# MOLECULAR WINE MICROBIOLOGY

# MOLECULAR WINE MICROBIOLOGY

*Edited by*

ALFONSO V. CARRASCOSA

ROSARIO MUÑOZ

RAMÓN GONZÁLEZ

AMSTERDAM • BOSTON • HEIDELBERG • LONDON
NEW YORK • OXFORD • PARIS • SAN DIEGO
SAN FRANCISCO • SINGAPORE • SYDNEY • TOKYO

Academic Press is an imprint of Elsevier

Academic Press is an imprint of Elsevier
32 Jamestown Road, London, NW1 7BY, UK
30 Corporate Drive, Suite 400, Burlington, MA 01803, USA
525 B Street, Suite 1800, San Diego, CA 92101-4495, USA

First edition 2011

Translation by Anne Murray and Iain Patten

**Notice**
No responsibility is assumed by the publisher for any injury and/or damage to persons or property as a matter of products liability, negligence or otherwise, or from any use or operation of any methods, products, instructions or ideas contained in the material herein. Because of rapid advances in the medical sciences, in particular, independent verification of diagnoses and drug dosages should be made

**British Library Cataloguing-in-Publication Data**
A catalogue record for this book is available from the British Library

**Library of Congress Cataloging-in-Publication Data**
A catalog record for this book is available from the Library of Congress

ISBN: 978-0-12-375021-1

For information on all Academic Press publications
visit our website at www.elsevierdirect.com

Typeset by TNQ Books and Journals

Printed and bound in the United States of America

11 12 13 14 15   10 9 8 7 6 5 4 3 1

Working together to grow
libraries in developing countries

www.elsevier.com | www.bookaid.org | www.sabre.org

ELSEVIER    BOOK AID International    Sabre Foundation

# Contents

# Preface

The publication of Louis Pasteur's *Mémoire sur la fermentation alcoolique* in 1857 has come to represent a milestone in the history of science and its applications, as it marked the beginning of a growing fascination with the biology of wine microorganisms among researchers worldwide. Since then, unprecedented improvements in winemaking processes have gone hand in hand with the development of modern microbiology, and it would now be impossible to understand the continuing progress made in the wine industry without taking into account the impact of advances in microbiological research.

A greater understanding of the microbiology of wine holds the key to critical issues affecting the industry, such as the management of safety and quality. For instance, by identifying and gaining a better understanding of the molecular mechanisms underlying the growth of microorganisms that cause wine spoilage or pose a threat to consumer health, winemakers will be better positioned to control and even eradicate them during the production process.

It is hoped that *Molecular Wine Microbiology* will be a useful tool for researchers and educators working in both the private and public sectors. Above all, however, it will be a valuable resource for those starting out on their fascinating journey through the world of wine microbiology.

Coordinated by Alfonso V. Carrascosa, Rosario Muñoz, and Ramón González from the Spanish National Research Council (CSIC), this book brings together contributions from a range of experts on the microbiology of wine working in universities, research centers, and industry.

Translation by Anne Murray and Iain Patten

The editors would like to acknowledge the excellent translation of the Spanish text. The translators have been able to capture all the nuances of the original, using accurate winemaking English terms.

# *Saccharomyces* Yeasts I: Primary Fermentation

*Agustín Aranda*[1], *Emilia Matallana*[1,2], *Marcel·lí del Olmo*[2]

[1] Departamento de Biotecnología, Instituto de Agroquímica y Tecnología de Alimentos, CSIC, Valencia, Spain and [2] Departament de Bioquímica i Biologia Molecular, Facultat de Ciències Biològiques, Universitat de València, Valencia, Spain

## OUTLINE

# 1. YEASTS OF INTEREST IN WINE PRODUCTION

## 1.1. Yeast Flora on the Grape, in the Winery, and in the Must

The fermentation of grape must is a complex microbiological process that involves interactions between yeasts, bacteria, and filamentous fungi (Fleet, 2007; Fugelsang & Edwards, 2007). Yeasts, which play a central role in the winemaking process, are unicellular fungi that reproduce by budding. Most yeasts belong to the phylum Ascomycota on the basis of their sexual development. In these organisms, the zygote develops within a sac-like structure, the ascus, while the nucleus undergoes two meiotic divisions, often followed by one or more mitotic divisions. A wall forms around each daughter nucleus and its surrounding cytoplasm to generate four ascospores within the ascus. The ascus then ruptures and releases the ascospores, which can germinate and produce new vegetative cells. Although thousands of yeast species have been identified, only 15 correspond to wine yeasts (Ribéreau-Gayon et al., 2006).

Traditionally, wine has been produced using yeast strains found on the surface of grapes and in the winery environment. The yeasts reach the grapes by wind and insect dispersal and are present on the wines from the onset of fruit ripening (Lafon-Lafourcade, 1983). The predominant species on the grape is *Kloeckera apiculata*, which can account for more than 50% of the flora recovered from the fruit (Fugelsang & Edwards, 2007). Other species of obligate aerobic or weakly fermentative yeasts with very limited alcohol tolerance may also be found in lesser proportions. These belong to the genera *Candida*, *Cryptococcus*, *Debaryomyces*, *Hansenula*, *Issatchenkia*, *Kluyveromyces*, *Metschnikowia*, *Pichia*, and *Rhodotorula* (Fleet & Heard, 1993; Ribéreau-Gayon et al., 2006). The fermentative species *Saccharomyces cerevisiae* and *Saccharomyces bayanus* are present in limited

numbers. This microflora can be affected by a wide variety of factors, principally temperature, rainfall, altitude, ripeness of the crop, and use of fungicides (Boulton et al., 1996). The flora associated with winery equipment is largely made up of *S. cerevisiae* (Fleet & Heard, 1993; Fleet, 2007; Martini & Vaughan-Martini, 1990), though species of the genera *Brettanomyces*, *Candida*, *Hansenula*, *Kloeckera*, *Pichia*, and *Torulaspora* have also been isolated.

The yeasts present in the must during the first few hours after filling the tanks belong to the same genera as those found on the grapes, predominantly *Hanseniaspora / Kloeckera*. In these spontaneous vinification conditions, *Saccharomyces* yeasts (mainly *S. cerevisiae*) begin to develop after around 20 h and are present alongside the grape-derived yeast flora. After 3 or 4 d of fermentation, *Saccharomyces* yeasts predominate and are ultimately responsible for alcoholic fermentation (Ribéreau-Gayon et al., 2006). This change in the yeast population is linked to the increasing presence of ethanol, the anaerobic conditions, the use of sulfites during harvesting and in the must, the concentration of sugar, and the greater tolerance of high temperatures shown by *S. cerevisiae* compared with other yeasts (Fleet & Heard, 1993; Fleet, 2007). *S. cerevisiae* comprises numerous strains with varying biotechnological properties (Ribéreau-Gayon et al., 2006). The importance of using genetic techniques to identify and characterize the different species and strains of yeast that participate in fermentation should not be underestimated. This is considered further in Chapter 5, which addresses the taxonomy of wine yeasts.

Currently, the usual strategy employed in winemaking involves inoculation of the must with selected yeasts in the form of active dried yeast. This practice, which emerged in the 1970s, shortens the lag phase, ensures rapid and complete fermentation of the must, and helps to create a much more reproducible final product (Bauer & Pretorius, 2000; Fleet & Heard,

1993). The selection of wine yeasts with specific genetic markers provides a system for the precise monitoring of the growth of particular strains during fermentation. Analyses of this type have shown that fermentation is driven mainly by inoculated yeasts (Delteil & Aizac, 1988), although these sometimes become only partially established (Esteve-Zarzoso et al., 1999). Given that the growth of the natural flora is not completely suppressed during the initial days of vinification, these strains can make substantial contributions to certain properties of the wine (Querol et al., 1992; Schütz & Gafner, 1993). Consequently, there is increasing interest in the use of mixed starter cultures in which non-Saccharomyces yeasts contribute desirable characteristics—particularly in terms of the organoleptic quality of the wine—that complement the fermentative capacity of Saccharomyces yeasts (Fleet, 2008).

The inoculated yeast strain must obviously be very carefully selected on the basis of certain necessary characteristics (Degré, 1993; Fleet, 2008). For instance, it must produce vigorous fermentation with short lag phases and little residual sugar, have reproducible fermentation characteristics, be tolerant of high pressure, ethanol, and suboptimal temperatures, and produce glycerol and β-glucosidases in adequate quantities to achieve a good aroma. Other valuable properties include fermentative capacity at low temperatures, low foaming, killer activity (Barre, 1980), certain levels of specific enzymatic activities (Darriet et al., 1988; Dubourdieu et al., 1988), and resistance to the adverse growth conditions present during winemaking (Zuzuarregui & del Olmo, 2004a). It is particularly important in the secondary fermentation of some sparkling wines for the yeast to be flocculent or easily separated from the medium (Degré, 1993; Zaworski & Heimsch, 1987). Autochthonous strains that meet these criteria have been increasingly used in recent years in an effort to obtain wines that maintain the sensory characteristics associated with specific wine-growing regions (Lafon-Lafourcade, 1983; Snow, 1983).

## 1.2. Morphology and Cellular Organization of Yeasts

Saccharomyces yeast cells have a rigid cell wall that allows them to resist the changes in osmotic pressure that can occur in the extracellular environment. Inside the cell wall, there is a periplasmic space and a plasma membrane surrounding the cytoplasm. Various transport mechanisms control the permeability of these structures and maintain their role as barriers.

Yeasts have multiple subcellular organelles characteristic of eukaryotic cells. These include a nucleus surrounded by a nuclear envelope, a smooth and a rough endoplasmic reticulum, a Golgi apparatus, mitochondria, and vacuoles. The cytoplasm contains numerous enzymes involved in the metabolic events described below, such as the enzymes responsible for alcoholic fermentation. Although some Saccharomyces strains lack mitochondria (respiration-deficient or "petite" mutants), these organelles play a fundamental role in metabolism. During fermentation, the high concentration of glucose in the medium inhibits synthesis of enzymes involved in the citric acid cycle and cytochromes from the respiratory chain through an effect known as glucose repression (Gancedo, 2008; Santangelo, 2006 and references therein). As a result, mitochondrial oxidative metabolism is limited under these conditions. However, aerobic metabolism, which is dependent upon mitochondria, does occur during the production of commercial yeasts for must inoculation and during some phases of the winemaking process.

Vacuoles are important for homeostasis, since enzymes that participate in the degradation and recovery of cell constituents are exclusively or predominantly localized to these structures. They also accumulate metabolites such as basic amino acids, S-adenosylmethionine, polyphosphates, allantoin, and allantoate at much higher concentrations than those found in the cytoplasm.

More in-depth reviews of the cellular organization of yeasts can be found in *The yeasts* (1991), edited by Rose and Harrison, and in *The molecular biology of the yeast Saccharomyces cerevisiae* (1982), edited by Strathern, Jones, and Broach.

## 1.3. Genetic Characteristics of Wine Yeasts

Unlike their counterpart laboratory strains, wine strains of *S. cerevisiae* are prototrophs, meaning that they do not require amino acids or nucleotides for their growth. This has important consequences for genetic manipulation, since genes conferring resistance to antibiotics, such as cycloheximide (del Pozo et al., 1991) or geneticin (Hadfield et al., 1990), must be used or auxotrophies introduced prior to transformation of these yeasts.

Wine strains of yeast are usually diploid, polyploid, or even aneuploid (Bakalinsky & Snow, 1990; Codón et al., 1995). Chromosome length in these yeasts is highly polymorphic (Bidenne et al., 1992; Rachidi et al., 1999), and this results in extensive variability in sporulation capacity and spore viability. This characteristic also influences the options for gene manipulation, since at least two copies of a gene need to be eliminated to obtain a deletion mutant. The ploidy of wine yeasts may provide them with advantages in adapting to changeable environments or, perhaps, represent a way of increasing the dose of genes that are important for fermentation (Bakalinsky & Snow, 1990; Salmon, 1997).

Finally, wine yeasts are predominantly homothallic (HO), meaning that following sporulation the daughter cells can change mating type, conjugate with a cell of the opposite mating type, and ultimately form a cell with 2n DNA content that is homozygous for all genes except the MAT locus (Thornton & Eschenbruch,

1976). In contrast, in heterothallic (ho) strains, the MAT locus is stable and cells remain in a haploid state until they encounter a cell of the opposite mating type with which to fuse (reviewed in Sprague, 1995). Wine strains also exhibit a high degree of heterozygosity (Barre et al., 1993; Codón et al., 1995), including for the HO locus (Guijo et al., 1997; Mortimer et al., 1994), and they can undergo mitotic recombination (Longo & Vézinhet, 1993; Puig et al., 2000), a characteristic that is not observed in haploid laboratory strains. This capacity for extensive genomic change means that wine yeasts do not display genetic stability (Pretorius, 2000; Snow, 1983). These factors and their relationship with evolutionary processes are discussed in detail by Pérez-Ortín et al. (2002) and are also considered in Chapter 6.

As in all eukaryotes, the mitochondria of *S. cerevisiae* have a circular mitochondrial DNA (mtDNA) (Christiansen & Christiansen, 1976; Hollenberg et al., 1970). This is usually located in the mitochondrial matrix but may occasionally be bound to the inner mitochondrial membrane. The mtDNA contains genes encoding proteins essential for mitochondrial function and, in yeasts, exhibits a high degree of polymorphism due to variability in the presence of certain introns and differences in the size of intergenic regions (Clark-Walker et al., 1981). This variability has been used in taxonomic studies, as discussed in Chapter 5.

# 2. GROWTH CHARACTERISTICS OF *SACCHAROMYCES* YEASTS DURING FERMENTATION

## 2.1. Must Composition

Grape must is a complex medium containing all of the nutrients necessary for the growth of *S. cerevisiae*. However, the varying composition of different musts, in addition to being crucial for the characteristics of the final product, influences the growth dynamics of the yeast. Vinification is a discontinuous, batch-type fermentation process in which all of the nutrients are present in the culture medium from the outset and the concentration of the nutrients declines as they are consumed by the yeast. As a result, the availability of some nutrients may act as a limiting factor for growth. Below we describe the main components of the must and their effect on the process of alcoholic fermentation.

### 2.1.1. Sugars

With the exception of water, monosaccharides are the most abundant component of grape must. Glucose and fructose are the main hexose sugars and are present in approximately equimolar concentrations. Other monosaccharides present as minor components include arabinose (0.2–1.5 g/L) and xylose (0.03–0.1 g/L); low concentrations of the disaccharide sucrose, which is generally hydrolyzed at the low pH found in must, are also present (Ough, 1992). Although polysaccharides such as pectins, gums, and dextrin are present at concentrations of around 3 to 5 g/L, they are not assimilable by wine yeasts. The total concentration of sugars is generally between 170 and 220 g/L (Ribéreau-Gayon et al., 2006). In musts with sugar concentrations of more than 200 g/L, there is a slowing of fermentation. Sugar concentrations between 250 and 300 g/L can inhibit yeast growth as a result of the high osmotic pressure and the elevated intracellular concentration of ethanol (Nishino et al., 1985). However, the low sugar concentrations typical of northerly wine-growing areas do not limit yeast growth and only affect the final alcohol concentration.

### 2.1.2. Organic Acids

The second most abundant compounds, organic acids, are present at concentrations of between 9 and 27 g/L (Ough, 1992). Tartaric and malic acid together account for 90% of the fixed acidity (Jackson, 1994); citric and ascorbic acid are found at lower concentrations. Tartaric acid predominates in must from warmer climates, where it reaches concentrations of 2 to 8 g/L, whereas, in cooler climates, malic acid concentrations may exceed those of tartaric acid, depending on the ripeness of the grapes. These acids have no direct effect on yeast growth but do play a decisive role in the pH of the must (see Section 2.2.3).

### 2.1.3. Nitrogenous Compounds

Nitrogen content is important since it tends to be limiting for the growth of S. cerevisiae (Ingledew & Kunkee, 1985) and the principal cause of stuck fermentation (Bisson, 1999). The concentration of soluble nitrogen varies between 0.1 and 1 g/L (Henschke & Jiranek, 1993). The composition of nitrogen sources in the must depends on a large number of factors, such as the grape variety, infection with Botrytis cinerea (which eliminates large quantities of the nutrients that can be assimilated by Saccharomyces yeasts), the timing of harvest, use of fertilizers, addition of supplements in the winery, and the extent of clarification of the musts, particularly in white grape musts (Lagunas, 1986). Variations in the quantity and form of the nitrogen sources in the must influence yeast cell growth, fermentation rate, and ethanol tolerance. The main compounds are ammonia (3–10%), amino acids (25–30%), polypeptides (25–40%), and proteins (5–10%). In addition, smaller quantities of nitrates, nucleotides, amines, and vitamins may be present. Nucleotides are only present at very low concentrations in the must (e.g., adenine and uracil nucleotides are found at concentrations of 4–15 mg/L and 4–8 mg/L, respectively). These are taken up by the yeast and incorporated into their nucleic acids, although yeast can also synthesize their own nucleotides (Monteiro & Bisson, 1992).

Saccharomyces yeasts cannot assimilate inorganic nitrogen sources such as nitrates and nitrites. They are also unable to assimilate proteins and polypeptides present in the medium, since they do not have a system for extracellular digestion of these types of compound. As a result, they are essentially dependent on the concentrations of ammonia and amino acids, their preferred nitrogen sources (Ough & Amerine, 1988). The most abundant amino acids in the must tend to be proline and arginine, and their concentrations vary in different musts. Proline cannot be metabolized by yeast under the low-oxygen conditions associated with alcoholic fermentation and should therefore not be taken into account when considering nitrogen availability. It has been reported that concentrations of assimilable nitrogen below 140 mg/L impair fermentation at normal sugar concentrations (Bely et al., 1990), and a concentration of ammonium ions below 25 mg/L is generally considered to be undesirable. However, outcomes can vary according to the individual strain. Cases have been described in which normal fermentation occurred in the presence of 120 mg/L assimilable nitrogen (Carrasco et al., 2003), while strains that require a minimum of 267 mg/L to complete the process have also been reported (Mendes-Ferreira et al., 2004). Since *Saccharomyces* yeasts can synthesize their own amino acids, the simplest solution to the problem of nitrogen deficiencies is to provide ammonium salt supplements, usually in the form of diammonium sulfate or phosphate (Ribéreau-Gayon et al., 2006). Addition of up to 30 g/hl of diammonium phosphate (DAP) is permitted in the European Union (EU), whereas in the United States up to 96 g/hl is allowed (Fugelsang & Edwards, 2007). Excessive nitrogen supplementation can alter the microbiological stability of the wine (providing nutrients for spoilage organisms) and its aroma (in many cases derived from deamination of amino acids). The timing of nitrogen addition is also important. Although reductions in fermentation time have been reported to occur independently of the timing of addition, better results are obtained when nitrogen is added during the exponential growth phase (Beltrán et al., 2005). Because ethanol impedes the uptake of nitrogenous compounds during later stages, it has been proposed that nitrogen should be added prior to or during the initial phases of fermentation, to coincide with aeration of the must (Sablayrolles et al., 1996). More recent data have shown poorer recovery of fermentation activity with addition of ammonia alone than with addition of amino acids or a combination of the two (Jiménez-Martí et al., 2007).

Among the nitrogenous compounds, vitamins deserve special mention. Wine yeasts are able to synthesize all of their own vitamins except for biotin (Ough, 1992) and nicotinic acid under anaerobic conditions (Panozzo et al., 2002), meaning that they are not as dependent as more complex organisms on the availability of these cofactors. Nevertheless, the presence of vitamins in the must stimulates the growth and metabolic activity of yeasts via the vitamins' participation as coenzymes in numerous biochemical reactions, and as a result they can be considered as growth factors (Ribéreau-Gayon et al., 2006). Must is generally rich in vitamins, but the concentrations of some are suboptimal. As a result, addition of vitamins can stimulate growth, particularly when the grapes have been subject to fungal infection, which always reduces the total concentration of vitamins. Thiamine is also an important component of the must. However, it is partially degraded by the sulfite added to prevent the appearance of spoilage organisms (Jackson, 1994) and is also consumed by the yeasts over the course of the fermentation. Consequently, it is advisable to add it to the must. The amount recommended by the EU is 50 mg/hl, whereas the maximum permitted level in the United States is 60 mg/hl (Fugelsang & Edwards, 2007). Deficiencies in other vitamins, such as pantothenic acid and pyroxidine, should also be avoided as they can lead to generation of undesirable compounds such as acetic acid and hydrogen sulfide (Wang et al., 2003).

### 2.1.4. Polyphenols

The many and varied phenolic compounds present in the must are essential elements in determining the organoleptic character of the wine (Waterhouse, 2002). Although it has been reported that the anthocyanins in red grape musts and the procyanidins in white grape musts can stimulate and inhibit growth, respectively (Cantarelli, 1989), these compounds have no relevant influence on the growth of wine yeasts. Their most noteworthy effect is as antioxidants, particularly in the case of quinones. It has recently been described that resveratrol and other polyphenols with recognized preventive effects in cardiovascular disease (Fremont, 2000) can extend the replicative lifespan of *Saccharomyces* yeasts (Howitz et al., 2003).

### 2.1.5. Mineral Salts

Inorganic elements are necessary for normal metabolism and maintenance of pH and ion balance in yeasts. Potassium, sodium, calcium, and magnesium are the predominant cations in the must, and chlorates, phosphates, and sulfates the main anions (Ough & Amerine, 1988). Must generally provides the inorganic elements required for yeast growth, but, if the concentration of one of these elements is limited, normal progression of fermentation can sometimes be affected (Bisson, 1999). Phosphate ions are particularly important given their vital metabolic role, as hexose sugars must be phosphorylated in order to be metabolized. Deficiencies in this anion can be compensated along with those of nitrogen by supplementation with DAP (see Section 2.1.3).

### 2.1.6. Lipids

Under the anaerobic conditions associated with wine fermentation, yeasts cannot synthesize sterols or long-chain unsaturated fatty acids. Synthesis of these compounds will only occur if oxygen is added during fermentation to increase yeast cell viability and the quality of fermentation (Sablayrolles, 1996). The lack of these types of lipid (especially ergosterol, the principal sterol in the plasma membrane of *Saccharomyces* yeasts) affects the structure and function of the plasma membrane and leads to increased effects of ethanol and poor glucose uptake (Jackson, 1994). These compounds are referred to as survival factors, since their presence is necessary for cell viability but their addition does not increase growth (Ribéreau-Gayon et al., 2006). Generally, the presence of these types of lipid in the must is guaranteed given their abundance in grape skins. Problems are only encountered in excessively clarified white wines, since up to 90% of unsaturated fatty acids may be lost under these conditions (Bertrand & Miele, 1984). In such situations, it is appropriate to supplement the must with yeast extract or lysed yeast (Muñoz & Ingledew, 1990). In other situations, the use of dried yeast grown under aerobic conditions usually guarantees the presence of sufficient lipids in the cell wall for fermentation of the must to take place.

### 2.1.7. Inhibitors

This section covers exogenous compounds added to the grapes and must to prevent the appearance of undesirable microorganisms that can also influence the growth of wine yeasts.

#### 2.1.7.1. SULFITES

Sulfites are added to control the appearance of spoilage organisms in the must. Industrial yeasts have been selected to be resistant to the quantities of sulfites used in wineries, and their growth is not usually affected by the concentrations of between 0.8 and 1.5 mg/L that are normally used. Concentrations above 1.5 mg/L, however, can inhibit growth (Sadraud & Chauvet, 1985). This inhibition is dependent upon the pH of the must; $SO_2$, the active molecular species, is generated at lower pH, and as a result the toxicity of a given concentration of the compound increases under those conditions

(Farkaš, 1988). Sulfite toxicity is also increased by the richness of methionine in the must, whereas it is reduced by higher concentrations of adenine (Aranda et al., 2006). Nevertheless, sulfite normally only delays the onset of fermentation and does not affect the rate or completion of the process.

### 2.1.7.2. PESTICIDES

Chemical compounds applied to the vines to prevent parasite infection can sometimes affect the growth of *Saccharomyces* yeasts during vinification. Folpet and captan, traditionally the most commonly used fungicides, have a substantial antiseptic effect on yeasts (Cabras & Angioni, 2000). New-generation fungicides are only marketed if they have been shown to have no effect on yeasts. For instance, metalaxyl, cymoxanil, famoxadone, fenhexamid, fluquinconazole, kresoxim-methyl, quinoxyfen, and trifloxystrobin have no effect on yeast growth (Jackson, 1994; Oliva et al., 2007). In addition, clarification of the must eliminates most of the pesticides present on the surface of the grapes, and many are degraded spontaneously under the acidic conditions of the must. As with sulfites, traces of fungicide in the must tend to inhibit the onset of fermentation rather than interfere with fermentation rate or completion.

## 2.2. Physical Parameters of Fermentation

The main physicochemical factors that affect the growth of *Saccharomyces* yeasts during alcoholic fermentation are described below.

### 2.2.1. Temperature

Temperature is the most important physical factor in the growth of yeasts and the progression of fermentation (Fleet & Heard, 1993). Although *S. cerevisiae* has an optimal growth temperature of around 30°C, it can adapt to a wide range of temperatures up to a maximum of 40°C, at which point viability begins to decline (Watson, 1987). Although there is a linear increase in the rate of fermentation between 10 and 32°C (doubling every 10°C), this does not mean that higher temperatures are the most appropriate for fermentation of the must. Ethanol toxicity increases with temperature, and higher temperatures lead to evaporation of ethanol and other volatile compounds that are essential to the organoleptic properties of the wine (Torija et al., 2003), particularly in the case of white wines. Excessively low temperatures are also not recommended, since they can cause stuck fermentation when yeast membrane fluidity begins to be affected (Bisson, 1999). It is also not economically viable to maintain fermentations under these conditions for extended periods. Consequently, controlling fermentation temperature is an essential element of modern wine production. White wines are generally fermented at between 10 and 18°C to improve the retention of aromas, whereas red wines tend to be fermented at higher temperatures (between 18 and 29°C) to achieve good extraction of phenolic compounds (Fugelsang & Edwards, 2007). Nevertheless, an initial fermentation temperature of 20°C is recommended in both cases in order to stimulate initiation of yeast growth (Jackson, 1994). Low temperatures may favor the growth of non-*Saccharomyces* yeasts during the initial stages of fermentation.

### 2.2.2. Aeration

*Saccharomyces* yeasts are facultative anaerobes, able to consume sugars in the absence of oxygen more effectively than non-*Saccharomyces* yeasts (Visser et al., 1990). In fact, excess oxygen can inhibit fermentation, a phenomenon known as the Pasteur effect. Nevertheless, a certain amount of oxygen is beneficial for the growth of wine yeasts since it is required for the synthesis of sterols (mainly ergosterol) and unsaturated fatty acids. A more oxygenated environment may be helpful in musts with nitrogen deficiencies, as this will allow the amino acid proline to be metabolized (Ingledew & Kunkee, 1985). It is also advisable to add exogenous nitrogen sources during aeration of the must (Sablayrolles et al., 1996). The oxygen captured by the must during pressing is usually sufficient to reach saturation, and is therefore generally adequate for normal progression of fermentation. In red wines, oxygen consumption due to oxidation of phenols is compensated by the aeration created during pump-over, resulting in oxygen concentrations of around 10 mg/L. This effect is most beneficial at the end of the exponential growth phase. Nevertheless, excessive aeration may lead to undesirable production of acetaldehyde and hydrogen sulfide, and reduced production of aromatic esters (Nykänen, 1986).

### 2.2.3. *pH*

The typical pH of grape must is between 2.75 and 4.2 (Heard & Fleet, 1988). These pH values do not have a negative effect on the growth of *Saccharomyces* yeasts, and problems only begin to present themselves at a pH below 2.8. The toxic effects of low pH are due to the increased effects of ethanol (Pampuhla & Loereiro-Dias, 1989) and sulfite (Farkaš, 1988). Tolerance of acidic pH depends on the abundance of potassium ions in the must (Kudo et al., 1998). Low pH favors the hydrolysis of disaccharides and, therefore, fermentation. In addition, the acidic character of the must prevents the appearance of spoilage microorganisms. Consequently, acids such as tartaric acid are sometimes added (addition of 1 g/L, for example, reduces the pH by 0.1 units). However, addition of excess tartaric acid can lead to undesirable precipitation.

### 2.2.4. *Clarification*

Elimination of solid particles from the must is an important element in the production of white wines. However, elimination of the nutrients that are associated with them, particularly nitrogenous compounds, can impair yeast growth (Ayestaran et al., 1995). Furthermore, solid particles act as nuclei for the formation of carbon dioxide bubbles and favor dissipation of the gas, which at high levels can inhibit the growth of *Saccharomyces* yeasts (Thomas et al., 1994). On the other hand, the final products obtained from clearer musts have better organoleptic characteristics. The extent of clarification must therefore be optimized to produce better wines without affecting the fermentation process.

### 2.2.5. *Carbon Dioxide*

Alcoholic fermentation of hexose sugars generates carbon dioxide, which can reach volumes equivalent to 56 times that of the fermented must (Boulton et al., 1996). The release of this gas contributes to the dissipation of some heat and produces convection currents within the must that aid the diffusion of nutrients. However, its evaporation also favors loss of ethanol and volatile compounds (Jackson, 1994). Furthermore, if produced in excess, carbon dioxide affects the viability of *Saccharomyces* yeasts, mainly due to membrane damage.

## 2.3. Yeast Growth and Fermentation Kinetics

Yeast growth during wine fermentation differs from that occurring in other industrial processes such as brewing, since the high concentration of sugars leads to the production of ethanol at concentrations that inhibit growth. Fermentation begins rapidly with inoculums containing approximately $10^6$ cells/mL. The typical growth cycle of *Saccharomyces* yeasts consists of three phases and begins following a short lag period (Lafon-Lafourcade, 1983). The first phase is the limited growth phase and lasts between 2 and 5 d, generating a population of up to $10^7$ or $10^8$ cells/mL. Fermentation during this phase occurs at a constant, maximal rate, and it tends to consume between a third and half of the initial sugar content (Castor & Archer, 1956). Next, growth enters a quasi-stationary phase that lasts around 8 d. During this time, there is no increase in the number of cells in the population. However, the cells are metabolically active and the rate of fermentation remains maximal. Finally, the culture enters the death phase, which is poorly characterized and highly variable. Whereas some authors claim that death does not occur until all of the sugars have been consumed (Boulton et al., 1996), others have assigned greater importance to this phase. According to this view, the death phase is estimated to be three or four times longer than the growth phase and still involves consumption of a considerable quantity of sugar (Ribéreau-Gayon et al., 2006). The loss of viability is accompanied by a reduction in the rate of fermentation, due not only to a reduction in the number of viable cells but also to inhibition of the metabolic activity of the nonproliferative cells. The loss of fermentative capacity of the cells in this final phase has been linked to the depletion of adenosine triphosphate (ATP) and the accumulation of ethanol, which have negative effects on membrane transport. It has been observed that, under these conditions, cellular enzyme systems are functional but the intracellular concentration of sugars decreases progressively.

Yeast growth is monitored by microscopic counts of the cells in diluted samples of fermenting must. The number of cells can also be estimated by measuring the optical density at 600 to 620 nm following the generation of standard curves for the inoculated strain. In both cases, estimations of the numbers of cells present in the fermenting must do not differentiate between viable and dead cells, a very important distinction when monitoring the progression of wine fermentation. To differentiate between the two, plate counts can be performed with solid nutrient media, on which only viable cells will be able to produce colonies; however, this type of analysis is slow, as the colonies take 3 to 4 d to grow. Other more rapid techniques based on the use of fluorescent reagents or bioluminescent quantification of ATP are available for estimation of the number of viable cells, but they are less reliable.

Another parameter that is analyzed in wine yeasts is vitality; that is, the capacity of the cells to achieve complete metabolic activity. There is a relationship between this metabolic activity and the time necessary to reach maximum fermentation rate. This is usually measured by indirect impedance; in other words, the reduction in impedance due to a solution of potassium hydroxide that reacts with the carbon dioxide produced by the metabolic activity of the yeast (Novo et al., 2007).

Because all of the methods for monitoring yeast growth are relatively difficult to implement in wineries, in practice, fermentation kinetics are analyzed using simpler techniques such as monitoring the reduction in sugar concentration, the increase in ethanol content, or the release of carbon dioxide (Ribéreau-Gayon et al., 2006). However, the simplest method to adapt to winery conditions is analysis of the density of the must, since measurement of the mass per unit volume provides an approximate measure of sugar content. During the course of fermentation, the sugar concentration decreases while ethanol content increases, and this leads to a reduction in density. The initial density of the must and the final density of the wine will depend on the initial sugar concentration, which will lead to a specific percentage of ethanol (approximately 1% [vol/vol] ethanol for every 17 g of sugars) (Ribéreau-Gayon et al., 2006).

## 2.4. Biochemistry of Fermentation

The biochemistry of wine production is also complex. The central metabolic process that takes place is alcoholic fermentation, a catabolic pathway involving the transformation of the hexose sugars present in the must into ethanol and carbon dioxide. Compounds are also generated that play a central role in yeast growth and in the organoleptic properties of the wine.

A more complete description of the biochemical processes that take place during wine production can be found elsewhere (Boulton et al., 1996; Rose & Harrison, 1991; Strathern et al., 1982). The aim here is to introduce readers to some of these pathways, in particular those that are most relevant in terms of yeast growth and the properties of the final product.

### 2.4.1. Alcoholic Fermentation

Carbon sources, in particular the hexose sugars glucose and fructose, allow cells to obtain energy by alcoholic fermentation. This metabolic pathway (Figure 1.1) occurs in the cytoplasm and can be expressed in terms of the following simplified equation:

$$C_6H_{12}O_6 + 2ADP + 2HPO_4^- \rightarrow 2C_2H_5OH + 2CO_2 + 2ATP + 2H_2O$$

Alcoholic fermentation involves the Embden-Meyerhof-Parnas (EMP) pathway, which was described by Embden, Meyerhof, and Parnas around 1940 and is also known as glycolysis. The pathway involves 10 reactions. The first five reactions correspond to the energy investment phase, in which sugars are metabolically activated by ATP-dependent phosphorylation to give rise to a six-carbon sugar, fructose-1,6-bisphosphate, which is cleaved to produce two moles of triose phosphate. During the energy generation phase (reactions 6 to 10), the triose phosphates are reactivated, generating two compounds with a high phosphate-transfer potential: firstly 1,3-bisphosphoglycerate and then phosphoenolpyruvate. Each of these compounds transfers a high-energy phosphate group to adenosine diphosphate (ADP), thus producing ATP in a process known as substrate-level phosphorylation. The chemical energy of ATP can be subsequently transformed in the cell into other forms of energy necessary for cell growth. The first reaction in this energy generation phase is an oxidation reaction catalyzed by the enzyme glyceraldehyde-3-phosphate dehydrogenase. This enzyme requires nicotinamide-adenine dinucleotide (NAD$^+$) as a coenzyme to accept the electrons from the substrate being oxidized. As a consequence, this coenzyme is reduced to NADH.

After glycolysis, alcoholic fermentation is completed with two additional reactions used to reoxidize NADH to NAD$^+$ to guarantee the continuation of glycolysis. In the first reaction, the resulting pyruvate is decarboxylated to acetaldehyde and carbon dioxide by the enzyme pyruvate decarboxylase, which requires thiamine pyrophosphate as a coenzyme. Finally, the acetaldehyde is reduced to ethanol by the

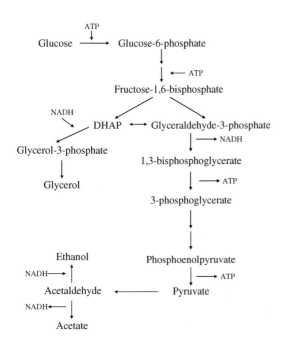

**FIGURE 1.1** Schematic diagram of the conversion of glucose into ethanol during alcoholic fermentation by the yeast *Saccharomyces cerevisiae*. The figure also shows the relationship between energy production in this pathway and the processes linked to the redox state of the coenzyme NAD$^+$/NADH. The reactions in which consumption or synthesis of ATP and NADH occur are indicated. DHAP = dihydroxyacetone phosphate. *Figure adapted from Norbeck and Blomberg (1997).*

enzyme alcohol dehydrogenase in a reaction involving oxidation of NADH to NAD$^+$.

The glycolysis pathway is not only involved in energy production for the yeast. It also generates metabolites that can be used as substrates for the biosynthesis of molecules linked to increased biomass (Figure 1.2). Glucose-6-phosphate can be directed towards the pentose phosphate pathway, which allows the formation of NADPH and ribose phosphate, molecules that are necessary for the biosynthesis of fatty acids and nucleotides, respectively. Pyruvate is also an important substrate for the synthesis of molecules such as oxaloacetate, succinate, organic acids, and amino acids. These molecules are produced at the beginning of vinification, when the activities of pyruvate decarboxylase and alcohol dehydrogenase are low. 3-Phosphoglycerate can also be diverted from glycolysis to participate in the synthesis of amino acids such as serine. Finally, dihydroxyacetone phosphate, one of the end products of the energy investment phase, is used to produce glycerol. This molecule has a powerful effect on the quality of the wine, participates in the biosynthesis of triacylglycerols, and is also the main compatible osmolyte that is produced by yeasts in response to the significant osmotic stress to which they

are exposed at the beginning of vinification (Blomberg & Adler, 1992).

The synthesis of glycerol also represents a mechanism for the oxidation of molecules of NADH generated during glycolysis that have not been reoxidized as a result of 1,3-bisphosphoglycerate or pyruvate being diverted towards products other than ethanol. It is therefore essential to maintain the redox balance in the cytoplasm. Higher alcohols are also produced during alcoholic fermentation. As discussed below, these compounds can also be generated from certain amino acids and are important in determining the aroma of wine.

### 2.4.2. Nitrogen Metabolism

S. cerevisiae is equally able to use amino acids, ammonia, uracil, proline derivatives, and urea as nitrogen sources (for a detailed review of the use of these compounds and their effect on yeast growth rate, see Cooper, 1982a). Among the nitrogenated components that can be found in the must, amino acids make the largest contribution to nitrogen provision for the synthesis of structural and functional proteins and the production of enzymes and transporters.

Figure 1.3 shows the uptake and use of nitrogen by yeasts when it is available in the

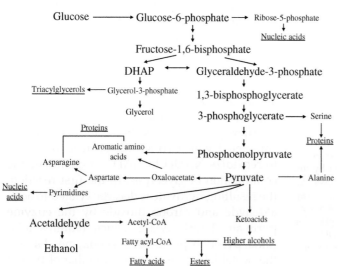

FIGURE 1.2 Biosynthetic precursors derived from alcoholic fermentation. DHAP = dihydroxyacetone phosphate. *Figure adapted from Henschke and Jiranek (1993).*

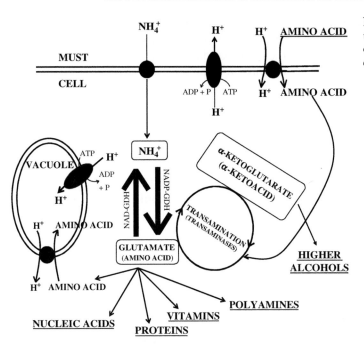

**FIGURE 1.3** Schematic diagram of the uptake and metabolism of nitrogenous compounds. *Reproduced from Llauradó i Reverchon (2002).*

medium. Most nitrogenous compounds are incorporated in the cell via active transport systems, specifically symport with ions, usually protons (Cooper, 1982b). In *S. cerevisiae*, a general amino acid permease (Gap1p) has been identified, and also specific permeases for different amino acids (Grenson et al., 1966; Horak, 1986). Ammonia undergoes active transport of the protonated species that requires the presence of glucose (Roon et al., 1977), and three systems have been identified involving the proteins Mep1p, Mep2p, and Mep3p (Marini et al., 1994; Marini et al., 1997). A detailed description of the uptake mechanisms of these different nitrogenous compounds is provided by Cartwright et al. (1989), Cooper (1982b), and Henschke and Jiranek (1993).

Nitrogenous compounds are assimilated during the first few hours of fermentation (Monteiro & Bisson, 1991) and degraded in a specific order that depends on factors such as the requirement for each compound in biosynthetic processes, efficiency of transport, and possible conversion into ammonia or glutamate without

releasing compounds that are toxic to the cell (Cooper, 1982a).

During yeast growth, more than half of the intracellular reserves of amino acids are in the vacuoles (Wiemken & Durr, 1974). This compartmentalization contributes to the regulation of the activity of various enzymes involved in their degradation (Sumrada & Cooper, 1982).

Ammonia and glutamate are central to all nitrogen metabolism in yeasts. Ammoniacal nitrogen is rapidly incorporated in the biosynthetic pathways through the activity of $NADP^+$-dependent glutamate dehydrogenase. In addition, it represents the end product of the catabolic pathways for nitrogenous compounds, in this case through the reaction catalyzed by $NAD^+$-dependent glutamate dehydrogenase. In turn, amino acids undergo interconversion processes via the transaminase system, in which glutamate plays an extremely important role as a donor and acceptor of amino groups.

Metabolism of nitrogenous compounds by yeasts also contributes to the formation of products that play an important role in the final quality of the wine by affecting its sensory properties.

## 2.5. The Importance of Yeast Metabolism in Wine Aroma

Wine aroma is generated by a series of aromatic and volatile compounds recognized by the senses of taste and smell. Some of these arise from the grapes and are responsible for what is known as varietal aroma. However, most arise from the fermentation process and their concentrations are essentially dependent on the yeasts that predominate during fermentation and the conditions under which fermentation takes place (Egli et al., 1998; Henick-Kling et al., 1998; Steger & Lambrechts, 2000). Vinification temperature plays a particularly important role, since more aromatic wines are produced when the process is carried out at temperatures close to or below 15°C (Bauer & Pretorius, 2000).

The main groups of aromatic compounds derived from yeast metabolism are organic acids, higher alcohols, esters, and, to a lesser extent, aldehydes (Rapp & Versini, 1991). Substances derived from fatty acids and from nitrogen- or sulfur-containing compounds also contribute (Boulton et al., 1996). Various studies undertaken in recent years show that the composition of the must and, in particular, the levels and nature of nitrogenous compounds present in the must or subsequently added to limiting fermentations play an important role in determining the organoleptic properties of the wine (see, for instance, Beltrán et al., 2005; Carrau et al., 2008; Jiménez-Martí et al., 2007; Mendes-Ferreira et al., 2009; Torija et al., 2003; Vilanova et al., 2007). Some of the compounds arising from metabolism have a negative contribution, as is the case for acetaldehyde, acetic acid, ethyl acetate, some higher alcohols when present at high concentrations, and, in particular, reduced sulfur compounds, organic sulfates, and thiols. Below we describe the origin of the main volatile compounds generated during yeast metabolism. Table 1.1 shows the compounds and their origins, their concentrations in wine, and the characteristics of the aroma produced. All of these elements are described in greater detail in the review by Lambrechts and Pretorius (2000).

The first group of compounds is the volatile fatty acids. These include acetic acid, long-chain fatty acids (C16 and C18), and short-chain fatty acids (C8, C10, and C12). The levels of acetic acid must be strictly controlled and should not exceed 1.0 to 1.5 g/L (Eglinton & Henschke, 1999). In yeast, fatty acids are generated from acetyl-coenzyme A (CoA) derived from oxidative decarboxylation of pyruvate. Their synthesis requires two enzyme systems: acetyl-CoA carboxylase and fatty acid synthase. For an extensive review of these pathways, see Paltauf et al. (1992) and Ratledge and Evans (1989).

Higher alcohols make the greatest contribution to wine aroma. At concentrations below 300 mg/L they introduce a desirable complexity, whereas at concentrations above 400 mg/L they have a negative effect on wine quality (Nykänen, 1986). The higher alcohols produced in the largest quantities are 1-propanol, 2-methyl-1-propanol, 2-methyl-1-butanol, 3-methyl-1-butanol, hexanol, and 2-phenylethanol (Henschke & Jiranek, 1993). One of the pathways through which these compounds are generated is the conversion of branched-chain amino acids (valine, leucine, isoleucine, and threonine); as a result, their accumulation depends on the quantity and type of nitrogen sources in the must (Giudici et al., 1993). However, most higher alcohols are synthesized de novo from sugars via the initial formation of the corresponding ketoacids.

Esters are largely responsible for the fruity and floral character of a number of wines. Acetate esters of higher alcohols (such as ethyl acetate, 2-phenylethanol acetate, or isoamyl acetate) and ethyl esters of medium-chain saturated fatty acids (such as ethyl hexanoate) make

**TABLE 1.1**    Principal Compounds Responsible for Wine Aroma

| | Compound | Concentration in wine (mg/L) | Aroma |
|---|---|---|---|
| Volatile fatty acids | Acetic acid | 150–900 | Vinegar |
| | Propionic acid | Trace | Rancid |
| | Butyric acid | Trace | Bitter |
| | Hexanoic acid | Trace to 37 | Rancid, vinegar, cheese |
| | Octanoic acid | Trace to 41 | Oily, rancid, sweet, buttery |
| | Decanoic acid | Trace to 54 | Unpleasant, rancid, bitter, phenolic |
| Higher alcohols | Propanol | 9–68 | Powerful |
| | Butanol | 0.5–8.5 | Petrol |
| | 2-methyl-1-butanol | 15–150 | Marzipan |
| | Isobutylic acid | 9–28 | Alcoholic |
| | Isoamyl alcohol | 45–490 | Marzipan |
| | Hexanol | 0.3–12 | Freshly mown grass |
| | 2-Phenylethanol | 10–180 | Floral, rose |
| Esters | Isoamyl acetate | 0.03–8.1 | Banana, pear |
| | 2-Phenylethyl acetate | 0.01–4.5 | Rose, honey, fruity, floral |
| | Ethyl acetate | 26–180 | Varnish, nail polish, fruity |
| | Isobutyl acetate | 0.01–0.8 | Banana |
| | Ethyl butanoate | 0.01–1.8 | Floral, fruity |
| | Ethyl hexanoate | Trace to 3.4 | Apple, banana, violet |
| | Ethyl octanoate | 0.05–3.8 | Pineapple, pear |
| | Ethyl decanoate | Trace to 2.1 | Floral |
| Carbonyl compounds | Acetaldehyde | 10–300 | Bitter, green pineapple |
| | Benzaldehyde | 0.003–4.1 | Bitter almond |
| | Diacetyl | 0.05–5 | Larder |
| Volatile phenols | 4-Vinylphenol | 0–1.15 | Medicinal |
| | 4-Vinyl guaiacol | 0–0.496 | Smoky, vanilla |
| | 4-Ethylphenol | 0–6.047 | Horse sweat |
| | 4-Ethyl guaiacol | 0–1.561 | Smoky, vanilla |

(*Continued*)

**TABLE 1.1**    Principal Compounds Responsible for Wine Aroma—*cont'd*

|  | Compound | Concentration in wine (mg/L) | Aroma |
|---|---|---|---|
| Sulfur compounds | Hydrogen sulfide | Trace to >0.080 | Rotten eggs |
|  | Dimethyl disulfide | Trace to 0.0016 | Boiled cabbage |
|  | Diethyl disulfide | Trace | Garlic, burnt rubber |
|  | Methyl mercaptan | Qualitative | Rotten eggs, cabbage |
|  | Ethyl mercaptan | Qualitative | Onion, rubber |

*Data taken from Cabanis et al. (1998) and Lambrechts and Pretorius (2000).*

the greatest contribution. According to the classification proposed by Baumes et al. (1986), all of these are included in the apolar group. The other group, the polar compounds, includes compounds found in greater quantities but that have more influence on the body than on the aroma of the wine. These include diethyl succinate, 2-ethyl-hydroxypropionate, diethyl malate, and ethyl-4-hydroxypropanoate. The principal ester is ethyl acetate, although concentrations above 170 mg/L in white wines and 160 mg/L in red wines are unacceptable (Corison et al., 1979). All these compounds are fundamentally derived from sugar metabolism. Acetate esters are synthesized in reactions between alcohol and acetyl-CoA catalyzed by alcohol acetyltransferases (Peddie, 1990). Fatty acid esters, on the other hand, are generated following activation of the corresponding fatty acid by CoA, catalyzed by an acyl-CoA synthase (Nordström, 1964a, 1964b, 1964c). Final concentrations of these compounds in the wine are affected by two factors: their hydrolysis during early phases of wine maturation (Ramey & Ough, 1980) and the extent to which they are transferred to the medium, which is reduced with increasing chain length and is influenced by the temperature at which fermentation takes place (Nykänen et al., 1977). Ethyl esters of amino acids have also been found in wines at concentrations of up to 58 mg/L (Herraiz & Ough, 1993; Heresztyn, 1984); these compounds

are mainly formed during the second half of fermentation, when the concentration of ethanol in the medium is high (Herraiz & Ough, 1993).

Volatile short-chain aldehydes are also important, and contribute in particular to pineapple and lemony aromas. Acetaldehyde and diacetyl account for around 90% of these compounds and have acceptable limits of 100 mg/L and 1 to 4 mg/L, respectively. These compounds are generated from two ketoacids derived from the synthesis or degradation of amino acids or higher alcohols.

Volatile phenols are important for flavor, color, and aroma of wines (Dubois, 1983). The most important are vinylphenols in white wines and ethylphenols in red wines (Chatonnet et al., 1997; Etievant, 1981; Singleton & Essau, 1969). These compounds are produced from the nonvolatile acids trans-ferulic and trans-p-cuomaric acid, essentially through the activity in red wines of contaminating yeasts belonging to the genera *Brettanomyces/Dekkera* (Chatonnet et al., 1997).

Finally, sulfur compounds make a significant contribution to the aroma of wine due to their high reactivity, although in some cases they are responsible for undesirable aromas. The main compound in this group is hydrogen sulfide, which has an acceptable limit of between 10 and 100 µ/L. This compound is essentially derived from sulfate in the medium and elemental sulfur introduced by fungicides, and its formation by yeasts is linked to nitrogen and sulfur metabolism (Henschke & Jiranek, 1993; Rauhut, 1993). In fact, it has been observed that deficiencies in easily assimilable sources of nitrogen are a major cause of hydrogen sulfide formation by yeasts (Stratford & Rose, 1986) and these levels can vary according to the initial concentration of nitrogen in the must and the strain under consideration (Mendes-Ferreira et al., 2009). Other sulfur compounds that contribute to wine aroma include methylmercaptan, ethylmercaptan, dimethyl disulfide, and diethyl disulfide (Rauhut & Kürbel, 1996).

## 3. GENE EXPRESSION DURING FERMENTATION

The capacity of yeasts to produce a wine with desirable properties must be related to the synthesis of specific molecules, proteins, and products of enzymatic reactions, and, consequently, substantial efforts have been made in recent years to investigate the molecular processes occurring during winemaking. Most studies of gene expression during winemaking have focused on the alcoholic fermentation phase, but reports have also been published on studies undertaken during the phases of industrial production and rehydration of commercial biomass, and also during aging. Initially, analyses focused on genes of interest, but more recent studies have analyzed global gene expression using DNA microarrays. In this chapter, we review the information available on the expression of particular genes (summarized in Table 1.2, which indicates their molecular function or the biological process in which they are implicated). The application of DNA microarrays to the understanding of gene expression in wine yeasts during the winemaking process will be discussed in Chapter 6.

Although in some cases gene expression has been analyzed in natural must fermentations, most of the studies that have been published

TABLE 1.2 *Saccharomyces cerevisiae* Genes Mentioned in This Chapter

| Gene | Molecular function or *biological process* |
|---|---|
| ADE4 | Phosphoribosyl pyrophosphate aminotransferase |
| ADH7 | Alcohol dehydrogenase |
| ALD2/3/4/6 | Aldehyde dehydrogenases |
| ATF1 | Alcohol o-acetyltransferase |
| ATH1 | Vacuolar acid trehalase |
| CAR1 | Arginase |
| COX6 | Cytochrome C oxidase |
| CUP1 | Metallothionein |
| CTT1 | Catalase |
| FBA1 | Fructose bisphosphate aldolase |
| GLK1 | Glucokinase |
| GLO1 | Glyoxylase |
| GPD1 | Glycerol-3-phosphate dehydrogenase |
| GPH1 | Glycogen phosphorylase |
| GPM1 | Phosphoglycerate mutase |
| GRE2 | Lactaldehyde dehydrogenase |
| GRX5 | Mitochondrial glutaredoxin V |
| GSH1 | Glutathione synthase I |
| GSY1/2 | Glycogen synthase |
| HOR7 | *Stress response* |
| HSP12 | *Stress response* |
| HSP26 | Molecular chaperone |
| HSP30 | *Stress response* |
| HSP78 | Molecular chaperone |
| HSP82 | Molecular chaperone |
| HSP104 | Molecular chaperone |
| HXK1,2 | Hexokinases |
| HXT1-18 | Glucose transporters |
| MET16 | Phosphoadenylylsulfate reductase |
| NTH1 | Cytosolic neutral trehalase |

*(Continued)*

**TABLE 1.2**  *Saccharomyces cerevisiae* Genes Mentioned in This Chapter—*cont'd*

| Gene | Molecular function or *biological process* |
|------|--------------------------------------------|
| *PAU3* | *Stress response* |
| *PGK1* | Phosphoglycerate kinase |
| *PMA1/2* | Proton transporters |
| *POT1* | Acetyl-CoA C-acyltransferase |
| *SNF3* | Glucose sensor |
| *SPI1* | *Stress response* |
| *SSA3* | *Stress response* |
| *SSA4* | Molecular chaperone |
| *STI1* | Molecular chaperone |
| *TDH2/3* | Glyceraldehyde-3-phosphate dehydrogenases |
| *THI4* | *Thiamine synthesis* |
| *TPS1* | Trehalose phosphate synthase |
| *TPS2* | Trehalose-6-phosphate phosphatase |
| *TRR1* | Thioredoxin reductase I |
| *TRX2* | Thiol-disulfide thioredoxin exchanger |
| *TSA1* | Thioredoxin peroxidase I |
| *UBI4* | Ubiquitin |
| *YGP1* | *Stress response* |

The molecular function of the gene product is shown, or, when this is not known, the biological process in which it is involved (in italics) according to the *Saccharomyces* Genome Database (http://www.yeastgenome.org) is shown.

are based on the use of synthetic musts with a defined chemical composition that mimics, among other characteristics, the composition of sugars and total nitrogen in natural musts. The advantage of synthetic media is that their defined composition allows reproducible experiments to be carried out even though the musts are prepared at different points in time. Furthermore, it is possible to analyze the effect that specific changes in must composition have on the fermentation process and the organoleptic properties of the final product.

## 3.1. Glycolytic Genes

The high concentration of sugars in the must represses the expression of genes encoding enzymes involved in mitochondrial respiration and also inhibits the activity of the expressed enzymes. As a result, the growth of *S. cerevisiae* during vinification largely involves fermentative metabolism. Gene expression analysis shows that this metabolic alternative involves increased flow through the glycolytic pathway. Northern blot analysis of the levels and accumulation of messenger RNA (mRNA) during fermentation of natural musts by wine yeasts has shown that, despite being expressed throughout fermentation, genes linked to the glycolytic pathway have specific, dynamic expression profiles during the different phases of the growth curve (Puig & Pérez-Ortín, 2000a). These studies analyzed the fermentation of both synthetic musts and natural musts derived from the grape varieties Bobal and Moscatel. The gene expression patterns observed in the different media led to certain general conclusions. Thus, the genes *TDH2/3*, which code for isoenzymes II and III of glyceraldehyde-3-phosphate dehydrogenase, have the highest expression levels in all of the musts analyzed. In addition, all of the glycolytic genes reach maximum expression levels during the first 24 to 48 h of fermentation, which coincides with the phase of cell growth and maximum fermentation rate. The extent of mRNA accumulation decreases progressively for all of the genes analyzed during the stationary phase in parallel with the slowing of fermentation. The results obtained by Puig and Pérez-Ortín (2000a) also indicated a significant difference between the musts analyzed. In the case of Bobal musts, increases in the expression of all genes were detected between 4 and 6 d after inoculation, a finding that did not occur in the other two musts. In addition, gene expression levels were generally higher in synthetic musts, and in some cases (*FBA1*, *TDH2/3*, and *GPM1*) lower in Moscatel varieties. These differences may be related to the way in which each of the musts was prepared and the consequent variations in their composition. In fact, the same group observed differences in the expression of other genes according to the must used (Puig & Pérez-Ortín, 2000a, 2000b). Comparison with the results obtained by Riou et al. (1997) using a synthetic must and a different strain highlights differences in the time course of stationary-phase mRNA levels for various genes, including the glycolytic gene *PGK1*. Other data on the expression of glycolytic genes in wine fermentations have been obtained from experiments designed to analyze global gene expression. The studies of Backhus et al. (2001), Erasmus et al. (2003), Jiménez-Marti and del Olmo (2008), Marks et al. (2003), Marks et al. (2008), and Rossignol et al. (2003) lead to the conclusion that transcriptional regulation of glycolytic genes during wine fermentation is affected under conditions that influence growth rate and, especially, fermentation rate.

### 3.2. Osmotic Stress-response Genes

The first few hours of fermentation constitute a critical period in which the capacity of the inoculated cells to adapt to the extremely high sugar concentration and initiate fermentation following a short lag phase is crucial if the inoculated strain is to dominate the fermentation at the expense of the natural flora of the must. Given that the inoculated yeast is derived from dried and rehydrated or precultured cells, contact with the must is likely to involve substantial reprogramming of gene expression in the yeast cells.

In a study designed to assess the effect of hyperosmolarity due to high concentrations of glucose in must, the expression of stress-response genes was analyzed over the course of the first day of fermentation in synthetic musts with differing sugar composition (Pérez-Torrado et al., 2002a). The results of that study

indicated that the molecular response to these conditions is complex and influenced by a number of factors. In response to hyperosmotic stress, *S. cerevisiae* accumulates glycerol, the main compatible osmolyte in yeasts. This is explained by the observation that the main change in the gene expression pattern in response to osmotic stress involves the induction of *GPD1*, which codes for glycerol-3-phosphate dehydrogenase, an enzyme involved in the synthesis of glycerol. Induction of the expression of this gene is a rapid phenomenon occurring in the initial phases of fermentations with high sugar concentrations (20%), and high levels of *GPD1* mRNA are detected 15 min after inoculation and reach a maximum induction approximately 1 h later. Analysis of the expression of other genes that act as markers for the general stress response and are dependent on Msn2/4p transcription factors, such as *HSP12* and *HSP104* (Martínez-Pastor et al., 1996), indicates that their expression is low following inoculation and their transcription begins to be induced some hours later, when the molecular response to osmotic stress has already finished. The gene-expression patterns observed during the first few hours of vinification depend on a variety of factors (Pérez-Torrado et al., 2002a; Zuzuarregui et al., 2005), including the metabolic status of the inoculum (differences are observed in the expression of *HSP* genes when rehydrated cells are used compared with that seen in cells derived from precultures), the nature of the osmolyte responsible for osmotic stress (glucose or glycerol), the pH, and the temperature (expression levels of *GPD1* and *HSP104* increase with reducing temperature between 15 and 28°C and with increasing pH between 3.0 and 3.6).

In our laboratory, a study was undertaken investigating the expression of 19 stress-response genes over the course of fermentations using commercial and noncommercial strains with different fermentation behavior (Zuzuarregui & del Olmo, 2004b). Aside from the differences between strains, the data obtained indicated that there was a significant reduction in mRNA levels between 1 and 6 h following inoculation of a synthetic must for the majority of the genes studied. This was the case for *HSP26*, *SSA3/4*, *STI1*, *HOR7*, *GRE2*, *SPI1*, *COX6*, *CAR1*, and *YGP1*, which are implicated in the response to different stress conditions and have specific biological roles and regulatory mechanisms. These results highlight the importance of genes that participate in cellular processes other than glycerol synthesis for the capacity of yeasts to overcome osmotic stress and initiate growth in must. More recent studies have shown that various signal transduction pathways (HOG, PKA, and TOR) are involved to a greater or lesser extent in the transcriptional response to high sugar concentrations, and that under these conditions the kinase Hog1p (essential for transcriptional activation under conditions of hyperosmolarity, reviewed in Hohmann & Mager, 2003) is phosphorylated in a manner similar to that seen in response to other types of osmotic stress (Jiménez-Martí et al., unpublished).

Recent studies of global gene expression during rehydration of dried yeast in different media and conditions and during subsequent inoculation into must (Novo et al., 2007; Rossignol et al., 2006) indicate that inoculation of must does not lead to a typical stress response, despite the hyperosmotic conditions that affect the yeast. The results show that the changes in gene expression, dependent on the presence of fermentable carbon sources, affect genes coding for proteins involved in fermentative metabolism, in the nonoxidative pentose phosphate pathway, and in ribosome biogenesis.

## 3.3. Genes Induced During the Stationary Phase

Analysis of gene expression in *Saccharomyces* yeasts during phases in which the yeast is not actively dividing is essential in order to understand the phenomenon of vinification, since, as mentioned in Section 2.3, approximately the last two thirds of the wine fermentation process occurs without cell division but in the presence of metabolically active yeasts that produce many enologically desirable compounds (Fuge et al., 1994). Termination of cell division essentially occurs as a result of the absence of one or more nutrients, leading to a series of physiological, biochemical, and morphological changes intended to ensure survival during periods of shortage, which characterize entry of cultures into the stationary phase (Herman, 2002; Werner-Washburne et al., 1996). Although under laboratory conditions the term "stationary phase" strictly applies to the termination of growth as a result of exhaustion of glucose (Werner-Washburne et al., 1996), in the case of wine fermentation, division can cease in the presence of high quantities of sugars (100–150 g/L), since other nutrients, primarily nitrogenous compounds, are usually the first to be consumed and cause termination of division (Fleet & Heard, 1993). At the end of wine fermentation, a high concentration of ethanol also affects cell metabolism (Jones, 1989). Ethanol toxicity acts via an effect on the fluidity and permeability of the plasma membrane (Alexandre et al., 1994). Both the absence of nutrients and the accumulation of ethanol coincide with the end of fermentation and their effects are synergistic and difficult to separate.

The molecular events associated with entry into the stationary phase have been described under laboratory conditions (Herman, 2002). Although winemaking conditions are not comparable to those used in the laboratory, much of the information that has been obtained can be extrapolated to the conditions currently used in wineries. For instance, Riou et al. (1997) studied the expression of 19 genes characterized as being associated with the stationary phase in laboratory strains. When those genes were analyzed under vinification conditions in an industrial strain of *S. cerevisiae*, the authors found that 60% of the genes displayed expression patterns similar to those obtained under laboratory conditions. In addition to metabolic genes used to respond to external deficiencies, many genes coding for heat shock proteins (*HSPs*) are expressed during the stationary phase. Expression of *HSPs* is typically associated with the response to stress and the gene products tend to function as molecular chaperones. These observations confirm the link between the absence of nutrients and other forms of stress. For instance, in addition to being strongly upregulated at the end of vinification, *HSP26* and *HSP30* expression is also induced by ethanol (Piper et al., 1994), and this may be responsible for the activation of their expression during the final phases. In studies of genes expressed late during microvinification, Puig et al. (1996) arrived at similar but not identical conclusions. The stress-response genes *SSA3*, *HSP12*, and *HSP26* were activated at the beginning of the stationary phase, but their levels were reduced at the end of the fermentation phase. *HSP104* and *POT1* (both identified as late-expressed genes under laboratory conditions), in contrast, were not expressed under winemaking conditions. These discrepancies may be explained by differences in the strains or musts used. The same authors identified a new gene of unknown function, *SPI1*, that is actively expressed during late phases under vinification conditions (Puig & Pérez-Ortín, 2000b). The peculiarities of the transcriptional control of this gene have allowed its promoter to be used to manipulate the expression of certain genes for biotechnological purposes (Cardona et al., 2006; Jiménez-Martí et al., 2009).

In another study, the use of Northern blotting to address global gene expression limited to the right arm of chromosome 3, Rachidi et al. (2000) identified two genes, *PAU3* (a member of a stress-response gene family of unknown function) and *ADH7* (a putative alcohol dehydrogenase), specifically expressed in the stationary phase under winemaking conditions. However, their role remains unclear. The gene *ATF1* encodes alcohol acetyltransferase, a key enzyme in the production of aromas (see Section 1.2.4). This gene also has a late expression profile. In a study by Lilly et al. (2000), it became detectable after 7 d of fermentation and reached maximal expression levels after 11 d. This expression pattern is of particular importance to the final product, since it indicates that aromas start to be produced at the end of fermentation. Expression of the gene for glycogen synthase (*GSY2*) is also increased with time of fermentation (Pérez-Torrado et al., 2002b), indicating that the accumulation of glycogen is a factor that defines entry into the stationary phase.

The expression of *YGP1* and *CAR1* has been studied under winemaking conditions with different quantities of nitrogen. Under these conditions, the expression patterns did not vary according to the quantity of nitrogen, whereas this was not the case under laboratory growth conditions (Carrasco et al., 2003). These results indicate that information obtained under controlled conditions in the laboratory cannot always be extrapolated to the conditions found in the winery. Subsequent studies have led to the identification of genes that are highly expressed under conditions in which nitrogen is limiting or absent (Jiménez-Martí et al., 2007; Mendes-Ferreira et al., 2007a, 2007b). Such genes could be used as markers to identify situations in which fermentation is limited by nitrogen deficiencies.

In our laboratory, we have analyzed the response to stress during the first half of vinification in seven industrial strains previously characterized in terms of their resistance to specific stress conditions (Zuzuarregui & del Olmo, 2004a). Coordinated expression of *SPI1*, *YGP1*, *CAR1*, and *COX6*, which are activated in response to a lack of nutrients, was detected in all strains during entry into the stationary phase after 149 h of culture (Zuzuarregui & del

Olmo, 2004b). Comparison of the expression of stress-response genes with fermentative behavior indicates that tight regulation of the response to various stress conditions may be fundamental to the adaptation of the yeast to the medium, although the absolute level of expression of the genes may sometimes be of lesser importance.

## 3.4. Gene Expression in Wine Yeasts Exposed to Specific Stress Conditions

Various stress conditions can affect the efficiency of yeasts during the different phases of wine production (Attfield, 1997; Bauer & Pretorius, 2000). Although it is essential to study the response of wine yeasts during real industrial processes, analysis of their behavior in response to specific individual stresses allows assessment of correlations with their adaptation to unfavorable environmental changes.

Ivorra et al. (1999) characterized the expression of typical stress-response genes (GPD1, HSP12, STI1, SSA3, and TRX2) to adverse conditions such as thermal shock, oxidative stress, hyperosmolarity, ethanol, and the absence of glucose. Although there was variability among different strains of wine yeast, comparison with a laboratory strain showed that the stress response mechanisms were essentially the same in all cases. HSP12, which encodes a small HSP, was the best marker for differences between strains, since it was expressed in all conditions and its levels of transcription in suboptimal conditions were lower in the strain displaying the greatest fermentative difficulty. This was the first indication of a link between defects in the stress response and stuck fermentation. That study was extended in the first systematic analysis of stress responses in 14 commercial strains, which analyzed the expression of the marker stress-response genes HSP12 and HSP104 in response to thermal shock (Carrasco et al., 2001). These two genes were expressed in all strains under conditions of

stress, and there was no clear correlation between the two variables. Resistance to stress is a complex phenomenon and the genes involved are likely to differ in importance according to specific conditions. Consequently, a more complete view of the process will require identification of a larger number of genes.

In our laboratory, we have also analyzed the effect of ethanol and, for the first time, acetaldehyde as isolated stress conditions in different wine and laboratory strains. Acetaldehyde accumulates within the cell during intense fermentation and can halt cell division (Stanley et al., 1993), and it has been proposed to be largely responsible for ethanol toxicity (Jones, 1989). Although both compounds generated transcriptional responses in which the expression of stress-response genes such as HSP12, HSP26, HSP82, and HSP104 was activated, acetaldehyde was a better transcriptional inducer than ethanol at lower concentrations (Aranda & del Olmo, 2004; Aranda et al., 2002). Those strains displaying greater induction in response to ethanol and acetaldehyde were more resistant to these stress conditions. The expression of genes encoding aldehyde dehydrogenases is also transcriptionally regulated by acetaldehyde and, to a lesser extent, by ethanol (Aranda & del Olmo, 2003).

The stress caused by addition of sulfite also has transcriptional effects. Maximal induction of the genes MET16 and ADE4 (biosynthesis of methionine and adenine) is delayed during vinification in the presence of $SO_2$ (Aranda et al., 2006). These transcriptional changes probably represent the molecular mechanism that links tolerance of sulfite to the metabolism of adenine and sulfur amino acids.

Other studies have revealed a correlation between resistance to the stress conditions relevant to wine production (oxidative stress and ethanol) and appropriateness for winemaking (Zuzuarregui & del Olmo, 2004a), a correlation that can be extended to the biological aging of certain wines, since flor strains with greater resistance to the substantial stress associated with this process (acetaldehyde, ethanol, moderate cold) are more abundant in *soleras* (Aranda et al., 2002).

It is increasingly apparent that post-transcriptional events such as the processing and transport of mRNA also influence gene expression. Thus, as the vinification process progresses, mRNA is observed to accumulate in the nucleus, and this is associated with the accumulation of ethanol at concentrations of more than 6% (Izawa et al., 2005).

Transcriptional analysis of the stress response has also been undertaken in laboratory-scale simulations of the industrial processes of biomass propagation and dehydration to obtain active dried yeast (Pérez-Torrado et al., 2005, 2009). Those studies have addressed the expression of a panel of genes that act as markers for different stresses (*TRX2, STI1, HSP12, GPD1, CUP1, GLO1, CTT1, GSH1, YGP1, GRE2, GRX5, TSA1, TRR1*). As a result, the response to oxidative stress has been identified as the most important in determining the fermentative efficiency of commercial inoculums during both yeast growth and dehydration (Garre et al., 2010; Pérez-Torrado et al., 2005, 2009).

## 4. GENETIC IMPROVEMENT OF YEAST EFFICIENCY DURING FERMENTATION

The use of pure strains for inoculation of industrial fermentations, together with a good understanding of the biochemistry, genetics, and molecular biology of *S. cerevisiae*, make it possible to develop strategies for the genetic improvement of their efficiency, taking into consideration the preservation of the genetic and genomic properties of natural strains in order to conserve their fermentative characteristics (Gimeno-Alcañiz & Matallana, 2001). Many aspects are open to improvement, in terms of both the fermentation process and the nutritional and organoleptic properties of the product (reviewed in Dequin, 2001; Pretorius & Bauer, 2002; Pretorius, 2003). Here we describe some strategies for the genetic improvement of the fermentation efficiency of yeasts. We will not discuss other types of genetic manipulation, as these will be described in Chapter 7.

*S. cerevisiae* is a highly efficient fermentative organism. However, on occasion, stuck fermentations or extended lag phases can prevent the yeasts from displacing the autochthonous flora. Different strategies have been developed to augment the fermentative capacity of active dried yeast; for instance, through accumulation of certain carbohydrate reserves implicated in stress resistance, such as glycogen and trehalose (Silljé et al., 1999). The accumulation of these metabolites can be modified by increasing the expression levels of genes that participate in their biosynthesis (*GSY1* and *GSY2* for glycogen [Farkas et al., 1991] and *TPS1* and *TPS2* for trehalose [González et al., 1992; Vuorio et al., 1993]) or by eliminating those involved in their mobilization (*GPH1* [Hwang et al., 1989] and *NTH1* and *ATH1* [Nwaka & Holzer, 1998; Nwaka et al., 1995]). In our laboratory, wine strains have been developed that show greater accumulation of glycogen as a result of regulated overexpression of the glycogen synthase encoded by the gene *GSY2*. The manipulated strain accumulates more glycogen during vinification in natural and synthetic musts, and under growth conditions similar to those used in industry to obtain yeast biomass (Pérez-Torrado et al., 2002b). It also showed greater viability when cells were recovered at the end of vinification following a period of 10 d in the finished wine, and this was accompanied by an increased fermentative capacity of the cells when reinoculated in fresh media. Other strategies that can increase the fermentative capacity are aimed at improving the efficiency of uptake and phosphorylation of sugars and have focused on the hexokinases *HXK1* and *HXK2*, the glucokinase *GLK1*, and the hexose transporters *HXT1-HXT18* and *SNF3* (Pretorius & Bauer, 2002). It would also be of interest to introduce heterologous genes coding for transporters and kinases that allow improvement in the utilization of fructose, a sugar that usually accumulates as a result of the preferential use of glucose by the yeasts (Pretorius & Bauer, 2002).

Modulation of the stress response might also contribute to improvements in the fermentative capacity of yeasts, given its relationship with the behavior of yeasts during winemaking (Aranda et al., 2002; Ivorra et al., 1999; Zuzuarregui & del Olmo, 2004a, 2004b). Recently, our group has developed strategies based on the modification of stress-response genes in wine yeasts that have led to an improvement in fermentative behavior (Cardona et al., 2006; Jiménez-Martí et al., 2009; Pérez-Torrado et al., 2009). The following genes were used: *TRX2* (encoding a cytoplasmic thioredoxin), *MSN2* (a transcription factor involved in the general stress response), *HSP26*, and *YHR087W* (a gene of unknown function that is induced under conditions of osmotic stress, including those generated by high concentrations of glucose; Jiménez-Martí, Zuzuarregui and del Olmo, unpublished results). These examples indicate that it is possible to obtain an improvement in the rate of consumption of sugars in different musts, under different conditions, and during specific stages of the vinification process by modifying stress-response genes, and this could help to facilitate the establishment of the inoculated strain and reduce fermentation time.

The low concentration of easily assimilable nitrogen sources in musts is a significant cause of loss of fermentative capacity. This problem may be alleviated through the use of modified wine yeasts able to use sources of nitrogen, such as proline, that are highly abundant in some musts but are not utilized during fermentation (Henschke, 1997).

Ethanol tolerance is very important for the capacity of inoculated yeasts to dominate the other microorganisms in the must. The accumulation of ethanol has negative consequences during fermentation because of its multiple toxic effects. The main effect is on membrane permeability, leading to loss of ions (principally magnesium and calcium) (Dombeck & Ingram, 1986; Nabais et al., 1988), and the passive diffusion of protons that alters the pH with increasing concentrations of ethanol (Cartwright et al., 1986; Leao & van Uden, 1984). These effects appear to be explained by changes in the activity of ATPase proton pumps (Cartwright et al., 1989; Rosa & Sá-Correia, 1991) and in membrane fluidity (Goldstein, 1987; Kunkee & Bisson, 1993; Sun & Sun, 1985). Strategies designed to improve ethanol tolerance should be directed towards the stimulation of sterol and long-chain unsaturated fatty acid metabolism to maintain membrane fluidity, and should also focus on genes coding for components of the membrane ATPase (*PMA1* and *PMA2*) to make it less sensitive to ethanol toxicity (Pretorius & Bauer, 2002). Wine strains have also been developed that produce more ethanol due to an increased tolerance of osmotic stress and ethanol (Hou et al., 2009). This was made possible by simultaneous overexpression of the general transcription factor Spt15p and the Spt3p subunit of SAGA complexes, implicated in the transcriptional activation of RNA-polymerase II-dependent genes.

The examples discussed in this chapter highlight how metabolic engineering can be exploited to improve the industrial applications of *S. cerevisiae*. Our increasing understanding of how this organism functions under wine fermentation conditions will lead to ever-greater opportunities for genetic improvement of the yeast strains employed.

## References

Alexandre, H., Rosusseaux, I., & Charpenter, C. (1994). Ethanol adaptation mechanisms in Saccharomyces cerevisiae. *Biotechnol. Appl. Biochem., 20*, 173–183.

Aranda, A., & del Olmo, M. (2003). Response to acetaldehyde stress in the yeast *Saccharomyces cerevisiae* involves a strain-dependent regulation of several *ALD* genes and is mediated by the general stress response pathway. *Yeast, 20,* 749–759.

Aranda, A., & del Olmo, M. (2004). Exposure to acetaldehyde in yeast causes an induction of sulfur amino acid metabolism and polyamine transporter genes, which depends on Met4p and Haa1p transcription factors respectively. *Appl. Environ. Microbiol., 70,* 1913–1922.

Aranda, A., Jiménez-Martí, E., Orozco, H., Matallana, M., & del Olmo, M. (2006). Sulfur and adenine metabolisms are linked and both modulate sulfite resistance in wine yeast. *J. Agric. Food. Chem., 54,* 5839–5846.

Aranda, A., Querol, A., & del Olmo, M. (2002). Correlation between acetaldehyde and ethanol resistance and expression of *HSP* genes in yeast strains isolated during the biological aging of sherry wines. *Arch. Microbiol., 177,* 304–312.

Attfield, P. V. (1997). Stress tolerance: The key to effective strains of industrial baker's yeast. *Nature Biotechnology, 15,* 1351–1357.

Ayestaran, B. M., Ancin, M. C., García, A. M., González, A., & Garrido, J. J. (1995). Influence of prefermentation clarification on nitrogenous contents of musts and wines. *J. Agric. Food Chem., 43,* 476–482.

Backhus, L. E., DeRisi, J., Brown, P. O., & Bisson, L. F. (2001). Functional genomic analysis of a commercial wine strain of *Saccharomyces cerevisiae* under differing nitrogen conditions. *FEMS Yeast Res., 1,* 111–125.

Bakalinsky, A. T., & Snow, R. (1990). The chromosomal constitution of wine strains of Saccharomyces cerevisiae. *Yeast, 6,* 367–382.

Barre, P. (1980). Role du facteur killer dans la concurrence entre souches de levures. *Bulletin de l'Office Internat. de la Vigne et du Vin, 593–594,* 560–572.

Barre, P., Vézinhet, F., Dequin, S., & Blondin, B. (1993). Genetic improvement of wine yeast. In G. H. Fleet (Ed.), *Wine Microbiology and Biotechnology* (pp. 421–447). UK: Harwood Academic, Reading.

Bauer, F. F., & Pretorius, I. S. (2000). Yeast stress response and fermentation efficiency: How to survive the making of wine. A review. *S. Afr. J. Enol. Vitic., 21,* 27–51.

Baumes, R., Cordonnier, R., Nitz, S., & Drawert, F. (1986). Identification and determination of volatile constituents in wines from different vine cultivars. *J. Sci. Food Agric., 37,* 927–943.

Beltrán, G., Esteve-Zarzoso, B., Rozés, N., Mas, A., & Guillamón, J. M. (2005). Influence of timing of nitrogen additions during synthetic grape must fermentations on fermentation kinetics and nitrogen consumption. *J. Agric. Food Chem., 53,* 996–1002.

Bely, M., Sablayrolles, J. M., & Barre, P. (1990). Automatic detection of assimilable nitrogen deficiencies during alcoholic fermentation in oenological conditions. *J. Ferment. Bioeng., 70,* 246–252.

Bertrand, A., & Miele, A. (1984). Influence de la clarification du mout de raisin sur sa teneeur en acides gras. *Connais Vigne Vin, 18,* 293–297.

Bidenne, C., Blondin, B., Dequin, S., & Vezinhet, F. (1992). Analysis of the chromosomal DNA polymorphism of wine strains of *Saccharomyces cerevisiae*. *Curr. Genet., 22,* 1–7.

Bisson, L. (1999). Stuck and sluggish fermentations. *Am. J. Enol. Viticult., 50,* 107–119.

Blomberg, A., & Adler, L. (1992). Physiology of osmotolerance in fungi. *Adv. Microbiol. Physiol., 33,* 145–212.

Boulton, R. B., Singleton, V. L., Bisson, L. F., & Kunkee, R. E. (Eds.), (1996). *Principles and practices of winemaking.* New York, NY: Chapman.

Cabanis, J. C., Cabanis, M. T., Cheynier, V., & Teissedre, P. L. (1998). Tables de composition. In E. Flanzy (Ed.), *Oenologie. Fondements scientifiques et technologiques* (pp. 315–336). Paris, France: Technique & Documentation.

Cabras, P., & Angioni, A. (2000). Pesticide residues in grapes, wines, and their processing productos. *J. Agric. Food Chem., 48,* 967–973.

Cantarelli, C. (1989). Phenolics and yeast: Remarks concerning fermented beberages. *Yeast, 5,* S53–61.

Cardona, F., Carrasco, P., Pérez-Ortín, J. E., del Olmo, M., & Aranda, A. (2006). Effect of a modification in the expression of MSN2 transcription factor in wine yeast on stress resistance and fermentative behaviour. *Int. J. Food Microbiol., 114,* 83–91.

Carrasco, P., Pérez-Ortín, J. E., & del Olmo, M. (2003). Arginase activity is a useful marker of nitrogen limitation during wine fermentations. *System. Appl. Microbiol., 26,* 471–479.

Carrasco, P., Querol, A., & del Olmo, M. (2001). Analysis of the stress resistance of commercial wine yeast strains. *Arch. Microbiol., 175,* 450–457.

Carrau, F. M., Medina, K., Farina, L., Boido, E., Henschke, P. A., & Dellacassa, E. (2008). Production of fermentation aroma compounds by *Saccharomyces cerevisiae* wine yeasts: Effects of yeast assimilable nitrogen on two model strains. *FEMS Yeast Res., 8,* 1196–1207.

Cartwright, C. P., Juroszek, J.-R., Beavan, M. J., Ruby, F. M. S., DeMorias, S. M. F., & Rose, A. H. (1986). Ethanol dissipates the proton-motive force across the plasma membrane of Saccharomyces cerevisiae. *J. Gen. Microbiol., 132,* 369–377.

Cartwright, C. P., Rose, A. H., Calderbank, J., & Keenan, H. J. (1989). Solute transport. In A. H. Rose & J. S. Harrison (Eds.), *The yeast. Metabolism and physiology of yeast* (pp. 5–49). London, UK: Academic Press.

Castor, J. G. B., & Archer, T. E. (1956). Amino acids in must and wines, praline, serine and threonine. *Am. J. Enol. Viticult., 7*, 19–25.

Chatonnet, P., Viala, C., & Dubourdieu, D. (1997). Influence of polyphenolic components of red wines on the microbial synthesis of volatile phenols. *Am. J. Enol. Viticult., 48*, 443–448.

Christiansen, G., & Christiansen, C. (1976). Comparison of the fine structure in the mitochondrial DNA from *Saccharomyces cerevisiae* and *S. carlsbergensis*: Electron microscopy of partially denatured molecules. *Nucleic Acids Res., 3*, 465–476.

Clark-Walker, G. D., McArthur, C. R., & Daley, D. J. (1981). Evolution of mitochondrial genomes in fungi. *Curr. Genet., 4*, 7–12.

Codón, A. C., Gasent-Ramírez, J. M., & Benítez, T. (1995). Factors which affect the frequency of sporulation and tetrad formation in *Saccharomyces cerevisiae* bakers yeasts. *Appl. Environ. Microbiol., 59*, 345–386.

Cooper, T. G. (1982a). Nitrogen metabolism in *Saccharomyces cerevisiae*. In J. N. Strathern, E. W. Jones & J. B. Broach (Eds.), *The molecular biology of the yeast Saccharomyces: Metabolism and gene expression* (pp. 39–99). Cold Spring Harbor, NY: Cold Spring Harbor Laboratory.

Cooper, T. G. (1982b). Transport in *Saccharomyces cerevisiae*. In J. N. Strathern, E. W. Jones, & J. B. Broach (Eds.), *The molecular biology of the yeast Saccharomyces: Metabolism and gene expression* (pp. 399–461). Cold Spring Harbor, NY: Cold Spring Harbor Laboratory.

Corison, C. A., Ough, C. S., Berg, H. W., & Nelson, K. E. (1979). Must acetic acid and ethyl acetate as mold rot indicators in grapes. *Am. J. Enol. Viticult., 30*, 130–134.

Darriet, P., Boldron, J. N., & Dubourdieu, D. (1988). L'hydrolyse des hétérosides terpéniques du Muscat àpetits grains par les enzymes périplasmiques de Saccharomyces cerevisiae. *Connaiss. Vigne Vin, 22*, 189–195.

Degré, R. (1993). Selection and commercial cultivation of wine yeast and bacteria. In G. H. Fleet (Ed.), *Wine microbiology and biotechnology* (pp. 421–447). Chur, Switzerland: Harwood Academic Publishers.

del Pozo, L. D., Abarca, D., Claros, M. G., & Jiménez, A. (1991). Cycloheximide resistance as a yeast cloning marker. *Curr. Genet., 19*, 353–358.

Delteil, D., & Aizac, T. (1988). Comparaison de différences techniques de levurage par suivi de l'implantation d'une souche de levure oenologique marquée. *Connaiss. Vigne Vin, 21*, 267–278.

Dequin, S. (2001). The potential of genetic engineering for improving brewing, wine-making and baking yeasts. *Appl. Microbiol. Biotechnol., 56*, 577–588.

Dombeck, K. M., & Ingram, L. O. (1986). Magnesium limitation and its role in apparent toxicity of ethanol during yeast fermentation. *Appl. Environ. Microbiol., 52*, 975–981.

Dubois, P. (1983). Volatile phenols in wine. In J. R. Piggot (Ed.), *Flavour of distilled beverages, origin and development* (pp. 110–119). Chichester, UK: Ellis Horwood.

Dubourdieu, D., Darriet, O., Ollivier, J. N., & Ribereau-Gayon, P. (1988). Role de la levure Saccharomyces cerevisiae dans l'hydrolyse enzymatique des héterosides terpéniques du jus de raisin. *C.R. Acad. Sci., 306*, 489–493.

Egli, C. M., Edinger, W. D., Mitrakul, C. M., & Henick-Kling, T. (1998). Dynamics of indigenous and inoculated yeast populations and their effect on the sensory character of Riesling and Chardonnay wines. *J. Appl. Microbiol., 85*, 779–789.

Eglinton, J. M., & Henschke, P. A. (1999). The occurrence of volatile acidity in Australian wines. *Australian Grapegrower and Winemaker, 426a*, 7–12.

Erasmus, D. J., van der Merwe, G. K., & van Vuuren, H. J. J. (2003). Genome-wide expression analyses: Metabolic adaptation of *Saccharomyces cerevisiae* to high sugar stress. *FEMS Yeast Res., 3*, 375–399.

Esteve-Zarzoso, B., Belloch, C., Uruburu, F., & Querol, A. (1999). Identification of yeast by RFLP analysis of the 5.8S rRNA gene and the two ribosomal internal transcribed spacers. *Int. J. Syst. Bacteriol., 49*, 329–337.

Etievant, P. X. (1981). Volatile phenol determination in wine. *J. Agric. Food Chem., 29*, 65–67.

Farkas, I., Hardy, T. A., Goebl, M. G., & Roach, P. J. (1991). Two glycogen synthase isoforms in *Saccharomyces cerevisiae* are coded by distinct genes that are differentially controlled. *J. Biol. Chem., 266*, 15602–15607.

Farkaš, J. (1988). *Technology and biochemistry of wine*. New York, NY: Gordon and Breach.

Fleet, G. H. (2007). Wine. In M. P. Doyle, & L. R. Beuchat (Eds.), *Food microbiology: Fundamentals and frontiers* (3rd ed.), (pp. 863–890) Washington, DC: ASM Press.

Fleet, G. H. (2008). Wine yeasts for the future. *FEMS Yeast Res., 8*, 979–995.

Fleet, G. H., & Heard, G. (1993). Yeast growth during fermentation. In G. Fleet (Ed.), *Wine microbiology and biotechnology* (pp. 27–57). Chur, Switzerland: Harwood Academic Publishers.

Fremont, L. (2000). Biological effects of resveratrol. *Life Sci., 66*, 663–673.

Fuge, E. K., Braun, E. L., & Werner-Washburne, M. (1994). Protein synthesis in long-term stationary-phase cultures of *Saccharomyces cerevisiae*. *J. Bacteriol., 176*, 5802–5813.

Fugelsang, K. C., & Edwards, C. G. (2007). *Wine microbiology: Practical applications and procedures*. New York, NY: Springer.

Gancedo, J. M. (2008). The early steps of glucose signalling in yeast. *FEMS Microbiol. Rev., 32*, 673–704.

Garre, E., Raginel, F., Palacios, A., Julien, A., & Matallana, E. (2010). Oxidative stress responses and lipid peroxidation damage are induced during dehydration in the production of dry active wine yeasts. *Int. J. Food Microbiol., 136*(3), 295–303.

Gimeno-Alcañiz, J. V., & Matallana, E. (2001). Performance of industrial strains of Saccharomyces cerevisiae during wine fermentation is affected by manipulation strategies based on sporulation. *System. Appl. Microbiol., 24,* 639–644.

Giudici, P., Zambonelli, C., & Kunkee, R. E. (1993). Increased production of n-propanol in wine by yeast strains having an impaired ability to form hydrogen sulphide. *Am. J. Enol. Vitic., 44,* 123–127.

Goldstein, D. (1987). Ethanol-induced adaptation in biological membranes. *Ann. N.Y. Acad. Sci., 492,* 103–111.

González, M. I., Stucka, R., Blázquez, M. A., Feldmann, H., & Gancedo, C. (1992). Molecular cloning of CIF1, a yeast gene necessary for growth on glucose. *Yeast, 8,* 183–192.

Grenson, M., Mousset, M., Wiame, J. M., & Bechet, J. (1966). Multiplicity of the amino acid permeases in Saccharomyces cerevisiae. I. Evidence for a specific arginine-transporting system. *Biochem. Biophys. Acta., 127,* 325–338.

Guijo, S., Mauricio, J. C., Salmon, J. M., & Ortega, J. M. (1997). Determination of the relative ploidy in different Saccharomyces cerevisiae strains used for fermentation and "flor" film ageing of dry sherry-type wines. *Yeast, 13,* 101–117.

Hadfield, C., Jordan, B. E., Mount, R. C., Pretorius, G. H., & Burak, E. (1990). G418-resistance as a dominant marker and reporter for gene expression in Saccharomyces cerevisiae. *Curr. Genet., 18,* 303–313.

Heard, G. M., & Fleet, G. H. (1988). The effects of temperature and pH on the growth of yeast species during the fermentation of grape juice. *J. Appl. Bact., 65,* 23–28.

Henick-Kling, T., Edinger, W., Daniel, P., & Monk, P. (1998). Selective effects of sulfur dioxide and yeast starter culture addition on indigenous yeast populations and sensory characteristics of wine. *J. Appl. Microbiol., 84,* 865–876.

Henschke, P. A. (1997). Wine yeasts. In F. K. Zimmermann & K.-D. Entian (Eds.), *Yeast sugar metabolism* (pp. 527–560). Lancaster, PA: Technomic Publishing Co.

Henschke, P. A., & Jiranek, V. (1993). Metabolism of nitrogen compounds. In G. H. Fleet (Ed.), *Wine microbiology and biotechnology* (pp. 77–164). Chur, Switzerland: Harwood Academic Publishers.

Heresztyn, N. (1984). Methyl and ethyl amino acid esters in wine. *J. Agric. Food Chem., 32,* 916–918.

Herman, P. K. (2002). Stationary phase in yeast. *Curr. Opin. Microbiol., 5,* 607–692.

Herraiz, T., & Ough, C. S. (1993). Formation of ethyl esters of amino acids by yeasts during the alcoholic fermentation of grape juice. *Am. J. Enol. Viticult., 44,* 41–48.

Hohmann, S., & Mager, W. H. (2003). *Yeast stress responses.* Berlin, Germany: Springer-Verlag.

Hollenberg, C. P., Borst, P., & van Bruggen, E. F. J. (1970). Mitochondrial DNA. A 25 micron closed circular duplex DNA molecule in wild type yeast mitochondria. Structure and genetic complexity. *Biochim. Biophys. Acta., 209,* 1–15.

Horak, J. (1986). Amino acid transport in eukaryotic microorganisms. *Biochim. Biophys. Acta., 864,* 223–256.

Hou, L., Cao, X., Wang, C., & Lu, M. (2009). Effect of overexpression of transcription factors on the fermentation properties of Saccharomyces cerevisiae industrial strains. *Lett. Appl. Microbiol., 49,* 14–19.

Howitz, K. T., Bitterman, K. J., Cohen, H. Y., Lamming, D. W., Lavu, S., Wood, J. G., et al. (2003). Small molecule activators of sirtuins extend Saccharomyces cerevisiae lifespan. *Nature, 425,* 191–196.

Hwang, P. K., Tugendreich, S., & Fletterick, R. J. (1989). Molecular analysis of GPH1, the gene encoding glycogen phosphorylase in Saccharomyces cerevisiae. *Mol. Cell. Biol., 9,* 1659–1666.

Ingledew, M., & Kunkee, R. E. (1985). Factors influencing sluggish fermentations of grape juice. *Am. J. Enol. Viticult., 36,* 65–76.

Ivorra, C., Pérez-Ortín, J. E., & del Olmo, M. (1999). An inverse correlation between stress resistance and stuck fermentations in wine yeasts. A molecular study. *Biotechnol. Bioeng., 64,* 698–708.

Izawa, S., Takemura, R., Miki, T., & Inoue, Y. (2005). Characterization of the export of bulk poly(A)$^+$ mRNA in Saccharomyces cerevisiae during the wine making process. *Appl. Environ. Microbiol., 71,* 2179–2182.

Jackson, R. S. (1994). *Wine science. Principles and applications.* San Diego, CA: Academic Press.

Jiménez-Martí, E., Aranda, A., Mendes-Ferreira, A., Mendes-Faia, A., & del Olmo, M. (2007). The nature of the nitrogen source added to nitrogen depleted vinifications conducted by a Saccharomyces cerevisiae strain in synthetic must affects gene expression and the levels of several volatile compounds. *Anton. Leeuw., 92,* 61–75.

Jiménez-Martí, E., & del Olmo, M. (2008). Addition of ammonia or amino acids to a nitrogen-depleted medium affects gene expression patterns in yeast cells during alcoholic fermentation. *FEMS Yeast Res., 8,* 245–256.

Jiménez-Martí, E., Zuzuarregui, A., Ridaura, I., Lozano, N., & del Olmo, M. (2009). Genetic manipulation of HSP26 and YHR087W stress genes may improve fermentative behaviour in wine yeasts under vinification conditions. *Int. J. Food Microbiol., 130,* 122–130.

Jones, R. P. (1989). Biological principles for the effects of ethanol. *Enzyme Microbiol. Technol., 11,* 130–153.

Kaeberlein, M., Andalis, A. A., Fink, G. R., & Guarente, L. (2002). High osmolarity extends life span in Saccharomyces cerevisiae by a mechanism related to calorie restriction. *Mol. Cell. Biol., 22,* 8056–8066.

Kudo, M., Vagnoli, P., & Bisson, L. F. (1998). Imbalance of potassium and hydrogen ion concentrations as a cause of stuck enological fermentations. *Am. J. Enol. Viticult., 49*, 295–301.

Kunkee, R. E., & Bisson, L. F. (1993). Wine-making yeats. In A. H. Rose & J. S. Harrison (Eds.), (2nd ed.), *The yeasts, Vol. 5* (pp. 69–127). London, UK: Academic Press.

Lafon-Lafourcade, S. (1983). Wine and brandy. In G. Reed (Ed.), *Biotechnology* (pp. 81–163). Heidelberg, Germany: Verlag-Chemie.

Lagunas, R. (1986). Misconceptions about the energy metabolism of *Saccharomyces cerevisiae*. *Yeast, 2*, 221–228.

Lambrechts, M. G., & Pretorius, I. S. (2000). Yeast and its importance to wine aroma – A review. *S. Afr. J. Enol. Vitic., 21*, 97–129.

Leao, C., & van Uden, N. (1984). Effects of ethanol and other alkanols on passive proton influx in the yeasts Saccharomyces cerevisiae. *Biochim. Biophys. Acta., 774*, 43–48.

Lilly, M., Lambretsch, M. G., & Pretorius, I. S. (2000). Effect of increased yeast alcohol acetyltransferase activity on flavour profiles of wine and distillates. *Appl. Environ. Microbiol., 66*, 744–753.

Llauradó i Reverchon, J.M. (2002). *Avaluació dels condicionants del most en el desenvolupament de la fermentació alcohòlica a baixes temperatures.* Doctoral thesis. Tarragona, Spain: Universitat Rovira i Virgili.

Longo, E., & Vézinhet, F. (1993). Chromosomal rearrangements during vegetative growth of a wild strain of Saccharomyces cerevisiae. *Appl. Environ. Microbiol., 59*, 322–326.

Marini, A. M., Soussi-Boudekou, S., Vissers, S., & Andre, B. (1997). A family of ammonium transporters in Saccharomyces cerevisiae. *Mol. Cell Biol., 17*, 4282–4293.

Marini, A. M., Vissers, S., Urrestarazu, A., & Andre, B. (1994). Cloning and expression of the MEP1 gene encoding an ammonium transporter in Saccharomyces cerevisiae. *EMBO J., 13*, 3456–3463.

Marks, V. D., Ho Sui, S. J., Erasmus, D., van der Merwe, G. K., Brumm, J., Wasserman, W. W., et al. (2008). Dynamics of the yeast transcriptome during wine fermentation reveals a novel fermentation stress response. *FEMS Yeast Res., 8*, 35–52.

Marks, V., van der Merwe, G., & van Vuuren, H. J. J. (2003). Transcriptional profiling of wine yeast in fermenting grape juice: Regulatory effect of diammonium phosphate. *FEMS Yeast Research, 3*, 269–287.

Martínez-Pastor, M. T., Marchler, G., Schuller, C., Marchler-Bauer, A., Ruis, H., & Estruch, F. (1996). The Saccharomyces cerevisiae zinc finger proteins Msn2p and Msn4p are required for transcriptional induction through the stress-response element (STRE). *EMBO J., 15*, 2227–2235.

Martini, A., & Vaughan-Martini, A. (1990). Grape must fermentation: Past and present. In J. F. T. Spencer &

D. M. Spencer (Eds.), *Yeast technology* (pp. 105–123). Berlin, Germany: Springer-Verlag.

Mendes-Ferreira, A., Barbosa, C., Falco, V., Leão, C., & Mendes-Faia, A. (2009). The production of hydrogen sulphide and other aroma compounds by wine strains of Saccharomyces cerevisiae in synthetic media with different nitrogen concentrations. *J. Int. Microbiol. Biotechnol., 36*, 571–583.

Mendes-Ferreira, A., del Olmo, M., García-Martínez, J., Jiménez-Martí, E., Leão, C., Mendes-Faia, A., et al. (2007a). Saccharomyces cerevisiae signatura genes for predicting nitrogen deficiency during alcoholic fermentation. *Appl. Env. Microbiol., 73*, 5363–5369.

Mendes-Ferreira, A., del Olmo, M., García-Martínez, J., Jiménez-Martí, E., Mendes-Faia, A., Pérez-Ortín, J. E., & Leão, C. (2007b). Transcriptional response of Saccharomyces cerevisiae to different nitrogen concentrations during alcoholic fermentation. *Appl. Env. Microbiol., 73*, 3049–3060.

Mendes-Ferreira, A., Mendes-Faia, A., & Leão, C. (2004). Growth and fermentation patterns of Saccharomyces cerevisiae under different ammonium concentrations and its implications in winemaking industry. *J. Appl. Microbiol., 97*, 540–545.

Monteiro, F., & Bisson, L. F. (1991). Amino acid utilization and urea formation during vinification fermentations. *Am. J. Enol. Viticult., 43*, 1–10.

Monteiro, F. F., & Bisson, L. F. (1992). Utilization of adenine by yeast during grape juice fermentation and investigation of the possible role of adenine as a precursor of urea. *Am. J. Enol. Viticult., 43*, 18–22.

Mortimer, R. K., Romano, P., Suzzi, G., & Polsinelli, P. (1994). Genome renewal: A new phenomenon revealed from a genetic study of 43 strains of Saccharomyces cerevisiae derived from natural fermentation of grape musts. *Yeast, 10*, 1543–1552.

Muñoz, E., & Ingledew, W. M. (1990). Yeast hulls in wine fermentation. A review. *J. Wine Res., 1*, 197–209.

Nabais, R. C., Sá-Correia, I., Viegas, C. A., & Novais, J. M. (1988). Influence of calcium ion on ethanol fermentation by yeasts. *Appl. Environ. Microbiol., 54*, 2439–2446.

Nishino, H., Miyazakim, S., & Tohjo, K. (1985). Effect of osmotic pressure on the growth rate and the fermentation activity of wine yeast. *Am. J. Enol.Viticult., 36*, 170–174.

Norbeck, J., & Blomberg, A. (1997). Metabolic and regulatory changes associated with growth of Saccharomyces cerevisiae in 1.4 M NaCl. *J. Biol. Chem., 272*, 5544–5554.

Nordström, K. (1964a). Formation of esters from acids by brewer's yeast. II. Formation from lower fatty acids. *J. Inst. Brew., 70*, 42–55.

Nordström, K. (1964b). Formation of esters from acids by brewer's yeast. IV. Effect of higher fatty acids and toxicity of lower fatty acids. *J. Inst. Brew., 70*, 233–242.

Nordström, K. (1964c). Formation of esters from alcohols by brewer's yeast. *J. Inst. Brew., 70*, 328–336.

Novo, M., Beltran, G., Rozes, N., Guillamon, J. M., Sokol, S., Leberre, V., et al. (2007). Early transcriptional response of wine yeast after rehydration: Osmotic shock and metabolic activation. *FEMS Yeast Res., 7*, 304–316.

Nwaka, S., & Holzer, H. (1998). Molecular biology of trehalose and trehalases in the yeast Saccharomyces cerevisiae. *Prog. Nucleic Acid Res. Mol. Biol., 58*, 197–237.

Nwaka, S., Mechler, B., Destreulle, M., & Holzer, H. (1995). Phenotypic features of trehalase mutants in Saccharomyces cerevisiae. *FEBS Lett., 360*, 286–290.

Nykänen, L. (1986). Formation and occurrence of flavour compounds in wine and distilled alcoholic beverages. *Am. J. Enol. Viticult., 37*, 84–96.

Nykänen, L. L., Nykänen, I., & Soumalainen, H. (1977). Distribution of esters produced during sugar fermentation between the yeast cell and the medium. *J. Inst. Brew., 83*, 32–34.

Oliva, J., Cayuela, M., Paya, P., Martinez-Cacha, A., Cámara, M. A., & Barba, A. (2007). Influence of fungicides on grape yeast content and its evolution in the fermentation. *Commun. Agric. Appl. Biol. Sci., 72*, 181–189.

Ough, C. S. (1992). *Winemaking basics*. Binghamton, NY: Haworth Press Inc.

Ough, C. S., & Amerine, M. A. (1988). *Nitrogen compounds. Methods for analysis of must and wines* (2nd ed.), University of California, CA: John Wiley & Sons.

Paltauf, F., Kohlwein, S. P., & Henry, S. A. (1992). Regulation and compartimentalization of lipid synthesis in yeast. In E. W. Jones, J. R. Pringle, & J. R. Broach (Eds.), *The molecular biology of the yeast Saccharomyces: Metabolism and gene expression* (pp. 415–500). Cold Spring Harbor, NY: Cold Spring Harbor Laboratory.

Pampuhla, M. E., & Loereiro-Dias, C. (1989). Combined effect of acetic acid, pH, and ethanol on intracellular pH of fermenting yeast. *Appl. Microbiol. Biotech., 31*, 547–550.

Panozzo, C., Nawara, M., Suski, C., Kucharczyka, R., Skoneczny, M., Becam, A. M., et al. (2002). Aerobic and anaerobic NAD$^+$ metabolism in. *Saccharomyces cerevisiae. FEBS Lett., 517*, 97–102.

Peddie, H. A. B. (1990). Ester formation in brewery fermentations. *J. Inst. Brew., 96*, 327–331.

Pérez-Ortín, J. E., García-Martínez, J., & Alberola, T. M. (2002). DNA chips for yeast biotechnology. The case of wine yeasts. *J. Biotechnol., 98*, 227–241.

Pérez-Torrado, R., Bruno-Barcena, J. M., & Matallana, E. (2005). Monitoring stress related genes during the process of biomass propagation of Saccharomyces cerevisiae strains used for wine making. *Appl. Environm. Microbiol., 71*, 6831–6837.

Pérez-Torrado, R., Carrasco, P., Aranda, A., Gimeno-Alcañiz, J., Pérez-Ortín, J. E., Matallana, E., et al. (2002a). Study of the first hours of microvinification by the use of osmotic stress-response genes as probes. *System. Appl. Microbiol., 25*, 153–161.

Pérez-Torrado, R., Gimeno-Alcañiz, J. V., & Matallana, E. (2002b). Wine yeast strains engineered for glycogen overproduction display enhanced viability under glucose deprivation conditions. *Appl. Environ. Microbiol., 68*, 3339–3344.

Pérez-Torrado, R., Gómez-Pastor, R., Larsson, C., & Matallana, E. (2009). Fermentative capability of dry active wine yeast requires a specific oxidative stress response during industrial biomass growth. *Appl. Microbiol. Biotechnol., 81*, 951–960.

Piper, P. W., Talreja, K., Panaretou, B., Moradas-Ferreira, P., Byrne, K., Praekelt, U. M., et al. (1994). Induction of major heat-shock proteins of Saccharomyces cerevisiae, including plasma membrane Hsp30, by ethanol levels above critical threshold. *Microbiology, 140*, 3031–3038.

Pretorius, I. S. (2000). Tailoring wine yeast for the new millennium: Novel approaches to the ancient art of winemaking. *Yeast, 16*, 675–729.

Pretorius, I. S. (2003). The genetic analysis and tailoring of wine yeasts. In J. H. de Winde (Ed.), *Functional genetics of industrial yeasts* (pp. 99–142). Berlin, Germany: Springer.

Pretorius, I. S., & Bauer, F. F. (2002). Meeting the consumer challenge through genetically customized wine-yeast strains. *Trends Biotechnol., 20*, 426–432.

Puig, S., & Pérez-Ortín, J. E. (2000a). Expression levels and patterns of glycolytic yeast genes during wine fermentation. *System. Appl. Microbiol., 23*, 300–303.

Puig, S., & Pérez-Ortín, J. E. (2000b). Stress response and expression patterns in wine fermentations of yeast genes induced at the diauxic shift. *Yeast, 16*, 139–148.

Puig, S., Querol, A., Barrio, E., & Pérez-Ortín, J. E. (2000). Mitotic recombination and genetic changes in Saccharomyces cerevisiae during wine fermentation. *Appl. Environ. Microbiol., 66*, 2057–2061.

Puig, S., Querol, A., Ramón, D., & Pérez-Ortín, J. E. (1996). Evaluation of the use of phase-specific gene promoters for the expression of enological enzymes in an industrial wine yeast strain. *Biotechnol. Lett., 18*, 887–892.

Querol, A., Barrio, E., Huerta, T., & Ramón, D. (1992). Molecular monitoring of wine fermentations conducted by active dry yeast strains. *Appl. Environ. Microbiol., 58*, 2948–2953.

Rachidi, N., Barre, P., & Blondin, B. (1999). Multiple Ty-mediated chromosomal translocations lead to karyotype changes in a wine strain of. *Saccharomyces cerevisiae. Mol. Gen. Genet., 261*, 841–850.

Rachidi, N., Barre, P., & Blondin, B. (2000). Examination of the transcriptional specificity of an enological yeast. A pilot experiment on the chromosome-III right arm. *Curr. Genet., 37*, 1–11.

Ramey, D. D., & Ough, C. S. (1980). Volatile ester hydrolysis on formation during storage of model solutions and wines. *J. Agric. Food Chem., 28*, 928–934.

Rapp, A., & Versini, G. (1991). Influence of nitrogen compounds in grapes on aroma compounds of wine. In J. M. Rantz (Ed.), *Proceedings of the International Symposium on Nitrogen in Grapes and Wines* (pp. 156–164). Davis, CA: American Society for Enology and Viticulture.

Ratledge, C., & Evans, C. T. (1989). In A. H. Rose, & J. S. Harrison (Eds.), *The yeasts. Lipids and their metabolism, Vol. 3* (pp. 367–455). London, UK: Academic Press.

Rauhut, D. (1993). Yeast-production of sulfur compounds. In G. H. Fleet (Ed.), *Wine microbiology and biotechnology* (pp. 183–223). Chur, Switzerland: Harwood Academic Publishers.

Rauhut, D., & Kürbel, H. (1996). Identification of wine aroma defects caused by sulfur-containing metabolites of yeast. In A. Lonvaud-Funel (Ed.), *Enologie 95, 5e Symposium International d'Oenologie* (pp. 515–519). London: Technique & Documentation.

Ribéreau-Gayon, P., Dubourdieu, D., Donèche, B., & Lonvaud, A. (2006). *Handbook of enology, Vol. 1. The microbiology of wine and vinifications* (2nd ed.), Chichester, UK: John Wiley & Sons.

Riou, C., Nicaud, J. M., Barre, P., & Gaillardin, C. (1997). Stationary-phase gene expression in *Saccharomyces cerevisiae* during wine fermentation. *Yeast, 13*, 903–915.

Roon, R. J., Levy, J. S., & Larimore, F. (1977). Negative interactions between amino acid and methylamine/ammonia transport systems of Saccharomyces cerevisiae. *J. Biol. Chem., 252*, 3599–3604.

Rosa, M. F., & Sá-Correia, I. (1991). *In vivo* activation by ethanol of plasma membrane ATPase of Saccharomyces cerevisiae. *Appl. Environ. Microbiol., 57*, 830–835.

Rose, A. H., & Harrison, J. S. (Eds.), (1991). *The yeasts* (2nd ed.), London, UK: Academic Press.

Rossignol, T., Dulau, L., Julien, A., & Blondin, B. (2003). Genome-wide monitoring of wine yeast gene expression during alcoholic fermentation. *Yeast, 20*, 1369–1385.

Rossignol, T., Postaire, O., Storaï, J., & Blondin, B. (2006). Analysis of the genomic response of a wine yeast to rehydration and inoculation. *Appl. Microbiol. Biotechnol., 71*, 699–712.

Sablayrolles, J. M., Dubois, C., Manginot, C., Roustan, J. L., & Barre, P. (1996). Effectiveness of combined ammoniacal nitrogen and oxygen additions for completion of sluggish and stuck fermentations. *J. Ferm. Bioeng., 82*, 377–381.

Sadraud, P., & Chauvet, S. (1985). Activite antilevure de l'anhydride sulfureux moleculaire. *Connais. Vigne Vin, 19*, 31–40.

Salmon, J. M. (1997). Enological fermentation kinetics of an isogenic ploidy series derived from an industrial *Saccharomyces cerevisiae* strain. *J. Ferment. Bioeng., 83*, 253–260.

Santangelo, G. M. (2006). Glucose signaling in Saccharomyces cerevisiae. *Microbiol. Mol. Biol. Rev., 70*, 253–282.

Schütz, M., & Gafner, J. (1993). Analysis of yeast diversity during spontaneous and induced alcoholic fermentations. *J. Appl. Bacteriol., 75*, 551–558.

Silljé, H. H. W., Paalman, J. W. G., Schure, E. G., Olsthoorn, S. Q. B., Verkleij, A. J., Boonstra, J., et al. (1999). Function of trehalose and glycogen in cell cycle progression and cell viability in Saccharomyces cerevisiae. *J. Bacteriol., 181*, 396–400.

Singleton, V. L., & Esau, P. (1969). Phenolic substances in grapes and wine and their significance. *Adv. Food Res., 1*, 1–282.

Snow, R. (1983). Genetic improvement of wine yeast. In J. F. T. Spencer, D. M. Spencer, & A. R. W. Smith (Eds.), *Yeast genetics. Fundamental and applied aspects* (pp. 439–459). New York, NY: Springer-Verlag.

Sprague, G. F. (1995). In A. H. Rose, A. E. Wheals, & J. S. Harrison (Eds.), *The yeasts*. London, UK: Academic Press.

Stanley, G. A., Douglas, N. G., Every, E. J., Tzanatos, T., & Pamment, N. B. (1993). Inhibition and stimulation of yeast growth by acetaldehyde. *Biotechnol. Lett., 15*, 1199–1204.

Steger, C. L. C., & Lambrechts, M. G. (2000). The selection of yeast strains for the production of premium quality South African brandy base products. *J. Ind. Microbiol. Biotech., 24*, 1–11.

Stratford, M., & Rose, A. H. (1986). Transport of sulphur dioxide by Saccharomyces cerevisiae. *J. Gen. Microbiol., 132*, 1–6.

Strathern, J. N., Jones, E. W., & Broach, J. R. (Eds.), (1982). *The molecular biology of the yeast Saccharomyces cerevisiae*. Cold Spring Harbor, NY: Cold Spring Harbor Laboratory.

Sumrada, R., & Cooper, T. G. (1982). Isolation of the *CAR1* gene from *Saccharomyces cerevisiae* and analysis of its expression. *Mol. Cell Biol., 2*, 1514–1523.

Sun, G. Y., & Sun, A. Y. (1985). Ethanol and membrane lipids. *Alcohol. Clin. Exp. Res., 9*, 164–180.

Thomas, K. C., Hynes, S. H., & Ingledew, W. M. (1994). Effects of particulate materials and osmoprotectants on very high gravity ethanolic fermentation by Saccharomyces cerevisiae. *Appl. Environ. Microbiol., 60*, 1519–1524.

Thornton, R. J., & Eschenbruch, R. (1976). Homothalism in wine yeasts. *Anton. Leeuw., 42*, 503–509.

Torija, M. J., Beltran, G., Novo, M., Poblet, M., Guillamón, J. M., Mas, A., et al. (2003). Effects of fermentation temperature and Saccharomyces species on the cell fatty acid composition and presence of volatile compounds in wine. *Int. J. Food Microbiol., 85*, 127–136.

Vilanova, M., Ugliano, M., Varela, M., Siebert, T., & Pretorius, I. S. (2007). Assimilable nitrogen utilisation and production of volatile and non-volatile compounds in chemically defined medium by *Saccharomyces cerevisiae* wine yeasts. *Appl. Microbiol. Biotechnol., 77*, 145–157.

Visser, W., Scheffers, W. A., Batenburg-van der Vegte, W. H., & van Dijken, J. P. (1990). Oxygen requirements of yeasts. *Appl. Environ. Microbiol., 56*, 3785–3792.

Vuorio, O. E., Kalkkinen, N., & Londesborough, L. (1993). Cloning of two related genes encoding the 56-kDa and 123-kDa subunits of trehalose synthase from the yeast Saccharomyces cerevisiae. *Eur. J. Biochem., 216*, 849–861.

Wang, X. D., Bohlscheid, J. C., & Edwards, C. G. (2003). Fermentative activity and production of volatile compounds by *Saccharomyces* grown in synthetic grape juice media deficient in assimilable nitrogen and/or pantothenic acid. *J. Appl. Microbiol., 94*, 349–359.

Waterhouse, A. L. (2002). Wine phenolics. *Ann. N.Y. Acad. Sci., 957*, 21–36.

Watson, K. (1987). Temperature relations. In A. H. Rose, & J. S. Harrison (Eds.) (2nd ed.), *The yeasts, Vol. 2* (pp. 41–47). London, UK: Academic Press.

Werner-Washburne, M., Braun, E. L., Crawford, M. E., & Peck, V. M. (1996). Stationary phase in Saccharomyces cerevisiae. *Mol. Microbiol., 19*, 1159–1166.

Wiemken, A., & Durr, M. (1974). Characterization of amino acid pools in the vacuolar compartment of Saccharomyces cerevisiae. *Arch. Microbiol., 101*, 45–57.

Zaworski, P. G., & Heimsch, R. C. (1987). The isolation and characterization of flocculent yeast. In G. G. Hiebsch (Ed.), *Biological research on industrial yeast III* (pp. 185–195). Boca Ratón, FL: CRC Press Inc.

Zuzuarregui, A., Carrasco, P., Palacios, A., Julien, A., & del Olmo, M. (2005). Analysis of the expression of some stress genes induced by stress in several commercial wine yeast strains at the beginning of vinification. *J. Appl. Microbiol, 98*, 299–307.

Zuzuarregui, A., & del Olmo, M. (2004a). Analyses of stress resistance under laboratory conditions constitute a suitable criterion for wine yeast selection. *Anton. Leeuw., 85*, 271–280.

Zuzuarregui, A., & del Olmo, M. (2004b). Expression of stress-response genes in wine strains with different fermentative behavior. *FEMS Yeast Res., 4*, 699–710.

# Saccharomyces Yeasts II: Secondary Fermentation

Alfonso V. Carrascosa[1], Adolfo Martinez-Rodriguez[1],
Eduardo Cebollero[2], Ramón González[3]

[1] Instituto de Investigación en Ciencias de la Alimentación (CIAL, CSIC-UAM),
Madrid, Spain, [2] Instituto de Fermentaciones Industriales (CSIC),
Madrid, Spain and [3] Instituto de Ciencias de la Vid y del Vino (CSIC-UR-CAR),
Logroño, Spain

## OUTLINE

# 1. SPARKLING WINES: TECHNOLOGY AND LEGISLATION

As seen in the previous chapter, *Saccharomyces* yeasts are the main microorganisms responsible for alcoholic fermentation in winemaking. Because this fermentation takes place in open-top tanks, the carbon dioxide generated is spontaneously released into the atmosphere. This release of carbon dioxide occurs during the production of most wines, generally referred to as still wines because of the small amounts of carbon dioxide they contain. Wines that contain more carbon dioxide are known as effervescent wines and include semi-sparkling and sparkling varieties. In natural semi-sparkling wines, the carbon dioxide that forms during alcoholic fermentation becomes trapped in the wine after bottling. When served, these wines produce bubbles that do not form a consistent or lasting mousse, unlike those of sparkling wines. Examples of semi-sparkling wines are the French Blanquette Méthode Ancestral and Blanquette de Limoux (http://www.limoux-aoc.com) wines and certain Asti spumante wines from the Italian Piedmont region (de Rosa, 1987). These wines are microbiologically stabilized prior to fermentation via cold processing, centrifugation, and/or pasteurization. When served, sparkling wines produce a persistent mousse and then gradually release bubbles; the production costs of these wines are higher than those of semi-sparkling wines. This chapter will look at the different chemical and microbiological phenomena that take place during the production of sparkling wines and analyze their impact on sensory quality. It will also provide a detailed explanation of events such as secondary fermentation and aging, which is when yeast autophagy and autolysis occur.

Most of the general microbiological aspects of *Saccharomyces* yeasts have already been discussed in Chapter 1 and many of the concepts are relevant, at least partly, to the understanding

and further analysis of the yeast strains considered in the current chapter. Chapter 5 discusses the molecular characterization of wine yeasts and several of the points covered in Chapters 1 and 6 will contribute to a better understanding of the principles of proteomics and genomics.

## 1.1. Sparkling Wines: Description and Classification

Traditional-method sparkling wines, which are made using particular varieties of grape, contain carbon dioxide gas as a natural consequence of the process used in their production. This gas is a byproduct of the secondary fermentation of natural or added sugars in the base wine. The fermentation takes place in closed vessels and the resulting wine has a minimum pressure of 4 atm at 20°C.

Sparkling wines, which are served as an aperitif or used to accompany meals or desserts, have varying sugar content and are generally acidic and white, although there are some rosé and a very small number of red varieties. Depending on the winemaking method used, sparkling wines are classified as tank-fermented wines (produced using large metallic tanks using the Charmat or the continuous method) or bottle-fermented wines (produced in the bottle using the transfer or the traditional [Champenoise] method) (Flanzy, 2000).

In the tank-fermented method, secondary fermentation takes place under isobaric conditions in a sealed tank with a capacity of tens of hectoliters. These tanks are equipped with stirring mechanisms that mix the yeast uniformly into the base wine. The minimum time a wine should remain in contact with the yeast before it can be sold is 21 d (Pozo-Bayón et al., 2003a). Sparkling wine produced using this method is bottled after clarification but cannot be labeled either Champagne or Cava, for example. In certain cases, the wine is pasteurized at temperatures of between 33 and 70°C for 2 to 5 d to induce yeast autolysis and improve the sensory quality of the final product. The method is attributed to Eugène Charmat, who, in 1916, designed a system for producing large quantities of sparkling wine. It is both simpler and cheaper than the traditional method and is used to produce low-cost sparkling wines. It is also suited to making wines from certain aromatic varieties of grape such as Muscat in which aging with yeast would mask the characteristic aromas of these grapes and detract from the wine's sensory quality. The Charmat method is used to produce sparkling wines from Asti and Trento in Italy.

In the continuous method, large tanks are used to reproduce the yeast autolysis that takes place in bottles in the traditional method (Flanzy, 2000). The process is conducted under isobaric conditions using a base wine to which 50 to 72 g/L of sugar is added. This is then pasteurized at 70°C to accelerate sucrose hydrolysis. The juice is then cooled, filtered, and inoculated with yeast ($\sim 10^6$ cells/mL). Secondary fermentation now takes place and steps are taken to reduce the yeast population and induce cell death and autolysis.

In the transfer method, the sparkling wine is produced in bottles, which are generally magnums measuring 1.5 or 2 L to minimize storage space requirements. It is then left to age on lees for at least 2 months, after which it is transferred to a tank maintained under isobaric conditions with carbon dioxide or nitrogen to prevent loss of the gas. The wine is then cold processed at −5°C, filtered, and sometimes transferred to a second tank, where the dosage, also known by the French term *liqueur d'expedition*, is added before rebottling. Another method involves filtering the wine in the tank after adding the dosage and before bottling. With this technique, disgorging is not required and certain advantage is taken of natural yeast autolysis, thus helping to keep production costs down. The label of these wines must state that the wine has been naturally fermented in a bottle.

## 1.2. Traditional-method Sparkling Wines

In traditional-method sparkling wines, secondary fermentation and the subsequent aging process both take place in the bottle that eventually reaches the consumer. It is an expensive, delicate procedure that is used to make high-quality, relatively expensive wines. In France, the wines in this group are known as Champagne and they are produced using base wines made with white Chardonnay grapes and red Pinot Noir and Pinot Meunier grapes. The minimum aging period is 12 months. Italian traditional-method sparkling wines are known as Talento and they are made with white Chardonnay or Pinot Bianco grapes or with red Pinot Nero or Pinot Meunier grapes. The minimum aging period is 15 months and the wines must be produced in the regions of Trento, Piamonte, Lombardía, el Trentino, el Alto Adige, Veneto, or Friuli.

In Spain, most sparkling wines produced using the traditional method are known as Cava. These wines must age on lees for at least 9 months. The authorized grape varieties are Macabeo, Xarel·lo, Parellada, Subirat (Malvasía riojana), and Chardonnay (white) and Garnacha tinta and Monastrell (red). Rosé Cava can be made using Pinot Noir or Trepat. The vast majority of Cava is produced in the Catalan region of El Penedès, with only around 1% of the total production coming from outside this region.

## 2. PRODUCTION OF SPARKLING WINE USING THE TRADITIONAL METHOD

The production of traditional-method sparkling wines involves two main stages: primary fermentation and secondary fermentation. The former converts the must into base wine and the latter creates the final product. Secondary fermentation is also known by the French term *prise de mousse* as it generates the carbon dioxide that forms the frothy mousse in the glass as the wine is served.

To produce the base wine, each variety of grape is fermented separately. The grapes are hand-picked and any leaves or spoiled berries removed. The grapes, still in bunches, are then transported to the press in shallow containers to protect the berries from bruising or damage and thus prevent the onset of premature fermentation.

To obtain the must, the grapes are pressed in several stages, with only the first press fractions used to make high-quality wines. Gentle crushing of the grapes ensures that the juice is extracted from the pulp of the berries without breaking the seeds or extracting compounds from the skin or stems that would increase sediment and colorant matter and add harsh vegetal aromas and flavors to the wine (Flanzy, 2000). The light pressure applied means the extraction is incomplete. The must should be transferred to the tanks as soon as possible after the grapes reach the press in order to minimize oxidation and thus safeguard against the development of flat aromas and browning.

The next stage involves addition of sulfites to the must. Sulfur dioxide is an important additive in wine because of its antioxidant and antimicrobial properties. Indeed, it plays a key role in determining which microorganisms participate in the initial phases of fermentation. The must is then clarified in large tanks at low temperatures. During this process, known as static clarification, the forces of gravity cause the solid particles suspended in the must to settle at the bottom of the tank. The liquid is then racked off the sediment and thus cleared of impurities.

## 2.1. Primary Fermentation

Primary fermentation to produce the base wine takes place in tanks at a controlled temperature of between 15 and 18°C. Because it is sometimes difficult to complete primary fermentation in these musts (due to extensive clarification, low pH, and the absence of grape skins), selected yeasts, generally *Saccharomyces cerevisiae*, are often added in proportions of approximately $10^6$ cells/mL to ensure even fermentation and prevent the formation of byproducts that would adversely affect the organoleptic characteristics of the base wine (Bidan et al., 1986; Martínez-Rodríguez et al., 2001a). These yeasts are commercialized as active dry yeast. Furthermore, by using specially selected, pure inoculums, winemakers can produce wines with distinctive characteristics using similar fermentation processes from one year to the next, thus circumventing the random effects of spontaneous fermentation. The maintenance of a low, stable temperature prevents the must from fermenting in an uncontrolled manner and thus protects against the loss of desirable aromas or the development of undesirable ones.

The base wine is racked off the sediment (solid particles that have settled at the bottom of the tank) and lees (yeast and adhered bentonite particles) and the sulfite level corrected. The level of free sulfites must be kept at under 15 mg/L to ensure normal yeast growth during secondary fermentation.

The next stage is the assembling or coupage stage. This consists of blending base wines made from different varieties of grape or identical varieties from different years, or sometimes even single varieties, in proportions that vary depending on the quality of the harvest.

The final step is tartrate stabilization, the aim of which is to prevent both the precipitation of potassium bitartrate in the bottle (as a result of the low storage temperatures and an increase in ethanol levels during secondary fermentation). One of the most common tartrate stabilization methods is to induce the formation of potassium bitartrate crystals by reducing the temperature of the base wine to $-4°C$. After several racking and filtration steps, all such crystals are removed from the base wine. In

**TABLE 2.1**  Optimal Characteristics of Base Wine Used to Make Cava

| Alcohol content | 9.5–11.5° |
| --- | --- |
| Total minimum acidity (tartaric acid) | 5.5 g/L |
| Nonreducing extract | 13–22 g/L |
| Maximum volatile acidity (acetic acid) | <0.60 g/L |
| Total sulfur dioxide | <140 mg/L |
| Ash | 0.70–2 g/L |
| pH | 2.8–3.3 |

addition to appropriate organoleptic characteristics, the final base wine must have analytical characteristics similar to those shown in Table 2.1. As the majority of sparkling wines are white, the rest of this chapter will refer to the production of sparkling wines made from white base wines.

## 2.2. *Prise de Mousse*

Once the base wine has been produced, the next stage is the *prise de mousse*, during which secondary fermentation, yeast autolysis, and possibly malolactic fermentation take place. Sparkling wines must remain in the bottle for a minimum number of months—stipulated by national legislation—before they can be sold. The main operations that take place during this period are tirage, stacking, riddling, disgorging, and dosage.

Tirage consists of filling the bottle with the base wine and the *liqueur de tirage*. The base wine receives no further treatment once it has been placed in the bottle. The *liqueur de tirage* is a suspension of yeast, sucrose (20–25 g/L), and a small quantity of bentonite (3 g/100 L) to aid flocculation and the subsequent removal of yeast cells. The amount of bentonite used is approximately 10 times less than that used in treatments aimed at removing proteins from wine. Interestingly, several authors have shown that the reduction of protein and/or peptide levels that occurs during aging (when the wine is left in contact with the yeast) may be due not only to precipitation caused by the increase in alcohol levels but also to adsorption to the bentonite contained in the *liqueur de tirage* (Luguera et al., 1997; Martinez-Rodriguez & Polo, 2003).

The bottles are then stacked horizontally in special aging rooms. Secondary fermentation, *prise de mousse*, aging, and yeast autolysis all occur when the bottle is in this position.

When aging is complete, the wines are riddled. This consists of gently shaking the bottles to direct the sediment (lees) formed by the yeast, bentonite, and any adhered substances towards the neck of the bottle. This used to be done manually by rotating the bottles one eighth of a turn every day for 15 d until they were practically perpendicular to the floor, but nowadays it is performed by more or less automated systems that can rotate large numbers of bottles simultaneously.

The next stage, known as disgorging, consists of removing the lees that have settled at the neck of the bottle. The lees are frozen by placing the neck of the bottle in a bath of freezing solution. The bottle is then placed in an upright position and the cork removed, and the ice plug containing the lees is expelled by the internal pressure in the bottle. Disgorging can be facilitated by using yeast immobilized on calcium alginate beads or enclosed in a special cartridge placed in the neck of the bottle.

Some liquid may be lost during disgorging but this is compensated for with the addition of dosage. This liqueur may be pure sparkling wine, sparkling wine containing sucrose, grape must, partially fermented grape must, grape must concentrate (which may or may not have been rectified), base wine, or a combination of all these. If necessary, wine distillates may also be added. The addition of dosage allows winemakers to give their sparkling wines a distinctive finish. Finally, the bottle is sealed with its definitive cork, which is held in place with a muzzle.

While in the bottle, the yeasts undergo or participate in a series of processes that are critical to the quality of traditional-method sparkling wines. These are secondary fermentation, followed by autophagy and autolysis.

## 3. SECONDARY FERMENTATION

Secondary fermentation begins after tirage. It starts with the inoculation of the base wine and ends when all fermentable sugars have been consumed. The yeasts used to make traditional-method sparkling wines must have certain, key, characteristics (Bidan et al., 1986). Specifically, they should

1. Have high resistance to ethanol (10−12°) as the base wines have an alcohol content of over 9.5°, which increases during secondary fermentation;
2. Display fermentation activity at low temperatures as, on occasion, the temperatures in cellars can be lower than 12°C;
3. Be resistant to pressure caused by carbon dioxide;
4. Be able to flocculate as this facilitates the subsequent elimination of lees during disgorging and prevents yeast deposits from adhering to the walls of the bottle (which is the bottle in which the wine is sold); and
5. Not produce unpleasant aromas as the aroma of sparkling wines is influenced not only by the grapes used but also by the metabolism of the yeast during primary and secondary fermentation and on-lees aging.

It was recently suggested that measurement of the autolytic and foaming capacity of yeast grown in synthetic media might be a valuable tool in the selection of strains for secondary fermentation (Martínez-Rodríguez et al., 2001a). Once a strain has been selected, the starter culture used in industrial processes is prepared following a general procedure that we will summarize in the following section using a practical example.

When dry active yeast is to be used, this should be rehydrated for 20 min (500 g of yeast in 5 L of water with 250 g of sucrose) at 35 to 40°C. When an agar slant culture of a selected strain is to be used, the strain should be grown in sterilized must or complete medium with sucrose until a volume of 5 L and a population of $10^8–10^9$ colony-forming units (CFU)/mL is achieved. The next stage is the conditioning phase, in which the yeast adapts to the alcohol environment. In our example, 600 L of wine, 645 L of water, 120 kg of sucrose, 500 g of yeast extract or 200 g of ammonium salts, 3 kg of tartaric acid, and 5 L of active biomass are added to a 2000 L tank, which is kept at 20°C for 3 to 4 d, until vigorous fermentation and a density of $1.000–1.002 \text{ kg/cm}^3$ is achieved. The next stage is the propagation phase. During this step, 12.5 kg of sucrose and 500 L of base wine are added to the mixture in the tank, which is kept at a temperature of 20°C until a density of $0.994–0.998 \text{ kg/cm}^3$ is reached (approximately 24 h). This culture can now be added to the base wine with 20 to 25 g/L of sugar and a fining agent in a proportion of between 8 and 10%. This produces a concentration of $8–12 \times 10^6 \text{ CFU/mL}$ and a sufficient amount with which to prepare between 20 000 and 26 000 bottles; proportional amounts can be used for other volumes. This operation is known as tirage.

Following the inoculation of the base wine, there is a short lag period in which the yeast adapts to the new substrate conditions. The growth pattern of *Saccharomyces* in secondary fermentation is similar to that in primary fermentation, although growth is generally slower as there are considerably fewer sources of carbon and nitrogen available. Other nutrients can become limiting as fermentation progresses, and increasing levels of ethanol and carbon dioxide build-up can also restrict growth. Our group found that a starter inoculum of approximately $10^6 \text{ cells/mL}$ produced a population of close to $10^7$ CFU/mL on termination of secondary fermentation (Martínez-Rodríguez et al., 2002). Secondary fermentation tended to take place in the first 15 to 20 d after tirage. After this, cell viability decreased slowly until it was no longer detectable (between days 60 and 90). Similar results have been reported by other groups (Feuillat & Charpentier, 1982).

Our group has analyzed morphological aspects of wine yeast viability and autolysis, including the presence/absence of budding, vacuole size, cell size, and cytoplasm separation from the cell wall (González et al., 2003; González et al., 2008; Martínez-Rodríguez et al., 2001b; Martínez-Rodríguez et al., 2004). In those studies, we compared viable cell counts and total microscopic cell count as we had observed an increasing concentration of dead yeast cells, ranging from $10^5$ cells/mL in the first week of incubation to $10^6$ cells/mL after day 20 of fermentation. Our analysis of the morphological characteristics of these cultures revealed the simultaneous presence of dead and live cells, indicating an overlap between secondary fermentation and autolysis (see Figure 2.1). The addition of bentonite

**Exponential phase**

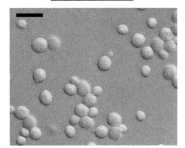

**Secondary fermentation in the bottle**

11°C                                    16 °C

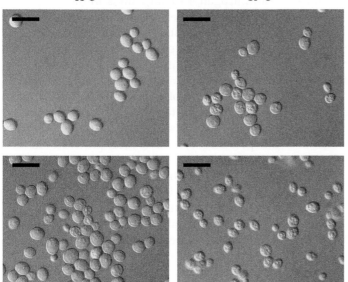

Day 32

Day 90

FIGURE 2.1 *Saccharomyces cerevisiae* EC1118 cells during exponential growth in synthetic medium and during secondary fermentation of base wine in the bottle at 11 and 16°C. Note the smaller cell size and presence of a granular cytoplasm typical of dead cells in the images of yeast cells fermented at 16°C. Images taken using Nomarski interference contrast microscopy; the bar corresponds to 10 μm.

does not appear to have a significant effect on variations in yeast concentrations (Martínez-Rodríguez et al., 2002).

Numerous variables have a considerable influence on the entire secondary fermentation process and on the viability of the yeasts that participate in this process. Temperature is one of the most critical variables, as was demonstrated in tests performed by our group, in which viability was seen to decrease dramatically after day 20 of fermentation in the bottle. At day 90 of fermentation, there was no evidence of viability at 16°C, but at 11°C levels were similar to those seen at the start of fermentation. Morphological and microscopic indicators of viability were visible in cultures that remained viable (Figure 2.1).

# 4. AGING

Traditional-method sparkling wines must be left to age once secondary fermentation is complete. The length of this aging period is regulated by national legislation and may vary from one country to the next. Champagne, for example, must be left to age for at least 11 months before it can be sold, whereas the minimum time stipulated for Cava by Spanish legislation is 9 months. The aim in all cases, however, is the same: to establish a minimum time during which the wine must remain in contact with the lees to guarantee a quality final product.

The most important biological process that takes place during aging is yeast autolysis, a phenomenon that causes the release of intracellular compounds into the wine. In the previous section, we saw how a study of yeast population dynamics indicated an overlap between secondary fermentation and autolysis. The term "autolysis" was first used in the scientific literature by Salkowsky at the end of the nineteenth century to refer to the self-degradation of cellular constituents that started after cell death. Autolysis has since been studied

by various authors, and, in recent years, numerous publications have contributed greatly to the understanding of this process and highlighted its importance in various types of vinification and related processes. In the next section, we will look at the most relevant aspects of autolysis from two perspectives: a biochemical perspective (dynamics and generation of different compounds released by yeast) and a microbiological perspective (addressing the mechanisms of autolysis based on changes occurring in the yeast cell).

## 4.1. Biochemical Changes During Aging

Sparkling wines contain a wide variety of organic compounds, including proteins, peptides, polysaccharides, monosaccharides, lipids, fatty acids, nucleic acids, and volatile components. Many of these compounds, or their precursors, can originate in either grapes or yeast. Nitrogenous compounds are the most abundant and as such have been studied in greatest detail. Indeed, they are considered by many authors to be the most important organic compounds in sparkling wines. The most abundant nitrogenous compounds in these types of wines are peptides and amino acids. Peptide levels rise at the start of fermentation (possibly due to the release of peptides as dead cells start to appear) and during autolysis (as mentioned in the previous section) and begin to fall towards the end of fermentation. This decrease has been attributed to the consumption of peptides by yeasts (Becker et al., 1973) and to the presence of active acid proteases in the wine (Lagace & Bisson, 1990). Because the protocols for isolating, analyzing, and characterizing peptides tend to be more complicated than those used for other nitrogenous compounds such as proteins and amino acids, it was only discovered in recent years, which is when most of the studies of these molecules were performed, that peptides account for the dominant fraction in autolysis.

In 1998, Moreno-Arribas et al. (1998), using high-performance liquid chromatography (HPLC), observed that Cava wines made with different grape varieties nevertheless had a similar peptide profile, leading them to suggest that the peptide composition at the end of secondary fermentation was more closely linked to yeast activity than to the initial composition of the must. Their hypothesis was confirmed in later studies, which demonstrated that the yeast strain used in secondary fermentation played a determining role in the final peptide fraction of wine (Martínez-Rodríguez & Polo, 2000). It was also seen that the addition of bentonite, a common operation in the making of sparkling wines, also influenced this fraction, even though only very small quantities of bentonite were used (Martínez-Rodríguez & Polo, 2003). Furthermore, the substance was seen to have an insignificant influence on cell viability. Although peptides are the predominant compounds in secondary fermentation, they are the best indicators of the dynamics of autolysis as they are released either directly by yeasts or indirectly from proteins and are simultaneously converted into free amino acids by enzymatic activity. Protease A is the best studied of all the proteases that participate in this conversion and it is also considered the most active as its optimum pH (3–3.5) is the same as that of wine.

The most abundant amino acids in base wine are proline, glutamic acid, lysine, leucine, arginine, and aspartic acid. Indeed, with the exception of lysine, these amino acids show the greatest decline during fermentation (Martínez-Rodríguez et al., 2002). The amino acid fraction of base wine is very important, as amino acids are the main source of nitrogen during fermentation. They also serve as precursors of aromatic compounds that contribute to the special characteristics of sparkling wines. The amino acids in sparkling wines are derived from various sources. Some come from the grapes used to make the base wines and are not metabolized by yeasts during growth, and others are released by yeasts either at the end of fermentation or during autolysis. The analysis of amino acids released by yeasts in both model systems and sparkling wines aged for varying lengths of time has shown that autolysis increases amino acid levels by just a few milligrams per liter. This may be because more peptides than amino acids are released during this process or because the amino acids released (primarily glutamic acid, arginine, and alanine) are converted through decarboxylation and deamination, resulting in a reduction of the final amino acid fraction. Nevertheless, although there are contrasting results regarding the specific behavior of certain amino acids, almost all authors agree that the concentrations of most amino acids decrease during secondary fermentation and increase again during aging, and that these amino acids act as important precursors of aromatic compounds (Charpentier & Feuillat, 1993).

Proteins levels, like peptide levels, increase at the start of fermentation (possibly because of the presence of dead cells) and during autolysis, and then decrease. This decrease is also influenced by the presence of bentonite (adsorption of proteins) and the increase in precipitation that occurs as a result of increasing alcohol content (Dizy & Polo, 1996). There is a progressive decrease in protein content during aging and autolysis as proteins are hydrolyzed into compounds with a lower molecular mass, which explains why traditional-method sparkling wines tend to have a lower protein fraction than the base wines used to make them.

The main sugar component of the polysaccharides in base wine is arabinose (66%), but this composition changes radically after secondary fermentation, with mannose and glucose becoming the dominant components (43 and 31%, respectively) (Núñez et al., 2005). This indicates that the polysaccharides present in sparkling wines after aging are primarily the result of the degradation of the yeast cell wall that takes place during autolysis. These polysaccharides are essentially glycoproteins with a sugar content of approximately 85 to 90% and a protein content of just 10 to 15% (Núñez et al., 2006). Glucan- and mannose-containing polysaccharides increase during fermentation and aging in a similar fashion to nitrogenous compounds. Their levels may therefore remain constant or decrease gradually if aging is prolonged. This decrease is primarily due to the activity of β-(1,3) glucanases, which are released by yeasts and remain active in the wine.

The lipids and fatty acids released by yeast during secondary fermentation and autolysis are very difficult to quantify as their levels are very low (~2—4 μm). Pueyo et al. (2000) developed an analytical method using HPLC and a light scattering detector to separate and quantify lipids by classes. They found that sterol esters were the most common type of lipid released during secondary fermentation (8.6%), followed by sterols (3.8%) and triglycerides (2%). As several authors have reported, lipid levels may decrease after an initial increase during secondary fermentation and at the beginning of aging because these compounds participate in the formation of esters, ketones, and aldehydes (Charpentier & Feuillat, 1993).

The volatile compounds responsible for the aroma of sparkling wines come from rather heterogeneous groups, including alcohols, aldehydes, ketones, esters, volatile acids, terpenes, and pyrazines, with levels varying considerably (from picograms to milligrams per liter). These compounds have various origins, and the aromas they generate can be classified into three groups according to whether they are derived from the grape, from fermentation, or from processes occurring during aging. The primary aroma is derived from substances in the grape and is also known as the varietal aroma as it is specific to the grape variety used (Codornnier & Bayonove, 1982). The secondary aroma, also known as the fermentation aroma, is generated by the metabolic activity of yeasts during fermentation. The most important compounds formed during this stage are alcohols, esters, fatty acids, and aldehydes. The tertiary aroma, or bouquet, evolves as the compounds generated during fermentation are transformed by aging. In a study conducted by our group, Pozo-Bayón et al. (2003b) confirmed that there was a close relationship between aging time and the final volatile compound fraction of sparkling wines, and explained that these compounds undergo both degradation and synthesis during aging.

Other compounds, such as nucleic acids, can also be found in traditional-method sparkling wines, albeit at very low levels (similar to those of lipids). In experiments performed using model systems, Hernawan and Fleet (1995) found that approximately 90% of RNA and 40% of cellular DNA was degraded during yeast autolysis and was soluble in wine.

## 4.2. Morphological Changes in Yeast Cells During Aging

Yeast autolysis, which is the dominant microbiological process in aging, takes place after cell death. As mentioned above, important morphological changes, which are visible by optical microscopy, occur in the yeast cell during secondary fermentation. These changes continue to occur during yeast autolysis throughout aging. Most of the changes affect the cell wall, which accounts for between 15 and 25% of the dry weight of the cell and is formed mostly by polysaccharides (80 to 90%), which are hydrolyzed during autolysis. These cell wall changes are visible only under an electron microscope. In a study conducted by our group, Martínez-Rodríguez et al. (2001b) observed the development of folds on the yeast cell wall caused by loss of volume during autolysis (see Figure 2.2), a finding that coincides with previous reports (Charpentier & Feuillat, 1993; González et al., 2008).

Not many studies have analyzed the changes that take place inside the cell during autolysis in

**FIGURE 2.2** *Saccharomyces cerevisiae* IFI-473 cells after 12 months of aging in the bottle. Note the wrinkles and folds on the wall. Images taken using low-temperature scanning electron microscopy.

winemaking conditions, but several groups have indicated that the disorganization of intracellular structures, together with the resulting release of hydrolytic enzymes, is the key step in the autolysis of yeast during the aging of traditional-method sparkling wines (Connew, 1998; Fornairon-Bonnefond et al., 2002).

Our group found morphological differences, visible by optical microscopy, on comparing cells that had undergone accelerated autolysis for a few hours and cells from wines that had been aged for 9 months (see Figure 2.3). The main difference was the presence of structures similar to the autophagosomes described in the cytoplasm of cells from bottle-aged wines. These structures will be studied in greater detail in Section 4.3.

## 4.3. The Genetics of Autolysis: Autophagy

It has traditionally been accepted that autolysis in winemaking involves the uncontrolled

**(a)**

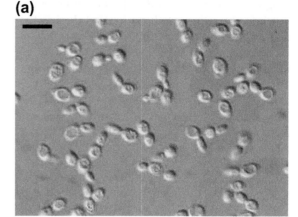

**(b)**

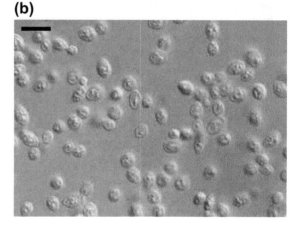

**FIGURE 2.3** *Saccharomyces cerevisiae* IFI-473 cells after 24 h of accelerated autolysis in wine medium (a) and after 9 months of aging in the bottle (b). Note the absence of autophagosomes after accelerated autolysis (a) and the presence of autophagosomes after 9 months of aging (b). Images taken using Nomarski contrast interference microscopy; the bar corresponds to 10 μm.

release of vacuolar enzymes; autophagy, in contrast, is an exquisitely organized and regulated process that occurs in response to the absence of essential nutrients and involves the trafficking of membranes and intracellular components. It is a catabolic process that has been conserved in all eukaryotic cells to degrade cytoplasmic material in the vacuole. In *S. cerevisiae*, thanks to autophagy, cells can survive for long periods of time in the absence of essential nutrients by using products formed during the degradation of cell constituents. The cytoplasm is transported to the vacuole through double-membrane vesicles known as autophagosomes, which, while being formed, sequester the surrounding cytoplasmic region in a nonspecific manner (Baba et al., 1994). Autophagosome formation can be induced in wine yeasts through prolonged nitrogen starvation (see Figure 2.4). The outer membrane of the autophagosome fuses with the vacuole, releasing a vesicle surrounded by a single membrane (autophagic body) into the lumen of the vacuole. This autophagic body is then digested by vacuolar enzymes (Takeshige et al., 1992). This digestion process can be experimentally interrupted by adding phenylmethylsulfonyl fluoride (PMSF), a protease inhibitor. The result is a vacuole full of autophagic bodies (see Figure 2.5) that is used as a marker of active autophagy.

Considerable advances have been made in our understanding of the molecular mechanisms underlying autophagy since the genes responsible for this process, *AUT* and *APG*, were first identified (Thumm et al., 1994; Tsukada & Ohsumi, 1993). (For a detailed review of the molecular aspects of autophagy, see Klionsky, 2005 and Nakatogawa et al., 2009.) Cloning of the genes required for autophagy revealed that most were involved in the cytosol-to-vacuole targeting (Cvt) pathway, a constitutive pathway for the transport of pro-aminopeptidase I to the vacuole that is morphologically and molecularly very similar to the autophagy pathway (Harding et al., 1996; Scott et al., 1996). To unify nomenclature, all genes, *APG*, *AUT*, and *Cvt* were renamed *ATG* (autophagy) genes (Klionsky et al., 2003). Sixteen *ATG* genes have been identified to date as playing a role in the formation of autophagosomes. Other genes involved in membrane fusion events and the degradation of vesicles in the vacuole have also been identified as necessary for autophagy and the normal

**(a)**

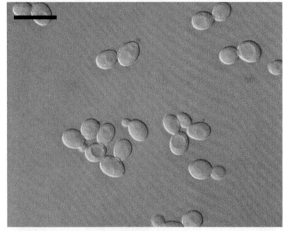

**(b)**

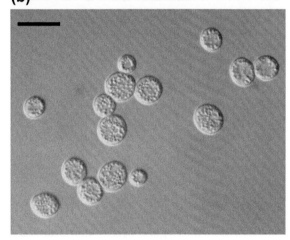

FIGURE 2.4 *Saccharomyces cerevisiae* IFI-473 cells during exponential growth (a) and after prolonged nitrogen starvation (b). Note the presence of budding and the absence of autophagosomes in the cytoplasm during the exponential phase (a) and the opposite after prolonged nitrogen starvation (b). Images taken using Nomarski contrast interference microscopy; the bar corresponds to 10 μm.

development of other vesicle transport pathways (reviewed in Levine & Klionsky, 2004).

Most of the Atg proteins involved in autophagosome formation localize to a single region adjacent to the vacuole. This region is called the

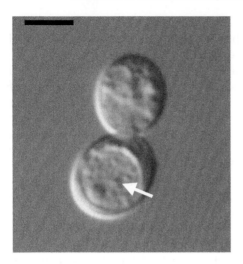

**FIGURE 2.5** Induction of autophagy in *Saccharomyces cerevisiae* IFI-473 strain by prolonged nitrogen starvation. Note the nondegraded autophagosomes in the vacuole (white arrow) caused by the interruption of the process using the protease inhibitor, phenylmethylsulfonyl fluoride. Images taken using Nomarski contrast interference microscopy; the bar corresponds to 5 μm.

pre-autophagosomal structure (PAS) and is thought to be a possible center for the formation of these vesicles (Kim et al., 2002; Suzuki et al., 2001). Analysis of the functional relationships between Atg proteins in the PAS has shed light on the cellular mechanism involved in autophagosome synthesis (Reggiori et al., 2004; Suzuki et al., 2004).

Because the Cvt pathway is a constitutive pathway and, as mentioned, shares most of its elements with the autophagy pathway, the induction of autophagy in starvation conditions may imply reorientation of a functional transport system. The main differences between the two pathways are vesicle size and cargo selectivity. Atg1 might play a key role in this transition by transducing the signal from other pathways such as the targets of rapamycin (TOR) pathway in response to starvation conditions (Kamada et al., 2000; Scott et al., 2000).

*ATG* gene products generally participate in the induction and control of autophagy via different mechanisms including the formation and degradation of protein complexes, enzymatic activities such as kinase activities (protein kinases, phosphatidylinositol kinases), and covalent modification of proteins through mechanisms similar to ubiquitination.

To determine whether or not autophagy could also take place in winemaking conditions, our group used a series of mutants that affect the Cvt and autophagy pathways, either simultaneously or independently, and analyzed the transport of aminopeptidase 1 (which can be transported by either pathway) to the vacuoles. We demonstrated that autophagy does indeed occur in conditions similar to those of secondary fermentation (Cebollero et al., 2005a). We later developed another strategy to analyze industrial strains (not necessarily mutant) in real winemaking conditions and concluded that, in sparkling wine production conditions, autolysis must be preceded by autophagy and the digestion of intracellular material in the vacuole has a clear influence on the nature and abundance of compounds released by yeasts into the wine during autolysis (Cebollero & González, 2006).

# 5. INFLUENCE OF AGING ON THE QUALITY OF TRADITIONAL-METHOD SPARKLING WINES

There is no question that yeast compounds released in wine during autolysis can considerably modify both the chemical composition and the sensory properties of the resulting wine. The interaction between these two aspects has been studied by various authors, who have shown not only that these compounds have a direct influence on the final quality of the wine but also that they can act as intermediaries in the formation of other substances that contribute to sensory quality. Mannoproteins, for example, which form part of the yeast cell wall, are released during autolysis and play a key role in the creation of a wine with small, lasting bubbles, two of the most desirable characteristics of sparkling wines (Brissonnet & Maujean, 1991; Núñez et al., 2005). Fatty acids, although found in low concentrations, are associated with the formation of esters, ketones, and aldehydes, all substances that have a very low sensory threshold, meaning that fatty acid concentrations can affect the flavor of wine (Charpentier & Feuillat, 1993). The situation is similar for nitrogenous compounds. Various aspects related to the sensory properties of sparkling wine and the quality of the mousse, in particular, have been associated with the nitrogen fraction. It is known, for example, that amino acids are precursors of aromatic compounds (Feuillat & Charpentier, 1982) and that the surfactant and sensory properties of proteins and peptides can influence the organoleptic properties of sparkling wines (Martínez-Rodríguez & Polo, 2003). In summary, in view of the importance of these aspects, it is clear why traditional-method sparkling wines are superior to those made using methods in which yeast autolysis has a lesser impact on the final product. Because autolysis in the bottle is a slow process requiring long aging times and considerable storage costs, being able to accelerate this process or achieve similar effects with the use of additives would represent a considerable improvement for makers of sparkling wines. The next section summarizes a range of strategies that have been explored with this goal in mind.

# 6. METHODS TO ACCELERATE YEAST AUTOLYSIS IN SPARKLING WINES AND IMPLICATIONS FOR THE PRODUCTION PROCESS

## 6.1. Increased Temperature and Addition of Autolysates

The temperature at which wine is aged is one of the main rate-limiting factors for autolysis. Consequently, increasing temperature was one of the first strategies for accelerating autolysis to be explored. Although the activity of enzymes involved in autolysis is known to increase with temperature (Fornairon-Bonnefond et al., 2002), increasing the storage temperature did not produce satisfactory results, as the flavor of the resulting wines was reported to be excessively yeasty or toasty. Another technique that has been explored involves the addition of yeast extracts to the base wine together with the *liquer de tirage* (containing sucrose and yeasts to trigger secondary fermentation). The result, however, was not satisfactory either, as the excessive proteolysis of the autolysates also resulted in undesirable flavors (e.g., toasty) and aromas (Peppler, 1982). To resolve this problem, a specific yeast autolysate preparation procedure was developed. The result was a lesser degradation of autolysates and, according to some authors, accelerated aging and improved aroma and bubble quality (Charpentier & Feuillat, 1993).

Another technique that has shown promising results in the laboratory is the use of mixed cultures of killer and sensitive strains of *S. cerevisiae*. Tests performed in synthetic media have shown that these mixed cultures had a 20 to 30% greater protein content than control cultures after 3 d of aging (Todd et al., 2000).

## 6.2. Genetic Improvements in Yeast

Because autolysis in wine is a lengthy process, which can last for years in some cases, it would be highly advantageous from a practical perspective to find yeast strains capable of autolysis in a shorter time. One way of achieving this would be through genetic improvements designed to create yeast strains with an accelerated autolytic capacity. These yeasts would help to accelerate autolysis without causing the problems associated with temperature increases and the addition of yeast autolysates. Such genetic improvements might indeed enhance the final quality of the wine without the need to alter the production process.

Genetic modification strategies include random mutagenesis and genetic engineering. While genetic engineering offers significant advantages, it also has several drawbacks. For example, it has limitations with respect to the alteration of complex genetic traits such as autolysis, it is subject to very strict legislative requirements, and it is currently negatively viewed by consumers (see Chapter 7). One of the advantages of random mutagenesis is that the system is relatively simple and an extensive knowledge of the targeted metabolic pathway is not required to create mutants. It is therefore likely to be much more widely accepted than genetic engineering at present and would also offer greater commercial potential. In a study performed by our group involving the creation of autolytic *S. cerevisiae* mutants by ultraviolet mutagenesis, the most promising mutant released greater amounts of nitrogenous compounds and amino acids in a model wine system at low temperatures, making it a potential candidate for use in secondary fermentation during sparkling wine production (González et al., 2003). The mutants with this capacity had a high level of cellular disorganization when viewed under an optical microscope (see Figure 2.6). More recently, Núñez et al. (2005) showed that one of the mutants created in the above study exhibited accelerated autolysis during the production of traditional-method sparkling wines and gave rise to quality wines, despite an aging time of just 6 months.

Following confirmation that autophagy also occurs during secondary fermentation, as was strongly suggested by our microscopic findings (see Figure 2.3), our group designed genetic engineering strategies aimed at accelerating or inhibiting the autophagy process. In the first case, using a gain-of-function allele of the *CSC1* gene called *CSC1-1*, we showed that overexpression of this gene was associated with accelerated autolysis, accelerated loss of viability, and more

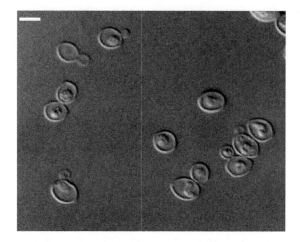

**FIGURE 2.6** Temperature-sensitive autolytic mutant of *Saccharomyces cerevisiae* IFI-473 strain in which autolysis was induced by incubation at 37°C for 24 h. Note the absence of cytoplasmic content in a large number of cells. Images taken using Nomarski contrast interference microscopy; the bar corresponds to 5 μm.

rapid release of nitrogenous compounds into the external environment; similar effects were observed in both laboratory strains and industrial strains used in secondary fermentation (Cebollero et al., 2005b; Cebollero et al., 2009). Strains in which autophagy was inhibited also showed an accelerated loss of viability, but this was associated with a more rapid release of intracellular material only in cases of alterations, with pleiotropic effects, in genes that both influence autophagy induction and participate in many cellular processes related to starvation response (Tabera et al., 2006).

## Acknowledgments

Work at the author's laboratories is funded by the Spanish Ministry for Science and Innovation (grants AGL2006-02558, AGL2009-07327, AGL2009-07894 and Consolider INGENIO2010 CSO2007-00063) as well as Comunidad de Madrid (CAM) (grant S2009/AGR-1469).

## References

Baba, M., Takeshige, K., Baba, N., & Ohsumi, Y. (1994). Ultrastructural analysis of the autophagic process in yeast: Detection of autophagosomes and their characterization. *J. Cell Biol., 124*, 903–913.

Becker, J. M., Naider, F., & Katchalski, E. (1973). Peptide utilization in yeast: Studies on methionine and lysine auxotrophs of Saccharomyces cerevisiae. *Biochim. Biophys. Acta, 291*, 388–397.

Bidan, P., Feuillat, M., & Moulin, J. P. (1986). Les vins mousseux. Rapport de la France. 65ème Asemblèe Génerale de L'OIV. *Bull. OIV, 59*, 563–626.

Brissonnet, F., & Maujean, A. (1991). Identification of some foam-active compounds in champagne base wines. *Am. J. Enol. Vitic., 42*, 97–102.

Cebollero, E., Carrascosa, A. V., & González, R. (2005a). Evidence for yeast autophagy during simulation of sparkling wine aging: A reappraisal of the mechanism of yeast autolysis in wine. *Biotechonol. Prog., 21*, 614–616.

Cebollero, E., & González, R. (2006). Induction of autophagy by second fermentation yeasts during elaboration of sparkling wines. *Appl. Environ. Microbiol., 72*, 4121–4127.

Cebollero, E., Gonzalez-Ramos, D., & Gonzalez, R. (2009). Construction of a recombinant autolytic wine yeast strain overexpressing the csc1–1 allele. *Biotechnol. Prog., 25*(6), 1598–1604.

Cebollero, E., Martínez-Rodríguez, A., Carrascosa, A. V., & González, R. (2005b). Overexpression of *csc1–1*. A plausible strategy to obtain wine yeast strains undergoing accelerated autolysis. *FEMS Microbiol. Lett., 246*, 1–9.

Charpentier, C., & Feuillat, M. (1993). Yeast autolysis. In G. H. Fleet (Ed.), *Wine Microbiology and Biotechnology* (pp. 225–242). Chur, Switzerland: Harwood Academic Publishers.

Codornnier, R., & Bayonove, C. (1982). Etude de la phase préfermentaire de la vinification: Extraction et formation de certains composes de l'arome; cas des terpenols, des aldehydes et des alcohols en $C_6$. *Connais. Vigne Vin, 15*, 269–286.

Connew, S. J. (1998). Yeast autolysis. A review of curret rese. *Arch. Wine Ind. J., 13*, 61–64.

de Rosa, T. (1987). La preparazione degli spumanti secondo il sistema dell' Asti spumante. In T. de Rosa (Ed.), *Tecnologia dei vini spumanti* (pp. 247–249). Brescia, Italy: AEB.

Dizy, M., & Polo, M. C. (1996). Changes of the concentration of nitrogenous compounds during fermentation of must at pilot-plant scale. *Food Sci. Technol. Int., 2*, 87–93.

Feuillat, M., & Charpentier, C. (1982). Autolysis of yeasts in champagne. *Am. J. Enol. Vitic., 33*, 6–13.

Flanzy, C. (2000). Los vinos espumosos. In A. Madrid Vicente (Ed.), *Enología: Fundamentos científicos y tecnológicos* (pp. 497–515). Madrid, Spain: Mundi-Prensa.

Fornairon-Bonnefond, C., Camarasa, C., Moutounet, M., & Salmon, J. M. (2002). New trends on yeast autolysis and wine aging on lees: A bibliographic review. *J. Int. Sci. Vigne Vin, 36*, 49–69.

González, R., Martínez-Rodríguez, A. J., & Carrascosa, A. V. (2003). Yeast autolytic mutants potentially useful for sparkling wine production. *Int. J. Food Microbiol., 84*, 21–26.

González, R., Vian, A., & Carrascosa, A. V. (2008). Morphological changes in *Saccharomyces cerevisiae* during the second fermentation of sparkling wines. *Food Sci. Technol. Int., 14*, 393–398.

Harding, T. M., Hefner-Gravink, A., Thumm, M., & Klionsky, D. J. (1996). Genetic and phenotypic overlap between autophagy and the cytoplasm to vacuole protein targeting pathway. *J. Biol. Chem., 271*, 17621–17624.

Hernawan, T., & Fleet, G. (1995). Chemical and cytological changes during the autolysis of yeast. *J. Ind. Microbiol., 14*, 440–450.

Kamada, Y., Funakoshi, T., Shintani, T., Pagano, K., Ohsumi, M., & Ohsumi, Y. (2000). Tor-mediated induction of autophagy via an Apg1 protein kinase complex. *J. Cell Biol., 150*, 1507–1513.

Kim, J., Huang, W.-P., Stromhaug, P. E., & Klionsky, D. J. (2002). Convergence of multiple autophagy and cytoplasm to vacuole targeting components to a perivacuolar membrane compartment prior to *de novo* vesicle formation. *J. Biol. Chem., 277*, 763–773.

Klionsky, D. J. (2005). The molecular machinery of autophagy: Unanswered questions. *J. Cell Sci., 118*, 7–18.

Klionsky, D. J., Cregg, J. M., Dunn, W. A., Emr, S. D., Sakai, Y., Sandoval, I. V., et al. (2003). A unified nomenclature for yeast autophagy-related genes. *Dev. Cell, 5*, 539–545.

Lagace, L. S., & Bisson, L. F. (1990). Survey of yeast proteases for effectiveness of wine haze reduction. *Am. J. Enol. Vitic., 41*, 147–155.

Levine, B., & Klionsky, D. J. (2004). Development by self-digestion: Molecular mechanism and biological functions of autophagy. *Dev. Cell, 6*, 463–477.

Luguera, C., Moreno-Arribas, V., Pueyo, E., & Polo, M. C. (1997). Capillary electrophoretic analysis of wine proteins. Modifications during the manufacture of sparkling wines. *J. Agric. Food Chem., 45*, 3766–3770.

Martínez-Rodríguez, A., Carrascosa, A. V., Barcenilla, J. M., Pozo-Bayón, M. A., & Polo, M. C. (2001a). Autolytic capacity and foam analysis as additional criteria for the selection of yeast strains for sparkling wine production. *Food Microbiol., 18*, 183–191.

Martínez-Rodríguez, A. J., Carrascosa, A. V., & Martin-Alvarez, P. J. (2002). Influence of the yeast strain on the changes of the amino acids, peptides and proteins during sparkling wine production by the traditional method. *Int. J. Microbiol., 29*, 314–322.

Martínez-Rodríguez, A. J., González, R., & Carrascosa, A. V. (2004). Morphological changes in autolytic wine yeast during aging in two model systems. *J. Food Sci., 69*, M233–M239.

Martínez-Rodríguez, A., & Polo, M. C. (2000). Characterization of the nitrogen compounds released during yeast autolysis in a model wine system. *J. Agric. Food Chem., 48*, 1081–1085.

Martínez-Rodríguez, A., Polo, M. C., & Carrascosa, A. V. (2001b). Structural and ultrastructural changes in yeast cells during autolysis in a model wine system and sparkling wines. *Int. J. Food Microbiol., 71*, 45–51.

Martínez-Rodríguez, A., & Polo, M. C. (2003). Effect of the addition of bentonite to the tirage solution on the nitrogen composition and sensory quality of sparkling wines. *Food Chem., 81*, 383–388.

Moreno-Arribas, V., Pueyo, E., Polo, M. C., & Martín-Álvarez, P. J. (1998). Changes in the amino acid composition of the different nitrogenous fractions during the ageing of wine with yeasts. *J. Agric. Food Chem., 46*, 4042–4051.

Nakatogawa, H., Suzuki, K., Kamada, Y., & Ohsumi, Y. (2009). Dynamics and diversity in autophagy mechanisms: Lessons from yeast. *Nature Reviews Molecular Cell Biology, 10*, 458–467.

Núñez, Y. P., Carrascosa, A. V., González, R., Polo, M. C., & Martínez-Rodríguez, A. (2005). Effect of accelerated autolysis of yeast on the composition and foaming properties of sparkling wines elaborated by a champenoise method. *J. Agric. Food Chem., 53*, 7232–7237.

Núñez, Y., Carrascosa, A. V., González, R., Polo, M. C., & Martínez-Rodríguez, A. J. (2006). Isolation and characterization of a thermally extracted yeast cell wall fraction potentially useful for improving the foaming properties of sparkling wines. *J. Agric. Food Chem., 54*, 7898–7903.

Peppler, H. J. (1982). In A. H. Rose (Ed.), *Economic microbiology. Yeast extracts, Vol. 7* (pp. 293–312). London, UK: Academic Press.

Pozo-Bayón, M. A., Martinez-Rodriguez, A. J., Moreno-Arribas, M. V., Pueyo, E., Martin-Alvarez, P. J., & Polo, M. C. (2003b). La elaboración de vinos espumosos y su repercusión en la composición química y la calidad. *Tecnología del Vino, 13*, 55–60.

Pozo-Bayón, M. A., Pueyo, E., Martín-Alvarez, P. J., Martínez-Rodríguez, A., & Y Polo, M. C. (2003a). Influence of yeast strain, bentonite addition, and aging time on volatile compounds of sparkling wines. *Am. J. Enol. Vitic., 54*, 273–278.

Pueyo, E., Martinez-Rodriguez, A., Polo, M. C., Santa-María, G., & Bartolomé, B. (2000). Release of lipids during yeast autolysis in a model wine system. *J. Agric. Food Chem., 48*, 116–122.

Reggiori, F., Tucker, K. A., Stromhaug, P. E., & Klionsky, D. J. (2004). The Atg1–Atg13 complex regulates Atg9 and Atg23 retrieval transport from the pre-autophagosomal structure. *Dev. Cell., 6*, 79–90.

Scott, S. V., Hefner-Gravink, A., Morano, K. A., Noda, T., Ohsumi, Y., & Klionsky, D. J. (1996). Cytoplasm to vacuole targeting and autophagy employ the same machinery to deliver proteins to the yeast vacuole. *Proc. Natl. Acad. Sci. USA., 93*, 12304–12308.

Scott, S. V., Nice, D. C., Nau, J. J., Weisman, L. S., Kamada, Y., Keizer-Gunnink, I., et al. (2000). Apg13p and Vac8p are part of a complex of phosphoproteins that are required for cytoplasm to vacuole targeting. *J. Biol. Chem., 275*, 25840–25849.

Suzuki, K., Kirisako, T., Kamada, Y., Mizushima, N., Noda, T., & Ohsumi, Y. (2001). The pre-autophagosomal structure organized by concertad functions of APG genes is essential for autophagosome formation. *EMBO J, 20*, 5971–5981.

Suzuki, K., Noda, T., & Ohsumi, Y. (2004). Interrelationships among Atg proteins during autophagy in. *Saccharomyces cerevisiae. Yeast, 21*, 1057–1065.

Tabera, L., Muñoz, R., & González, R. (2006). Deletion of BCY1 from the *Saccharomyces cerevisiae* genome is semidominant and induces autolytic phenotypes suitable for improvement of sparkling wines. *Appl. Environ. Microbiol., 72*, 2351–2358.

Takeshige, K., Baba, M., Tsuboi, S., Noda, T., & Ohsumi, Y. (1992). Autophagy in yeast demonstrated with proteinase-deficient mutants and conditions for its induction. *J. Cell Biol., 119*, 301–311.

Thumm, M., Egner, R., Koch, B., Schlumpberger, M., Straub, M., Veenhuis, M., et al. (1994). Isolation of autophagocytosis mutants of. *Saccharomyces cerevisiae. FEBS Lett., 349*, 275–280.

Todd, B. E. N., Fleet, G. H., & Henschke, P. A. (2000). Promotion of autolysis through the interaction of killer and sensitive yeast: Potential application in sparkling wine production. *Am. J. Enol. Vitic., 51*, 65–72.

Tsukada, M., & Ohsumi, Y. (1993). Isolation and characterization of autophagy-defective mutants of Saccharomyces cerevisiae. *FEBS Lett., 333*, 169–174.

CHAPTER

# 3

# Yeasts Used in Biologically Aged Wines

*Tahía Benítez, Ana M. Rincón, Antonio C. Codón*

Departamento de Genética, Facultad de Biología, Universidad de Sevilla, Sevilla, Spain

OUTLINE

## 1. INTRODUCTION

Most wines that undergo biological aging are known in Spanish as *vinos generosos* (literally, generous wines). They are fortified with grape spirit at the end of fermentation until an alcohol content of at least 16% is reached, and then aged under a film of yeast known as *velo de flor* (literally, flower veil) or simply *flor*. In Spain, most biologically aged wines are produced in the areas of Jerez, Montilla-Moriles, Condado de Huelva, Aljarafe, and Rueda (Benítez et al., in press). Outside Spain, the *flor* aging method is used mostly in France (*vins jaunes*, or yellow wines, from the Jura region), Sardinia, California, South Africa, Australia, and certain areas of Europe (mainly the Hungarian region of Tokaj, which produces botrytized wines) (Charpentier et al., 2009; Fleet, 2007; Kovacs et al., 2008; Pirino et al., 2004; Sipiczki, 2008).

The best-known biologically aged wines are from the Spanish region of Jerez, which boasts a unique mix of soil, elevation, climate, and flora. Other characteristic features of this region include its planting and cultivation methods, the architecture of the cellars in which the wines are aged, and its unique winemaking and aging techniques (Bravo-Abad, 1986). Thanks to their special climate, geography, and topography, the Guadalquivir valley and Jerez region were highly successful producers and traders of wine between the first century BC and the first century AD (Celestino-Peréz, 1999). The existence of vineyards and wines in the area has been documented as far back as Roman times, and there are also records of the wineries of Marco de Jerez in the urban land registers for the city of Jerez de la Frontera undertaken at the order of Alfonso X the Wise in 1264 (García del Barrio, 1995). Since alcohol is added at the end of fermentation, most biologically aged wines are dry. The high alcohol content of these wines is achieved with the addition of grape spirit and the resulting wines are renowned for their extraordinary finesse. The wines are made from white grapes, mainly of the varieties Palomino (95%) and Pedro Ximénez and Moscatel (5%), and produced under the appellation of Jerez-Xérèz-Sherry (see Table 3.1). The characteristic white soils of the area, known as *albarizas*, are rich in limestone and retain water. Given the nature of this soil, the local climate, and the predilection for Palomino grapes, one would expect Jerez wines to be relatively insubstantial, but the unique winemaking process that characterizes the area produces extraordinarily unique and flavorful wines (Martínez-Llopis et al., 1992).

The juice is extracted from the grapes in different press fractions. The resulting musts are then separated according to quality and used to produce different types of wines (García Maiquez, 1995) (see Table 3.1). This extremely delicate pressing system is a key step in a carefully controlled production process. Clusters of grapes from healthy vines are harvested at the end of August or at the beginning of September and vacuum-pressed to prevent contamination of the must by skins, stems, pips, or similar material. This first press fraction is known by the Spanish term *primera yema*. The grapes are then vacuum-pressed a second time, using greater pressure, to produce the second *yema*. The remaining grape products are generally ground and fermented and then distilled to produce the grape spirit used to fortify the wines at the end of fermentation. The grape skins and seeds are then separated from the must by static sedimentation. The first *yema* is used to produce *Fino* wines (see Figure 3.1). This superior must is normally inoculated with a *Saccharomyces cerevisiae* yeast strain from the winery. This locally occurring inoculum is known by the name of *pie de cuba*. Fermentation traditionally took place in oak barrels measuring 500 to 600 L but nowadays it is carefully conducted in stainless steel tanks with close monitoring of temperature and other parameters (Suárez-Lepe, 1997). Fermentation ends at the beginning of December and the wine is then stored until the solid particles have settled.

Around January or February, the fermented musts may undergo a second selection process, after which they are destined to produce *Fino* wines (biological aging), *Amontillado* wines (biological aging followed by oxidative aging), or *Oloroso* wines (oxidative aging only). The fermented musts used for *Finos*, which have an alcohol content of 10 to 12% vol/vol, are fortified with grape spirit to a strength of 15.5 to 16% and then transferred to oak butts (barrels), which are filled to five sixths of their capacity (see Figure 3.1). Shortly afterwards, a film of yeast (the *flor*), formed mostly (>95%) by strains from different races of *S. cerevisiae*, starts to grow

**TABLE 3.1** Grape Varieties, Types of Wine, and Aging Methods in Jerez-Sanlúcar and Montilla-Moriles Wines

| Appellation | Jerez-Xérèz-Sherry and Manzanilla-Sanlúcar | | Montilla-Moriles | |
|---|---|---|---|---|
| **REGULATIONS** | | | | |
| | Spanish Ministerial Order 2-V-77 | | Spanish Ministerial Order 12-XII-85 | |
| **AUTHORIZED GRAPE VARIETIES** | | | | |
| | Palomino de Jerez | | Airén | |
| | Palomino Fino | | Baladí-Verdejo = Jaén blanco | |
| | Pedro Ximénez | | Moscatel | |
| | Moscatel | | Pedro Ximénez | |
| **TYPES OF WINE** | | | | |
| | Fortified wines | Alcohol strength | Aged fortified wines | Alcohol strength |
| | *Fino* | 15° | *Fino* | 14–17.5° |
| | *Amontillado* | 16–18° | *Amontillado* | 16–22° |
| | *Oloroso* | 18–20° | *Oloroso* | 16–20° |
| | *Palo Cortado and Raya* | 18–20° | *Palo Cortado* | 16–18° |
| | *Manzanilla* | 15° | *Raya* | 16–20° |
| | Natural sweet wine | | Natural sweet wine | |
| | *Pedro Ximénez* | | *Pedro Ximénez* | |
| | | | White | |
| | | | Without aging | 10–12° |
| | | | With aging | Min. 13° |
| | | | Without aging | |
| | | | *Ruedos* | Min. 14° |
| **AGING METHOD** | | | | |
| | In oak butts using the traditional *criaderas* and *soleras* method. All the wines sold under this appellation must be at least 3 years old. | | In oak butts with a maximum capacity of 1000 L for at least 2 years using the traditional *criaderas* and *soleras* method. | |

on the surface of the wine. These yeasts are responsible for the biological aging of the wine (Martínez et al., 1995). At this stage of the process, alcohol is the only available carbon source for the yeasts as all the fermentable sugars in the wine have already been metabolized. The combined effect of the oxidative metabolism of the *flor* yeasts and the physical barrier they form on the surface of the wine creates reducing conditions that are responsible

**FIGURE 3.1** Fermentation and biological aging of Jerez wines. The musts are fermented until a dry wine with an alcohol content of 10 to 12% is obtained (white vinification). Following an initial selection, the less delicate wines are fortified to 18% alcohol and left to undergo oxidative aging to produce *Oloroso* wines. The paler and more delicate wines are fortified to a strength of 15.5% and left to undergo biological aging. Following a second selection, some of these more delicate wines, used to produce *Fino*, are placed in oak butts in which they continue to age under a layer of yeasts. The other wines are fortified to a strength of 17.5% and used to produce *Amontillado* wines following a process of oxidative aging. Wines are aged for 1 to 3 years (*añadas*) in a static system known as *sobretablas* before undergoing further aging in a dynamic system known as *soleras* and *criaderas*. In this system, a certain amount of wine is taken from the oldest butts (*soleras*) to be bottled and is then replaced with a younger wine from the first *criadera*. This, in turn, is replaced by wine from the second *criadera* and so on until the youngest wine is replaced with wine from the *añadas* (*sobretablas*) system (Benítez & Codón, 2005).

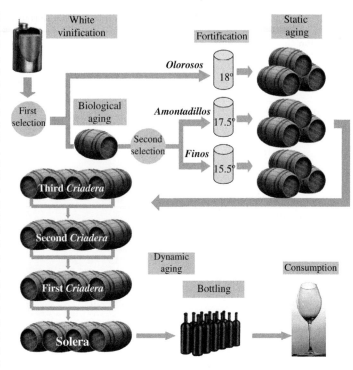

for the pale color and many of the organoleptic characteristics of the final product (Suárez-Lepe, 1997). The *flor* acts as an insulating layer between the wine and the surrounding oxygen, which is continuously consumed by the strongly oxidative metabolism of the yeast. These reducing conditions, together with the products of yeast metabolism, influence the aroma, flavor, and color of the resulting *Fino* wine, which is renowned for its pale, straw-gold color; intense yet delicate aroma that is slightly reminiscent of hazelnuts; and rich, dry feel on the palate.

Biologically aged wines produced in Sanlúcar de Barrameda are called *Manzanillas*. They are made with the grape variety Listán, which is a precursor of the Palomino Fino grape. The influence of the nearby Atlantic endows *Manzanillas* with iodine aromas and greater acidity.

The musts used to produce *Oloroso* wines are fortified to a strength of 18 to 20%. At such high alcohol concentrations, yeasts are incapable of growing and forming a *flor*. The absence of a biofilm during oak aging results in an oxidized wine with a characteristic color and aroma. *Olorosos*, for example, are very dark and have an intense aroma and full body, and are either dry or tending towards medium sweet. In the combined aging system, wines are aged biologically and, following a second selection process (see Figure 3.1), fortified to 16 to 22% alcohol, causing the *flor* to disappear. They are then left to undergo oxidative aging, producing *Amontillados* in the case of *Finos*, and *Manzanilla Pasada* wines in the case of *Manzanillas* from Sanlúcar. These amber-colored wines are smooth and dry and have a hazelnut aroma (see Table 3.1) (García Maiquez, 1995).

Certain wines aged using the combined system develop the distinctive aroma of *Amontillados* and flavor of *Olorosos* (see Table 3.1). These are known as *Palo Cortado* wines (see Table 3.1). Fortified wines fermented on lees that undergo oxidative aging have less intense aromas than *Olorosos* and are called *Raya* wines (see Table 3.1) (Benítez et al., 2009).

Aging takes place in two phases: a static phase known as *sobretablas*, which takes place in oak butts, and a dynamic phase using a system known as *criaderas* and *soleras*. In the second system, the butts are stacked on top of each other to form what is known as a scale. The bottom row, nearest the floor, is called the *solera* and contains the oldest wine (see Figure 3.1). The row immediately above this is called the *primera criadera* (literally, first nursery) and contains the second-oldest wine. Above this come the second and the third *criaderas*, and so on until a height of five or six rows is reached. The wines in the scales are mixed and standardized using a system known as *saca* and *rocío*, where wine taken from the bottom row to be bottled (twice a year) is replaced with an equal volume of wine from the row above. This occurs successively up through the rows until the youngest wine (the topmost *criadera*) is replaced by a *sobretabla* wine, which has been aged using the static system for 1 to 3 years (*añadas*, or vintages). The wines taken from the *soleras* are standardized, stabilized, filtered, and bottled (Martínez et al., 1995). They are also mixed with wines from other *soleras* from the same winery in suitable proportions to guarantee a final product with consistent properties.

Jerez also produces sweet wines, made from grapes that have been exposed to a lot of sunlight (mainly Pedro Ximénez and Moscatel grapes). After a partial fermentation phase, the wines are fortified to halt fermentation and retain the sweetness of the wine. They then undergo oxidative aging, just like *Olorosos*, until a velvety, mahogany-colored wine is obtained (Martínez-Llopis et al., 1992; Martínez et al., 1995).

Certain wines from nearby areas (Montilla-Moriles in the province of Córdoba, and Huelva) also undergo biological aging under a layer of *flor*, but they have certain distinctive characteristics that set them apart from Jerez wines. The grapes used to produce Montilla-Moriles wines are grown in limestone soils and have a very high sugar content. Most (90%) are of the variety Pedro Ximénez but Airén, Moscatel, and Baladí grapes are also used (see Table 3.1). While Pedro Ximénez grapes are perfectly adapted to the local climate, characterized by long, dry summers, Palomino grapes, which produce excellent wines in the neighboring Jerez, cannot withstand the heat. Because the wines destined to produce *Finos* reach an alcohol content of 15% during fermentation, they do not need to be fortified. These natural *Finos*, with their distinctive, unique aroma and flavor, are considered among the best in the world and their production has been documented as far back as 700 BC (Bravo-Abad, 1986). White *Finos* that undergo biological aging have complex aromas and almond flavors and can reach an alcohol content of 17.5%. *Amontillados*, which are dry on the palate and have hazelnut notes, can reach an alcohol content of 22%, while the amber-colored, velvety *Olorosos* can reach levels of 18 to 20%. Montilla also has young wines with a low alcohol content that are covered by the appellation (see Table 3.1). This is not the case in Jerez, however, where certain young wines do not meet the requirements of the appellation.

The wines from Huelva are made with grapes grown in well-drained, sandy soils. The grape of choice is Zalema, although other varieties are used, including Listán, Moscatel, Palomino, and Garrido Fino. Wines from this area that undergo biological aging are straw colored and have an alcohol content of 14 to 17%, although there is also a mahogany-colored wine called *Condado Viejo* that is aged using the oxidative system and can reach alcohol levels of 23% (Martínez-Llopis et al., 1992). The sensory properties of these wines are similar to those of Montilla and Jerez. Young wines with a low alcohol content are covered by the Huelva appellation. The first wines to reach the new world following the discovery of America were from this region.

The area of Rueda, in the province of Valladolid, also produces biologically aged wines, made with the grape variety Viura. While this grape has a good set of primary aromas, it lacks the complexity conferred by other varieties such as Verdejo. Until recently, the majority of wines produced in Rueda were dry, fortified wines, generally produced using oxidative aging, similar to the wines of Montilla. The *Pálido Rueda* wines from the same region are biologically aged and fortified with grape spirit to a strength of 15%. They are fragrant, dry wines that are aged for at least 4 years. For the last 3 years, they are aged in oak barrels. The other fortified wine produced in the region, known as *Dorado Rueda*, undergoes oxidative aging and has an alcohol content of 15% (Martínez-Llopis et al., 1992).

Outside Spain, Australia produces flor sherry, a wine aged using the continuous method through a column filled with wood shavings. In South Africa and California, *flor* yeast strains from Spain (*S. cerevisiae beticus* race) have been used to seed fermented musts following a process similar to that used for Montilla wines (Benítez et al., 2009). In France, the *vins jaunes* of the Jura region are fermented until they reach an alcohol content of 14 to 15% and subsequently aged for at least 6 years under a yeast film that forms on the surface (Charpentier et al., 2009). The wine is characterized by its yellow color and persistent walnut flavor, derived from its high acetaldehyde content (Suárez-Lepe, 1997). In other areas of Europe, particularly in the Hungarian region of Tokaj, *flor* yeasts are used to age wines made from grapes infected with the noble rot (*Botrytis cinerea*) (Kovacs et al., 2008; Sipiczki, 2008). In Sardinia, fortified wines are produced

using natural fermentation, followed by aging under a *flor* identical to the age-old method used by the wineries in Jerez (Zara et al., 2008). None of these wines are fortified as they naturally reach an alcohol content of at least 15% through fermentation.

In addition to the controlled aging of white wines already discussed, new methods are being developed to age red wines using selected *flor* yeasts (Suárez-Lepe, 1997). The resulting wines have the characteristic organoleptic properties of biologically aged *Finos*.

# 2. CHARACTERISTICS OF YEASTS USED IN BIOLOGICALLY AGED WINES

## 2.1. Physiological Characteristics of Fermentation and Aging Yeasts

The majority of yeasts responsible for fermentation and aging correspond to strains of *S. cerevisiae*. While yeasts that participate in fermentation are found on the vine and later in fermentation tanks for about just 4 weeks a year, those that participate in aging are constantly present in aging barrels. Interestingly, wine strains cannot be isolated in the vineyards in the weeks immediately before or after harvesting (Mortimer, 2000). One possible explanation is that these yeasts are carried to the vineyard by insects when the grapes are almost ripe. If this is the case, insects could be considered the natural reservoir of these yeasts. In the majority of today's wineries, the main fermentation yeasts are inoculated using selected strains with desirable characteristics, although the indigenous microflora also makes an important contribution to the organoleptic properties of the final product. The yeast film that forms on the surface of fortified wines after fermentation is the result of colonization by naturally occurring yeasts in the wineries. Although these yeasts also belong to the species

*S. cerevisiae*, they do not participate in fermentation and are completely different to those that do in terms of metabolic, physiological, and genetic characteristics (Esteve-Zarzoso et al., 2001). They are autochthonous strains in wineries that produce biologically aged wines. They remain in the aging butts all year round and displace fermentation yeasts once the wine has been fortified to high alcohol levels (Infante et al., 2003). During aging, they release acetaldehyde, consume glycerol and ethanol, reduce volatile acidity, and increase concentrations of higher alcohols (see Table 3.2). The cost of producing wines using this system, however, is increased by the long aging period required and the need to periodically replace

the alcohol consumed by the yeasts (which can be as high as 7.5–9 L per 500 L barrel per year of aging).

A wide variety of fungi, yeasts, and bacteria exist alongside *S. cerevisiae* throughout the fermentation, production, and aging of wines produced using the biological aging method, but this microflora is reduced to just a few yeasts by increasing the alcohol content of the wine, adjusting the pH level to between 3.0 and 3.5, and adding sulfur (to a level of between 100 and 130 mg/L) (Campo et al., 2008). The musts used to create Jerez wines contain fungi such as *Mucor*, *Rhizopus*, and *Aspergillus* species at the beginning of fermentation. These are all natural components of the grape microflora

**TABLE 3.2** Basic Characteristics of Jerez Wines

| Parameters that determine the character of the wine | Wines | | |
|---|---|---|---|
| | *Finos* | *Olorosos* | *Pedro Ximénez* |
| Soil | *Albariza* | *Albariza* | *Albariza* |
| Grape variety | Palomino | Palomino | Pedro Ximénez |
| Vinification (pressure of must [atm]) | <0.5 | 0.5–1.5 | >5 |
| Wine fortified to (degrees) | 15.5 | 18 | 10.0–15.5 |
| *AGING* | | | |
| Method | Biological | Biological/oxidative | Oxidative |
| Average age (years) | 3–5 | 8–10 | 8–25 |
| *AVERAGE ANALYTICAL VALUES* | | | |
| pH | 2.9–3.3 | 3.1–3.5 | 3.6–4.1 |
| Alcohol (degrees) | 15.5–17.0 | 18–21 | 15.5 |
| Total acidity (g/L of tartaric acid) | 3.7–5.2 | 4.5–6.0 | 5.2–7.1 |
| Volatile acidity (g/L of acetic acid) | <0.3 | <0.8–1.2 | <0.8–1.3 |
| Acetaldehyde (mg/L) | 200–400 | 60–80 | 150–200 |
| Glycerol (g/L) | <1.0 | 5–8 | 3–5 |
| Malic acid (mg/L) | 134–268 | 335–603 | 2500 |
| Lactic acid (mg/L) | <900 | <720 | <400 |
| Total polyphenols (mg/L) | 250 | 275–350 | 500 |

*Reproduced from García Maiquez (1995).*

but do not have a defined role in fermentation. Other microorganisms present at this stage are bacteria from the genera *Acetobacter, Pediococcus,* and *Lactobacillus*—which can alter the wine but are mostly eliminated when acidity levels are corrected—and yeasts from the genera *Hanseniaspora, Kloeckera, Candida, Pichia, Hansenula, Saccharomycodes,* and *Saccharomyces*. The dominant yeasts at the end of fermentation, however, are *Saccharomyces* yeasts. In Montilla-Moriles wines, new batches of must are added during fermentation to prevent excessive glycerol and acidity levels caused by the high sugar content of the Pedro Ximénez grapes used to make these wines. Thanks to this process, the must retains a relatively high alcohol content throughout fermentation, explaining why the majority of yeasts isolated during this period are *S. cerevisiae* (Sancho et al., 1986). Indeed, *S. cerevisiae* is the predominant species at the end of fermentation in the majority of biologically aged wines (Charpentier et al., 2009; Martínez et al., 1995). Exceptions are wines from Sardinia, in which *Saccharomyces prostoserdovii* dominates (Fatichenti et al., 1983), and the botrytized wines of Tokaj (Sipiczki, 2003) and the Sauternes region in France (Naumov et al., 2000), where *Saccharomyces bayanus* var. *uvarum* dominates. *S. bayanus* seems to develop in musts fermented at low temperatures (Naumov et al., 2000, 2002).

The value of non-*Saccharomyces* yeasts in the production of sweet wines is a topic of debate (Urso et al., 2008). Generally speaking, they produce low levels of higher alcohols and ethyl esters compared to *Saccharomyces* yeasts. *Pichia* yeast strains, for example, are undesirable because they produce ethyl acetate (García Maiquez, 1995), and *Hanseniaspora* and *Kloeckera* strains have been found to produce high, undesirable levels of acetate, acetaldehyde, ethyl acetate, and acetoin. Nonetheless, there have also been reports of strains from the genera *Hansenula, Kloeckera, Candida,* and *Pichia*, and other strains with low fermentation activity exerting a seemingly favorable effect on wine aroma,

above all due to the release of metabolites through the activity of enzymes produced by these strains (proteases, lipases, esterases, and pectinases) (Esteve-Zarzoso et al., 1998). Other yeasts belonging to the genera *Kluyveromyces, Torulaspora,* and *Saccharomyces* make an enormous contribution to the final aroma of wine thanks to their ability to convert monoterpene alcohols during fermentation.

Wineries are currently working towards controlling the fermentation process by inoculating musts with a selected yeast strain such as the *pie de cuba* to shorten fermentation time and reduce but not completely eliminate the number of other microorganisms present. Although most of the glucose and fructose present in must is converted to ethanol, small quantities undergo glycerol-pyruvic fermentation, giving rise to the release of glycerol and pyruvate (Martínez et al., 1998). Later, during biological aging, the *flor* yeasts convert ethanol to acetaldehyde and acetate through oxidation; they also consume glycerol, organic acids (acetic, lactic, citric, and succinic acid), and amino acids (including proline), and produce higher alcohols (isobutanol and isoamyl alcohol), acetaldehyde, and acetoin (see Table 3.2) (Martínez et al., 1998; Muñoz et al., 2006).

The musts are clarified prior to fermentation to prevent the premature alteration of the properties of the wine (Roldán et al., 2006). Interestingly, the fermentation of musts with a high solid content gives rise to *Finos* with a lower volatile acidity and greater concentrations of acetaldehyde, higher alcohols, and glycerol. Clarification may remove fatty acids and sterols, which would then need to be produced by yeasts through acetyl-coenzyme A (CoA) (Roldán et al., 2006). Microaerobic conditions inhibit the synthesis of fatty acids, and the subsequent hydrolysis of acetyl-CoA increases volatile acidity (Martínez et al., 1998; Zara et al., 2009). Indeed, the inhibition of phospholipid and sterol biosynthesis in microaerobic conditions often causes stuck fermentation, even before high alcohol concentrations are reached (Mauricio et al., 1990). Oxygen is also necessary for the consumption of proline, which is the main source of nitrogen in the must (Ingledew et al., 1987). If proline levels become depleted or there is a shortage of oxygen for proline consumption, other compounds that make an important contribution to the aroma of the wine may be consumed (Berlanga et al., 2001; Gómez et al., 2004). In an environment with limited oxygen supply, yeasts can also release amino acids (threonine, methionine, cysteine, tryptophan), which are synthesized de novo from ethanol to restore redox potential; these amino acids serve as electron acceptors to oxidate excess nicotinamide adenine dinucleotide (Berlanga et al., 2001; Mauricio et al., 2001; Vriesekoop et al., 2009).

Analysis of *flor* films formed in Jerez and Sanlúcar wines during biological aging has revealed the presence of a relatively complex microflora. In some wineries, over 95% of the flora has been found to be formed by *S. cerevisiae*, with two predominant races—*S. cerevisiae beticus* and *S. cerevisiae montuliensis*—and two minority races—*S. cerevisiae cheresiensis* and *S. cerevisiae rouxii*. These races can be distinguished on the basis of their metabolic characteristics (Martínez et al., 1997a). The remaining population (approximately 4%) is composed of yeasts from the genera *Debaryomyces*, *Pichia*, *Hansenula*, and *Candida*. Just a single dominant race—*S. cerevisiae beticus*—in coexistence with minority *Dekkera* and *Brettanomyces* strains (which might be responsible for sporadic spoilage of *Fino* wines due to acidification) has been isolated in certain wineries, and molecular analyses have even shown a single *S. cerevisiae beticus* population per butt (Ibeas et al., 1996). In other wineries, the microflora has been found to be mainly composed of *beticus* and *cheresiensis* races, with *Pichia* species existing as minority yeasts (Mesa et al., 1999). The literature also contains reports of wineries with minority populations of *Candida*, *Dekkera*, *Hanseniaspora*, *Zygosaccharomyces*, and *Metschnikowia* species

during fermentation that are replaced by a single dominant population belonging to the *beticus* race during aging (Esteve-Zarzoso et al., 2001). The *flor* that forms on the French *vins jaunes* from the Jura region features *S. cerevisiae* of the *beticus*, *montuliensis*, and *cheresiensis* races (Charpentier et al., 2009). In Sardinia, the *flor* is formed by *S. prostoserdovii* and *S. bayanus* in addition to *S. cerevisiae* (Zara et al., 2008). Differences in *flor* composition have been partly attributed to strain differences in sensitivity to compounds such as acetaldehyde, leading to a natural evolution towards strains with greater tolerance of those compounds (Martínez et al., 1997b). Other authors have attributed these differences to the displacement of sensitive populations by killer strains (Mesa et al., 1999). In some wineries, all the strains analyzed have been found to be sensitive to K1 but resistant to K2, with the majority of strains not producing toxins (Ibeas & Jiménez, 1996). Other studies, in contrast, have found all *beticus* and *cheresiensis* races to be resistant to K1 and K2 toxins and *montuliensis* and *rouxii* races to be sensitive to both. None of the strains were found to produce toxins and the authors reported a balance between the four races in the static and dynamic aging systems that they attributed to speed of *flor* formation and resistance to acetaldehyde rather than to the killer character of the strains (see Figure 3.2) (Martínez et al., 1997a). The enormous variations that seem to exist in microflora composition from one winery to the next would explain why wines from different wineries in geographically close areas often have very different organoleptic properties, though made using similar processes (Budroni et al., 2005; Mérida et al., 2005).

*Finos*, *Amontillados*, and *Olorosos* all extract tannins, phenols, and other compounds from the wood of the butts in which they are aged. In the case of *Finos* and *Amontillados*, the metabolic activities of the *flor* yeasts also lead to enrichment of 3-methylbutanal, phenylacetaldehyde, methional, and sotolon, as well as methyl

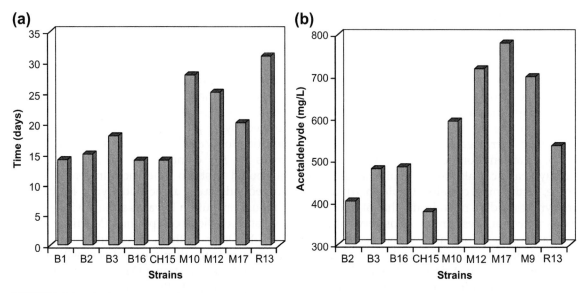

**FIGURE 3.2** *Flor* formation speed (a) and acetaldehyde production (b) in different strains of *Saccharomyces cerevisiae* in young sherry wine (*sobretabla*). Strains from the same race can have different characteristics. *S. cerevisiae beticus* (B) and *S. cerevisiae cheresiensis* (CH), for example, are faster at forming a *flor* than *S. cerevisiae montuliensis* (M) or *S. cerevisiae rouxii* (R), but they are also characterized by less acetaldehyde production and tolerance. *Reproduced from Martínez et al. (1997a).*

esters derived from methylpentanoic acids. *Finos*, in particular, are especially rich in acetaldehyde, diacetyl, ethyl esters of branched aliphatic acids, and 4-ethylguaiacol (Campo et al., 2008; Peinado et al., 2004). The characteristic flavor of *Fino* wines is principally derived from acetaldehyde, but diacetylene and acetoin also have a role (Muñoz et al., 2006). Concentrations of ethanol, glycerol, acetaldehyde, acetic acid, and nitrogenous and volatile compounds change continually throughout aging (Martínez et al., 1998). While considerable amounts of ethanol, glycerol, organic acids, and amino acids are consumed during biological aging (Martínez et al., 1998), this consumption is not continuous. Ethanol, for instance, is consumed in the greatest quantities during the formation of the *flor*, while glycerol, organic acids, and amino acids are consumed once the film has been established. The *flor* also contains bacterial populations, most of which are species of *Lactobacillus*, associated with varying populations of

*S. cerevisiae* (Kawarai et al., 2007). These bacteria in the *flor* play a key role in the consumption of organic acids (gluconic, malic, and lactic acid). Lactic acid bacteria are responsible for malolactic fermentation during the aging of younger wines, normally in the fourth or fifth *criadera* rows. These reactions have been detected in wines from both Jerez and Montilla-Moriles (Bravo-Abad, 1986; Mérida et al., 2005; Peinado et al., 2004). While certain races of *S. cerevisiae* (*beticus*, *cheresiensis*) are more efficient than others (*montuliensis*, *rouxii*) at reducing volatile acidity, they are less efficient when it comes to consuming alcohol and producing acetaldehyde (Martínez de la Ossa et al., 1987b; Martínez et al., 1993, 1998). Even within the same race, some strains are much more efficient than others at reducing volatile acidity. Variations in volatile concentrations depend not only on the yeast strains that form the *flor* but also on aging conditions, number of *criadera* levels, butt replenishment methods, vineyard density, and climate

(Martínez et al., 1993). In *Fino* wines, the dry extract decreases to under 15 g/L, mainly because of the consumption of glycerol by the *flor* yeasts, while in *Oloroso* wines this extract can exceed levels of 22 g/L due to the concentration of compounds in the wine produced by evaporation (Martínez de la Ossa et al., 1987a) (Table 3.2).

### 2.1.1. Flor *Formation*

The *flor* is a structure of cells that forms a thick, white, rough film that floats on the surface of the wine. The yeasts that form this film are responsible for the biological aging of *Fino* and *Amontillado* wines and can survive in alcohol concentrations of around 16%. Other hostile conditions that these yeasts have to tolerate are high concentrations of acetaldehyde, oxidative stress due to the metabolism of nonfermentable carbon sources, water stress, and, often, high levels of metals (e.g., copper) and nitrogen sources that are difficult to assimilate (e.g., proline). The *flor* is thus considered to be an adaptive mechanism in which yeast cells change their size, shape, and hydrophobicity in response to different stresses. Hydrophobicity, which is imparted by the presence of specific surface proteins (including Flo11 [Muc1], which will be discussed in Section 3.3,

causes the cells to aggregate. The resulting aggregate adheres to the gas bubbles generated during respiration and floats on the surface (see Figure 3.3). It has been suggested that the *flor* is the result of ethanol-induced lipogenic activity (Bravo-Abad, 1986). Later studies, however, reported that, while the addition of oleic acid or ergosterol did not affect *flor* formation, the addition of different proteases disintegrated the *flor* and reduced hydrophobicity (Martínez et al., 1997c), indicating that *flor* formation depends on the presence of hydrophobic cell-surface proteins. More recent studies have assigned a key role to mannoproteins containing over 90% mannose in hydrophobicity and *flor* formation (Caridi, 2006), and described how yeast cells secrete glucose and mannose polysaccharides that surround cell aggregates (Beauvais et al., 2009).

While the presence of fermentable carbon sources or ammonium salts generally inhibits *flor* formation, both proline and ethanol appear to activate the process (Fidalgo et al., 2006; Martínez et al., 1997b). Indeed, hydrophobicity increases and films become more compact as alcohol levels increase. Nonetheless, yeasts have also been found to form *flors* in sweet botrytized wines (Kovacs et al., 2008). According to some authors, *flor* formation is very

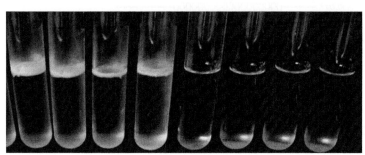

**A9　A9-F1　A9-F2 Ar5-H12 Ar5-H12　W3　W3-F1　W3-F2**
　　　　　　　　　　　　　　　-F1

**FIGURE 3.3** *Flor* formation by wild-type strains and their Δ*flo11* disruptants. Different yeast strains were cultivated overnight and inoculated into *flor* SD medium with 3% ethanol (vol/vol) and a pH of 3.5. The tubes containing each culture were photographed after 3 d of static incubation at 30°C. The data indicate that *FLO11* is the primary factor in *flor* formation but that there are other genes involved. A9 is a diploid *flor* strain containing two functional copies of *FLO11*; A9-F1 is a strain derived from A9 with a disrupted copy of *FLO11*; and A9-F2, also derived from A9, has two disrupted copies of *FLO11*. Ar5-H12 is a *flor* strain with a single functional copy of *FLO11*; its derivative Ar5-H12-F1 has an undisrupted copy of *FLO11*. W3 is a wine strain incapable of forming a *flor*, as are its derivatives W3-F1 and W3-F2, which have one and two disrupted *FLO11* alleles, respectively. *Reproduced from Ishigami et al. (2006).*

positively influenced by the presence of polyphenol compounds (Budroni et al., 1995) and biotin (Bravo-Abad, 1986). Others, in contrast, have found that the process requires pantothenic acid in addition to the oxidative metabolism (Martínez et al., 1997b). These differences, however, may simply be due to differences in the composition of the wine (Charpentier et al., 2009).

Several authors have likened *flor* formation to a form of flocculation and to pseudohyphal development and invasive growth (Budroni et al., 1995; Lambrechts et al., 1996). The similarity between *flor* formation and flocculation and filamentation lies in the notable increase in cell hydrophobicity that occurs in all these cases (Straver & Kijne, 1996) and the activation of these processes in environments with limited nitrogen supply (Douglas et al., 2007; Ma et al., 2007), as will be discussed at the end of this section.

The genes responsible for flocculation (*FLO* genes) form subtelomeric families that include both functional genes and pseudogenes (Teunissen & Steensma, 1995; van Mulders et al., 2009). They encode cell-surface glycosylphosphatidylinositol (GPI)-linked proteins that are covalently bound to glucans in the cell wall (Beauvais et al., 2009; Verstrepen & Klis, 2006). These cell-surface proteins, known as adhesins (Huang et al., 2009), are composed of repeating motifs organized in a characteristic fashion from the plasma membrane, through the cell wall, to the cell surface (Douglas et al., 2007; van Mulders et al., 2009) (see Figure 3.4). Recombination of the internal repeats results in an increase or decrease in protein size, which, in turn, would alter phenotypes such as adherence, *flor* formation, and flocculation (Rando & Verstrepen, 2007; Verstrepen et al., 2005; Verstrepen et al., 2004). One of the genes

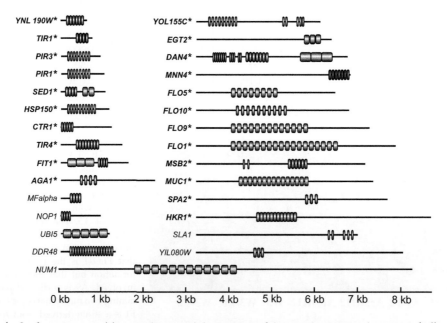

**FIGURE 3.4** *Saccharomyces cerevisiae* proteins containing conserved intragenic repeats. A screen of all open reading frames in the *S. cerevisiae* genome revealed 29 genes with large repeats (>40 nucleotides). Some of the repeats (vertical boxes) showed variations in size from one strain to the next. The majority of repeats occurred in cell-surface proteins. The names of the genes encoding proteins of this type are shown with asterisks. *Reproduced from Verstrepen et al. (2005).*

**(a)**

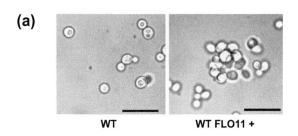

**WT**  **WT FLO11 +**

**(b)**

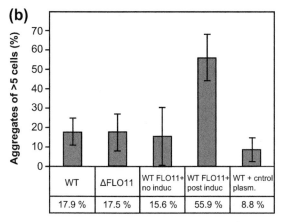

| WT | ΔFLO11 | WT FLO11+ no induc | WT FLO11+ post induc | WT + cntrol plasm. |
|---|---|---|---|---|
| 17.9 % | 17.5 % | 15.6 % | 55.9 % | 8.8 % |

**(c)**

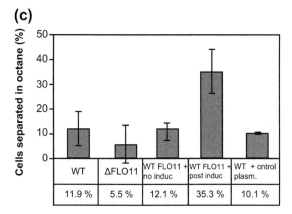

| WT | ΔFLO11 | WT FLO11 + no induc | WT FLO11 + post induc | WT + cntrol plasm. |
|---|---|---|---|---|
| 11.9 % | 5.5 % | 12.1 % | 35.3 % | 10.1 % |

responsible for cell flocculation, *FLO11*, regulates pseudomycelium formation, *flor* formation, and invasive growth in yeasts (Barrales et al., 2008; Lambrechts et al., 1996; Lo & Dranginis, 1996; Palecek et al., 2000; Tamaki et al., 2000). All of these phenomena are characterized by an increase in cell hydrophobicity (Purevdorj-Gage et al., 2007) (see Figure 3.5). Another of the genes responsible for flocculation, *FLO1*, is involved in protecting cell aggregates from a range of hostile conditions such as the presence of ethanol and certain antimicrobial compounds (Beauvais et al., 2009; Smukalla et al., 2008), which is precisely the environment the *flor* yeasts have to endure (see Figure 3.6). Adverse conditions such as the presence of ethanol and acetaldehyde, low pH, and a lack of nutrients activate the genes encoding adhesins (Barrales et al., 2008; van Dyk et al., 2005; Verstrepen & Klis, 2006) (see Figure 3.7). Nonetheless, synergistic effects between different hostile conditions (e.g., high temperature and ethanol) can lengthen the time required for the *flor* to form and even cause it to disappear (Ibeas & Jiménez, 1997) (see Figure 3.8).

The Flo11 adhesin is regulated via three main signaling pathways: the mitogen-activated protein kinase (MAPK)-dependent pathway (which is regulated by nitrogen depletion at least) and two pathways regulated by glucose (Figure 3.7). Furthermore, other activators involved in chromatin remodeling (e.g., Msn1) and pH response pathways (e.g., Rim20) also control the synthesis of Flo11 in these pathways (see Figure 3.9). These observations, together

**FIGURE 3.5** *Flo11* overexpression confers greater ability to coaggregate and greater hydrophobicity. Micrograph of wild-type (WT) strain (left) and strain overexpressing *FLO11* under the control of the *GAL1* promoter (WT FLO11+) (right), both during exponential growth (a). Percentage of cells in multiple cell aggregates (b). Hydrophobicity (c). The percentage of cells partitioned in octane was measured using an aqueous-hydrocarbon biphasic assay. Optical density (OD)$_{600nm}$ was measured after placing a volume of octane on the surface of the aliquots, vortexing the tubes, and leaving the phases to separate. The difference between the OD$_{600nm}$ before and after the addition of octane was used to determine the hydrophobicity of the culture. The greater the percentage of partitioned cells, the greater the hydrophobicity. No induc = *GAL1* not induced; post-induc = *GAL1* induced; ΔFLO11 = strain with disrupted *FLO11* gene; WT+cntrol plasm = wild-type transformed with empty plasmid. *Reproduced from Purevdorj-Gage et al. (2007).*

FIGURE 3.6 Presence of Flo1 protein and flocculation capacity confer resistance to different stresses. *Saccharomyces cerevisiae* cells with (KV210) and without (KV22) *FLO1* expression were subjected to various stress treatments, after which the percentage of surviving cells was measured. Asterisks indicate statistically significant differences between flocculent and nonflocculent cultures ($\alpha = 0.05$); error bars correspond to standard deviation. *Reproduced from Smukalla et al. (2008).*

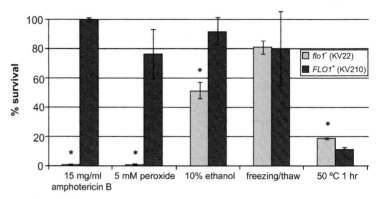

with the role attributed to *FLO* genes and the possible relationship between flocculation, *flor* formation, and pseudomycelium growth (Ishigami et al., 2006), are further supported by the fact that high levels of these genes (mainly of *FLO11*) have been detected during the *flor* formation phase (Infante et al., 2003).

## 2.2. Genetic Characteristics of Fermentation and Aging Yeasts

The emergence of molecular biology techniques has permitted a more accurate classification of wine yeasts and revealed enormous variability among the different strains of *S. cerevisiae* (Fernández-Espinar et al., 2003; Martorell et al., 2005). While traditional metabolic methods were successfully used to distinguish between different races of *S. cerevisiae* in the fermentation and *flor* microflora of biologically aged wines, they were unable to unequivocally distinguish between different populations of the same race (Martínez et al., 1995). The emergence of molecular techniques, however, has greatly improved the genetic characterization of both fermentation and *flor* yeasts (Benítez et al., 1996). The use of these techniques has shed light on the DNA content of yeast strains (flow cytometry), chromosome numbers and size (pulsed-field gel electrophoresis [PFGE]), homology with genes from other yeasts (hybridization with specific probes), and mitochondrial

DNA (mtDNA) polymorphisms (restriction fragment length polymorphism [RFLP] analysis) (Martínez et al., 1995). Research performed since these techniques became available has confirmed that the fermentation and *flor* strains of *S. cerevisiae* in biologically aged wines display a high degree of genetic variability, not only in terms of DNA content (variations of 1.3 to almost 4.0 n) but also in the number and size of nuclear chromosomes and mtDNA restriction fragments (Martínez et al., 1995). Electrophoretic karyotyping has also revealed important differences between the size and number of chromosomes in both fermentation and *flor* yeast strains in biologically aged wines (Valero et al., 2007). Furthermore, *flor* yeasts as a whole seem to have a different chromosome pattern to other wine yeasts.

Even greater variations, however, seem to exist in mtDNA (Martínez et al., 1995). mtDNA restriction analysis is, thus, a sufficiently simple, rapid, and unequivocal method for studying the yeast populations involved in fermentation and wine aging and monitoring the development of inoculated strains to determine whether or not they displace indigenous populations (Esteve-Zarzoso et al., 2001).

The first genetic characterization studies of yeast strains in biologically aged wines were conducted by Sancho et al. (1986) in Montilla-Moriles wines. The authors isolated and characterized strains during fermentation and

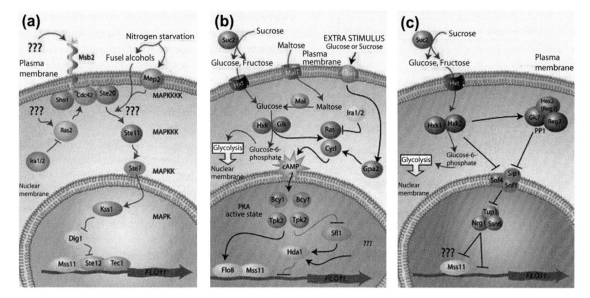

**FIGURE 3.7** Signaling cascades that regulate *Flo11*. (a) The MAPK-dependent filamentous growth pathway. The core of this pathway in *Saccharomyces cerevisiae* is formed by the central kinases Ste11 (MAPKKK) and Ste7 (MAPKK). These kinases are shared by other MAPK signaling cascades, such as the mating response pathway and the high osmolarity glycerol (HOG) pathway. Msb2 is thought to function as a sensor at the top of the pathway, but the conditions that trigger Msb2 have not yet been characterized. Other known triggers of FLO11 that (at least partially) act through MAPK signaling include nitrogen starvation (which might be sensed through the ammonium permease Mep2) and elevated concentrations of certain fusel alcohols such as butanol. The specific downstream part of the MAPK-dependent filamentous growth pathway includes the MAPK Kss1 and the transcriptional regulators Dig1, Ste12, and Tec1. (b) The Ras/AMPc/PKA pathway. The Ras/cAMP/PKA pathway responds to the presence of glucose or sucrose in the medium. The pathway is activated by two independent triggers. First, the intracellular phosphorylation of glucose enhances the activity of adenylate cyclase Cyr1. Second, a G protein-coupled receptor system, consisting of the receptor Gpr1 and the Gα protein Gpa2, senses extracellular glucose and sucrose. Activation of the Gpr1/Gpa2 complex causes a further increase in Cyr1 activity, resulting in a transient cAMP peak. Subsequently, cAMP activates the protein kinase A complex (PKA), resulting in the dissociation of the Bcy1 subunits from the Tpk catalytic subunits of PKA. The three different Tpk subunits, Tpk1, Tpk2, and Tpk3, have been shown to have distinct roles in *FLO11* regulation: Tpk2 mostly acts as an activator, while Tpk1 and Tpk3 function as inhibitors. Once released from the inhibitory Bcy1 subunits, the free Tpk2 kinase inactivates Sfl1 (suppressor of flocculation) and activates the positive regulator Flo8. (c) The main glucose repression pathway. The hexose transporters (Hxt) allow glucose uptake from the medium. Once inside the cell, glucose is phosphorylated to glucose-6-phosphate by one of the hexokinases (Hxk). This phosphorylation process and/or the depletion of AMP due to the increase in ATP production inactivate(s) the central Snf1 protein kinase. Inactivation of Snf1 allows the regulatory proteins Mig1 and Nrg1 to bind to the *FLO11* promoter and recruit the general repressors Tup1 and Ssn6, resulting in repression of *FLO11*. Questions marks indicate unknown mechanism. *Reproduced from Verstrepen and Klis (2006).*

biological aging and found that all the yeasts belonged to different races of *S. cerevisiae*. More interestingly, however, they found that the races did not mix, which suggests strong sexual isolation between the populations to prevent the random distribution of metabolic characteristics. Jiménez and Benítez (1988) confirmed this sexual isolation in a later study

of *flor* strains in which they discovered heterozygotic lethal recessive alleles. This indicated that these strains never sporulated or at least that sporulation occurred less frequently than mutations in lethal alleles. More recently, Puig et al. (2000) attributed genetic variability in wine strains to mitotic recombination, repair, and gene conversion during growth and not to

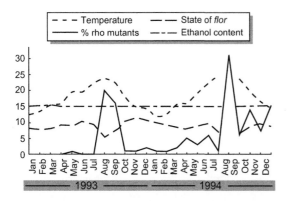

**FIGURE 3.8** Alcohol content (% vol/vol), temperature (°C), rho (petite) mutants (%), and state of the *flor* during the biological aging of Jerez wines. Synergistic effects of alcohol and temperature may give rise to a high rate of mutants with nonfunctional mitochondria (petite mutants) and the deterioration or disappearance of the *flor* film. *Reproduced from Ibeas et al. (1997).*

sporulation and meiotic recombination. There have, however, been reports of sporadic intra- and interspecific crosses between *Saccharomyces* strains from wine and other enological sources (González et al., 2008).

Mesa et al. (1999) found enormous genetic variation between *beticus* and *cheresiensis* races of *flor* yeasts isolated in a winery in terms of both chromosome patterns and mtDNA profiles. They also reported preferential associations between chromosome and mitochondrial patterns. In a subsequent study, the same group found a correlation between strains with specific patterns and different aged wines, with certain patterns found only in *soleras* and others found only at *criadera* levels (Mesa et al., 2000). In an earlier study, Nadal et al. (1996) had observed a strong association between specific mtDNA patterns and fermentation strains with high tolerance of ethanol and temperature. Shortly afterwards, Martínez et al. (1997a) found that different aged wines in the *solera* and *criadera* system were correlated with races rather than with specific mtDNA profiles. They proposed that *beticus* races were more abundant in younger wines because they

were able to form a *flor* film more quickly and that *montuliensis* races were more abundant in older wines because of their greater production and tolerance of acetaldehyde (see Figure 3.2). Other authors, on analyzing the molecular profiles of the *flor* in other wineries, found a single dominant strain (Esteve-Zarzoso et al., 2001; Ibeas et al., 1997).

Because the majority of *flor* strains never sporulate (or sporulate only poorly) and meiotic products are often inviable (Martínez et al., 1995), chromosome constitution has been established by studying spores produced using mass mating methods and analyzing the segregation frequencies of markers in the chromosomes after the sporulation of industrial–laboratory hybrids (Bakalinsky & Snow, 1990). This method can be used to distinguish between disomic, trisomic, and tetrasomic complements in parent strains. The study by Bakalinsky and Snow showed numerous cases of aneuploidy in the strains analyzed but of particular interest was the large number of extra copies of chromosomes V, VII, and XIII detected. Chromosome XIII contains the genes for alcohol and aldehyde dehydrogenases, which play a key role in the production and consumption of ethanol and acetaldehyde. Using similar methods, Guijo et al. (1997) analyzed the chromosome structure of wine strains involved in the fermentation and aging of Montilla-Moriles wines. They detected aneuploidy in all of the strains analyzed as well as a high rate of polysomy for chromosome XIII. Considerable genomic plasticity has also been detected for chromosomes IV, VIII, and XII in fermentation strains, including those that participate in biologically aged wines (Infante et al., 2003; Puig et al., 2000). These variations have been linked to chromosome translocations due to homologous, asymmetric, or ectopic recombination between sequences of *Ty*, *δ*, or *Y′* elements that resulted in strains that were better adapted to specific industrial conditions. Variability in chromosome patterns has been attributed to recombination events that occur during

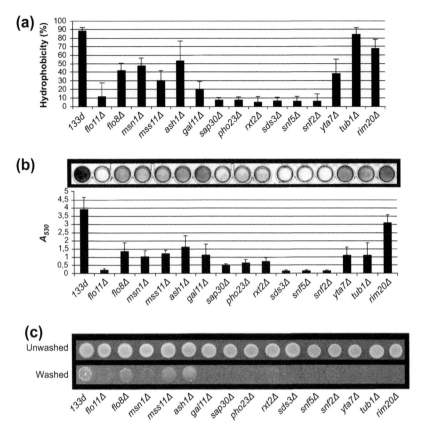

**FIGURE 3.9** *FLO11*-related phenotypes in mutants with different gene deletions. In addition to the regulatory function described in Figure 3.6, proteins involved in chromatin remodeling and pH response can regulate the synthesis of Flo11 protein. Hydrophobicity (a) was measured using an aqueous-hydrocarbon biphasic assay (Purevdorj-Gage et al., 2007). *Flor* film on solid surface (b). Exponentially growing cells were placed in microtiter plate wells and incubated for 1 h at 28°C. The cells were then stained with crystal violet and the wells washed repeatedly with water and photographed. For biofilm quantification, the crystal violet was solubilized using SDS and optical density$_{530nm}$ was measured. Invasive growth (c). Exponentially growing cells were spotted on yeast extract peptone dextrose (YPED) solid medium and photographed before (unwashed) and after (washed) washing. 133d = wild-type strain; ahs1 = transcription factor; flo11 = adhesin; flo8 = transcription factor; gal11 = mediator complex component; msn1 = transcriptional activator; mss11 = transcription factor; pho23 = component of Rpd3 histone deacetylase complex; rim20 = protein involved in proteolytic activation of Rim101 in response to alkaline pH; rxt2 = subunit of histone deacetylase complex Rpd3L; sap30 = subunit of histone deacetylase complex; sds3 = component of Rpd3p/Sin3p deacetylase complex; snf2 = catalytic subunit of SWI/SNF chromatin remodeling complex; snf5 = subunit of SWI/SNF chromatin remodeling complex; tup1 = general transcriptional repressor; yta7 = protein of unknown function. *Reproduced from Barrales et al. (2008).*

vegetative growth due to the hostile environment the yeasts have to endure (Infante et al., 2003). Several chromosomal rearrangements appear to be mediated by *Ty1* elements, and yeasts that participate in biological aging have been found to possess *Ty1* elements located exclusively on chromosome XII. Whether the severe restrictions affecting *Ty1* mobility and expression in laboratory strains also apply to wine strains is not known (Nyswaner et al., 2008; Wu & Jiang, 2008). Hybridization experiments based on DNA microarrays used to

compare, gene by gene, the genomes of *flor* and laboratory yeasts have revealed the existence of chromosomes, chromosomal regions, and genes that affect events such as aneuploidy, amplification, and deletion and are partly responsible for the genetic variability detected and the enological properties of yeast strains (Hu et al., 2007; Infante et al., 2003). High expression levels of genes involved in the biosynthesis of amino acids and the metabolism of nitrogen and sulfur, for example, have been found in wine strains only, as has the

**TABLE 3.3**    Open Reading Frames (ORFs)[1] Included in Genomic Regions Amplified in *Saccharomyces cerevisiae* Flor Yeast Strain 11.3 That Have Been Found Overexpressed in This Strain With Respect to *S. cerevisiae* X2180 Strain During Growth Under Enological-like Conditions

| ORF | Name | Chromosome | Characteristics of gene product |
|---|---|---|---|
| YBL092W | *RPL32* | II (20–82 kb) | 60S large subunit ribosomal protein |
| YBR089C-A | *NHP6B* | II (427–436 kb) | Regulation of transcription (chromatin architecture) |
| YCL018W | *LEU2* | III (76–105 kb) | 3-Isopropylmalate dehydrogenase |
| YCL050C | *APA1* | III (3.5–70 kb) | ATP adenyltransferase |
| YDL198C | *YHM1* | IV (0–116 kb) | Mitochondrial carrier protein (maintenance of mitochondrial genome) |
| YEL017C-A | *PMP2* | V (30–128 kb) | Plasma membrane H+-ATPase regulator |
| YER044C[a] | *ERG28* | V (196–313 kb) | Involved in ergosterol biosynthesis |
| YER163C | | V (488–554 kb) | Biological process/function unknown |
| YGR234W[a] | *YHB1* | VII (697–1095 kb) | Flavohemoglobin (cell protection against nitrosylation) |
| YHR053C | *CUP1-1* | VIII (208–217 kb) | Copper-binding (metallothionein) protein |
| YHR055C | *CUP1-2* | VIII (208–217 kb) | Copper-binding (metallothionein) protein |
| YHR096C | *HXT5* | VIII (285–320 kb) | Hexose transporter |
| YHR162W[a] | | VIII (320–481 kb) | Biological process/function unknown |
| YIL065C | *FIS1* | IX (232–243 kb) | Involved in mitochondrial fission |
| YIL155C | *GUT2* | IX(18–57 kb) | Glycerol 3-phosphate dehydrogenase (mitochondrial) |
| YIR019C | *MUC1* | IX (312–425 kb) | Cell surface glycoprotein involved in biofilm formation |
| YIR037W | *HYR1* | IX (312–425 kb) | Glutathione peroxidase |
| YMR009W | | XIII (196–427 kb) | Biological process/function unknown |
| YPL092W | *SSU1* | XVI (374–590 kb) | Sulfite transport (sulfite resistance) |
| YPR099C | | XVI (729–825 kb) | Biological process/function unknown |

[1]*ORFs with significant log ratios, which indicate a higher copy in strain 11.3, but included within a chromosomal region with equal copy number in both 11.3 and 1.28 strains. The majority of ORFs correspond to proteins of enological interest.*
The limits of the genomic regions in each chromosome (from left telomere) are indicated.
*Reproduced from Infante et al. (2003).*

overexpression of genes linked to tolerance of high levels of sulfur dioxide (Backhus et al., 2001). Furthermore, in the study by Backhus et al., genes that regulate the change from fermentative to aerobic metabolism and are repressed by glucose in laboratory strains were induced in wine strains under low-nitrogen conditions, despite high levels of glucose.

Analysis of *flor* strains has revealed aneuploidy of chromosomes I, III, and IV and chromosomes X and XII in *beticus* and *montuliensis* races, respectively (Infante et al., 2003). In the same study, variations were also found in the copy number of 38% of the open reading frames (ORFs) that make up the genome (see Table 3.3). The amplified regions appeared to be associated with the presence of *Ty* transposons and long terminal repeat elements. Moreover, it was found that, of the genes overexpressed during the *flor* phase that are of enological interest and involved in *flor* formation, sulfite tolerance, ergosterol synthesis, and the metabolism of glutathione (GSH) and other substances (see Table 3.3), 37% were amplified genes, pointing to a possible association between gene rearrangements and adaptation to specific conditions (Infante et al., 2003). There have also been reports of ribosomal DNA (rDNA) rearrangements involving chromosome XII in wine yeasts not mediated by either homologous recombination or *Ty1* elements (Blake et al., 2006). Several authors have suggested that more than one gene may be involved in the formation and maintenance of the *flor*. Jiménez and Benítez (1988) and Castrejón (2000), for example, using meiotic analysis of *flor* yeast strains found that *flor* formation was regulated by more than one gene. Those authors also found variations in aneuploidy.

Restriction analysis of the intergenic region of 5.8S rDNA has identified a 24-base-pair deletion in over 150 *S. cerevisiae flor* strains analyzed; this deletion was not found in fermentation strains, providing further evidence that the two types of strain are different (Esteve-Zarzoso et al., 2001). The 24-base-pair deletion was also found in strains of *Saccharomyces aceti* and *Saccharomyces gaditensis* isolated from *flor* films in Jerez wines and in strains of *S. prostoserdovii* isolated from *flor* films in Vernaccia de Oristano wines (Fernadez-Espinar et al., 2000). In more recent studies, the same deletion was detected in *flor* yeasts (all *S. cerevisiae*) isolated in different *añadas*, *soleras*, and *criaderas* in Jerez and Montilla-Moriles wines (Naumova et al., 2005) and in the majority of strains isolated from *flors* in botrytized wines (Kovacs et al., 2008). The deletion was not detected in *S. cerevisiae flor* strains from the French *vins jaunes* (Charpentier et al., 2009), indicating that *flor* yeasts possibly have different phylogenic origins.

Although each race of *S. cerevisiae* confers specific characteristics to wine, the four majority races of *flor* yeasts have the same restriction pattern in the intergenic region of 5.8S rDNA (Esteve-Zarzoso et al., 2004). This means that metabolic differences are not detectable at the molecular level. Furthermore, strains can vary even within the same race. This explains the existence of phenotypes characterized by low, moderate, and high acetaldehyde production (550–800 mg/L for *montuliensis* strains and 350–450 mg/L for *beticus* strains) and low, moderate, or high rate of *flor* formation (15–20 d for *beticus* strains and 25–35 d for *montuliensis* strains) (See Figure 3.2). These variations mean that a very large number of individuals must be analyzed before a property can be assigned to a particular race.

# 3. INFLUENCE OF ENVIRONMENTAL FACTORS ON THE CHARACTERISTICS OF YEASTS INVOLVED IN BIOLOGICAL AGING

## 3.1. Influence on Mitochondria

Yeasts responsible for the biological aging of wines (basically *flor* yeasts) show enormous variability in terms of both chromosome and mtDNA restriction fragment patterns (Martínez et al., 1995). mtDNA variability could be a result of mutations induced by the high mutagenic concentrations of alcohol the yeasts are exposed to. The frequency of spontaneous mutants with nonfunctional mitochondria (petite mutants) has been reported to increase ten-fold at concentrations of 24% ethanol (Bandas & Zakharov, 1980; Castrejón et al., 2002) (see Figure 3.8). The mitochondrial genome is responsible for cell viability in high-alcohol environments (Jiménez & Benítez, 1988). In particular, the mitochondria of *flor* yeasts remain functional under these conditions thanks to their exceptional resistance

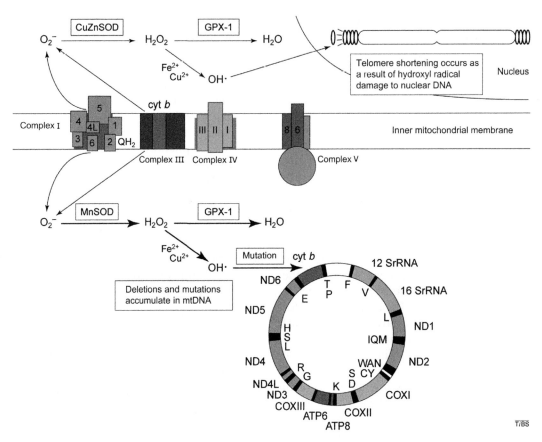

**FIGURE 3.10** Generation of reactive oxygen species (ROS) in the mitochondrion and role of superoxide dismutases. Mitochondrial DNA is the target of ROS. The increased resistance of *flor* yeasts to ROS and ethanol is largely due to a greater efficiency of the mitochondrial manganese superoxide dismutase (MnSOD). *Reproduced from Raha and Robinson (2000).*

to ethanol. Because *flor* yeasts have an oxidative metabolism, it has been suggested that ethanol might induce a very high rate of mtDNA mutation but that the requirement for functional mitochondria would lead to elimination of mutants with nonfunctional mitochondrial genomes (Martínez et al., 1995).

It has also been reported that ethanol causes loss of mtDNA (Ibeas & Jiménez, 1997). Because ethanol is a membrane solvent, its mutagenic effect has been attributed to alterations in the mitochondrial membrane that lead to the loss of mtDNA. Nevertheless, both ethanol and acetaldehyde cause breaks in chromosomal DNA

(Ristow et al., 1995). Ethanol toxicity has been correlated with the production of mitochondrial reactive oxygen species (ROS) (see Figure 3.10) (Abbott et al., 2009; Costa et al., 1997; Du & Takagi, 2007; Gales et al., 2008; Landolfo et al., 2008; Longo et al., 1996; Piper, 1999; Raha & Robinson, 2000), although data are unavailable on the effect of ROS on mtDNA. The generation of superoxide radicals probably causes high rates of mutagenesis in mtDNA as a result of oxidative stress, and the tolerance shown by yeast mitochondria to these mutagenic effects could be thanks to their ability to prevent damage by these radicals (Costa et al., 1997).

Castrejón et al. (2002) reported that both ethanol and acetaldehyde induced petite mutants in populations of *flor* yeasts and demonstrated that the induction mechanism initially involved mtDNA damage. Indeed, mtDNA is lost after prolonged incubations. Acetaldehyde and ethanol, thus, both cause irreversible changes in mtDNA and alter restriction patterns. While these mitochondrial alterations eventually lead to complete DNA loss, in *Fino* wines, mutant populations do not survive due to the lack of fermentable carbon sources (Castrejón et al., 2002). Petite mutants nevertheless remain detectable temporarily. Ibeas et al. (1997) found that ethanol and temperature exerted a synergistic effect on the *flor* layer that led to the formation of petite mutants (20–30%) (see Figure 3.8). Esteve-Zarzoso et al. (2001), in turn, reported the coexistence of large cells with functional mitochondria and petite mutants (a minority) with different mtDNA restriction patterns. Finally, the greatest mtDNA polymorphism has been found in yeasts that generate wine with higher levels of acetaldehyde.

## 3.2. Influence on Chromosomes

As mentioned, ethanol and acetaldehyde have been found to cause breaks in yeast chromosomal DNA (Ristow et al., 1995). In *Fino* wines with high concentrations of ethanol and acetaldehyde, the chromosomal DNA sequences of *flor* yeasts in all probability undergo numerous changes due to errors generated during the repair of DNA breaks by recombination. This would explain the high frequency of chromosomal rearrangements seen in such cases and also the differences in chromosomal organization between laboratory strains and wine and *flor* strains (Infante et al., 2003; Pirino et al., 2004; Puig et al., 2000; Zara et al., 2008).

Large differences in gene order due to translocations have also been described in the mtDNA of wine yeasts (Cardazzo et al., 1998).

The preferential mutagenic action of acetaldehyde and ethanol on mtDNA is due, on the one hand, to the fact that mtDNA is very close to the production site of ROS, and, on the other, to the fact that mtDNA polymerase lacks repair ability. *Flor* yeasts possibly have more resistant mitochondria because of the greater efficiency of mitochondrial manganese superoxide dismutase (MnSOD) (see Figure 3.10 and Table 3.4) and the enzymes associated with the metabolism of GSH and other compounds that protect against oxidative damage (see Figure 3.11). Indeed, *SOD* genes are expressed much more during the *flor* phase than during other growth phases (Castrejón et al., 2002; Infante et al., 2003). In a study of 176 mutants with an ethanol-sensitivity phenotype (each with a deletion in a different gene), almost all of the genes analyzed encoded proteins involved in respiration and mitochondrial adenosine triphosphate (ATP) synthesis (Kumar et al., 2008).

Ethanol might also induce chromosome loss due to its role as a membrane solvent. This would explain the high frequency of aneuploidy found in *flor* yeasts (Guijo et al., 1997; Martínez

**TABLE 3.4** Sensitivity to Ethanol in Respiration-deficient Mutants in Cytoplasmic Superoxide (CuZnSOD), Encoded by *SOD1*, and in Mitochondrial SOD (MnSOD[1]), Encoded by *SOD2*

| | Viability (%) | | | |
|---|---|---|---|---|
| | Diauxic shift | | Post-diauxic | |
| *Saccharomyces cerevisiae* | 14% EtOH | 20% EtOH | 14% EtOH | 20% EtOH |
| aBR10 (wild-type) | 90 (3) | 56 (8) | 79 (7) | 82 (2) |
| sod1 | 83 (10) | 44 (15) | 87 (11) | 79 (8) |
| sod2 | 72 (3) | 0 | 76 (3) | 9 (3) |

[1]*Believed to play a key role in resistance to ethanol and oxidative stress.*
Values are means (*SD*) of five independent experiments.
*Reproduced from Costa et al. (1997). EtOH = ethanol.*

FIGURE 3.11 Resistance to oxidative stress in different *Saccharomyces cerevisiae* mutants with deletion of genes involved in glutathione (GSH) metabolism and response to oxidative stress. Reduced GSH plays an important role in resistance to oxidative stress, mainly in cells in the *flor* phase. In this assay, cells were taken during exponential or stationary growth phases or during *flor* formation (mats, or colonies of cells). Growth inhibition assays were performed in Petri

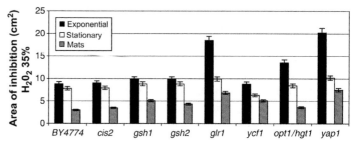

dishes following exposure to 35% $H_2O_2$. cis2 = γ-glutamyl transpeptidase; glr1 = glutathione reductase; gsh1 = γ-glutamylcysteine ligase; gsh2 = glutathione synthethase; opt1/hgt1 = cell-surface transporter of oligopeptides and GSH; yap1 = stress-induced transcription factor; ycf1 = vacuolar transporter of GSH and GSH-X. *Reproduced from Gales et al. (2008).*

et al., 1995; Mesa et al., 2000; Naumova et al., 2005). Finally, the maintenance of aneuploidy and chromosomal rearrangements is advantageous in the conditions that *flor* yeasts have to endure. It also plays a key role in sexual isolation, which, in turn, prevents the random distribution of favorable characteristics, which is why the majority of *flor* yeasts do not sporulate. Sporulation does, however, increase cell resistance to hostile conditions, which explains why a small percentage of wine (not *flor*) yeasts are apomictic, meaning that they do not complete the first or second meiotic divisions and that the two spores in the ascus have an identical genetic structure to that of the parent cells (Castrejón et al., 2004). Apomixis is less dependent on environmental conditions and therefore much more common than meiotic sporulation. It always takes place in unfavorable conditions and prevents recombination and the loss of optimal genotypes.

## 3.3. Influence on the Membrane and Cell Wall

Yeast cell membranes undergo major changes when exposed to high concentrations of ethanol. The yeasts synthesize lipids enriched in C18:1 to compensate for the decrease in palmitic acid. There is also a general increase in the proportion of ergosterol, unsaturated fatty acids, and

phospholipids and a decrease in the sterol/protein ratio (van Uden, 1989). Supplementation with ergosterol and unsaturated fatty acids increases ethanol tolerance in yeasts, strongly indicating that the cell membrane is the main target of ethanol-induced toxicity. The transfer of mitochondria from strains with high ethanol tolerance to laboratory strains has been shown to considerably increase ethanol tolerance and thermotolerance in the receptor strains (Jiménez & Benítez, 1988). The study by Jiménez and Benítez also showed loss of ethanol-induced mitochondrial functions following transfer, indicating that the mitochondrial genome is partly responsible for tolerance. This tolerance is partly due to the role of aerobic metabolism in the biosynthesis of ergosterol and unsaturated fatty acids, which are essential membrane components and key determinants of ethanol tolerance (Sulo et al., 2003; Zara et al., 2009). Indeed, there have been reports of considerable improvements in ethanol tolerance during the fermentation of Montilla-Moriles wines following aeration, with direct correlations between oxygen concentrations, cell membrane sterol, and phospholipid content (Mauricio et al., 1990).

Ethanol also induces the expression of antioxidant proteins (Piper, 1999) and proteins that protect against oxidative stress (e.g., SOD, catalase, or enzymes associated with GSH metabolism) (Abbott et al., 2009; Du & Takagi, 2007; Gales et al., 2008; Infante et al., 2003; Land-olfo et al., 2008), as indicated previously. Both ethanol and temperature induce the expression of proteins involved in the metabolism of treha-lose (a protective agent in cell membranes) and stress-response proteins such as Hsp104 that contribute to thermotolerance and ethanol toler-ance. Resistance to osmotic shock, ethanol, cold, oxidative stress, and acetaldehyde in *flor* strains has been associated with the synthesis of heat shock proteins other than Hsp104, such as Hsp12, Hsp82, and Hsp26 (Aranda et al., 2002). Hsp12, which confers protection to membrane liposomes, has also been associated with *flor* strains (Zara et al., 2002). This protein, however, does not appear to play a determining role in ethanol tolerance, as deletion of the corre-sponding gene prevented film formation but did not increase cell sensitivity to ethanol.

As has already been indicated, high concen-trations of ethanol also induce the production of adhesins and other hydrophobic cell-surface proteins with internal repeats (Pir) (Huang et al., 2009) (see Figure 3.4), explaining why cell hydrophobicity has been found to double during *flor* formation (Barrales et al., 2008; Rando & Verstrepen, 2007; van Mulders et al., 2009) (see Figure 3.5). The synthesis of hydro-phobic proteins, particularly those encoded by *FLO* genes, leads to increased resistance to ethanol, temperature, pH, and mechanical stress (Beauvais et al., 2009; Castrejón, 2000; Smukalla et al., 2008). Furthermore, it has been seen (in *flor* yeasts only) that a deletion in the repression domain of the *FLO11* promoter considerably increased *FLO11* gene expression levels and that rearrangements in the repeat domains of the coding region increased both the number and hydrophobicity of the corresponding proteins (Barrales et al., 2008; Fidalgo et al., 2006).

Many attempts have been made to identify other proteins responsible for both *flor* forma-tion and increased tolerance of a range of hostile conditions. In addition to isolating proteins encoded by the *FLO1* and *FLO11* genes (Barrales et al., 2008; Smukalla et al., 2008), studies of *flor* yeasts have also uncovered a strongly hydro-phobic 49 kDa cell surface mannoprotein of as-yet-undetermined function (Alexandre et al., 2000). Kovacs et al. (2008) isolated and compared another cell-surface Pir protein, Hsp150 (see Figure 3.4), in numerous fermenta-tion and film-forming yeast strains from different geographical areas. It is noteworthy that all the flor strains analyzed were lacking three of the 11 protein repeat regions found in other strains, but whether or not this deletion is relevant to *flor* formation or maintenance is not known. Hydrophobin isolation protocols have been used to isolate fungal proteins on the surface of *flor* yeasts that protect against desiccation and other stress conditions (Castre-jón, 2000). The function of these proteins, however, has yet to be fully elucidated.

# 4. EVOLUTION OF CELLULAR GENOMES AND YEAST POPULATIONS IN BIOLOGICALLY AGED WINES

The chromosomal constitution of a popula-tion of wine yeasts is determined by a combina-tion of the variability introduced by mutation and meiotic or mitotic recombination and the adaptive selection that occurs in specific fermentation or aging conditions. There is evidence that interspecific hybridization occurs between different *Saccharomyces* wine strains. Examples include brewing strains that are hybrids of *S. cerevisiae* and *Saccharomyces pastor-ianus* and cider strains with nuclear chromo-somes from three different species of *Saccharomyces* (Belloch et al., 2008; de Barros Lopes et al., 2002; González et al., 2007; Sipiczki, 2008). No evidence of barriers to interspecies conjugation has been found. *Flor* yeasts, however, have less chromosome polymorphism than other wine yeasts, possibly because they have to endure very hostile conditions that tend to select for a practically unique karyotype (Martínez et al., 1995; Naumova et al., 2005). This relative uniformity is favored by sexual isolation during fermentation and aging (Sancho et al., 1986) (which would prevent meiotic recombination), a shortage of *Ty1* elements (Ibeas & Jiménez, 1996; Nyswaner et al., 2008), poor mobility of *Ty2* elements (Rachidi et al., 1999; Wu & Jiang, 2008) (which would notably reduce rearrangements and chromosome changes during mitosis), and frequent aneuploidy (Martínez et al., 1995) (which would render the majority of meiotic products inviable). The chromosome changes observed in wine strains during fermentation have been attributed to mitotic recombination and gene conversion processes designed to eliminate potentially deleterious alleles (Carro & Pina, 2001; Puig et al., 2000). Cells might use recombination to repair the continual damage caused by ethanol and acetaldehyde—the main sources of variability—during both fermentation and aging (Benítez & Codón, 2002; Castrejón et al., 2002). There have also been reports of rearrangements caused by reciprocal chromosomal translocation, which in all cases has been found to be the result of ectopic recombination between *Ty* elements (Nyswaner et al., 2008; Rachidi et al., 1999; Wu & Jiang, 2008). Several authors have identified *flor* strains with an identical karyotype but different mtDNA RFLPs and vice versa, and attributed this to conjugation and recombination processes between cells (Cardazzo et al., 1998; Ibeas & Jiménez, 1996). The conditions that develop during fermentation, and above all during aging, determine whether or not changes in the genome of the new recombinants will be maintained. This explains the high copy number of certain chromosomes or genes that have been observed in such cases (Guijo et al., 1997) (see Table 3.3). It also explains the associations detected between specific genotypes and both winemaking conditions (Nadal

et al., 1996) and the different rows in the dynamic *solera* and *criadera* aging system (Infante et al., 2003).

Chromosomal constitution is thus influenced by variations that can occur in the genome of a yeast strain during fermentation or aging and the selection of specific genotypes in certain environmental conditions. In the case of biologically aged wines, it has been suggested that *beticus* and *cheresiensis* races predominate in younger wines because of their ability to form a *flor* much more quickly than other races (Martínez et al., 1997c). Likewise, it is thought that *montuliensis* and *rouxii* races predominate in older wines because of their greater ability to tolerate and produce acetaldehyde (see Figure 3.2). As explained previously, in the dynamic biological aging system, young wine from the upper rows is added to the older wines below. *Beticus* and *cheresiensis* races would thus gain access to these older wines, in which they would initially proliferate because of their superior ability to repair and re-form the *flor*. As the concentrations of acetaldehyde in the wine increased, those yeasts would be gradually displaced by *montuliensis* and *rouxii* races until younger wine was added again (Martínez et al., 1997c). The association detected in some wineries between specific genotypes and different levels in the aging scale has been attributed to the speed with which the *flor* forms in younger wines and, in older wines, the tolerance to acetaldehyde (Infante et al., 2003). A comparison between strains capable of accelerated *flor* formation and strains with a greater acetaldehyde production capacity revealed copy number differences in 38% of their genes, as well as differences in aneuploidy (detected in chromosomes I, III, and VI in the former group and in chromosomes X and XII in the latter group) (Mesa et al., 1999, 2000) (see Table 3.3). Ethanol tolerance alleles have also been localized to some of these chromosomes (VI, VII, IX, and XII) (Hu et al., 2007). In the presence of very dominant phenotypes, a population can be reduced to a single race, or almost even a single genotype if the dominant strain succeeds in completely displacing the other populations (Ibeas et al., 1997).

## 5. GENETIC IMPROVEMENT OF WINE YEASTS

### 5.1. Improving the Characteristics of Fermentation Yeasts

Fermentation yeasts must be able to favor a quick onset of fermentation, convert all sugars to ethanol, and produce low levels of undesirable byproducts such as volatile acids. Rapid onset of fermentation is important as it eliminates or reduces the contaminants in this phase, while total consumption of sugars is important as it prevents contamination during aging, particularly by strains of *Brettanomyces*, which have a major impact on the organoleptic quality of biologically aged wines (Barata et al., 2008; Vigentini et al., 2008). The key elements in fermentation are sugar permeases and certain glycolytic enzymes such as hexokinase 2 (Berthels et al., 2008). Over 20 *HXT* genes have been implicated in hexose transport in *S. cerevisiae*. *HXT2* overexpression, in particular, has been reported to considerably reduce the lag phase at the start of fermentation, although this effect has not been observed in the fermentation of musts from Palomino grapes. Because most wine strains of *S. cerevisiae* are more efficient transporters of glucose than of fructose, the residual sugar is formed by fructose in musts with a glucose to fructose ratio of 1:1 at the start of fermentation. The expression of Fsy1, a specific fructose transporter found in *S. pastorianus*, complements an *hxt* null mutation in *S. cerevisiae* (Rodrigues de Sousa et al., 2004) and increases the efficiency with which Palomino grapes are fermented. The resulting residual sugar level in such cases is under 0.5 g/L, contrasting with the level of 2 to 5 g/L

found in musts fermented with strains that do not express this transporter.

Modulation of the glycerol-to-ethanol ratio during fermentation has been the focus of recent investigation. Low-alcohol wines can be obtained by increasing glycerol content and consequently decreasing ethanol content. Glycerol is also important in wines lacking body as it enhances their sweetness and masks possible acidity. Wine strains capable of producing more glycerol than ethanol have been created by overexpression of the gene encoding glycerol-3-phosphate dehydrogenase (Michnick et al., 1997). The wines produced using these strains are characterized not only by greater glycerol content but also by small variations in other metabolites (Remize et al., 2003). Similar designs used with brewing strains have succeeded in increasing glycerol concentrations five- to six-fold. Considerable increases in glycerol production have also been seen in response to ethanol stress (Vriesekoop et al., 2009). Nonetheless, in the case of biologically aged wines, the fermentation of musts made with Palomino grapes containing high sugar concentrations (particularly when partially dehydrated grapes that have been exposed to many hours of sun are used) increases both glycerol levels and volatile acidity. Yeasts synthesize glycerol to combat osmotic stress (Remize et al., 2003) and acetic acid (responsible for volatile acidity) to maintain the redox balance during glycerol production. Wine yeast strains expressing bacterial genes linked to malolactic fermentation have also been engineered to reduce the acidity of wines with a high concentration of malic acid (Williams et al., 1984).

Terpenes are characteristic substances found in different grape varieties. They are present in sugar-bound forms, which are transported through the plant, and as free terpenes, which have a substantial impact on the organoleptic properties of wine. Wine yeasts expressing heterologous genes encoding hydrolases capable of breaking the glycosidic bonds between sugars and terpenes have also been obtained, as have yeasts that hydrolyze plant cell materials that would otherwise retain aromas produced by the vine, resulting thus in more aromatic wines (Villanueva et al., 2000). Constructs in which heterologous genes are under the control of regulatable promoters have been used to control the release of aromas at the end of fermentation or at any other desired time (Puig & Perez-Ortín, 2000). Nonetheless, in one study, the addition of enzyme preparations during the fermentation of musts from Palomino grapes did not improve the organoleptic quality of the resulting wines (Roldán et al., 2006). This indicates the scarcity of primary aromas in this variety of grape and suggests that the sensory properties of biologically aged wines are mainly derived from secondary and tertiary aromas generated by yeast metabolism (García Maiquez, 1995; Roldán et al., 2006).

In 1989, musts from the French Midi region inoculated with Champagne or Burgundy yeasts produced wines with aromas reminiscent of the wines from those regions (Suárez-Lepe, 2002). These aromas are derived from the production of higher alcohols, esters, fatty acids, aldehydes, sulfur compounds, phenols, and terpenes by yeasts. All of these compounds, but particularly esters and alcohols, are formed as secondary metabolites during glycerol-pyruvic fermentation (Cordente et al., 2009; Linderholm et al., 2008; Suárez-Lepe, 2002). Considering that yeasts use some amino acids as precursors of higher alcohols and esters, it might be possible to enhance the fruity aroma of wine by increasing amino acid content (Carrau et al., 2008; Muñoz et al., 2006; Thibon et al., 2008; Torrea-Goñi & Ancín-Azpilicueta, 2001). The addition of amino acid solutions (particularly threonine and serine) to must made from 11 varieties of grape has been found to increase the concentration of higher alcohols, some of their acetates, ethyl butyrate, and some acids, as well as the aromatic character of the

wine (Hernández-Orte et al., 2002). Strains that overproduce leucine have been successfully used to increase levels of isobutyl and isoamyl alcohol (Watanabe et al., 1990); likewise, phenylethyl alcohol levels have been increased using strains that overproduce aromatic amino acids (Fukuda et al., 1990, 1991a, 1991b). The alcohol acetyltransferase gene family has been implicated in the synthesis of esters during alcoholic fermentation (Mason & Dufour, 2000), and, as a consequence, attempts have been made to increase alcohol activity transferase and hence increase ester levels while reducing those of alcohols. It has been reported that acetate ester concentration is increased by increasing alcohol acetyltransferase activity in yeast strains used to make sake (Fukuda et al., 1998), beer (Fujii et al., 1994; Verstrepen et al., 2003a, 2003b), and wine and spirits (Dequin, 2001; Lilly et al., 2000). The organoleptic properties of wine could also be improved through the manipulation of Adh6 and Adh7 (alcohol dehydrogenase enzymes involved in the synthesis of higher alcohols) (de Smidt et al., 2008) or the Pdr12 transporter (involved in the secretion of fusel acids) (Hazelwood et al., 2006).

Finally, efforts are also being made in the area of "healthy" wines. The phytoalexin resveratrol, a member of the stilbene family, has been identified in numerous plants, including vines. In Palomino grapes, for example, levels of this phenolic compound vary from 2 to 7 mg/L in the juice, and from 14 to 64 mg/L in the skin (Roldán et al., 2003). These variations are related to numerous factors such as climate and the health of the grapes. Resveratrol is also found in wine and has been associated with beneficial health effects in the areas of cancer and heart disease (Giménez-García, 2000). Growing efforts are thus being made to increase resveratrol concentrations in wine by improving extraction of this compound from grapes, analyzing factors that influence resveratrol levels, and inducing the expression of stilbene synthases in wine strains (Becker et al., 2003).

## 5.2. Improving the Characteristics of Aging Yeasts

The synergistic action of ethanol, acetaldehyde, low pH, water stress, oxidative stress, and temperature can contribute to the loss of the *flor* film (Martínez et al., 1997c) (see Figure 3.8). The use of strains capable of forming a stronger film in less time would thus enhance the quality of *Fino* wines and shorten the aging time required. While hybrids of nonisogenic strains capable of forming a *flor* faster than parental strains have been developed (Castrejón, 2000), they do not confer the same organoleptic properties as their progenitors.

Both *flor* formation and resistance to hostile conditions are largely determined by the proteins on the cell surface of the yeasts. Changes in external pH induce the synthesis of GPI-linked proteins anchored to the membrane (Barrales et al., 2008) (see Figure 3.4). Many of these proteins are dependent on the high-osmolarity glycerol response (osmotic stress) (see Figure 3.7), which increases yeast resistance to enzymatic lysis (Barrales et al., 2008). Sedi1, another GPI-linked membrane protein, is synthesized only in the stationary phase, and its absence increases cell sensitivity to enzymatic lysis. Reynolds and Fink (2001) described a family of GPI-linked membrane glycoproteins similar to adhesins that were present in the yeast cell wall. The corresponding genes are expressed in carbon- or nitrogen-starvation conditions and the main function of the proteins is to adhere to inert surfaces or other cells. An example of one of these proteins is that encoded by the *SPI1* gene, regulated by Msn2 and Msn4 (Puig & Perez-Ortín, 2000) (see Figures 3.7 and 3.9). SPI1 confers resistance to 2,4 D and β-1,3-glucanase and controls the pH gradient across the membrane in adverse conditions. Overexpression of this protein could improve a strain's ability to form a film that is resistant to hostile conditions because weak organic acids alter the transmembrane proton gradient, exert considerable oxidative stress, and are mutagens for mtDNA (Piper, 1999). Another of these proteins, Flo11 (described in Section 3.3), is responsible for invasive growth (van Dyk et al., 2005), filamentation (Ma et al., 2007; Palecek et al., 2000; Tamaki et al., 2000), flocculation (Guo et al., 2000), and *flor* formation (Guo et al., 2000; Ishigami et al., 2006; Reynolds & Fink, 2001) (see Figure 3.3). It also appears to be responsible for the greater resistance shown by cells to temperature variations, high alcohol levels, low pH levels, and mechanical stress during the *flor* formation phase (Barrales et al., 2008; Castrejón, 2000). Its overexpression might, thus, contribute to the formation of more stable *flors* in a shorter time.

Another class of cell-surface protein involved in resistance to stress is highly glycosylated and attached to the cell wall via disulfide bridges (Caridi, 2006). Thus, lower glycosylation is associated with greater sensitivity to hostile conditions. An example of such a protein is Hsp150, which is strongly induced by osmotic stress or low pH (Moukadiri & Zueco, 2001) and gives rise to increased stress resistance. Attempts to correlate differential expression of other *HSP* genes (mainly *HSP12*, *HSP123*, *HSP82*, *HSP26*, and *HSP104*) with resistance to cold, osmotic stress, oxidative stress, and acetaldehyde- or ethanol-induced stress in *flor* strains have not succeeded in confirming that these genes are directly involved in either *flor* formation or resistance to hostile conditions (Aranda et al., 2002; Zara et al., 2002). Products of the *HSP70* gene family, however, in association with other factors but independently of Flo11, have been directly implicated in *flor* formation (Martineau et al., 2007). Hsp70 proteins are chaperones involved in protein folding and transport through the endoplasmic reticulum and above all through the mitochondrial membrane.

High expression levels of other genes associated with ethanol metabolism, redox potential, glycerol uptake, and oxidative stress have also been observed in *flor* yeasts (Infante et al.,

2003; Longo et al., 1996). Ethanol and acetaldehyde toxicity have been associated with the production of mitochondrial ROS (Landolfo et al., 2008) (see Figure 3.10), leading to the accumulation of mutations and deletions in mtDNA (Raha & Robinson, 2000). Enzymes such as N-acetyltransferase (Du & Takagi, 2007), catalase (Abbott et al., 2009), and above all SOD (Costa et al., 1997), in contrast, protect against this toxicity (see Table 3.4 and Figure 3.10). Resistance to oxidative stress has also been found to be increased by proteins associated with GSH metabolism (Gales et al., 2008) (see Figure 3.11). Improved strains could also be obtained by overexpressing the SOD enzymes Sod1 and Sod2, among others. The role of these two enzymes in ethanol tolerance, for example, has been well established (Costa et al., 1997) (see Figure 3.10 and Table 3.4) and the *SOD1* gene is strongly expressed in the stationary phase and above all in the *flor* phase (Infante et al., 2003). Finally, because the Sod1 protein is involved in molecular crosslinking, it has been possible to purify this protein in the *flor* using a purification protocol that favors aggregation at the liquid–air interface (Castrejón, 2000).

# 6. CONCLUSIONS

*S. cerevisiae* strains that participate in the production of biologically aged wines have a highly heterogeneous genetic profile. They are easily distinguishable from each other because of the enormous differences in their DNA content, mtDNA restriction profiles, and chromosomal patterns. Strains isolated in different wineries or even at different levels of aging scales within the same winery also display genetic variability. Exploitation of differences between *S. cerevisiae* races or strains with varying genetic profiles associated with specific metabolic characteristics will allow targeted strategies aimed at producing biologically

aged wines with desirable characteristics such as high levels of acetaldehyde and higher alcohols, and low volatile acidity. Furthermore, research efforts have uncovered genes that appear to be involved in *flor* formation (*FLO11* gene) and tolerance of the hostile conditions that characterize aging (*SOD* genes). The use of yeasts overexpressing these genes would allow more stable films to be formed and shorten the aging period required. Finally, the development of fermentation strains that overproduce amino acids and produce higher levels of alcohol transferase, or that express hydrolases or stilbene synthase, could give rise to more aromatic, healthier young wines.

# Acknowledgments

During the writing of this chapter, our work was supported by funds from the Spanish Ministry for Science and Innovation (project AGL2006-03947), the Autonomous Government of Andalusia, Spain (Proyecto de Excelencia, PO6-CVI-01646), the Interministerial Committee for Scientific and Technology (CICYT) (Proyecto TRACE, PET 2008_0283), and the company Bean Global, SA, in Jerez de la Frontera.

# References

Abbott, D. A., Suir, E., Duong, G. H., de Hulster, E., Pronk, J. T., & van Maris, A. J. (2009). Catalase overexpression reduces lactic acid-induced oxidative stress in Saccharomyces cerevisiae. *Appl. Environ. Microbiol.*, 75, 2320–2325.

Alexandre, H., Blanchet, S., & Charpentier, C. (2000). Identification of a 49-kDa hydrophobic cell wall mannoprotein present in velum yeast which may be implicated in velum formation. *FEMS Microbiol. Lett.*, 185, 147–150.

Aranda, A., Querol, A., & del Olmo, M. (2002). Correlation between acetaldehyde and ethanol resistance and expression of HSP genes in yeast strains isolated during the biological aging of sherry wines. *Arch. Microbiol.*, 177, 304–312.

Backhus, L. E., DeRisi, J., & Bisson, L. F. (2001). Functional genomic analysis of a commercial wine strain of Saccharomyces cerevisiae under differing nitrogen conditions. *FEMS Yeast Res.*, 1, 111–125.

Bakalinsky, A. T., & Snow, R. (1990). The chromosomal constitution of wine strains of Saccharomyces cerevisiae. *Yeast*, 6, 367–382.

Bandas, E. L., & Zakharov, I. A. (1980). Induction of rho-mutations in yeast *Saccharomyces cerevisiae* by ethanol. *Mutat. Res., 71*, 193–199.

Barata, A., González, S., Malfeito-Ferreira, M., Querol, A., & Loureiro, V. (2008). Sour rot-damaged grapes are sources of wine spoilage yeasts. *FEMS Yeast Res., 8*, 1008–1017.

Barrales, R. R., Jiménez, J., & Ibeas, J. I. (2008). Identification of novel activation mechanisms for *FLO11* regulation in Saccharomyces cerevisiae. *Genetics, 178*, 145–156.

Beauvais, A., Loussert, C., Prevost, M. C., Verstrepen, K., & Latge, J. P. (2009). Characterization of a biofilm-like extracellular matrix in *FLO1*-expressing *Saccharomyces cerevisiae* cells. *FEMS Yeast Res., 9*, 411–419.

Becker, J. V., Armstrong, G. O., van der Merwe, M. J., Lambrechts, M. G., Vivier, M. A., & Pretorius, I. S. (2003). Metabolic engineering of *Saccharomyces cerevisiae* for the synthesis of the wine-related antioxidant resveratrol. *FEMS Yeast Res., 4*, 79–85.

Belloch, C., Orlic, S., Barrio, E., & Querol, A. (2008). Fermentative stress adaptation of hybrids within the Saccharomyces sensu stricto complex. *Int. J. Food Microbiol., 122*, 188–195.

Benítez, T., & Codón, A. C. (2002). Genetic diversity of yeasts in wine production. In G. G. Khachatourians & D. K. Arora (Eds.), *Applied mycology and biotechnology, Vol. 2* (pp. 19–44). Amsterdam, the Netherlands: Elsevier Science B.V.

Benítez, T., & Codón, A. C. (2005). Levaduras. *Saccharomyces* III. Levaduras de vinos de crianza biológica. In A. V. Carrascosa, R. Muñoz, & R. Gónzález (Eds.), *Microbiología del vino* (pp. 78–113). Madrid, Spain: AMV Ediciones.

Benítez, T., Martínez, P., & Codón, A. C. (1996). Genetic constitution of industrial yeast. *Microbiologia, 12*, 371–384.

Benítez, T., Rincón, A.M., & Codón, A.C. (in press). Yeasts associated with production of fortified wines. In A. Romano & G. H. Fleet (Eds.), *Yeasts in the production of wines*. Springer.

Berlanga, T. M., Atanasio, C., Mauricio, J. C., & Ortega, J. M. (2001). Influence of aeration on the physiological activity of flor yeasts. *J. Agric. Food Chem., 49*, 3378–3384.

Berthels, N. J., Cordero Otero, R. R., Bauer, F. F., Pretorius, I. S., & Thevelein, J. M. (2008). Correlation between glucose/fructose discrepancy and hexokinase kinetic properties in different *Saccharomyces cerevisiae* wine yeast strains. *Appl. Microbiol. Biotechnol., 77*, 1083–1091.

Blake, D., Luke, B., Kanellis, P., Jorgensen, P., Goh, T., Penfold, S., et al. (2006). The F-box protein Dia2 overcomes replication impedance to promote genome stability in *Saccharomyces cerevisiae*. *Genetics, 174*, 1709–1727.

Bravo-Abad, F. (1986). Crianza biológica de vino: Procedimiento tradicional de vinos finos de D.O. Jerez y D.O. Montilla-Moriles. *Enología y Enotecnia, 1*, 15–19.

Budroni, M., Roggio, T., Pinna, G., Pretti, L., & Farris, G. A. (1995). Which factors favour the formation of a biofilm by *Saccharomyces cerevisiae*. In *XII International Conference on Yeast Genetics and Molecular Biology*. Lisbon, Portugal: J.W.S. Ltd. June 10–16.

Budroni, M., Zara, S., Zara, G., Pirino, G., & Mannazzu, I. (2005). Peculiarities of flor strains adapted to Sardinian sherry-like wine ageing conditions. *FEMS Yeast Res., 5*, 951–958.

Campo, E., Cacho, J., & Ferreira, V. (2008). The chemical characterization of the aroma of dessert and sparkling white wines (Pedro Ximenez, Fino, Sauternes, and Cava) by gas chromatography-olfactometry and chemical quantitative analysis. *J. Agric. Food Chem., 56*, 2477–2484.

Cardazzo, B., Minuzzo, S., Sartori, G., Grapputo, A., & Carignani, G. (1998). Evolution of mitochondrial DNA in yeast: Gene order and structural organization of the mitochondrial genome of *Saccharomyces uvarum*. *Curr. Genet., 33*, 52–59.

Caridi, A. (2006). Enological functions of parietal yeast mannoproteins. *Anton. Leeuw., 89*, 417–422.

Carrau, F. M., Medina, K., Farina, L., Boido, E., Henschke, P. A., & Dellacassa, E. (2008). Production of fermentation aroma compounds by *Saccharomyces cerevisiae* wine yeasts: Effects of yeast assimilable nitrogen on two model strains. *FEMS Yeast Res., 8*, 1196–1207.

Carro, D., & Pina, B. (2001). Genetic analysis of the karyotype instability in natural wine yeast strains. *Yeast, 18*, 1457–1470.

Castrejón, F. (2000). Mejora de cepas de *Saccharomyces cerevisiae* utilizables en la elaboración de vinos de Jerez. In *Genética*. Seville, Spain: University of Seville.

Castrejón, F., Codón, A. C., Cubero, B., & Benítez, T. (2002). Acetaldehyde and ethanol are responsible for mitochondrial DNA (mtDNA) restriction fragment length polymorphism (RFLP) in flor yeasts. *Syst. Appl. Microbiol., 25*, 462–467.

Castrejón, F., Martínez-Force, E., Benítez, T., & Codón, A. C. (2004). Genetic analysis of apomictic wine yeasts. *Curr. Genet., 45*, 187–196.

Celestino-Peréz, S. (Ed.) (1999). El vino en la antigüedad romana. In *Simposio Arqueología del vino, Jerez. Consejo Regulador de las Denominaciones de Origen Jerez-Xeres-Sherry y manzanilla-San-lúcar de Barrameda y Universidad Autónoma de Madrid, Madrid*. October 2–4.

Cordente, A. G., Heinrich, A., Pretorius, I. S., & Swiegers, J. H. (2009). Isolation of sulfite reductase variants of a commercial wine yeast with significantly reduced hydrogen sulfide production. *FEMS Yeast Res., 9*, 446–459.

Costa, V., Amorim, M. A., Reis, E., Quintanilha, A., & Moradas-Ferreira, P. (1997). Mitochondrial superoxide dismutase is essential for ethanol tolerance of Saccharomyces cerevisiae in the post-diauxic phase. Microbiology, 143, 1649–1656.

Charpentier, C., Colin, A., Alais, A., & Legras, J. L. (2009). French Jura flor yeasts: Genotype and technological diversity. Anton. Leeuw., 95, 263–273.

de Barros Lopes, M., Bellon, J. R., Shirley, N. J., & Ganter, P. F. (2002). Evidence for multiple interspecific hybridization in Saccharomyces sensu stricto species. FEMS Yeast Res., 1, 323–331.

de Smidt, O., du Preez, J. C., & Albertyn, J. (2008). The alcohol dehydrogenases of Saccharomyces cerevisiae: A comprehensive review. FEMS Yeast Res., 8, 967–978.

Dequin, S. (2001). The potential of genetic engineering for improving brewing, wine-making and baking yeasts. Appl. Microbiol. Biotechnol., 56, 577–588.

Douglas, L. M., Li, L., Yang, Y., & Dranginis, A. M. (2007). Expression and characterization of the flocculin Flo11/Muc1, a Saccharomyces cerevisiae mannoprotein with homotypic properties of adhesion. Eukaryot Cell, 6, 2214–2221.

Du, X., & Takagi, H. (2007). N-Acetyltransferase Mpr1 confers ethanol tolerance on Saccharomyces cerevisiae by reducing reactive oxygen species. Appl. Microbiol. Biotechnol., 75, 1343–1351.

Esteve-Zarzoso, B., Fernández-Espinar, M. T., & Querol, A. (2004). Authentication and identification of Saccharomyces cerevisiae "flor" yeast races involved in sherry ageing. Anton. Leeuw., 85, 151–158.

Esteve-Zarzoso, B., Manzanares, P., Ramón, D., & Querol, A. (1998). The role of non- Saccharomyces yeasts in industrial winemaking. Int. Microbiol., 1, 143–148.

Esteve-Zarzoso, B., Peris-Torán, M. J., García Maiquez, E., Uruburu, F., & Querol, A. (2001). Yeast population dynamics during the fermentation and biological aging of sherry wines. Appl. Environ. Microbiol., 67, 2056–2061.

Fatichenti, F., Farris, G. A., & Deiana, P. (1983). Improved production of a Spanish-type sherry by using selected indigenous film-forming yeasts as starters. Am. J. Enol. Vitic., 34, 216–220.

Fernández-Espinar, M. T., Barrio, E., & Querol, A. (2003). Analysis of the genetic variability in the species of the Saccharomyces sensu stricto complex. Yeast, 20, 1213–1226.

Fernádez-Espinar, M. T., Esteve-Zarzoso, B., Querol, A., & Barrio, E. (2000). RFLP analysis of the ribosomal internal transcribed spacers and the 5.8S rRNA gene region of the genus Saccharomyces: A fast method for species identification and the differentiation of flor yeasts. Anton. Leeuw., 78, 87–97.

Fidalgo, M., Barrales, R. R., Ibeas, J. I., & Jiménez, J. (2006). Adaptive evolution by mutations in the FLO11 gene. Proc. Natl. Acad. Sci. U.S.A., 103, 11228–11233.

Fleet, G. (2007). Wine. In M. Doyle, & L. Beuchat (Eds.), Food microbiology: Fundamentals and frontiers (pp. 863–890). Washington, DC: ASM Press.

Fujii, T., Nagasawa, N., Iwamatsu, A., Bogaki, T., Tamai, Y., & Hamachi, M. (1994). Molecular cloning, sequence analysis, and expression of the yeast alcohol acetyltransferase gene. Appl. Environ. Microbiol., 60, 2786–2792.

Fukuda, K., Watanabe, M., Asano, K., Ouchi, K., & Takasawa, S. (1991a). Isolation and genetic study of p-fluoro-DL-phenylalanine-resistant mutants overproducing beta-phenethyl-alcohol in Saccharomyces cerevisiae. Curr. Genet., 20, 449–452.

Fukuda, K., Watanabe, M., Asano, K., Ouchi, K., & Takasawa, S. (1991b). A mutated ARO4 gene for feedback-resistant DAHP synthase which causes both o-fluoro-DL-phenylalanine resistance and beta-phenethyl-alcohol overproduction in Saccharomyces cerevisiae. Curr. Genet., 20, 453–456.

Fukuda, K., Watanabe, M., Asano, K., Ueda, H., & Ohta, S. (1990). Breeding of brewing yeast producing a large amount of β-phenylethyl alcohol and β-phenylethyl acetate. Agric. Biol. Chem., 54, 269–271.

Fukuda, K., Yamamoto, N., Kiyokawa, Y., Yanagiuchi, T., Wakai, Y., Kitamoto, K., et al. (1998). Balance of activities of alcohol acetyltransferase and esterase in Saccharomyces cerevisiae is important for production of isoamyl acetate. Appl. Environ. Microbiol., 64, 4076–4078.

Gales, G., Penninckx, M., Block, J. C., & Leroy, P. (2008). Role of glutathione metabolism status in the definition of some cellular parameters and oxidative stress tolerance of Saccharomyces cerevisiae cells growing as biofilms. FEMS Yeast Res., 8, 667–675.

García del Barrio, I. (1995). Las bodegas del vino de Jerez (Historia, microclima y construcción). In J. J. Iglesias-Rodríguez (Ed.), Historia y cultura del vino en Andalucía (pp. 141–178). Sevilla, Spain: Secretariado de Publicaciones de la Universidad de Sevilla.

García Maiquez, E. (1995). Sherry wine microorganisms. Microbiologia, 11, 51–58.

Giménez-García, J. M. (2000). Contenido en resveratrol en la uva de variedad Bobal. In Uva y Vino: Los valores saludables, (pp. 49–54). Valencia, Spain: Fundación Valenciana de Estudios Avanzados and Generalitat Valenciana, Consellería de Agricultura, Pesca y Alimentación.

Gómez, M. E., Igartuburu, J. M., Pando, E., Luis, F. R., & Mourente, G. (2004). Lipid composition of lees from Sherry wine. J. Agric. Food. Chem., 52, 4791–4794.

González, S. S., Barrio, E., & Querol, A. (2007). Molecular identification and characterization of wine yeasts isolated from Tenerife (Canary Island, Spain). J. Appl. Microbiol., 102, 1018–1025.

González, S. S., Barrio, E., & Querol, A. (2008). Molecular characterization of new natural hybrids of Saccharomyces *cerevisiae* and *S. kudriavzevii* in brewing. *Appl. Environ. Microbiol., 74,* 2314–2320.

Guijo, S., Mauricio, J. C., Salmon, J. M., & Ortega, J. M. (1997). Determination of the relative ploidy in different *Saccharomyces cerevisiae* strains used for fermentation and "flor" film ageing of dry sherry-type wines. *Yeast, 13,* 101–117.

Guo, B., Styles, C. A., Feng, Q., & Fink, G. R. (2000). A *Saccharomyces* gene family involved in invasive growth, cell–cell adhesion, and mating. *Proc. Natl. Acad. Sci. U.S.A., 97,* 12 158–12 163.

Hazelwood, L. A., Tai, S. L., Boer, V. M., de Winde, J. H., Pronk, J. T., & Daran, J. M. (2006). A new physiological role for Pdr12p in *Saccharomyces cerevisiae*: Export of aromatic and branched-chain organic acids produced in amino acid catabolism. *FEMS Yeast Res., 6,* 937–945.

Hernández-Orte, P., Cacho, J. F., & Ferreira, V. (2002). Relationship between varietal amino acid profile of grapes and wine aromatic composition. Experiments with model solutions and chemometric study. *J. Agric. Food Chem., 50,* 2891–2899.

Hu, X. H., Wang, M. H., Tan, T., Li, J. R., Yang, H., Leach, L., et al. (2007). Genetic dissection of ethanol tolerance in the budding yeast *Saccharomyces cerevisiae*. *Genetics, 175,* 1479–1487.

Huang, G., Dougherty, S. D., & Erdman, S. E. (2009). Conserved WCPL and CX4C domains mediate several mating adhesin interactions in *Saccharomyces cerevisiae*. *Genetics, 182,* 173–189.

Ibeas, J. I., & Jiménez, J. (1996). Genomic complexity and chromosomal rearrangements in wine-laboratory yeast hybrids. *Curr. Genet., 30,* 410–416.

Ibeas, J. I., & Jiménez, J. (1997). Mitochondrial DNA loss caused by ethanol in *Saccharomyces* flor yeasts. *Appl. Environ. Microbiol., 63,* 7–12.

Ibeas, J. I., Lozano, I., Perdigones, F., & Jiménez, J. (1996). Detection of *Dekkera- Brettanomyces* strains in sherry by a nested PCR method. *Appl. Environ. Microbiol., 62,* 998–1003.

Ibeas, J. I., Lozano, I., Perdigones, F., & Jiménez, J. (1997). Effects of ethanol and temperature on the biological aging of Sherry wines. *Am. J. Enol. Vitic., 48,* 71–74.

Infante, J. J., Dombek, K. M., Rebordinos, L., Cantoral, J. M., & Young, E. T. (2003). Genome-wide amplifications caused by chromosomal rearrangements play a major role in the adaptive evolution of natural yeast. *Genetics, 165,* 1745–1759.

Ingledew, W. M., Magnus, C. A., & Sosulski, F. W. (1987). Influence of oxygen on proline utilization during the wine fermentation. *Am. J. Enol. Vitic., 38,* 246–248.

Ishigami, M., Nakagawa, Y., Hayakawa, M., & Iimura, Y. (2006). *FLO11* is the primary factor in flor formation caused by cell surface hydrophobicity in wild-type flor yeast. *Biosci. Biotechnol. Biochem., 70,* 660–666.

Jiménez, J., & Benítez, T. (1988). Yeast cell viability under conditions of high temperature and ethanol concentrations depends on the mitochondrial genome. *Curr. Genet., 13,* 461–469.

Kawarai, T., Furukawa, S., Ogihara, H., & Yamasaki, M. (2007). Mixed-species biofilm formation by lactic acid bacteria and rice wine yeasts. *Appl. Environ. Microbiol., 73,* 4673–4676.

Kovacs, M., Stuparevic, I., Mrsa, V., & Maraz, A. (2008). Characterization of Ccw7p cell wall proteins and the encoding genes of *Saccharomyces cerevisiae* wine yeast strains: Relevance for flor formation. *FEMS Yeast Res., 8,* 1115–1126.

Kumar, G. R., Goyashiki, R., Tramakrishnan, V., Karpel, J. E., & Bisson, L. F. (2008). Genes required for ethanol tolerance and utilization in Saccharomyces cerevisiae. *Am. J. Enol. Vitic., 59,* 401–411.

Lambrechts, M. G., Bauer, F. F., Marmur, J., & Pretorius, I. S. (1996). Muc1, a mucin-like protein that is regulated by Mss10, is critical for pseudohyphal differentiation in yeast. *Proc. Natl. Acad. Sci. U.S.A., 93,* 8419–8424.

Landolfo, S., Politi, H., Angelozzi, D., & Mannazzu, I. (2008). ROS accumulation and oxidative damage to cell structures in *Saccharomyces cerevisiae* wine strains during fermentation of high-sugar-containing medium. *Biochim. Biophys. Acta, 1780,* 892–898.

Lilly, M., Lambrechts, M. G., & Pretorius, I. S. (2000). Effect of increased yeast alcohol acetyltransferase activity on flavor profiles of wine and distillates. *Appl. Environ. Microbiol., 66,* 744–753.

Linderholm, A. L., Findleton, C. L., Kumar, G., Hong, Y., & Bisson, L. F. (2008). Identification of genes affecting hydrogen sulfide formation in *Saccharomyces cerevisiae*. *Appl. Environ. Microbiol., 74,* 1418–1427.

Lo, W. S., & Dranginis, A. M. (1996). FLO11, a yeast gene related to the STA genes, encodes a novel cell surface flocculin. *J. Bacteriol., 178,* 7144–7151.

Longo, V. D., Gralla, E. B., & Valentine, J. S. (1996). Superoxide dismutase activity is essential for stationary phase survival in *Saccharomyces cerevisiae*. Mitochondrial production of toxic oxygen species in vivo. *J. Biol. Chem., 271,* 12 275–12 280.

Ma, J., Jin, R., Jia, X., Dobry, C. J., Wang, L., Reggiori, F., et al. (2007). An interrelationship between autophagy and filamentous growth in budding yeast. *Genetics, 177,* 205–214.

Martineau, C. N., Beckerich, J. M., & Kabani, M. (2007). Flo11p-independent control of "mat" formation by hsp70 molecular chaperones and nucleotide exchange factors in yeast. *Genetics, 177,* 1679–1689.

Martínez, P., Codón, A. C., Pérez, L., & Benítez, T. (1995). Physiological and molecular characterization of flor yeasts: Polymorphism of flor yeast populations. *Yeast, 11,* 1399–1411.

Martínez, P., Pérez, L., & Benítez, T. (1997a). Evolution of flor yeast population during the biological aging of fino sherry wine. *Am. J. Enol. Vitic., 48,* 160–168.

Martínez, P., Pérez, L., & Benítez, T. (1997b). Factors which affect velum formation by flor yeast isolated from sherry wine. *Syst. Appl. Microbiol., 20,* 154–157.

Martínez, P., Pérez, L., & Benítez, T. (1997c). Velum formation by flor yeasts isolated from sherry wine. *Am. J. Enol. Vitic., 48,* 55–62.

Martínez, P., Valcárcel, M. J., González, P., Benítez, T., & Pérez, L. (1993). Consumo de etanol, glicerina y aminoácidos totales en vinos finos durante la crianza biológica bajo "velo de flor." *Alimentación: Equipos y tecnología, 2,* 61–65.

Martínez, P., Valcárcel, M. J., Pérez, L., & Benítez, T. (1998). Metabolism of *Saccharomyces cerevisiae* flor yeasts during fermentation and biological aging of fino sherry: By-products and aroma compounds. *Am. J. Enol. Vitic., 49,* 240–250.

Martínez de la Ossa, E., Caro, I., Bonat, M., Pérez, L., & Domecq, B. (1987a). Dry extract in sherry and its evolution in the aging process. *Am. J. Enol. Vitic., 38,* 321–325.

Martínez de la Ossa, E., Pérez, L., & Caro, I. (1987b). Variation of the major volatiles trough aging of sherry. *Am. J. Enol. Vitic., 38,* 293–297.

Martínez-Llopis, M., Mijares, M. I., & Chirbes, R. (1992). *Historia del vino español.* Madrid, Spain: Vinoselección.

Martorell, P., Querol, A., & Fernández-Espinar, M. T. (2005). Rapid identification and enumeration of *Saccharomyces cerevisiae* cells in wine by real-time PCR. *Appl. Environ. Microbiol., 71,* 6823–6830.

Mason, A. B., & Dufour, J. P. (2000). Alcohol acetyltransferases and the significance of ester synthesis in yeast. *Yeast, 16,* 1287–1298.

Mauricio, J. C., Arroyo, M., Millán, C., & Ortega, J. M. (1990). Relationship between the phospholipid and sterol contents in *Saccharomyces cerevisiae* and *Torulaspora delbrueckii,* and their permanence during fermentation of musts from grapes of the Pedro Ximenez variety. *Biotechnol. Lett., 12,* 265–270.

Mauricio, J. C., Valero, E., Millán, C., & Ortega, J. M. (2001). Changes in nitrogen compounds in must and wine during fermentation and biological aging by flor yeasts. *J. Agric. Food Chem., 49,* 3310–3315.

Mérida, J., López-Toledano, A., Márquez, T., Millán, C., Ortega, J. M., & Medina, M. (2005). Retention of browning compounds by yeasts involved in the winemaking of sherry type wines. *Biotechnol. Lett., 27,* 1565–1570.

Mesa, J. J., Infante, J. J., Rebordinos, L., & Cantoral, J. M. (1999). Characterization of yeasts involved in the biological aging of sherry wines. *Lebensm. Wiss. Technol., 32,* 114–120.

Mesa, J. J., Infante, J. J., Rebordinos, L., Sánchez, J. A., & Cantoral, J. M. (2000). Influence of the yeast genotypes on enological characteristics of sherry wines. *Am. J. Enol. Vitic., 51,* 15–21.

Michnick, S., Roustan, J. L., Remize, F., Barre, P., & Dequin, S. (1997). Modulation of glycerol and ethanol yields during alcoholic fermentation in *Saccharomyces cerevisiae* strains overexpressed or disrupted for *GPD1* encoding glycerol 3-phosphate dehydrogenase. *Yeast, 13,* 783–793.

Mortimer, R. K. (2000). Evolution and variation of the yeast (*Saccharomyces*) genome. *Genome Res., 10,* 403–409.

Moukadiri, I., & Zueco, J. (2001). Evidence for the attachment of Hsp150/Pir2 to the cell wall of *Saccharomyces cerevisiae* through disulfide bridges. *FEMS Yeast Res., 1,* 241–245.

Muñoz, D., Peinado, R. A., Medina, M., & Moreno, J. (2006). Higher alcohols concentration and its relation with the biological aging evolution. *Eur. Food Res. Technol., 222,* 629–635.

Nadal, D., Colomer, B., & Pina, B. (1996). Molecular polymorphism distribution in phenotypically distinct populations of wine yeast strains. *Appl. Environ. Microbiol., 62,* 1944–1950.

Naumov, G. I., Masneuf, I., Naumova, E. S., Aigle, M., & Dubourdieu, D. (2000). Association of *Saccharomyces bayanus* var. *uvarum* with some French wines: Genetic analysis of yeast populations. *Res. Microbiol., 151,* 683–691.

Naumov, G. I., Naumova, E. S., Antunovics, Z., & Sipiczki, M. (2002). *Saccharomyces bayanus* var. *uvarum* in Tokaj wine-making of Slovakia and Hungary. *Appl. Microbiol. Biotechnol., 59,* 727–730.

Naumova, E. S., Ivannikova Iu, V., & Naumov, G. I. (2005). Genetic differentiation of the sherry yeasts *Saccharomyces cerevisiae. Prikl. Biokhim. Mikrobiol., 41,* 656–661.

Nyswaner, K. M., Checkley, M. A., Yi, M., Stephens, R. M., & Garfinkel, D. J. (2008). Chromatin-associated genes protect the yeast genome from *Ty1* insertional mutagenesis. *Genetics, 178,* 197–214.

Palecek, S. P., Parikh, A. S., & Kron, S. J. (2000). Genetic analysis reveals that *FLO11* upregulation and cell polarization independently regulate invasive growth in *Saccharomyces cerevisiae. Genetics, 156,* 1005–1023.

Peinado, R. A., Moreno, J., Medina, M., & Mauricio, J. C. (2004). Changes in volatile compounds and aromatic series in sherry wine with high gluconic acid levels subjected to aging by submerged flor yeast cultures. *Biotechnol. Lett., 26,* 757–762.

Piper, P. W. (1999). Yeast superoxide dismutase mutants reveal a pro-oxidant action of weak organic acid food preservatives. *Free Radic. Biol. Med., 27,* 1219—1227.

Pirino, G., Zara, S., Pinna, G., Farris, G. A., & Budroni, M. (2004). Diversity of Y region at *HML* locus in a *Saccharomyces cerevisiae* strain isolated from a Sardinian wine. *Anton. Leeuw., 85,* 29—36.

Puig, S., & Perez-Ortín, J. E. (2000). Stress response and expression patterns in wine fermentations of yeast genes induced at the diauxic shift. *Yeast, 16,* 139—148.

Puig, S., Querol, A., Barrio, E., & Pérez-Ortín, J. E. (2000). Mitotic recombination and genetic changes in *Saccharomyces cerevisiae* during wine fermentation. *Appl. Environ. Microbiol., 66,* 2057—2061.

Purevdorj-Gage, B., Orr, M. E., Stoodley, P., Sheehan, K. B., & Hyman, L. E. (2007). The role of *FLO11* in *Saccharomyces cerevisiae* biofilm development in a laboratory based flow-cell system. *FEMS Yeast Res., 7,* 372—379.

Rachidi, N., Barre, P., & Blondin, B. (1999). Multiple *Ty*-mediated chromosomal translocations lead to karyotype changes in a wine strain of *Saccharomyces cerevisiae. Mol. Gen. Genet., 261,* 841—850.

Raha, S., & Robinson, B. H. (2000). Mitochondria, oxygen free radicals, disease and ageing. *Trends Biochem. Sci., 25,* 502—508.

Rando, O. J., & Verstrepen, K. J. (2007). Timescales of genetic and epigenetic inheritance. *Cell, 128,* 655—668.

Remize, F., Cambon, B., Barnavon, L., & Dequin, S. (2003). Glycerol formation during wine fermentation is mainly linked to Gpd1p and is only partially controlled by the HOG pathway. *Yeast, 20,* 1243—1253.

Reynolds, T. B., & Fink, G. R. (2001). Bakers' yeast, a model for fungal biofilm formation. *Science, 291,* 878—881.

Ristow, H., Seyfarth, A., & Lochmann, E. R. (1995). Chromosomal damages by ethanol and acetaldehyde in *Saccharomyces cerevisiae* as studied by pulsed field gel electrophoresis. *Mutat. Res., 326,* 165—170.

Rodrigues de Sousa, H., Spencer-Martins, I., & Gonçalves, P. (2004). Differential regulation by glucose and fructose of a gene encoding a specific fructose/$H^+$ symporter in *Saccharomyces sensu stricto* yeasts. *Yeast, 21,* 519—530.

Roldán, A., Palacios, V., Caro, I., & Pérez, L. (2003). Resveratrol content of Palomino fino grapes: Influence of vintage and fungal infection. *J. Agric. Food Chem., 51,* 1464—1468.

Roldán, A., Palacios, V., Peñate, X., Benítez, T., & Pérez, L. (2006). Use of *Trichoderma* enzymatic extracts on vinification of Palomino fino grapes in the sherry region. *J. Food Eng., 75,* 375—382.

Sancho, E. D., Hernández, E., & Rodríguez-Navarro, A. (1986). Presumed sexual isolation in yeast populations during production of sherrylike wine. *Appl. Environ. Microbiol., 51,* 395—397.

Sipiczki, M. (2003). *Candida zemplinina* sp. nov., an osmotolerant and psychrotolerant yeast that ferments sweet botrytized wines. *Int. J. Syst. Evol. Microbiol., 53,* 2079—2083.

Sipiczki, M. (2008). Interspecies hybridization and recombination in *Saccharomyces* wine yeasts. *FEMS Yeast Res., 8,* 996—1007.

Smukalla, S., Caldara, M., Pochet, N., Beauvais, A., Guadagnini, S., Yan, C., et al. (2008). *FLO1* is a variable green beard gene that drives biofilm-like cooperation in budding yeast. *Cell, 135,* 726—737.

Straver, M. H., & Kijne, J. W. (1996). A rapid and selective assay for measuring cell surface hydrophobicity of brewer's yeast cells. *Yeast, 12,* 207—213.

Suárez-Lepe, J. A. (1997). *Levaduras vínicas: Funcionalidad y uso en bodega.* Madrid, Spain: Mundi-prensa.

Suárez-Lepe, J. A. (2002). Impacto de levaduras y bacterias en los aromas vínicos fermentativos. In *I Encuentro internacional de ciencias sensoriales y de la percepción* (pp. 43—45). Barcelona, Spain: Rubes Editorial, CS2002.

Sulo, P., Spirek, M., Soltesova, A., Marinoni, G., & Piskur, J. (2003). The efficiency of functional mitochondrial replacement in Saccharomyces species has directional character. *FEMS Yeast Res., 4,* 97—104.

Tamaki, H., Miwa, T., Shinozaki, M., Saito, M., Yun, C. W., Yamamoto, K., et al. (2000). *GPR1* regulates filamentous growth through *FLO11* in yeast *Saccharomyces cerevisiae. Biochem. Biophys. Res. Commun., 267,* 164—168.

Teunissen, A. W., & Steensma, H. Y. (1995). Review: The dominant flocculation genes of *Saccharomyces cerevisiae* constitute a new subtelomeric gene family. *Yeast, 11,* 1001—1013.

Thibon, C., Marullo, P., Claisse, O., Cullin, C., Dubourdieu, D., & Tominaga, T. (2008). Nitrogen catabolic repression controls the release of volatile thiols by *Saccharomyces cerevisiae* during wine fermentation. *FEMS Yeast Res., 8,* 1076—1086.

Torrea-Goñi, D., & Ancín-Azpilicueta, C. (2001). Influence of yeast strain on biogenic amines content in wines: Relationship with the utilization of amino acids during fermentation. *Am. J. Enol. Vitic., 52,* 185—190.

Urso, R., Rantsiou, K., Dolci, P., Rolle, L., Comi, G., & Cocolin, L. (2008). Yeast biodiversity and dynamics during sweet wine production as determined by molecular methods. *FEMS Yeast Res., 8,* 1053—1062.

Valero, E., Cambon, B., Schuller, D., Casal, M., & Dequin, S. (2007). Biodiversity of *Saccharomyces* yeast strains from grape berries of wine-producing areas using starter commercial yeasts. *FEMS Yeast Res., 7,* 317—329.

van Dyk, D., Pretorius, I. S., & Bauer, F. F. (2005). Mss11p is a central element of the regulatory network that controls *FLO11* expression and invasive growth in *Saccharomyces cerevisiae. Genetics, 169,* 91—106.

van Mulders, S. E., Christianen, E., Saerens, S. M., Daenen, L., Verbelen, P. J., Willaert, R., et al. (2009). Phenotypic diversity of Flo protein family-mediated adhesion in *Saccharomyces cerevisiue. FEMS Yeast Res., 9,* 178–190.

van Uden, N. (1989). *Alcohol toxicity in yeasts and bacteria.* Boca Raton, FL: CRC Press.

Verstrepen, K. J., Derdelinckx, G., Dufour, J. P., Winderickx, J., Pretorius, I. S., Thevelein, J. M., et al. (2003a). The *Saccharomyces cerevisiae* alcohol acetyl transferase gene *ATF1* is a target of the cAMP/PKA and FGM nutrient-signalling pathways. *FEMS Yeast Res., 4,* 285–296.

Verstrepen, K. J., Jansen, A., Lewitter, F., & Fink, G. R. (2005). Intragenic tandem repeats generate functional variability. *Nat. Genet., 37,* 986–990.

Verstrepen, K. J., & Klis, F. M. (2006). Flocculation, adhesion and biofilm formation in yeasts. *Mol. Microbiol., 60,* 5–15.

Verstrepen, K. J., Reynolds, T. B., & Fink, G. R. (2004). Origins of variation in the fungal cell surface. *Nat. Rev. Microbiol., 2,* 533–540.

Verstrepen, K. J., van Laere, S. D., Vanderhaegen, B. M., Derdelinckx, G., Dufour, J. P., Pretorius, I. S., et al. (2003b). Expression levels of the yeast alcohol acetyl-transferase genes *ATF1, Lg-ATF1,* and *ATF2* control the formation of a broad range of volatile esters. *Appl. Environ. Microbiol., 69,* 5228–5237.

Vigentini, I., Romano, A., Compagno, C., Merico, A., Molinari, F., Tirelli, A., et al. (2008). Physiological and oenological traits of different *Dekkera/Brettanomyces bruxellensis* strains under wine-model conditions. *FEMS Yeast Res., 8,* 1087–1096.

Villanueva, A., Ramón, D., Valles, S., Lluch, M. A., & MacCabe, A. P. (2000). Heterologous expression in *Aspergillus nidulans* of a *Trichoderma longibrachiatum* endoglucanase of enological relevance. *J. Agric. Food Chem., 48,* 951–957.

Vriesekoop, F., Haass, C., & Pamment, N. B. (2009). The role of acetaldehyde and glycerol in the adaptation to ethanol stress of *Saccharomyces cerevisiae* and other yeasts. *FEMS Yeast Res., 9,* 365–371.

Watanabe, M., Fukuda, K., Asano, K., & Ohta, S. (1990). Mutants of bakers' yeasts producing a large amount of isobutyl alcohol or isoamyl alcohol, flavour components of bread. *Appl. Microbiol. Biotechnol., 34,* 154–159.

Williams, S. A., Hodges, R. A., Strike, T. L., Snow, R., & Kunkee, R. E. (1984). Cloning the gene for the malolactic fermentation of wine from *Lactobacillus delbrueckii* in *Escherichia coli* and yeasts. *Appl. Environ. Microbiol., 47,* 288–293.

Wu, X., & Jiang, Y. W. (2008). Overproduction of non-translatable mRNA silences. The transcription of *Ty1* retrotransposons in *S. cerevisiae* via functional inactivation of the nuclear cap-binding complex and subsequent hyperstimulation of the TORC1 pathway. *Yeast, 25,* 327–347.

Zara, G., Angelozzi, D., Belviso, S., Bardi, L., Goffrini, P., Lodi, T., et al. (2009). Oxygen is required to restore flor strain viability and lipid biosynthesis under fermentative conditions. *FEMS Yeast Res., 9,* 217–225.

Zara, G., Mannazzu, I., Sanna, M. L., Orro, D., Farris, G. A., & Budroni, M. (2008). Exploitation of the semi-homothallic life cycle of *Saccharomyces cerevisiae* for the development of breeding strategies. *FEMS Yeast Res., 8,* 1147–1154.

Zara, S., Antonio Farris, G., Budroni, M., & Bakalinsky, A. T. (2002). *HSP12* is essential for biofilm formation by a Sardinian wine strain of *S. cerevisiae. Yeast, 19,* 269–276.

# Non-*Saccharomyces* Yeasts in the Winemaking Process

*Paloma Manzanares, Salvador Vallés, Fernando Viana*

Departamento de Biotecnología de Alimentos, Instituto de Agroquímica y Tecnología de Alimentos, Consejo Superior de Investigaciones Científicas, Valencia, Spain

## 1. INTRODUCTION

Numerous studies designed to isolate and identify yeasts present on the surface of grapes and winery equipment (Barata et al., 2008b; Beltrán et al., 2002; Combina et al., 2005; Martini & Vaughan-Martini, 1990; Martini et al., 1996; Raspor et al., 2006; Sabaté et al., 2002), along with quantitative and qualitative analyses of the yeast species present during alcoholic fermentation (Esteve-Zarzoso et al., 2001; Fleet & Heard, 1993; Guillamón et al., 1998; Renouf et al., 2007; Schütz & Gafner, 1993; Urso et al., 2008) have confirmed that in traditional winemaking processes (without the use of starter cultures) grape must is transformed into wine through the sequential activity of different yeast species. Under these conditions, fermentation generally begins with the growth of weakly fermentative yeast species belonging to the genera *Candida*, *Debaryomyces*, *Dekkera*, *Hanseniaspora*, *Metschnikowia*, *Pichia*, *Torulaspora*, and *Zygosaccharomyces* (Heard & Fleet, 1985). The growth of these species, known collectively as non-*Saccharomyces* yeasts, is limited to the first 2 or 3 d of fermentation, after which they die as a result of ethanol toxicity. As these yeasts disappear, highly fermentative strains of the species *Saccharomyces cerevisiae* begin to multiply until they become solely responsible for alcoholic fermentation. Despite growing only during the first few days of fermentation, non-*Saccharomyces* yeasts produce a large number of compounds that can have a significant influence on the quality of the wine (reviewed in Fleet, 2003). This chapter will describe the role of non-*Saccharomyces* yeasts in the winemaking process and their contribution to the final product, both in terms of their influence on wine aroma and their role in the vinification process itself. We will begin by discussing traditional methods for identification of yeasts before addressing the methods used to isolate and identify non-*Saccharomyces* species. We will then describe the role of non-*Saccharomyces* yeasts in the vinification process. Finally, we will introduce the design of mixed starter cultures containing non-*Saccharomyces* yeasts and discuss how they might be used to exploit the positive characteristics of these yeasts while minimizing their possible negative effects.

## 2. ISOLATION, ENUMERATION, AND IDENTIFICATION OF NON-SACCHAROMYCES YEASTS

Before we can exploit the potential of non-*Saccharomyces* yeasts and understand their contribution to the vinification process, we must be able to isolate and correctly identify them using appropriate techniques.

### 2.1. Isolation and Enumeration

Isolation and enumeration of yeast from grapes, must, wine, and wineries has traditionally involved plate counts. While spread-plate or pour-plate methods can be used, spread plates seem to yield better results (King et al., 1986). Other systems for the enumeration of yeasts include membrane filtration, direct microscopic counts, dye reduction tests, and the most probable number method (reviewed in Jay, 1994).

Various nutrient agars have been described for the isolation of wine yeasts. Examples include those based on grape must or malt agar. In general, these are complex, nutritionally rich media containing an energy source (e.g., glucose, fructose, or sucrose), a hydrolyzed protein (e.g., peptone, tryptone, or casitone), a complex vitamin supplement such as yeast or malt extract, antibiotics to prevent bacterial growth (oxytetracycline, chlorotetracycline, or chloramphenicol), and compounds to inhibit fungal growth (e.g., rose bengal or dichloran) (reviewed in Beuchat, 1998). However, these nonselective media allow growth of all species of yeast associated with the winemaking process, and this limits their usefulness. For instance, when the sample contains a mixture of very different populations derived from a range of yeast species, most of the colonies on the plate belong to the predominant species and impede recognition of colonies from minority species. Selective and differential methods have been developed to circumvent these limitations. Media for use in enology must allow *S. cerevisiae*, spoilage yeasts belonging to the genera *Saccharomyces*, and non-*Saccharomyces* yeasts to be distinguished from each other. The best example of a selective medium is lysine agar, which allows *S. cerevisiae* to be distinguished from non-*Saccharomyces* yeasts. Its selectivity is based on its failure to support the growth of *S. cerevisiae*, which is unable to use lysine as a nitrogen source. Heard and Fleet (1986a) were the first to demonstrate the effectiveness of lysine agar for the selective isolation and enumeration of *Kloeckera apiculata* and *Candida stellata* populations during fermentation. This is the medium mainly used in the brewing industry to analyze contamination by non-*Saccharomyces* yeasts. Nutrient media containing ethanol and sodium metabisulfite have been described for the selective isolation and enumeration of *Saccharomyces* species during winemaking (Kish et al., 1983). In this medium, non-*Saccharomyces* yeasts, particularly *Kloeckera/Hanseniaspora* species, do not grow because of their lower tolerance of sulfur and ethanol. Media containing high concentrations of sorbates and benzoates have also been described for the selective isolation of *Zygosaccharomyces bailii*, a typical wine spoilage yeast (Pitt & Hocking, 1985), and media containing sulfite and bismuth have been used to analyze the presence of yeasts that produce hydrogen sulfide (Jiranek et al., 1995; Rupela & Tauro, 1984). A differential medium containing glucose and formic acid and supplemented with an indicator was recently described by Schuller et al. (2000) for the enumeration of *Z. bailii* in wine samples. This medium allowed *Z. bailii* to be distinguished from other spoilage yeasts. The same medium has been successfully used to isolate *Z. bailii* and *Zygosaccharomyces bisporus* from damaged grapes, despite the low numbers of these yeasts present (Barata et al., 2008a). Rodrigues et al. (2001) described a medium for the detection of yeasts belonging to the genera *Dekkera/Brettanomyces* based on the use of ethanol as a carbon source and supplemented with cycloheximide, bromocresol green, and ρ-cumaric acid. It has also been possible to demonstrate the presence of *Dekkera bruxellensis* on the surface of grape berries thanks to the development of an enrichment medium (EBB) comprising must, ethanol, malt and yeast extracts, ammonium sulfate, and Tween-80 (Renouf & Lonvaud-Funel, 2007).

Although yeast plate counts are ideal for ecologic studies, obtaining reliable results usually requires incubation of the plates for up to 4 d, and this makes the approach too slow for use in quality-control procedures. Consequently, more rapid, automated methods based on impedance, adenosine triphosphate (ATP) measurements, or fluorescence microscopy have been developed for the enumeration of yeast populations (reviewed in Lightfoot & Maier, 2002). Flow cytometry using fluorescent dyes has also been employed to obtain rapid estimates of the number of yeast and bacteria present in wine and in samples obtained during fermentation (Malacrino et al., 2001). Connell et al. (2002) have developed a filter-based chemiluminescent in situ hybridization method for the detection, identification, and enumeration of *Brettanomyces* species in winery air samples.

## 2.2. Identification

Standard identification criteria classify yeasts on the basis of their morphological, physiological, and biochemical characteristics (Barnett et al., 1990; Kreger-van Rij, 1984; Kurtzman & Fell, 1998). Physiological characteristics serve mainly to describe and identify species and, to a lesser extent, genera. The most commonly used tests for routine identification are fermentation capacity on different carbon sources, growth on different carbon and nitrogen sources, vitamin requirements, growth at different temperatures and in media containing high concentrations of sugar or sodium chloride, urea hydrolysis, and antibiotic resistance (Yarrow, 1998). However, Yarrow (1998) points out that, in the absence of a standardized method for identification tests, the results will depend on the technique used. In addition, the results can vary in different strains of the same yeast species, and this can lead to identification errors. Alongside these drawbacks, traditional identification methods require evaluation of 60 to 90 tests. This complicated and laborious process can take 2 to 3 weeks to complete. Obviously, this methodology cannot be routinely applied in the food industry or indeed in wineries. Consequently, efforts have been made to develop simplified and shortened identification techniques based on the response of yeasts to a series of carefully selected tests (Deák & Beuchat, 1996). Velázquez et al. (2001) developed a kit comprising a series of 24 physiological and biochemical tests along with software to analyze the data obtained. Those authors had previously proposed a kit involving 10 tests for the identification of wine yeasts (Velázquez et al., 1993). However, these simplified techniques are based on the same principles as the more traditional methods. Consequently, although the process may be automated or computerized, the time required is the same and identification is often incorrect (Loureiro & Querol, 1999).

Recently, new molecular biological techniques have been developed for the identification and characterization of yeasts. These techniques, which will be described in detail in Chapter 5, include restriction fragment length polymorphism (RFLP) analysis of mitochondrial DNA (mtDNA), electrophoretic separation of chromosomes, restriction analysis of ribosomal DNA (rDNA), and random amplification of polymorphic DNA (RAPD). Other techniques designed to overcome the problems of routine identification involve analyzing the profile of proteins and long-chain fatty acids. We describe these methods in Sections 2.2.1 and 2.2.2.

### 2.2.1. Chromatographic Analysis of Long-chain Fatty Acid Profile

This technique involves the extraction of yeast fatty acids and the analysis of sample composition by gas chromatography. In the past, a number of problems have prevented this technique from being successfully applied to the identification of yeasts in the food industry. These include the dependence of fatty acid composition on the growth conditions of the yeast, the lack of differentiation between the fatty acid composition of different yeasts, the high cost, the time required to obtain the results, and the need for skilled personnel. Nevertheless, some of these problems can be overcome, for instance by standardizing the growth conditions to minimize as far as possible any variation in lipid composition or by using solid media, which allow results to be obtained within 2 d of isolating and purifying the unknown yeast (Malfeito-Ferreira et al., 1997). Furthermore, the technique is no more expensive than most available rapid-identification techniques. According to Malfeito-Ferreira et al. (1997), analysis of the fatty acid profile can be used to identify most of the yeasts that are important in the food industry, classify the species according to their potential for spoilage of a foodstuff, and identify sources of contamination in wineries and bottling plants. The yeasts typically associated with foodstuffs can be separated into three groups according to their fatty acid composition. The yeasts with the highest potential for spoilage, such as *Dekkera anomala*, *D. bruxellensis*, and various species of the genus *Zygosaccharomyces*, can be identified by the presence of significant quantities of linoleic acid (C18:2) and the absence of linolenic acid (C18:3). The presence of both fatty acids is typical of yeast species belonging to the genera *Candida*, *Cryptococcus*, *Debaryomyces*, *Kluyveromyces*, and *Pichia*, which are associated with poor hygiene during the processing of food-stuffs. The lack of C18 polyunsaturated fatty acids suggests the presence of fermentative strains that may cause spoilage, such as those belonging to the genus *Saccharomyces*. A combination of fatty acid profile and typical PCR methods has also been used as a rapid system for the detection of contaminating yeasts (Sancho et al., 2000). The main drawback of this technique in the food industry is the absence of a database of fatty acid profiles for yeasts typically found in foodstuffs. However, various research laboratories now have access to such databases, meaning that it is simply a question of time before they become available in the food industry (Loureiro & Querol, 1999). In some cases, a degree of overlap has been found between the fatty acid profiles of different yeasts, as was observed for the profile of *Kluyveromyces*, which overlaps with that of *Saccharomyces* (Augustyn et al., 1992).

### 2.2.2. *Electrophoretic Analysis of Protein Profile*

Analysis of the electrophoretic profile of extracellular and/or intracellular proteins or of isoenzymes allows differentiation of strains. In this technique, extracted proteins are separated by polyacrylamide gel electrophoresis. The appearance of specific bands for a given strain is used as a criterion for differentiation. The technique has been used with various yeast genera and its taxonomic validity has been confirmed. Strains of industrial yeast belonging to the genus *Saccharomyces* have been differentiated using this method (Degré et al., 1989; Duarte et al., 1999; Guillamón et al., 1993; van Vuuren & van der Meer, 1987). Duarte et al. (1999) analyzed isoenzyme profile and grouped 35 strains of yeast in the four recognized species of *Saccharomyces sensu stricto* (*S. cerevisiae*, *Saccharomyces bayanus*, *Saccharomyces pastorianus/Saccharomyces carlsbergensis*, and *Saccharomyces paradoxus*). The results of that study confirm the validity of isoenzyme profile as a criterion

for identification and highlight the usefulness of the approach as a rapid and sensitive method of identifying strains of this important group of industrial yeasts. Electrophoretic analysis of protein profiles has also been used for taxonomic purposes in the genera *Candida* (Vancanneyt et al., 1991) and *Zygosaccharomyces* (Duarte et al., 2004). In addition, the technique has been used in a clinical context to identify different strains of *Candida albicans* (Boriollo et al., 2003).

Electrophoretic analysis of protein profile is sensitive, does not require expensive or sophisticated equipment, and can be completed in 48 h following isolation of the yeast strain. Consequently, it should be used more extensively in the food industry. However, as in the case of long-chain fatty acid profiles, representative databases of isoenzyme profiles must first be developed for yeasts used in the wine industry.

# 3. THE ROLE OF NON-SACCHAROMYCES YEASTS IN VINIFICATION

The microflora present at the beginning of fermentation is derived solely from the grapes and essentially comprises species belonging to genera with very limited ethanol tolerance, such as *Hanseniaspora/Kloeckera*, *Hansenula*, *Metschnikowia*, and *Candida*, and strains of *S. cerevisiae*, which are more ethanol tolerant but only represent a very small proportion of the microflora at this stage. This proportion depends on a wide variety of factors, such as the harvesting method, type of transport, fermentation temperature, and quantity of sulfur added. The ratio of non-*Saccharomyces* to *Saccharomyces* yeasts can contribute to accentuating the chemical and sensory changes that take place during fermentation, with clear consequences for the quality of the wine obtained. Consequently, irrespective of whether or not the fermentations are inoculated with *Saccharomyces* yeast, wineries can consider enhancing the non-*Saccharomyces* flora in order to take advantage of its fermentative characteristics.

Numerous studies have characterized the non-*Saccharomyces* yeast strains found on different varieties of grape. In an effort to exploit the putative causal relationship between the presence of these strains and the type of wine produced, fermentations have been attempted with starter cultures that include both *S. cerevisiae* and yeasts belonging to the genera *Kloeckera*, *Cryptococcus*, *Hanseniaspora*, *Candida*, *Pichia*, and *Hansenula* (Fleet & Heard, 1993). The wines produced with these mixed starter cultures differ significantly in both chemical composition and sensory characteristics (Egli et al., 1998). By further analyzing the outcome of these fermentations, greater insight may be gained into the particular characteristics of non-*Saccharomyces* yeasts that affect the type of wine produced.

One characteristic that is thought to differ between non-*Saccharomyces* yeasts and *Saccharomyces* species is the production of enzymes (esterases, glycosidases, lipases, β-glucosidases, proteases, cellulases, etc.). By interacting with substrates in the medium, these enzymes can improve particular phases of the process (such as maceration, filtration, or clarification), increase yield and color extraction, and enhance the characteristics of the wine, especially the aroma (Charoenchai et al., 1997). Since grapes produce only small quantities of enzymes with very limited activity, exogenous enzymes are often introduced during vinification. However, if instead we exploit the contribution of enzymes from yeasts involved in the vinification process, we may be able to produce a more natural product and at the same time improve both the vinification process and the sensory attributes of the wine. In-depth studies are still required to both assess the nature of these yeast-derived enzymes and determine how they favor vinification. However, various studies have already shown that, unlike

*S. cerevisiae* (McKay, 1990), non-*Saccharomyces* yeasts are notable producers of extracellular enzymes (Dizy & Bisson, 2000; Lagace & Bisson, 1990; Strauss et al., 2001).

## 3.1. Influence of Non-*Saccharomyces* Yeasts on the Winemaking Process

Grapes, the raw material for winemaking, contain numerous different compounds, notably phenols, aromatic precursors, enzymes, and structural components. The structural components include pectins, cellulose, glycans, hemicelluloses, proteins, and lignin. The enzymatic degradation of this structure can improve the different stages of vinification, for instance by enhancing the yield and clarification of the must, increasing color extraction, and improving filtration of the wine. Not all of the enzyme activities of interest can be obtained from the grape and those enzymes that are present may not be fully effective under vinification conditions. Therefore, it is of particular interest to control the development of non-*Saccharomyces* yeasts as sources of these enzymes.

The group of secreted enzymes involved in hydrolysis of structural components are referred to as macerating enzymes, and include pectolytic, proteolytic, cellulolytic, and hemicellulolytic enzymes.

### 3.1.1. Pectolytic Enzymes

Pectolytic enzymes cleave long-chain pectins to generate shorter, more soluble chains. This plays an important role in the changes that occur during grape ripening. Later, during winemaking, it facilitates grape pressing and contributes to clarification of the must. The presence of pectolytic enzymes can also improve filtration of the wines and increase the extraction of substances that contribute to color and aroma while the must remains in contact with the grape skins.

The most notable pectolytic enzymes are the pectinases. These act on pectins, the major constituents of the primary cell wall in higher plants. Pectins are heteropolysaccharides with a backbone made of repeating $\alpha$-1,4-D galacturonic acid units. These galacturonic acid units are periodically replaced by $\beta$-1,2- and $\beta$-1,4-linked L-rhamnose units (approximately one per 25 galacturonic acid units), from which a series of side chains of varying length and composition branch off. Although these rhamnogalacturonans are the most common pectins, highly branched arabinogalactans also exist.

Pectinases are mainly classified according to their mechanism of action. Pectin methylesterases, for instance, act via de-esterification, releasing methanol and reducing the degree of methoxylation of the pectin, whereas the polygalacturonases, pectin lyases, and pectate lyases act through depolymerization. The polygalacturonases are the most important from a wine-making perspective. There are two types: the exopolygalacturonases, which hydrolyze the terminal groups and reduce the chain length only slightly, and the endogalacturonases, which act at random. The endogalacturonases alter the dimensions of the molecules more rapidly, reducing the viscosity of the pulp and generally improving some phases of the vinification process, such as clarification.

Studies published to date indicate that non-Saccharomyces yeasts secrete polygalacturonase and pectin methylesterase. Polygalacturonase is produced by species of the genera Candida, Pichia, and Kluyveromyces, whereas pectin methylesterase is produced by Candida, Debaryomyces, and Pichia (Table 4.1).

Since it is difficult to perform experiments in natural conditions, in most cases activity has been detected by growing the microorganisms on plates. Consequently, little information is available on the effect that components of the media might have on the induction or inhibition of enzyme production, and, as a result, a definitive relationship between the presence of the yeast and the secreted activity cannot be assumed. For instance, glucose in the must is thought to inhibit polygalacturonase production in non-Saccharomyces yeasts but induce it in S. cerevisiae (Strauss et al., 2001). Non-Saccharomyces yeasts could be used to reinforce the production of polygalacturonases by S. cerevisiae (McKay, 1990). The combined activity of these enzymes and pectin methylesterase could improve the degradation of pectins in the medium during fermentation.

### 3.1.2. Proteolytic Enzymes

Proteins are present in varying quantities in the grape and, along with polysaccharides, are responsible for increasing must and wine turbidity. Although these proteins can be eliminated with bentonite, this nonspecific process also leads to loss of aromas and compounds that influence flavor. Protease treatment, on the other hand, specifically hydrolyzes proteins and improves the clarity and stability of the wine. The smaller, more soluble peptides and amino acids generated by this enzymatic hydrolysis are also nitrogen-containing compounds. Consequently, in addition to improving clarification and stabilization, protease treatment helps to prevent stuck fermentation caused by a lack of assimilable nitrogen in the must.

Yeast proteases play an important role in the process of autolysis during on-lees aging of wines and in the development of protein haze (protein degradation), especially in white wines. However, not all proteases are active under the particular conditions found in wine. Analysis of protease activity from non-Saccharomyces yeasts has revealed the importance of nitrogen sources in the production of the extracellular enzymes (Charoenchai et al., 1997).

As with pectinases, the main problem with proteases is their weak activity under the particular conditions found in wine. As a result, tests to detect proteolytic activity tend to be done on plates or using a model wine, even though these options do not ensure that the proteases detected will be active under vinification conditions. Nevertheless, the use of proteases from

**TABLE 4.1**   Macerating Enzymes Produced by Non-*Saccharomyces* Yeasts

| | Macerating enzymes | | | | | | |
|---|---|---|---|---|---|---|---|
| Yeasts | PG | PME | CEL | GLU | XYL | PR | Reference |
| *Candida albicans* | | | X | | | | (7) |
| *Candida flavus* | | | | | | X | (1) |
| *Candida hellenica* | | | X | X | | | (1) |
| *Candida krusei* | | X | | | | | (3) |
| *Candida lambica* | | | | X | | | (1) |
| *Candida lipolytica* | | | | | | X | (4) |
| *Candida norvegensis* | X | | | | | | (1) |
| *Candida olea* | | | | | | X | (1),(4) |
| *Candida oleophila* | X | | | | X | | (1) |
| *Candida pelliculosa* | | | | | X | | (1) |
| *Candida pulcherrima* | X | | X | X | X | X | (1),(4) |
| *Candida silvae* | X | | | | | | (1) |
| *Candida sorbosa* | | | | X | | | (1) |
| *Candida stellata* | X | | X | X | X | X | (1),(6) |
| *Candida tropicalis* | X | | | | | | (2) |
| *Candida valida* | X | | | | | | (1) |
| *Candida wickerhamii* | | | X | | | | (8) |
| *Cryptococcus sp.* | | | X | | | | (11) |
| *Cryptococcus albidus* | X | | | | | | (12) |
| *Debaryomyces hansenii* | | | | | | X | (1) |
| *Debaryomyces membranaefaciens* | | X | | | | | (3) |
| *Hanseniaspora guilliermondii* | | | | | | X | (5) |
| *Kloeckera apiculata* | | | | | | X | (1),(4),(5) |
| *Kloeckera thermotolerans* | X | | | | | | (6) |
| *Kluyveromyces marxianus* | X | | | | | | (10) |
| *Metschnikowia pulcherrima* | | | | | | X | (1),(6) |
| *Pichia anomala* | X | | | | | | (6) |
| *Pichia guilliermondii* | X | | | | | | (6) |

*(Continued)*

**TABLE 4.1** Macerating Enzymes Produced by Non-*Saccharomyces* Yeasts—*cont'd*

| | Macerating enzymes | | | | | | |
|---|---|---|---|---|---|---|---|
| Yeasts | PG | PME | CEL | GLU | XYL | PR | Reference |
| *Pichia kluyveri* | X | X | | | | | (9) |
| *Pichia membranaefaciens* | X | | | | | | (6) |

CEL = cellulase; GLU = β-glucanase; PG = polygalacturonase; PME = pectin methylesterase; PR = protease; XYL = xylanase. (1) Strauss et al. (2001); (2) Luh and Phaff (1951); (3) Bell and Etchells (1956); (4) Lagace and Bisson (1990); (5) Dizy and Bisson (2000); (6) Fernández et al. (2000); (7) Chambers et al. (1993); (8) LeClerc et al. (1984); (9) Masoud and Jespersen (2006); (10) Serrat et al. (2004); (11) Thongekkaew et al. (2008); (12) Servili et al. (1990).

non-*Saccharomyces* yeasts in the vinification process has been investigated. Specifically, the addition of proteases from *K. apiculata* has been used successfully to degrade some of the protein in Chenin Blanc and Chardonnay wines. It has even been demonstrated that proteases from *Candida olea*, *Candida lipolytica*, *Candida pulcherrima*, and *K. apiculata* can produce a moderate reduction in wine turbidity (Lagace & Bisson, 1990). Dizy and Bisson (2000) demonstrated that some species belonging to the genus *Kloeckera/Hanseniaspora* are the highest producers of proteolytic activity in the must and affect the protein profile of the finished wines.

Despite these positive properties, in some cases proteolytic activity does not significantly reduce the temperature-related turbidity of the wines and may even increase it in fermentations with a high proteolytic activity (Strauss et al., 2001). Table 4.1 shows the species with proteolytic activity.

### 3.1.3. Cellulolytic and Hemicellulolytic Enzymes

Given that cellulose and hemicellulose are the main structural polysaccharides of the plant cell wall, their enzymatic degradation will allow extraction and release of pigments and aromas from the grape skins. Treatment with these enzymes reduces the maceration time, as the desired results are achieved sooner.

When working with grapes infected with the fungus *Botrytis cinerea*, the clarification and filtration process is also impaired by high-molecular-weight (1,3)-β-D-glucans, even though they are only present at low concentrations. These polysaccharides can be eliminated by enzymatic treatment. As in the case of the pectins, these compounds are degraded by a series of enzymes, including cellulolytic (endoglucanases, exoglucanases, cellobiases, and β-glucanases) and hemicellulolytic (xylanases) enzymes.

During cellulose degradation, the endoglucanases and exoglucanases act at random and from the ends, respectively, generating a mixture of oligosaccharides, predominantly cellobiose, which is hydrolyzed to glucose by cellobiase. The β-glucanases are specific enzymes for the hydrolysis of the β-glucans mentioned above.

Unlike cellulose, the hemicelluloses, notably xylane, are branched heteropolymers, and as a result of their complexity must be degraded by multiple enzymes, such as the xylanases, galactanases, and mannanases.

To date, the only non-*Saccharomyces* yeasts that have been described as producers of cellulolytic or hemicellulolytic enzymes are *Candida* and *Cryptococcus*, as shown in Table 4.1.

## 3.2. Influence on Aroma

Aroma is one of the organoleptic characteristics that determine the quality of a wine. As with many other foodstuffs, wine aroma is determined by hundreds of different compounds with concentrations that can vary between $10^{-1}$ and $10^{-10}$ g/kg (Rapp & Mandery, 1986).

The concentration of these compounds depends on factors such as grape variety, climate, soil, rainfall, and time of harvesting, as well as numerous variables relating to the fermentation process (pH, temperature, nutrients, and microflora) and the operations that it encompasses (filtration, clarification, etc.). The aromatic quality of the wine is determined by the balance and interaction of these compounds.

It is important to distinguish between three different types of wine aroma: the varietal or primary aroma, determined by the grape variety; the fermentation or secondary aroma; and the bouquet or tertiary aroma resulting from the transformation of aromas during aging. Non-*Saccharomyces* yeasts can influence both the primary and secondary aroma, as described below.

### 3.2.1. *Primary Aroma*

The varietal aroma is mainly determined by the quantity and chemical nature of the volatile secondary metabolites present in the grape (van Rensburg & Pretorius, 2000). Of these, the terpenes have the greatest influence on flavor and aroma, particularly in wines derived from Moscatel grapes but also in other less aromatic varieties (Marais, 1983; Rapp & Mandery, 1986).

Terpenes are volatile compounds that are present in the grape as free molecules or non-aromatic glycosylated precursors. In general, the precursors are glycosides formed from a disaccharide and a terpene, with $\alpha$-L-arabinofuranosyl-$\beta$-D-glucopyranosides, $\alpha$-L-rhamnopyranosyl-$\beta$-D-glucopiranosides, and $\beta$-D-apiosyl-$\beta$-D-glucopyranosides of geraniol, nerol, and linalool among the most abundant (Gunata et al., 1988). The hydrolysis that releases the volatile aromatic compounds occurs in two steps (Figure 4.1). Firstly, the glycosidic bonds are cleaved to release specific sugars according to the substrate and the enzyme involved. For instance, arabinose, rhamnose, and apiose are released by $\alpha$-L-arabinofuranosidase, $\alpha$-L-rhamnosidase, and $\beta$-D-apiosidase enzymes,

respectively. In the next step, $\beta$-D-glucosidase activity releases terpenes from the glucosides generated in the first step.

Suboptimal conditions (pH and temperature) or inhibition by glucose and ethanol nevertheless result in reduced activity of hydrolytic enzymes derived from the grapes or from *S. cerevisiae*. Consequently, these precursors are commonly hydrolyzed in very small proportions during fermentation (Gunata, 1984). The degree of inhibition depends on the species and strains of organism involved (Aryan et al., 1987; Delcroix et al., 1994; LeClerc et al., 1987; Rosi et al., 1994). For instance, Grossman et al. (1987) showed that the $\beta$-glucosidase of *Hansenula* sp., isolated from fermented must, was able to release aromatic substances when added to the wine but was less effective in the must.

Studies of yeast glycosidases indicate that some specific enzymes can influence the varietal aroma of the wine (Laffort et al., 1989), especially when fermentation is carried out under natural conditions (Fugelsang, 1997), where non-*Saccharomyces* yeasts predominate during the initial stages. This apparent influence of non-*Saccharomyces* yeasts may be explained by their marked hydrolytic activity, which is absent in most *Saccharomyces* strains (Charoenchai et al., 1997; Fernández et al., 2000; Gunata et al., 1994; Manzanares et al., 1999, 2000; Mendes-Ferreira et al., 2001; Strauss et al., 2001; Úbeda & Briones, 2000; Zoecklein et al., 1997).

Rosi et al. (1994) showed that yeasts of the genera *Candida, Debaryomyces, Hansenia-spora/Kloeckera, Kluyveromyces, Metschnikowia, Pichia, Saccharomycodes, Schizosaccharomyces,* and *Zygosaccharomyces* can produce $\beta$-glucosidase, and this was later confirmed by other authors (Charoenchai et al., 1997; Manzanares et al., 2000; McMahon et al., 1999; Strauss et al., 2001). However, the analogues used as substrates for the selection of $\beta$-glucosidase can equally detect exoglucanase, and the activities of these enzymes can therefore be confused (Strauss et al., 2001).

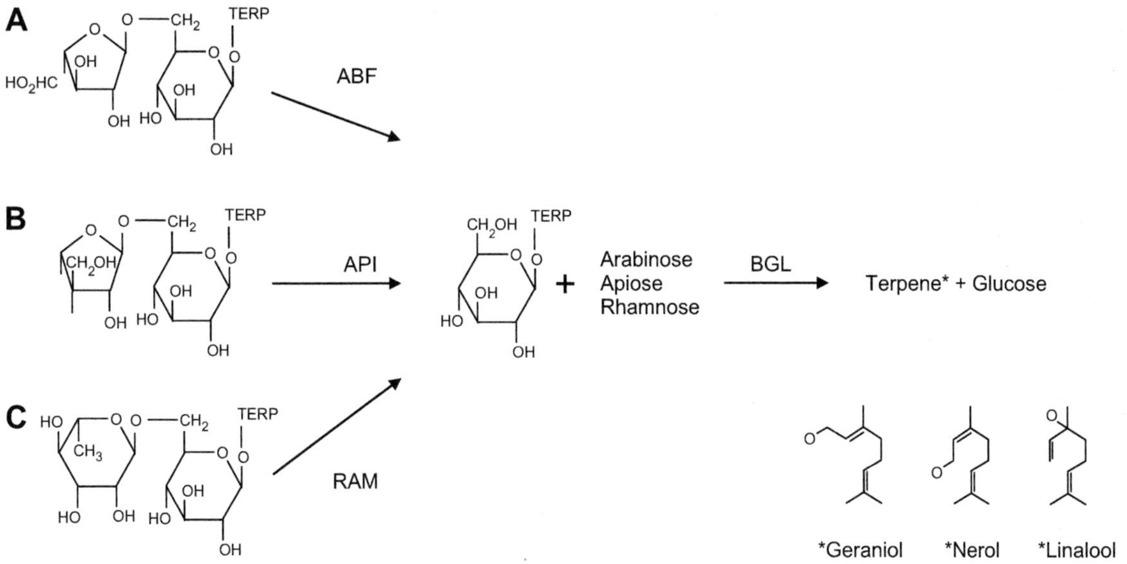

**FIGURE 4.1** Enzymatic hydrolysis of glycosylated precursors. A = α-L-arabinofuranosyl-(1,6)-β-D-glucopyranoside; B = β-D-apiosyl-(1,6)-β-D-glucopyranoside; C = α-L-rhamnopyranosyl-(1,6)-β-D-glucopyranoside; ABF = α-L-arabinofuranosidase; API = β-D-apiosidase; BGL = β-D-glucosidase; RAM = α-L-rhamnosidase.

As a result, the putative production of β-glucosidase by those yeasts should be analyzed in greater detail. Likewise, natural substrates such as precursor extracts should be used, since the β-glucosidases selected as a result of their activity with artificial substrates could prove ineffective when it comes to hydrolyzing aromatic precursors in grape must. This is the case for the β-glucosidase from *Brettanomyces bruxellensis*, which was found to be unable to hydrolyze an extract of precursors from grape must (Mansfield et al., 2002). The potential effectiveness of yeast-derived β-glucosidases is even further reduced in most cases by the fact that the enzymes are intracellular and released only in very small amounts into the culture medium (McMahon et al., 1999).

Another limitation of these enzymes is their very weak activity in the presence of glucose in the must or wine, making it especially necessary to analyze their inhibition by this sugar. *Candida*, *Debaryomyces*, *Kluyveromyces*, and *Pichia* species produce extracellular β-glucosidases that are not inhibited by glucose; in particular, *Candida peltata* β-glucosidase remains unaffected by glucose at concentrations of up to 250 mg/mL (Saha & Bothast, 1996).

In contrast to the limited information available on macerating enzymes, β-glucosidases are quite well characterized. In some cases these enzymes have even been purified; for example, in *Debaryomyces hansenii* strains (Riccio et al., 1999; Yanai & Sato, 1999). *D. hansenii* β-glucosidase maintains its activity in the presence of ethanol concentrations of up to 15% (vol/vol) and releases terpenes, not only from extracts of glycosylated precursors but also when added to the must during fermentation, where it increases the concentrations of linalool and nerol by 90 and 116%, respectively (Yanai & Sato, 1999). Extracellular β-glucosidase has also been purified from *Debaryomyces vanrijiae* and found to be active in the presence of 200 mM glucose (80% activity) and 15%

(vol/vol) ethanol (64% residual activity). Its addition to Moscatel grape must during fermentation also increases the concentration of terpenes (Belancic et al., 2003). A β-glucosidase from *Candida molischiana* has also been shown to release terpenes and alcohols both from a Moscatel glycoside extract and from the wine itself (Genovés et al., 2003).

β-D-xylosidase is also involved in releasing aromas, although data are limited on its hydrolytic capacity. Manzanares et al. (1999) selected eight yeast strains belonging to the genera *Hanseniaspora* and *Pichia* as the best producers of β-D-xylosidase from a total of 54 species of wine yeast. Of these two genera, only the species *Hanseniaspora osmophila*, *Hanseniaspora uvarum*, and *Pichia anomala* exhibited β-D-xylosidase activity. Although this is not a new observation for *Pichia* species, which have previously been described to hydrolyze xylane (Lee et al., 1986), it has not been reported previously for *Hanseniaspora* species. Another genus that is able to produce this enzyme is *Candida*. *Candida utilis* produces a β-D-xylosidase that can hydrolyze precursors and increase the concentration of terpenes following addition of the purified enzyme to Moscatel grape must during fermentation (Yanai & Sato, 2001).

There is only one purified α-L-arabinofuranosidase from *Pichia capsulata* that effectively releases terpenes from precursors obtained from Moscatel grape must. Its main characteristics are that it is not inhibited by glucose and its activity is stimulated by the presence of ethanol (Yanai & Sato, 2000). Another species from this genus, *P. anomala*, has been described as a producer of α-L-arabinofuranosidase activity (Spagna et al., 2002).

In a recent study addressing glycosidase activities, a strain of *Candida guilliermondii* was found to produce α-L-rhamnosidase (Rodríguez et al., 2004). Table 4.2 shows the glycosidase activities produced by different species of non-*Saccharomyces* yeast.

### 3.2.2. Secondary Aroma

Yeasts are responsible for the secondary or fermentation aroma of the wine. This aroma arises during alcoholic fermentation and is determined by compounds produced as part of wine yeast metabolism. Although ethanol, glycerol, and carbon dioxide are quantitatively the most abundant of these compounds and play a fundamental role in wine aroma, their contribution to the secondary aroma is relatively limited. Volatile fatty acids, higher alcohols, esters, and, to a lesser extent, aldehydes, have a greater contribution to secondary aroma (Rapp & Versini, 1991). These compounds, shown in Table 4.3, are generated through the conversion of directly fermentable sugars and also of long-chain fatty acids and nitrogenated and sulfur compounds, among others. These components of the must are able to penetrate the cell wall and participate in a variety of chemical reactions that generate a range of volatile compounds (Boulton et al., 1996).

Although *S. cerevisiae* is the wine yeast *par excellence* and the main yeast responsible for the fermentation products, the contribution of non-*Saccharomyces* yeasts should not be forgotten either in natural fermentation or when a commercial strain of *S. cerevisiae* is inoculated. In the latter case, the influence of non-*Saccharomyces* yeasts is reduced, although it has been shown that the use of starter cultures does not prevent the growth or metabolic activity of other natural strains of *S. cerevisiae* or of *K. apiculata*, *H. uvarum*, *C. stellata*, or *Torulaspora delbrueckii* (Egli et al., 1998; Heard & Fleet, 1986b, 1985; Henick-Kling et al., 1998; Lema et al., 1996).

Below, we describe the influence of the different yeast species on the formation of the main aromatic compounds originating from wine yeast metabolism that determine the secondary aroma. The biosynthesis of these compounds has been reviewed by Lambrechts and Pretorius (2000).

**TABLE 4.2**　Glycosidases Produced by Non-*Saccharomyces* Yeasts

| Yeasts | Glycosidases | | | | |
|---|---|---|---|---|---|
| | BGL | XYL | RAM | ARA | Reference |
| *Brettanomyces bruxellensis* | X | | | | (1) |
| *Candida stellata* | X | | | | (2),(3) |
| *Candida pulcherrima* | X | | | | (3),(4) |
| *Candida cacaoi* | X | | | | (20) |
| *Candida cantarelli* | X | | | | (5) |
| *Candida colliculosa* | X | | | | (3) |
| *Candida dattila* | X | | | | (5) |
| *Candida domerquiae* | X | | | | (5) |
| *Candida famata* | X | | | | (3) |
| *Candida guilliermondii* | X | | X | | (4),(6) |
| *Candida hellenica* | X | | | | (2) |
| *Candida krusei* | X | | | | (3) |
| *Candida molischiana* | X | | | | (7),(21) |
| *Candida parapsilosis* | X | | | | (6) |
| *Candida peltata* | X | | | | (8) |
| *Candida utilis* | | X | | | (9) |
| *Candida vinaria* | X | | | | (5) |
| *Candida vini* | X | | | | (5) |
| *Candida wickerhamii* | X | | | | (21) |
| *Cryptococcus albidus* | | X | | | (19) |
| *Debaryomyces hansenii* | X | | | | (10),(11) |
| *Debaryomyces vanrijiae* | X | | | | (12),(13) |
| *Hanseniaspora guilliermondii* | X | | | | (5) |
| *Hanseniaspora osmophila* | X | X | | | (5),(14) |
| *Hanseniaspora uvarum* | X | X | | | (5),(14),(15) |
| *Kloeckera apiculata* | X | | | | (2),(3),(4),(6) |
| *Metschnikowia pulcherrima* | X | | | | (5),(6) |
| *Pichia anomala* | X | X | | X | (3),(5),(14),(16) |
| *Pichia capsulata* | | | | X | (17) |

(*Continued*)

**TABLE 4.2**   Glycosidases Produced by Non-*Saccharomyces* Yeasts—*cont'd*

| Yeasts | Glycosidases | | | | |
| --- | --- | --- | --- | --- | --- |
| | BGL | XYL | RAM | ARA | Reference |
| *Pichia membranaefaciens* | X | | | | (5) |
| *Pichia stipitis* | | X | | | (18) |
| *Zygosaccharomyces bailii* | X | | | | (5) |
| *Zygosaccharomyces mellis* | X | | | | (5) |
| *Zygosaccharomyces rouxii* | X | | | | (5) |

ARA = α-arabinofuranosidase; BGL = β-glucosidase; RAM = α-rhamnosidase; XYL = β-xylosidase. (1) Mansfield et al. (2002); (2) Strauss et al. (2001); (3) Charoenchai et al. (1997); (4) Rodríguez et al. (2004); (5) Manzanares et al. (2000); (6) McMahon et al. (1999); (7) Genovés et al. (2003); (8) Saha and Bothast (1996); (9) Yanai and Sato (2001); (10) Yanai and Sato (1999); (11) Riccio et al. (1999); (12) Belancic et al. (2003); (13) García et al. (2002); (14) Manzanares et al. (1999); (15) Fernández-González et al. (2003); (16) Spagna et al. (2002); (17) Yanai and Sato (2000); (18) Lee et al. (1986); (19) Peciarová and Biely (1982); (20) Drider et al. (1993); (21) Gunata et al. (1990).

### 3.2.2.1. VOLATILE FATTY ACIDS

Acetic acid is responsible for 90% of the volatile acidity of wines (Radler, 1993). The remaining fatty acids, such as propanoic and butanoic acid, are present in small quantities as products of yeast metabolism. It should be remembered that the production of these fatty acids is also associated with bacterial growth (Riberau-Gayon et al., 1998).

Long-chain fatty acids (C16 and C18) are essential precursors for the synthesis of many

**TABLE 4.3**   Principal Volatile Fatty Acids, Higher Alcohols, Esters, and Carbonyl Compounds Produced During Alcoholic Fermentation

| Volatile fatty acids | Higher alcohols | Esters | Carbonyl compounds |
| --- | --- | --- | --- |
| **Acetic acid** | Propanol | **Ethyl acetate** | **Acetaldehyde** |
| Butyric acid | Butanol | **2-Phenylethyl acetate** | Benzaldehyde |
| Formic acid | Isobutyl alcohol | **Isoamyl acetate** | Butanal |
| Isobutyric acid | Amyl alcohol | **Isobutyl acetate** | Diacetyl |
| Isovaleric acid | **Isoamyl alcohol** | Hexyl acetate | Propanal |
| Propionic acid | Hexanol | Ethyl butanoate | Isobutanal |
| Valeric acid | **Phenylethanol** | Ethyl caprate | Pentanal |
| Hexanoic acid | | Ethyl caprylate | Isovaleraldehyde |
| Heptanoic acid | | Ethyl caproate | 2-Acetyl tetrahydropyridine |
| Octanoic acid | | Ethyl isovalerate | |
| Nonanoic acid | | Ethyl 2-methylbutanoate | |
| Decanoic acid | | | |
| Tridecanoic acid | | | |

The most abundant compounds in wines are shown in boldface.

lipid compounds in yeast. These fatty acids occur in the cell membrane as esters, specifically palmitoleic and oleic acids, which constitute 70% of the cell membrane in *S. cerevisiae* (Ratledge & Evans, 1989). In general, these acids do not appear in wines, but they are found in products distilled in the presence of yeast lees. In contrast, intermediate-chain fatty acids (C8, C10, and C12) do appear alongside their ethyl esters as components of wine.

Although acetic and lactic acid bacteria can generate high levels of acetic acid, yeasts are also involved in its production. Delfini and Cervetti (1991) classified *Saccharomyces* yeast strains into three groups according to their production of acetic acid: low (0–0.3 g/L), intermediate (0.31–0.60 g/L), and high (>0.61 g/L). Studies of acetic acid production in non-*Saccharomyces* yeasts have generated highly variable results, and the concentrations reached may be greater than or less than those produced by *S. cerevisiae*. For instance, *K. apiculata* produces between 1 and 2.5 g/L, *C. stellata* between 1 and 1.3 g/L, *Metschnikowia pulcherrima* between 0.1 and 0.15 g/L, *Candida krusei* 1 g/L, *Hansenula anomala* between 1 and 2 g/L, and *T. delbrueckii* between 0.01 and 1.07 g/L (Fleet & Heard, 1993; Renault et al., 2009).

In general, C8 and C10 fatty acids and their esters are produced in lower quantities by non-*Saccharomyces* yeasts than by *S. cerevisiae* (Herraiz et al., 1990; Ravaglia & Delfini, 1993; Renault et al., 2009; Rojas et al., 2001; Viana et al., 2008). It is notable that the concentrations of short-chain fatty acids produced by non-*Saccharomyces* yeasts are substantially below the levels that can inhibit the growth of *S. cerevisiae* and stop fermentation (Edwards et al., 1990).

### 3.2.2.2. HIGHER ALCOHOLS

The term "higher alcohol" encompasses those alcohols with more than two carbon atoms and a higher molecular weight and boiling point than ethanol. They are the largest group of aromatic compounds (Amerine et al., 1980). At the concentrations normally present in wine (<300 mg/L), they contribute to the aromatic complexity of the product. When their concentrations exceed 400 mg/L, they are considered to have a negative effect on aroma (Rapp & Mandery, 1986). Among the aliphatic higher alcohols, the most predominant is isoamyl alcohol, although in this group propanol, isobutyl alcohol, and amyl alcohol are also produced. The aromatics form another class of higher alcohols, notable among which is 2-phenylethanol (Nykänen et al., 1977). The importance of these compounds is also related to their role as precursors for the formation of esters (Soles et al., 1982), compounds that are very important in wine aroma.

Production of higher alcohols is a strain-specific characteristic and can be used to select yeasts for industrial applications (Giudici et al., 1990, 1993). The proportions of isoamyl and amyl alcohol, isobutanol, and propanol (Herraiz et al., 1990), and the production of dodecanol and tetradecanol (Longo et al., 1992), are also specific to each strain. In general, studies of higher alcohol production in non-*Saccharomyces* yeasts highlight the influence that these yeasts can have on the chemical composition and quality of the wine (Gil et al., 1996; Herraiz et al., 1990; Longo et al., 1992; Mateo et al., 1991). In fermented musts, the total production of higher alcohols by pure cultures of non-*Saccharomyces* yeasts is lower than that found with *S. cerevisiae* (Moreira et al., 2008; Rojas et al., 2003; Viana et al., 2008, 2009). However, when non-*Saccharomyces* yeasts are used in mixed cultures, the difference is reduced and the total quantity of higher alcohols is similar in all wines.

### 3.2.2.3. ESTERS

Esters are the most abundant compounds found in wine, with around 160 identified to date. In general, the concentration of esters in wine is above the perception threshold (Salo,

1970a, 1970b), and some of the sensory descriptors used in wine evaluation correspond to ester aromas (Etievant, 1991; Maarse & Visscher, 1989). The fresh, fruity aroma of young wines, for instance, is due to a mixture of esters generated during fermentation, in particular acetate esters (Ferreira et al., 1995; Marais, 1990).

Although various esters can be formed during fermentation, the most abundant are those derived from acetic acid and higher alcohols (ethyl acetate, isoamyl acetate, isobutyl acetate, and 2-phenylethyl acetate) and ethyl esters of saturated fatty acids (ethyl butanoate, ethyl caproate, ethyl caprilate, and ethyl caprate).

Various genera of non-*Saccharomyces* yeasts have been described as good producers of esters (Table 4.4). *Candida*, *Hansenula*, and *Pichia* species have a greater capacity to produce ethyl acetate than wine strains of *S. cerevisiae* (Nykänen, 1986; Ough et al., 1968). The genus *Rhodotorula* has been reported to produce isoamyl acetate (Suomalainen & Lehtonen, 1979), whereas the genus *Hanseniaspora*, specifically the species *H. uvarum*, is reported to be a good producer of esters in general (Mateo et al., 1991; Romano et al., 1997; Sponholz, 1993). In a recent study, Viana et al. (2008) grouped ester production according to the yeast genus and demonstrated the capacity of the genus *Hanseniaspora* to produce acetate esters, particularly

**TABLE 4.4** Non-*Saccharomyces* Yeast Genera That Produce Esters

| Genus | Ester produced | | | |
| | Ethyl acetate | Isoamyl acetate | 2-Phenylethyl acetate | Ethyl caproate |
| --- | --- | --- | --- | --- |
| *Candida* | + | | | |
| *Hanseniaspora* | + | + | + | |
| *Pichia* | + | + | | |
| *Rhodotorula* | | + | | |
| *Torulaspora* | | | | + |

2-phenylethyl acetate. In addition to its production of ethyl acetate, the genus *Pichia* also stands out as a good producer of isoamyl acetate. In terms of the ethyl esters, the genus *Saccharomyces* was the best producer of ethyl caproate, whereas the genus *Torulaspora* was the strongest producer of ethyl caprylate.

The species and strain of yeast are, among other variables, determinants of the levels of esters produced (Lambrechts & Pretorius, 2000). For instance, species of the genus *Hanseniaspora* (*H. guilliermondii*, *H. osmophila*, and *H. uvarum*) produce significant quantities of 2-phenylethyl acetate and isoamyl acetate (Moreira et al., 2005, 2008; Plata et al., 2003; Rojas et al., 2001, 2003; Viana et al., 2008, 2009), although there are notable differences among strains (Viana et al., 2008).

### 3.2.2.4. CARBONYL COMPOUNDS

Due to their low perception threshold and the characteristics that they confer on the wine (apple, lemon, and nutty aromas), volatile aldehydes are among the most interesting carbonyl compounds. Acetaldehyde constitutes more than 90% of the total aldehyde content of wines (Nykänen et al., 1977). Other carbonyl compounds of interest include diacetyl, which indicates growth of lactic acid bacteria when present at high concentrations (1−4 mg/L) (Sponholz, 1993), and the tetrahydropyridines responsible for the acetamide (mousy) aroma and tightly linked to the growth of lactic acid bacteria and *Brettanomyces* (Heresztyn, 1986; Rapp, 1998).

Data are available on the effect of non-*Saccharomyces* yeasts on the total concentration of aldehydes in wine. The species *K. apiculata*, *C. krusei*, *C. stellata*, *H. anomala*, and *M. pulcherrima* produce quantities ranging from undetectable to 40 mg/L, whereas *S. cerevisiae* produces between 6 and 190 mg/mL (Fleet & Heard, 1993; Then & Radler, 1971). In a study describing the aromatic profile of different species of yeast, Romano et al. (2003) found little variation in the production of acetaldehyde by 52 strains of *S. cerevisiae* (with a mean of approximately 50 mg/L), whereas there were significant differences among the 59 strains of *H. uvarum* studied (mean acetaldehyde concentration of approximately 25 mg/L). Data have also been reported on the maximum production of benzaldehyde (1200 mg/L) achieved by *Schizosaccharomyces* and *Zygosaccharomyces* species (Delfini et al., 1991).

### 3.2.2.5. VOLATILE PHENOLS AND SULFUR COMPOUNDS

From a quantitative point of view, volatile phenols and sulfur compounds (Table 4.5) make a lesser contribution to wine aroma than the compounds described above. However, qualitatively they are very important, since their perception thresholds are very low and in general they have a negative contribution to wine aroma.

Volatile phenols are generated by microbiological transformation of hydroxycinnamic acids (trans-ferulic, trans-p-coumaric, and

caffeic acid) present in the grapes through the sequential action of two enzymes. First, hydroxycinnamate decarboxylase converts hydroxycinnamic acids into vinylphenols (4-vinyl guaiacol and 4-vinylphenol), and these are then reduced to ethylphenols (4-ethylguaiacol and 4-ethylphenol) by vinylphenol reductase. Their concentrations vary between 0 and 6047 µg/L, and when they exceed the perception threshold they are responsible for the phenolic aroma of wines. The presence of volatile phenols is always undesirable, since even at concentrations below the perception threshold they are reported to mask the fruity notes of white wines (Chatonnet et al., 1992; Dubois, 1983).

Although it was thought that only species of the genera *Brettanomyces/Dekkera* were able to transform hydroxycinnamic acids into ethylphenols, more recent studies have identified other non-*Saccharomyces* yeast strains with this capacity, although only some strains of *Pichia guilliermondii* displayed the same conversion capacity as *Dekkera* species (Dias et al., 2003; Renault et al., 2009; Shinohara et al., 2000). However, the initial decarboxylation step of hydroxycinnamic acids into vinylphenols is much more common in both the non-*Saccharomyces* yeasts found in wine (e.g., *Hanseniaspora*, *Pichia*, and *Zygosaccharomyces* species) and in wine strains of *S. cerevisiae* (Chatonnet et al., 1992). Table 4.6 shows the main species of yeast that have been identified as producers of 4-ethylphenol, although it should be remembered that this is a strain-dependent characteristic. The production of the ethylphenols found in wine has been reviewed by Suárez et al. (2007).

The sulfur compounds present in wine can be divided into various groups according to their chemical structure: sulfides, heterocyclic polysulfide compounds, thioesters, and thiols. The sensory properties of these compounds vary extensively, and, although most of them are associated with negative aromatic descriptors, they can have a positive contribution to wine

**TABLE 4.5** Principal Phenolic and Sulfur Compounds Produced During Alcoholic Fermentation

| Phenolic compounds | Sulfur compounds |
| --- | --- |
| 2-Vinylphenol | Hydrogen sulfide |
| 4-Vinyl guaiacol | Dimethyl sulfide |
| 4-Ethylphenol | Diethyl sulfide |
| 4-Ethyl guaiacol | Dimethyl disulfide |
| | Diethyl disulfide |
| | Methyl mercaptan |
| | Ethyl mercaptan |
| | S-methyl thioacetate |
| | S-ethyl thioacetate |
| | 4-Mercapto-4-methylpentan-2-one (4MMP) |
| | 3-Mercaptohexan-1-ol (3MH) |
| | 3-Mercaptohexyl acetate (3MHA) |

**TABLE 4.6** Species of Yeast that Produce 4-Ethylphenol

| Yeasts | 4-Ethylphenol |
|---|---|
| *Brettanomyces lambicus* | ++ |
| *Candida cantarelli* | + |
| *Candida halophila* | + |
| *Candida mannitofaciens* | + |
| *Candida versatilis* | + |
| *Candida wickerhamii* | + |
| *Debaryomyces hansenii* | + |
| *Dekkera anomala* | ++ |
| *Dekkera bruxellensis* | ++ |
| *Dekkera intermedia* | ++ |
| *Kluyveromyces lactis* | + |
| *Pichia guilliermondii* | ++ |
| *Torulaspora delbrueckii* | + |

+ = weak producer; ++ = strong producer.

aroma through the introduction of fruity notes (reviewed in Swiegers et al., 2005).

The most extensively studied sulfur compound is hydrogen sulfide, since it often occurs in musts low in nitrogen. Almost all studies have focused on *S. cerevisiae* and very limited information is available on the production of hydrogen sulfide by non-*Saccharomyces* yeasts. Strauss et al. (2001) included this characteristic, along with the production of extracellular enzymes, in a study characterizing non-*Saccharomyces* yeasts isolated from musts and grapes from South Africa. Almost all of the strains studied produced hydrogen sulfide, with *Candida* species displaying the highest production. Species of the genus *Hanseniaspora* have also been reported to produce hydrogen sulfide, as has *T. delbrueckii* (Renault et al., 2009; Viana et al., 2008).

Recently, the capacity of *H. uvarum* and *H. guilliermondii* to produce heavy sulfur compounds has been evaluated (Moreira et al.,

2008) and, although there were significant differences between the two species, the concentrations produced were similar to those of *S. cerevisiae*.

# 4. DESIGN OF MIXED STARTER CULTURES

The variability in the yeast flora present in grape musts can be controlled by routine inoculation with cultures that predominate and therefore standardize the initial flora, leading to homogeneous fermentation year after year. This is the reason why nowadays fewer wineries produce their wines by natural or spontaneous fermentation and instead tend to induce fermentation of the must with selected strains of *S. cerevisiae*. The use of selected yeasts is common in large wineries and in the most technologically advanced appellations, since it ensures rapid initiation of fermentation and, as a result, reduces the risk of oxidation and contamination. Prior selection of yeasts also allows wines with improved organoleptic properties to be produced. It should not be forgotten, then, that non-*Saccharomyces* yeasts can make a positive contribution to the winemaking process. These yeast species could be used as part of a strategy to obtain different types of wines, especially in terms of aromatic profile. The combination of different *S. cerevisiae* and non-*Saccharomyces* strains in starter cultures could be used to produce wines with unique aromatic characteristics. Thus, the presence of *S. cerevisiae* would prevent premature termination of fermentation and non-*Saccharomyces* yeast species would introduce aromatic complexity. This proposal has been supported by a number of authors in studies addressing the effect of non-*Saccharomyces* yeasts on the organoleptic characteristics of wines (Egli et al., 1998; Gil et al., 1996; Henick-Kling et al., 1998; Lema et al., 1996; Mateo et al., 1991; Moreno et al., 1991; Romano et al., 1997; Zironi et al., 1993). In two of those studies, emphasis was placed on the sensory characteristics of the wines obtained. In a study involving wines produced from Riesling grapes, Henick-Kling et al. (1998) concluded that the intensity of the fruity aroma generated with non-*Saccharomyces* yeasts was greater than that obtained with *S. cerevisiae* starter cultures. In a similar study involving Riesling and Chardonnay grapes, Egli et al. (1998) analyzed the organoleptic properties of wines produced by spontaneous fermentation in the presence of non-*Saccharomyces* yeasts or with starter cultures from two strains of *S. cerevisiae*. According to a panel of tasters, the wines of both varieties achieved a higher score as a result of the greater intensity of their floral and fruity aromas when produced by spontaneous fermentation.

In contrast to these spontaneous fermentations, recent studies have addressed the effects of mixed starter cultures on the aroma and structure of the wines. Soden et al. (2000) reported that wines produced using sequential fermentations with *C. stellata* AWRI 1159 and *S. cerevisiae* AWRI 838 had higher concentrations of succinic and acetic acid, glycerol, and ethyl acetate and lower concentrations of acetaldehyde and ethanol than wines derived from pure cultures of *S. cerevisiae*. The wines produced with the mixed starter cultures also had a distinct aromatic character. Similar results were obtained in pilot-scale fermentations with immobilized *C. stellata* and subsequent inoculation with *S. cerevisiae* (Ferraro et al., 2000).

The effect of mixed and sequential starter cultures of *C. cantarellii* and *S. cerevisiae* has also been analyzed during fermentation of musts from the Syrah grape variety (Toro & Vazquez, 2002). The main differences in the analytic profile of these wines, compared with those obtained using *S. cerevisiae*, were in the quantities of acetoin, propanol, and succinic acid, as well as the higher concentrations of ethanol and glycerol (between 7.8 and 10% and between 44.3 and 52.8%, respectively). Zohre and Erten (2002)

described the influence of *K. apiculata* and *C. pulcherrima* in mixed starter cultures with *S. cerevisiae*. They concluded that the wines obtained had a different aromatic profile to those obtained with *S. cerevisiae* alone and that none of the compounds produced had a negative influence on the organoleptic quality of the wine. Similar studies using musts with high sugar contents indicated that wines obtained with mixed starter cultures of *H. uvarum*, *T. delbrueckii*, or *Kluyveromyces thermotolerans* and *S. cerevisiae* had comparable or improved analytic profiles compared to those obtained with *S. cerevisiae* alone (Ciani et al., 2006). However, these same yeasts halted fermentation when inoculated sequentially.

As with *S. cerevisiae* strains currently used for controlled fermentations, the most rational approach with the greatest likelihood of success would be to select non-*Saccharomyces* yeasts on the basis of their production both of enzymes relevant for the winemaking process and of metabolites that influence the quality of the wine. Selection on the basis of enzyme production has been analyzed in studies addressing the terpene fraction of a Moscatel wine produced with a mixed culture of *S. cerevisiae* and *D. vanrijiae*, which was chosen for its β-glucosidase activity. It was found that the wines obtained with a mixed culture differed in their concentrations of certain volatile compounds, in particular geraniol (Belancic et al., 2003; García et al., 2002). To address selection on the basis of metabolite production, Rojas et al. (2003) studied the effect of mixed starter cultures using *S. cerevisiae* and non-*Saccharomyces* yeast strains selected for their capacity to produce 2-phenylethyl acetate (*H. guilliermondii* CECT 11104) and ethyl acetate (*P. anomala* CECT 10590) in microbiological culture medium. However, when tested under winemaking conditions, 2-phenylethyle acetate was produced by *H. guilliermondii* but isoamyl acetate was not produced by *P. anomala*. In addition, although the concentration of ethyl acetate in the wines obtained with mixed cultures was lower than in those obtained with pure cultures of non-*Saccharomyces* yeasts, it was nevertheless excessive and surpassed acceptable limits. Based on the balanced production of secondary metabolites and the organoleptic characteristics of the wines, Mingorance-Cazorla et al. (2003) selected a strain of *Pichia fermentans* as a good candidate for use in mixed starter cultures. Subsequent studies showed that it was effective in musts sequentially inoculated with *S. cerevisiae* and gave rise to wines with higher levels of esters, alcohol, and glycerol (Clemente-Jiménez et al., 2005).

A simple solution to avoid the possible negative effects of non-*Saccharomyces* yeasts is to include both positive and negative features in the selection criteria. Viana et al. (2008) included in the selection criteria not only the formation of acetate esters with positive effects on wine aroma (2-phenylethyl acetate and isoamyl acetate) but also the excessive formation of ethyl acetate as a negative characteristic. In addition, hydroxycinnamate decarboxylase activity and the production of hydrogen sulfide, acetaldehyde, acetic acid, and short-chain fatty acids were included as negative selection criteria. These criteria allowed selection of *H. osmophila* 1471 as a strain for use with *S. cerevisiae* in mixed starter cultures. In a later study, the same authors showed that wines with increased levels of 2-phenylethyl acetate could be produced using the mixed starter culture and that the concentrations of the ester could be controlled by changing the proportions of the two yeasts in the starter culture (Viana et al., 2009).

The species *T. delbrueckii*, which is characterized by its low production of acetaldehyde, acetoin, acetic acid, and ethyl acetate (Cabrera et al., 1988; Herraiz et al., 1990), has been proposed for inclusion in a mixed starter culture for fermentation of musts with a high sugar content (Bely et al., 2008). The problem with these musts,

which are obtained from grapes infected with *B. cinerea* and can reach sugar concentrations of up to 450 g/L, is that conventional yeasts produce excessive concentrations of acetic acid. Bely et al. (2008) have shown that a mixed culture of *T. delbrueckii*—*S. cerevisiae* at a 20:1 ratio is the most appropriate for improving the analytic profile of these sweet wines. The wines obtained have approximately half the volatile acidity and acetaldehyde concentration of sweet wines produced with *S. cerevisiae* alone. That study also addressed the influence of combined or sequential inoculation of the two yeasts and the proportions of the species in the starter culture.

Studies of this type confirm the potential offered by selecting non-*Saccharomyces* yeasts for use in mixed starter cultures. Taking into account their general characteristics, it would be possible to design mixed starter cultures based on non-*Saccharomyces* yeasts that produce macerating enzymes (with possible consequences on the technological aspects of winemaking) and/or glycosidases and acetate esters (with possible effects on wine aroma). Exploiting the enzymatic potential of non-*Saccharomyces* yeasts could even represent an alternative to the use of exogenous enzymes in the winemaking process, currently a common practice in many wineries. Good selection of non-*Saccharomyces* yeasts based on enological criteria may therefore help in the design of optimized mixed starter cultures without compromising wine quality.

## 5. FINAL CONSIDERATIONS

It is clear that non-*Saccharomyces* yeasts have considerable unexploited potential as microorganisms able to influence both wine aroma and the winemaking process itself. More in-depth studies are now required to select those with characteristics that exert a positive influence on winemaking and to minimize negative effects. The use of non-*Saccharomyces* yeasts in mixed starter cultures could extend the range of the current market by allowing wines with defined organoleptic characteristics to be produced. Finally, as non-*Saccharomyces* yeasts are natural components of the grape microflora, these proposals are in line with the current trend in the food industry to use as few additives as possible and satisfy consumer demand for natural products.

## Acknowledgments

The work of the authors in the areas discussed in this chapter has been supported by grants VIN01-018-C2-1 and VIN03-007-C2-1 from the Spanish National Institute for Agricultural and Food Science Research and Technology (Instituto Nacional de Investigación y Tecnología Agraria y Alimentaria [INIA]) and by grants AGL2003-01295 and AGL2004-00978/ALI from the Spanish Ministry of Science and Education-FEDER.

## References

Amerine, A. M., Berg, H. V., Kunkee, R. E., Ough, C. S., Singleton, V. L., & Webb, A. D. (1980). *The technology of winemaking* (4th ed.). Wesport, CT: AVI Technical Books Inc.

Aryan, A. P., Wilson, B., Strauss, C. R., & Williams, P. J. (1987). The properties of glycosidases of *Vitis vinifera* and a comparison of their β-glucosidase activity with that of exogenous enzymes. An assessment of possible applications in enology. *Am. J. Enol. Vitic., 38*, 182–188.

Augustyn, O. P. H., Kock, J. L. F., & Ferreira, D. (1992). Differentiation between yeast species, and strains within a species by cellular fatty acid analysis. 5. A feasible technique? *Syst. Appl. Microbiol., 15*, 105–115.

Barata, A., González, S., Malfeito-Ferreira, M., Querol, A., & Loureiro, V. (2008b). Sour rot-damaged grapes are sources of wine spoilage yeasts. *FEMS Yeast Res., 8*, 1008–1017.

Barata, A., Seborro, F., Belloch, C., Malfeito-Ferreira, M., & Loureiro, V. (2008a). Ascomycetus yeast species recovered from grapes damaged by honeydew and sour rot. *J. Appl. Microbiol., 104*, 1182–1191.

Barnett, J. A., Payne, R. W., & Yarrow, D. (1990). *Yeasts: Characteristics and identification* (2nd ed.). Cambridge, UK: Cambridge University Press.

Belancic, A., Gunata, Z., Vallier, M.-J., & Agosin, E. (2003). β-glucosidase from the grape native yeast *Debaryomyces vanrijiae*: Purification, characterization, and its effect on monoterpene content of a Muscat grape juice. *J. Agric. Food Chem., 51*, 1453–1459.

Bell, T. A., & Etchells, J. L. (1956). Pectin hydrolysis by certain salt-tolerant yeasts. *Appl. Microbiol., 4*, 196–202.

Beltrán, G., Torija, M. J., Novo, M., Ferrer, N., Poblet, M., Guillamón, J. M., et al. (2002). Analysis of yeast populations during alcoholic fermentation: A six year follow-up study. *Syst. Appl. Microbiol., 25*, 287–293.

Bely, M., Stoeckle, P., Masneuf-Pomarède, I., & Dubourdieu, D. (2008). Impact of mixed Torulaspora delbrueckii-Saccharomyces cerevisae culture on high-sugar fermentation. *Int. J. Food Microbiol., 122*, 312–320.

Beuchat, L. R. (1998). Progress in conventional methods for the detection and enumeration of foodborne yeasts. *Food Technol. Biotechnol., 36*, 267–272.

Boriollo, M. F. G., Rosa, E. A. R., Bernardo, W. L. D. C., Gonçalves, R. B., & Höfling, J. F. (2003). Electrophoretic protein patterns and numerical analysis of *Candida albicans* from the oral cavities of healthy children. *Rev. I. Med. Trop., 45*, 249–257.

Boulton, R. B., Singleton, V. H., Bisson, L. F., & Kunkee, R. E. (1996). *Principles and practices of winemaking*. New York, NY: Chapman & Hall.

Cabrera, M. J., Moreno, J., Ortega, J. M., & Medina, M. (1988). Formation of ethanol, higher alcohols, esters and terpenes by five yeast strains in must from Pedro Ximenez grapes in various degrees of ripeness. *Am. J. Enol. Vitic., 39*, 283–287.

Chambers, R. S., Broughton, M. J., Cannon, R. D., Carne, A., Emerson, G. W., & Sullivan, P. A. (1993). An exo-β-(1, 3)-glucanase of *Candida albicans*: Purification of the enzyme and molecular cloning of the gene. *J. Gen. Microbiol., 139*, 325–334.

Charoenchai, C., Fleet, G. H., Henschke, P. A., & Todd, B. E. N. (1997). Screening of non-Saccharomyces wine yeasts for the presence of extracellular hydrolytic enzymes. *Aust. J. Grape Wine Res., 3*, 2–8.

Chatonnet, P., Dubourdieu, D., Boidron, J., & Pons, M. (1992). The origin of ethylphenols in wines. *J. Sci. Food Agric., 60*, 178–184.

Ciani, M., Beco, L., & Comitini, F. (2006). Fermentation behaviour and metabolic interactions of multistarter wine yeast fermentations. *Int. J. Food Microbiol., 108*, 239–245.

Clemente-Jiménez, J. M., Mingorance-Cazorla, L., Martínez-Rodríguez, S., Las Heras-Vázquez, F. J., & Rodríguez-Vico, F. (2005). Influence of sequential yeast mixtures on wine fermentation. *Int. J. Food Microbiol., 98*, 301–308.

Combina, M., Mercado, L., Borgo, P., Elia, A., Jofre, V., Ganga, A., et al. (2005). Yeasts associated to Malbec grape berries from Mendoza, Argentina. *J. Appl. Microbiol., 98*, 1055–1061.

Connell, L., Stender, H., & Edwards, C. G. (2002). Rapid detection and identification of Brettanomyces from winery air samples based on peptide nucleic acid analysis. *Am. J. Enol. Vitic., 53*, 322–324.

Deák, T., & Beuchat, L. R. (1996). *Handbook of food spoilage yeasts.* Boca Raton, FL: CRC Press.

Degré, R., Thomas, D. Y., Ash, J., Mailhiot, K., Morin, A., & Dubord, C. (1989). Wine yeast strain identification. *Am. J. Enol. Vitic., 40*, 309–315.

Delcroix, A., Gunata, Z., Sapis, J. C., Salmon, J. M., & Bayonove, C. (1994). Glycosidase activities of three enological yeast strains during winemaking: Effect on the terpenol content of Muscat wine. *Am. J. Enol. Vitic., 45*, 291–296.

Delfini, C., & Cervetti, F. (1991). Metabolic and technological factors affecting acetic acid production by yeasts during alcoholic fermentation. *Vitic. Enol. Sci., 46*, 142–150.

Delfini, C., Gaia, P., Bardi, L., Mariscalco, G., & Conteiro, M. (1991). Production of benzaldehyde, benzyl alcohol and benzoic acid by yeasts and *Botrytis cinerea* isolated from grape musts and wines. *Vitis, 30*, 253–263.

Dias, L., Dias, S., Sancho, T., Stender, H., Querol, A., Malfeito-Ferreira, M., et al. (2003). Identification of yeasts isolated from wine-related environments and capable of producing 4-ethylphenol. *Food Microbiol., 20*, 567–574.

Dizy, M., & Bisson, L. F. (2000). Proteolytic activity of yeasts strains during grape juice fermentations. *Am. J. Enol. Vitic., 51*, 155–167.

Drider, D., Chemardin, P., Arnaud, A., & Galzy, P. (1993). Isolation and characterization of the exocellular β-glucosidase of *Candida cacaoi*: Possible use in carbohydrates degradation. *Lebensm. Wiss. U. Technol., 26*, 198–204.

Duarte, F. L., Pais, C., Spencer-Martins, I., & Leão, C. (1999). Distinctive electrophoretic isoenzyme profiles in Saccharomyces cerevisiae sensu stricto. *Int. J. System. Bacterial, 49*, 1907–1913.

Duarte, F. L., Pais, C., Spencer-Martins, I., & Leão, C. (2004). Isoenzyme patterns: A valuable molecular tool for the differentiation of *Zygosaccharomyces* species and detection of misidentified isolates. *Syst. Appl. Microbiol., 27*, 436–442.

Dubois, P. (1983). Volatile phenols in wine. In J. R. Piggot (Ed.), *Flavour of distilled beverages, origin and development* (pp. 110–119). Chichester, UK: Ellis Horwood.

Edwards, C. G., Beelman, R. B., Bartley, C. E., & McConnell, A. L. (1990). Production of decanoic acid and other volatile compounds on the growth of yeast and malolactic bacteria during vinification. *Am. J. Enol. Vitic., 41*, 48–56.

Egli, C. M., Edinger, W. D., Mitrakul, C. M., & Henick-Kling, T. (1998). Dynamics of indigenous and inoculated yeast populations and their effect on the sensory character of Riesling and Chardonnay wines. *J. Appl. Microbiol., 85*, 779–789.

Esteve-Zarzoso, B., Peris-Torán, M. J., García-Maiquez, E., Uruburu, F., & Querol, A. (2001). Yeast population dynamics during the fermentation and biological aging of sherry wines. *Appl. Environ. Microbiol., 67*, 2056–2061.

Etievant, P. X. (1991). Wine. In H. Maarse (Ed.), *Volatile compounds in foods and beverages* (pp. 483–587). New York, NY: Marcel Dekker Inc.

Fernández, M., Úbeda, J. F., & Briones, A. I. (2000). Typing of non- Saccharomyces yeasts with enzymatic activities. *Int. J. Food Microbiol., 59*, 29–36.

Fernández-González, M., di Stefano, R., & Briones, A. (2003). Hydrolysis and transformation of terpene glycosides from Muscat must by different yeast species. *Food Microbiol., 20*, 35–41.

Ferraro, L., Fatichenti, F., & Ciani, M. (2000). Pilot scale vinification process using immobilized *Candida stellata* cells and Saccharomyces cerevisiae. *Process Biochem., 35*, 1125–1129.

Ferreira, V., Fernández, P., Peña, C., Escudero, A., & Cacho, J. F. (1995). Investigation on the role played by fermentation esters in the aroma of young Spanish wines by multivariate analysis. *J. Sci. Food Agric., 67*, 381–392.

Fleet, G. H. (2003). Yeast interactions and wine flavour. *Int. J. Food Microbiol., 86*, 11–22.

Fleet, G. H., & Heard, G. M. (1993). Yeast-growth during winemaking. In G. H. Fleet (Ed.), *Wine microbiology and biotechnology* (pp. 27–54). Chur, Switzerland: Harwood Academic Publishers.

Fugelsang, K. C. (1997). *Wine microbiology.* New York, NY: Chapman & Hall.

García, A., Carcel, C., Dulau, L., Samson, A., Aguera, E., Agosin, E., & Gunata, Z. (2002). Influence of a mixed culture with *Debaryomyces vanriji* and *Saccharomyces cerevisiae* on the volatiles of a Muscat wine. *J. Food Sci., 67*, 1138–1143.

Genovés, S., Gil, J. V., Manzanares, P., Aleixandre, J. L., & Vallés, S. (2003). *Candida molischiana* β-glucosidase production by *Saccharomyces cerevisiae* and its application in winemaking. *J. Food Sci., 68*, 2096–2100.

Gil, J. V., Mateo, J. J., Jiménez, M., Pastor, A., & Huerta, T. (1996). Aroma compounds in wine as influenced by apiculate yeasts. *J. Food Sci., 61*, 1247–1249.

Grossmann, M. K., Rapp, A., & Rieth, W. (1987). Enzymatische freisetzung gebundener aromastoffe in wein. *Deut. Lebensm.-Rundsch, 83*, 7–12.

Guillamón, J. M., Querol, A., Jiménez, M., & Huerta, T. (1993). Phylogenetic relationships among wine yeast strains based on electrophorectic whole-cell protein patterns. *Int. J. Food Microbiol., 18*, 115–125.

Guillamón, J. M., Sabaté, J., Barrio, E., Cano, J., & Querol, A. (1998). Rapid identification of wine yeast species based on RFLP analysis of the ribosomal internal transcribed spacer (ITS) region. *Arch. Microbiol., 169,* 387–392.

Guiudici, P., Romano, P., & Zambonelli, C. (1990). A biometric study of higher alcohol production in Saccharomyces cerevisiae. *Can. J. Microbiol., 36,* 60–64.

Guiudici, P., Zambonelli, C., & Kunkee, R. E. (1993). Increased production of n-propanol in wine yeast strains having an impaired ability to form hydrogen sulphide. *Am. J. Enol. Vitic., 44,* 123–127.

Gunata, Y. Z. (1984). *Récherches sur la fraction liee de nature glycosidique de l'arome du raisin: Importance des terpenyl-glycosides, actino des glycosidases.* Montpellier, France: Doctorial thesis, Universite des Sciences et Techniques du Languedoc.

Gunata, Y., Bitteur, S., Brillouet, J. M., Bayonove, C. L., & Cordonnier, R. E. (1988). Sequential enzymatic hydrolysis of potentially aromatic glycosides from grape. *Carbohydr. Res., 184,* 139–149.

Gunata, Z., Bayonove, C. L., Cordonnier, R. E., Arnaud, A., & Galzy, P. (1990). Hydrolysis of grape monoterpenyl glycosides by Candida molischiana and Candida wickerhamii β-glucosidases. *J. Sci. Food Agric., 50,* 499–506.

Gunata, Y. Z., Dugelay, I., Sapis, J. C., Baumes, R., & Bayonove, C. (1994). Role of enzymes in the use of the flavor potential from grape glycosides in winemaking. In P. Schreier, & P. Winterhalter (Eds.), *Progress in flavor precursor studies.* Wurzburg, Germany: Proceedings of the International Conference.

Heard, G. M., & Fleet, G. H. (1985). Growth of natural yeast flora during the fermentation of inoculated wines. *Appl. Environ. Microbiol., 50,* 727–728.

Heard, G. M., & Fleet, G. H. (1986a). Evaluation of selective media for enumeration of yeasts during wine fermentation. *J. Appl. Bacteriol., 60,* 477–481.

Heard, G. M., & Fleet, G. H. (1986b). Occurrence and growth of yeast species during the fermentation of some Australian wines. *Food Technol. Australia, 38,* 22–25.

Henick-Kling, T., Edinger, W., Daniel, P., & Monk, P. (1998). Selective effects of sulfur dioxide and yeast starter culture addition on indigenous yeast populations and sensory characteristics of wine. *J. Appl. Microbiol., 84,* 865–876.

Heresztyn, T. (1986). Formation of substituted tetrahydropyridines by species Brettanomyces and Lactobacillus isolated from mousy wines. *Am. J. Enol. Vitic., 37,* 127–132.

Herraiz, T., Reglero, G., Herraiz, M., Martín-Alvarez, P. J., & Cabezudo, M. D. (1990). The influence of the yeast and type of culture on the volatile composition of wines fermented without sulphur dioxide. *Am. J. Enol. Vitic., 41,* 313–318.

Jay, J. M. (1994). Identificación de los microorganismos y/o de sus productos metabólicos en los alimentos. In *Microbiología moderna de los alimentos* (3rd ed.). Zaragoza, Spain: Editorial Acribia, SA. pp. 113–219.

Jiranek, V., Langride, P., & Henschke, P. A. (1995). Validation of bismuth-containing indicator media for predicting $H_2S$ producing potential of Saccharomyces cerevisiae wine yeast under enological conditions. *Am. J. Enol. Vitic., 46,* 269–273.

King, A. P., Pitt, J. I., Beuchat, L. R., & Corry, J. (1986). *Methods for the mycological examination of food.* New York, NY: Plenum Press.

Kish, S., Sharf, R., & Margalith, P. (1983). A note on a selective medium for wine yeasts. *J. Appl. Bacteriol., 55,* 177–179.

Kreger-van Rij, N. J. W. (1984). *The yeasts: A taxonomic study* (3rd ed.). Amsterdam, the Netherlands: Elsevier Science B.V.

Kurtzman, C. P., & Fell, J. W. (1998). *The yeasts: A taxonomic study* (4th ed.). Amsterdam, the Netherlands: Elsevier Inc.

Laffort, J. F., Romat, H., & Darriet, P. (1989). Les levures et lèxpression aromatique des vins blancs. *Revue des Oenologues, 53,* 9–12.

Lagace, L. S., & Bisson, L. F. (1990). Survey of yeast acid protease for effectiveness of wine haze reduction. *Am. J. Enol. Vitic., 41,* 147–155.

Lambrechts, M. G., & Pretorius, I. S. (2000). Yeast and its importance to wine aroma. A review. *S. Afr. J. Enol. Vitic., 21,* 97–129.

LeClerc, M., Arnaud, A., Ratomahenina, R., & Galzy, P. (1987). Yeast β-glucosidases. *Biotechnol. Genet. Engin. Rev., 5,* 269–295.

LeClerc, M., Arnaud, A., Ratomahenina, R., Galzy, P., & Nicolas, M. (1984). The enzyme system in a strain of Candida wickerhamii Meyer and Yarrow participating in the hydrolysis of cellodextrins. *J. Gen. Appl. Microbiol., 30,* 509–521.

Lee, H., Biely, P., Latta, R. K., Barbosa, M. F. S., & Schneider, H. (1986). Utilization of xylan by yeasts and its conversion to ethanol by Pichia stipitis strains. *Appl. Environ. Microbiol., 52,* 320–324.

Lema, C., Garcia-Jares, C., Orriols, I., & Angulo, L. (1996). Contribution of Saccharomyces and non-Saccharomyces populations to the production of some components of Albariño wine aroma. *Am. J. Enol. Vitic., 47,* 206–216.

Lightfoot, N. F., & Maier, E. A. (2002). Métodos cuantitativos y evaluación de procedimientos. In *Análisis microbiológico de alimentos y aguas. Directrices para el aseguramiento de la calidad.* Zaragoza, Spain: Editorial Acribia SA. pp. 95–133.

Longo, E., Velazquez, J. B., Siereiro, C., Cansado, J., Calo, P., & Villa, T. G. (1992). Production of higher alcohols, ethyl acetate, acetaldehyde and other compounds by 14 S. cerevisiae strains isolated from the same region (Salnes, NW Spain). *World J. Appl. Microbiol. Biotech., 8,* 539–541.

Loureiro, V., & Querol, A. (1999). The prevalence and control of spoilage yeasts in foods and beverages. *Trends Food Sci. Technol., 10*, 356−365.

Luh, B. S., & Phaff, H. J. (1951). Studies on polygalacturonase of certain yeast. *Arch. Biochem. Biophys., 33*, 212−227.

Maarse, H., & Visscher, C. A. (1989). *Volatile compounds in alcoholic beverages −qualitative and quantitative data*. Zeist, the Netherlands: TNO-CIVO Food Analysis Institute.

Malacrino, P., Zapparoli, G., Torriani, S., & Dellaglio, F. (2001). Rapid detection of viable yeasts and bacteria in wine by flow cytometry. *J. Microbiol. Methods, 45*, 127−134.

Malfeito-Ferreira, M., Tareco, M., & Loureiro, V. (1997). Fatty acid profiling: A feasible typing system to trace yeast contamination in wine bottling plants. *Int. J. Food Microbiol., 38*, 143−155.

Mansfield, A. K., Zoecklein, B. W., & Whiton, R. S. (2002). Quantification of glycosidase activity in selected strains of *Brettanomyces bruxellensis* and Oenococcus oeni. *Am. J. Enol. Vitic., 53*, 303−307.

Manzanares, P., Ramón, D., & Querol, A. (1999). Screening of non-Saccharomyces wine yeasts for the production of β-D-xylosidase activity. *Int. J. Food Microbiol., 46*, 105−112.

Manzanares, P., Rojas, V., Genovés, S., & Vallés, S. (2000). A preliminary search for anthocyanin-β-D-glucosidase activity in non- Saccharomyces wine yeasts. *Int. J. Food Sci. Technol., 35*, 95−103.

Marais, J. (1983). Terpenes in the aroma of grapes and wines. A review. *S. Afr. J. Enol. Vitic., 4*, 49−58.

Marais, J. (1990). Wine aroma composition. *Food Rev., 17*, 18−21.

Martini, A., Ciani, M., & Scorzetti, G. (1996). Direct enumeration and isolation of wine yeasts from grapes surfaces. *Am. J. Enol. Vitic., 47*, 435−440.

Martini, A., & Vaughan-Martini, A. V. (1990). Grape must fermentation, past and present. In J. F. T. Spencer, & D. M. Spencer (Eds.), *Yeast technology* (pp. 105−123). Berlin, Germany: Springer-Verlag.

Masoud, W., & Jespersen, L. (2006). Pectin degrading enzymes in yeasts involved in fermentation of Coffea arabica in East Africa. *Int. J. Food Microbiol., 110*, 291−296.

Mateo, J. J., Jiménez, M., Huerta, T., & Pastor, A. (1991). Contribution of different yeasts isolated from musts of Monastrell grapes to the aroma of wine. *Int. J. Food Microbiol., 14*, 153−160.

McKay, A. M. (1990). Degradation of polygalacturonic acid by Saccharomyces cerevisiae. *Lett. Appl. Microbiol., 11*, 41−44.

McMahon, H., Zoecklein, B. W., Fugelsang, K., & Jasinski, Y. (1999). Quantification of glycosidase activities in selected yeasts and lactic acid bacteria. *J. Ind. Microbiol. Biotechnol., 23*, 198−203.

Mendes-Ferreira, A., Clímaco, M. C., & Mendes Faia, A. (2001). The role of non-*Saccharomyces* species in realising glycosidic fraction of grape aroma components. A preliminary study. *J. Appl. Microbiol., 91*, 67−71.

Mingorance-Cazorla, L., Clemente-Jiménez, J. M., Martínez-Rodríguez, S., Las Heras-Vázquez, F. J., & Rodríguez-Vico, F. (2003). Contribution of different natural yeasts to the aroma of two alcoholic beverages. *World J. Microbiol. Biotechnol., 19*, 297−304.

Moreira, N., Mendes, F., Guedes de Pinho, P., Hogg, T., & Vasconcelos, I. (2008). Heavy sulphur compounds, higher alcohols and esters production profile of Hanseniaspora uvarum and Hanseniaspora guilliermondii grown as pure and mixed cultures in grape must. *Int. J. Food Microbiol., 124*, 231−238.

Moreira, N., Mendes, F., Hogg, T., & Vasconcelos, I. (2005). Alcohols, esters and heavy sulphur compounds production by pure and mixed cultures of apiculate wine yeasts. *Int. J. Food Mocrobiol., 103*, 285−294.

Moreno, J. J., Millán, C., Ortega, J. M., & Medina, M. (1991). Analytical differentiation of wine fermentations using pure and mixed yeast cultures. *J. Indust. Microbiol., 7*, 181−190.

Nykänen, L. (1986). Formation and occurrence of flavour compounds in wine and distilled alcoholic beverages. *Am. J. Enol. Vitic., 37*, 84−96.

Nykänen, L., Nykänen, I., & Suomalainen, H. (1977). Distribution of esters produced during sugar fermentation between the yeast cell and the medium. *J. Inst. Brew., 83*, 32−34.

Ough, C. S., Cook, J. A., & Lider, L. A. (1968). Rootstock-scion interactions concerning wine making. II. Wine compositional and sensory changes attributed to rootstock and fertilizer differences. *Am. J. Enol. Vitic., 19*, 254−265.

Peciarová, A., & Biely, P. (1982). β-xylosidases and a nonspecific wall-bound β-glucosidase of the yeast *Cryptococcus albidus*. *Biochim. Biophys. Acta (BBA) − General Subjects, 716*, 391−399.

Pitt, J. I., & Hocking, A. D. (1985). *Fungi and food spoilage*. Sydney, Australia: Academic Press.

Plata, C., Millán, C., Mauricio, J. C., & Ortega, J. M. (2003). Formation of ethyl acetate and isoamyl acetate by various species of wine yeasts. *Food Microbiol., 20*, 217−224.

Radler, F. (1993). Yeasts-metabolism of organic acids. In G. H. Fleet (Ed.), *Wine microbiology and biotechnology* (pp. 165−182). Chur, Switzerland: Harwood Academic Publishers.

Rapp, A. (1998). Volatile flavour of wine. Correlation between instrumental analysis and sensory perception. *Nahrung, 42*, 351−363.

Rapp, A., & Mandery, H. (1986). Wine aroma. *Experientia, 42*, 873−884.

Rapp, A., & Versini, G. (1991). Influence of nitrogen compounds in grapes on aroma compounds of wine. In J. M. Rantz (Ed.), In: *Proceedings of the international symposium on nitrogen in grapes and wines* (pp. 156–164). Davis, CA: American Society for Enology and Viticulture.

Raspor, P., Milek, D. M., Polanc, J., Mozina, S. S., & Cadez, N. (2006). Yeasts isolated from three varieties of grapes cultivated in different locations of the Dolenjska vine-growing region, Slovenia. *Int. J. Food Microbiol., 109*, 97–102.

Ratledge, C., & Evans, C. T. (1989). Lipids and their metabolism. In *The yeasts, Vol 3*. London, UK: Academic Press. pp. 367–455.

Ravaglia, S., & Delfini, C. (1993). Production of medium chain fatty acids and their ethyl esters by yeast strains isolated from musts and wines. *Ital. J. Food Sci., 5*, 21–36.

Renault, P., Miot-Sertier, C., Marullo, P., Hernández-Orte, P., Lagarrigue, L., Lonvaud-Funel, A., et al. (2009). Genetic characterization and phenotypic variability in Torulaspora delbrueckii species: Potencial applications in the wine industry. *Int. J. Food Microbiol., 134*, 201–210.

Renouf, V., Claisse, O., & Lonvaud-Funel, A. (2007). Inventory and monitoring of wine microbial consortia. *Appl. Microbiol. Biotechnol., 75*, 149–164.

Renouf, V., & Lonvaud-Funel, A. (2007). Development of an enrichment medium to detect *Dekkera/Brettanomyces bruxellensis*, a spoilage wine yeast, on the surface of grape berries. *Microbiol. Res., 162*, 154–167.

Ribereau-Gayon, P., Glories, Y., Maugean, A., & Dubordieu, D. (1998). *Chimie du vin, stabilisation et traitements, Vol. 2*. Paris, France: Dunod.

Riccio, P., Rossana, R., Vinella, M., Domizio, P., Zito, F., Sansevrino, F., et al. (1999). Extraction and immobilization in one step of two β-glucosidases released from a yeast strain of debaryomyces hansenii. *Enzyme Microb. Technol., 24*, 123–129.

Rodrigues, N., Gonçalves, G., Pereira-da-Silva, M., Malfeito-Ferreira, M., & Loureiro, V. (2001). Development and use of a new medium to detect yeasts of the genera *Dekkera/Brettanomyces*. *J. Appl. Microbiol., 90*, 588–599.

Rodríguez, M. E., Lopes, C. A., van Broock, M., Vallés, S., Ramón, D., & Caballero, A. C. (2004). Screening and typing of Patagonian wine yeasts for glycosidase activities. *J. Appl. Microbiol., 96*, 84–95.

Rojas, V., Gil, J. V., Piñaga, F., & Manzanares, P. (2001). Studies on acetate ester formation by non-Saccharomyces wine yeasts. *Int. J. Food Microbiol., 70*, 283–289.

Rojas, V., Gil, J. V., Piñaga, F., & Manzanares, P. (2003). Acetate ester formation in wine by mixed cultures in laboratory fermentations. *Int. J. Food Microbiol., 86*, 181–188.

Romano, P., Fiore, C., Paraggio, M., Caruso, M., & Capece, A. (2003). Function of yeast species and strains in wine flavour. *Int. J. Food Microbiol., 86*, 169–180.

Romano, P., Suzzi, G., Comi, G., Zironi, R., & Maifreni, M. (1997). Glycerol and other fermentation products of apiculate wine yeasts. *J. Appl. Microbiol., 82*, 615–618.

Rosi, I., Vinella, M., & Domizio, P. (1994). Characterization of β-glucosidase activity in yeasts of oenological origin. *J. Appl. Bacteriol., 77*, 519–527.

Rupela, O. P., & Tauro, P. (1984). Isolation and characterization of low hydrogen sulfide producing wine yeast. *Enzyme Microb. Technol., 6*, 419–421.

Sabaté, J., Cano, J., Esteve-Zarzoso, B., & Guillamon, J. M. (2002). Isolation and identification of yeasts associated with vineyard and winery by RFLP analysis of ribosomal genes and mitocondrial DNA. *Microbiol. Res., 157*, 267–274.

Saha, B. C., & Bothast, R. J. (1996). Production, purification, and characterization of a highly glucose-tolerant novel β-glucosidase from Candida peltata. *Appl. Environ. Microbiol., 62*, 3165–3170.

Salo, P. (1970a). Determining the odor thresholds for some compounds in alcoholic beverages. *J. Food Sci., 35*, 95–99.

Salo, P. (1970b). Variability of odour thresholds for some compounds in alcoholic beverages. *J. Sci. Food Agric., 21*, 597–600.

Sancho, T., Giménez-Jurado, G., Malfeito-Ferreira, M., & Loureiro, V. (2000). Zymological indicators: A new concept applied to the detection of potential spoilage yeast species associated with fruit pulps and concentrates. *Food Microbiol., 17*, 613–624.

Schuller, P., Corte-Real, M., & Leao, C. (2000). A differential medium for the enumeration of the spoilage yeast *Zygosaccharomyces bailii* in wine. *J. Food Protect, 63*, 1570–1575.

Schütz, M., & Gafner, J. (1993). Analysis of yeast diversity during spontaneous and induced alcoholic fermentations. *J. Appl. Bacteriol., 75*, 551–558.

Serrat, M., Bermudez, R. C., & Villa, T. G. (2004). Polygalacturonase and ethanol production in *Kluyveromyces marxianus*: Potential use of polygalacturonase in foodstuffs. *Appl. Biochem. Biotechnol., 117*, 49–64.

Servili, M., Begliomini, A. L., Montedoro, G. G., Petruccioli, M., & Federico, F. (1990). Use of an endo-polygalacturonase preparation from *Cryptococcus albidus var. albidus* in manufacture of red wines. *Industrie delle Bevande., 19*, 332–336.

Shinohara, T., Kubodera, S., & Yanagida, F. (2000). Distribution of phenolic yeasts and production of phenolic off-flavours in wine fermentation. *J. Biosci. Bioeng., 90*, 90–97.

Soden, A., Francis, I. L., Oakey, H., & Henschke, P. A. (2000). Effects of co-fermentation with Candida stellata and Saccharomyces cerevisiae on the aroma and composition of Chardonnay wine. *Aust. J. Grape Wine Res., 6*, 21–30.

Soles, R. M., Ough, C. S., & Kunkee, R. E. (1982). Ester concentration differences in wine fermented by various species and strains of yeasts. *Am. J. Enol. Vitic., 33,* 94–98.

Spagna, G., Barbagallo, R. N., Palmeri, R., Restuccia, C., & Giudici, P. (2002). Properties of endogenous β-glucosidase of a *Pichia anomala* strain isolated from Sicilian musts and wines. *Enzyme Microb. Technol., 31,* 1036–1041.

Sponholz, W. R. (1993). Wine spoilage by microorganisms. In G. H. Fleet (Ed.), *Wine microbiology and biotechnology* (pp. 395–420). Chur, Switzerland: Harwood Academic Publishers.

Strauss, M. L. A., Jolly, N. P., Lambrechts, M. G., & van Rensburg, P. (2001). Screening for the production of extracellular hydrolytic enzymes by non-*Saccharomyces* wine yeasts. *J. Appl. Microbiol., 91,* 182–190.

Suárez, R., Suárez-Lepe, J. A., Morata, A., & Calderón, F. (2007). The production of ethylphenols in wine by yeasts of the genera *Brettanomyces* and *Dekkera*: A review. *Food Chem., 102,* 10–21.

Suomalainen, H., & Lehtonen, M. (1979). The production of aroma compounds by yeast. *J. Inst. Brew., 85,* 149–156.

Swiegers, J. H., Bartowsky, E. J., Henschke, P. A., & Pretorius, I. S. (2005). Yeast and bacteria modulation of wine aroma and flavour. In R. J. Blair, M. E. Francis, & I. S. Pretorius (Eds.), *Advances in wine science* (pp. 159–187). Urrbrae, Australia: The Australian Wine Research Institute.

Then, R., & Radler, F. (1971). Vergleichende untersuchung der acetaldehydbilding bei der aeroben vergärung von glucose bei verschiedenen stämmen von *Saccharomyces cerevisiae* und Saccharomyces carlsbergensis. *Mschr. Brauerei, 24,* 127–130.

Thongekkaew, J., Ikeda, H., Masaki, K., & Iefuji, H. (2008). An acidic and thermostable carboxymethyl cellulase from the yeast *Cryptococcus* sp. S-2: Purification, characterization and improvement of its recombinant enzyme production by high cell-density fermentation of Pichia pastoris. *Protein Express. Purif., 60,* 140–146.

Toro, M. E., & Vazquez, F. (2002). Fermentation behaviour of controlled mixed and sequential cultures of *Candida cantarellii* and *Saccharomyces cerevisiae* wine yeasts. *World J. Microbiol. Biotechnol., 18,* 347–354.

Úbeda, J., & Briones, A. (2000). Characterization of differences in the formation of volatiles during fermentation within synthetic and grape must by wild *Saccharomyces* strains. *Lebensm.-Wiss. Technol., 33,* 408–414.

Urso, R., Rantsiou, K., Dolci, P., Rolle, L., Comi, G., & Cocolin, L. (2008). Yeast biodiversity and dynamics during sweet wine production as determined by molecular methods. *FEMS Yeast Res., 8,* 1053–1062.

van Rensburg, P., & Pretorius, I. S. (2000). Enzymes in winemaking: Harnessing natural catalysts for efficient biotransformations. A review. *S. Afr. J. Enol. Vitic., 21,* 52–73.

van Vuuren, H. J. J., & van der Meer, L. (1987). Fingerprinting of yeasts by protein electrophoresis. *Am. J. Enol. Vitic., 38,* 49–53.

Vancanneyt, M., Pot, B., Hennebert, G., & Kersters, K. (1991). Differentiation of yeast species based on electrophoretic whole-cell protein patterns. *System. Appl. Microbiol., 14,* 23–32.

Velázquez, E., Cruz-Sánchez, J. M., Mateos, P. F., Monte, E., & Chordi, A. (1993). Prolewine: A program for winemaking yeast identification. *Binary-Comput. Microb., 5,* 58–61.

Velázquez, E., Cruz-Sánchez, J. M., Rivas-Palá, T., Zurdo-Piñeiro, J. L., Mateos, P. F., Monte, E., et al. (2001). YeastIdent-Food/ProleFood, a new system for the identification of food yeasts based on physiological and biochemical tests. *Food Microbiol., 18,* 637–646.

Viana, F., Gil, J. V., Genovés, S., Vallés, S., & Manzanares, P. (2008). Rational selection of non-*Saccharomyces* wine yeasts for mixed starters based on ester formation and enological traits. *Food Microbiol., 25,* 778–785.

Viana, F., Gil, J. V., Vallés, S., & Manzanares, P. (2009). Increasing the levels of 2-phenylethyl acetate in wine through the use of a mixed culture of Hanseniaspora osmophila and Saccharomyces cerevisiae. *Int. J. Food Microbiol., 135,* 68–74.

Yanai, T., & Sato, M. (1999). Isolation and properties of β-glucosidase produced by *Debaryomyces hansenii* and its application in winemaking. *Am. J. Enol. Vitic., 50,* 231–235.

Yanai, T., & Sato, M. (2000). Purification and characterization of a novel α-L-arabinofuranosidase from Pichia capsulata X91. *Biosci. Biotechnol. Biochem., 64,* 1181–1188.

Yanai, T., & Sato, M. (2001). Purification and characterization of a β-D-xylosidase from *Candida utilis* IFO 0639. *Biosci. Biotechnol. Biochem., 65,* 527–533.

Yarrow, D. (1998). Methods for the isolation, maintenance and identification of yeasts. In C. P. Kurtzman & J. W. Fell (Eds.), *The yeasts. A taxonomic study* (4th ed.). San Diego, CA: Elsevier Inc. pp. 77–100.

Zironi, R., Romano, P., Suzzi, G., Battistutta, F., & Comi, G. (1993). Volatile metabolites produced in wine by mixed and sequential cultures of *Hanseniaspora guilliermondii* or *Kloeckera apiculata* and Saccharomyces cerevisiae. *Biotechnol Lett., 15,* 235–238.

Zoecklein, B. W., Marcy, J. E., Williams, J. M., & Jasinski, Y. (1997). Effect of native yeasts and selected strains of *Saccharomyces cerevisiae* in glycosyl glucose, potential volatile terpenes, and selected aglycones of White Riesling (*Vitis vinifera* L.) wines. *J. Food Composition Anal., 10,* 55–65.

Zohre, D. E., & Erten, H. (2002). The influence of *Kloeckera apiculata* and *Candida pulcherrima* yeasts on wine fermentation. *Process Biochem., 38,* 319–324.

CHAPTER

# 5

# Molecular Identification and Characterization of Wine Yeasts

M. Teresa Fernández-Espinar[1], Silvia Llopis[1], Amparo Querol[1], Eladio Barrio[2]

[1] Departamento de Biotecnología, Instituto de Agroquímica y Tecnología de Alimentos (CSIC), Valencia, Spain and [2] Unitat de Genètica Evolutiva, Institut "Cavanilles" de Biodiversitat i Biologia Evolutiva, Universitat de València, Edificio de Institutos, Valencia, Spain

OUTLINE

## 1. INTRODUCTION

The transformation of grape must into wine is a complex microbiological process involving yeasts and lactic acid bacteria, though only yeasts of the genus *Saccharomyces* (principally *Saccharomyces cerevisiae*) are responsible for alcoholic fermentation. Wine has traditionally been produced by natural fermentation caused by the growth of yeasts derived from the grapes and winery environment. The composition of the microflora on the surface of the grape is affected by a variety of factors, including temperature, rainfall, and other climatic variables (Longo et al., 1991; Querol et al., 1990); the ripeness of the crop (Martínez et al., 1989; Rosini et al., 1982); the use of fungicides (Bureau et al., 1982); physical damage caused by fungi, insects, etc. (Longo et al., 1991); and the grape variety. The surfaces of winery equipment (presses, tanks, fermenters, pumps, etc.) are another source of yeasts as they come into contact with the grape must. Apiculate yeasts of the genera *Kloeckera* and *Hanseniaspora* (predominant species on the surface of the grape that account for 50% to 70% of the total yeast population) and anaerobic yeast of the genera *Candida* (*Candida stellata* and *Candida pulcherrima*), *Cryptococcus*, *Hansenula*, *Kluyveromyces*, *Pichia*, and *Rhodotorula* grow during the initial phases of fermentation, but increasing alcohol concentration and anaerobic conditions later favor the growth of yeasts belonging to the genus *Saccharomyces*, specifically *S. cerevisiae*, which are responsible for alcoholic fermentation. Although *S. cerevisiae* is only found at low levels on grapes, it multiplies rapidly and displaces other microorganisms present in the grape must. As a result of its ability to tolerate high concentrations of alcohol and to thrive at higher temperatures than other yeasts, *S. cerevisiae* comes to dominate the fermentation environment. During vinification, *Saccharomyces* strains can survive and continue fermentation at temperatures of up to 38°C, whereas growth of most of the flora present in the grape must is inhibited at temperatures close to 25°C. Despite these characteristics of *S. cerevisiae*, fermentation often stops at high temperatures.

Although *S. cerevisiae* is the most common species in wine fermentations and has been the subject of most of the studies performed to date, other species belonging to the so-called *Saccharomyces sensu stricto* complex, due to their phylogenetic proximity to *S. cerevisiae*, may also be present during alcoholic fermentation and even become the predominant species. For instance, *Saccharomyces bayanus* predominates in wines from regions with a continental climate and *Saccharomyces paradoxus* has been described recently to predominate in Croatian wines (Redzepovic et al., 2002). Furthermore, although *Saccharomyces* species form the majority of the flora resident in the winery (Fleet & Heard, 1993; Martini & Vaughan-Martini, 1990), species belonging to the genera *Brettanomyces*, *Candida*, *Hansenula*, and *Pichia* have also been isolated in the winery environment and in finished wines. These may be responsible for organoleptic changes that result in wine spoilage, as has been observed for species belonging to the genera *Pichia* and *Brettanomyces* (Dias et al., 2003).

Many of the studies designed to identify different species of yeast and different strains within the same species have been based on morphological and physiological criteria (Barnett et al., 1990; Kreger-van Rij, 1984). Examples of the characteristics that distinguish the main yeast species are shown in Table 5.1. These characteristics can vary according to the culture conditions (Scheda & Yarrow, 1966, 1968; Yamamoto et al., 1991), and species are sometimes defined by a single physiological character that may even be controlled by a single gene. Consequently, identification depends upon the physiological state of the yeast. An example is seen in galactose fermentation, which has traditionally been used by enologists to differentiate between the species *S. cerevisiae* and *S. bayanus* (Kurtzman & Phaff, 1987; Price et al., 1978). More recently, methods have been developed to differentiate between yeasts based

on analysis of total cell proteins (Vacanneyt et al., 1991; van Vuuren & van der Meer, 1987), isoenzyme profiles (Duarte et al., 1999), and analysis of fatty acids by gas chromatography (Cottrell et al., 1986; Moreira da Silva et al., 1994; Tredoux et al., 1987). The reproducibility of these techniques, however, is somewhat questionable, since in many cases they depend on the physiological state of the yeasts (Golden et al., 1994). Molecular biological techniques circumvent these difficulties by allowing direct analysis of the genome, irrespective of the physiological state of the cell. Many such techniques have now been developed and successfully applied to the identification and molecular characterization of yeasts. In this chapter, we describe the main techniques that have been used for the analysis of wine yeasts. The principles on which some of them are based have been described previously (Giudici & Pulvirenti, 2002). Although some molecular studies of non-*Saccharomyces* wine yeasts have been performed (see Tables 5.1–5.8), research has focused mainly on yeasts of the genus

**TABLE 5.1** The Most Relevant Physiological and Morphological Characteristics for the Identification of Predominant Species in Winemaking

| | Morphology | Assimilation | | | | | | | Fermentation | | | | | | |
|---|---|---|---|---|---|---|---|---|---|---|---|---|---|---|---|
| | | Gal | Glc | Lac | Mal | Raf | Sac | Tre | Gal | Glc | Lac | Mal | Raf | Sac | Tre |
| *Candida stellata* | Globose/ovoid | − | + | − | − | +/− | + | − | − | + | − | − | + | + | − |
| *Dekkera bruxellensis* | Ellipsoid − oval and elongated | − | + | − | + | − | + | | − | | | + | +w | + | + |
| *Hansiniaspora uvarum* | Lemon-shaped | − | + | − | − | | − | | − | | | − | | − | |
| *Metschnikowia pulcherrima* | Globose/ellipsoid | + | | − | + | − | + | + | +/− | + | − | − | − | − | |
| *Saccharomyces cerevisiae* | Ovoid | v | | − | v | v | v | v | v | + | − | v | | v | |
| *Zygosaccharomyces bailli* | Ellipsoid/ovoid | v | | − | − | −(+) | v | v | − | + | − | − | | v | |

+ = positive; − = negative; v = variable; Gal = galactose; Glc = glucose; Lac = lactose; Mal = maltose; Raf = raffinose; Sac = saccharose; Tre = trehelose.

**TABLE 5.2**    Studies Using Restriction Analysis of the 5.8S-ITS Ribosomal DNA Region for Species Identification

| Species studied | Reference | Application | Observations |
|---|---|---|---|
| Non-*Saccharomyces* yeasts; *Saccharomyces cerevisiae* | Constantí et al. (1998) | Population dynamics, controlled fermentation (effect of sulfite dioxide and inoculum) | Other techniques used: mtDNA |
| Non-*Saccharomyces* yeasts; *S. cerevisiae* | Guillamón et al. (1998) | Identification of collection strains | |
| Non-*Saccharomyces* yeasts; *S. cerevisiae* | Granchi et al. (1999) | Ecological study of spontaneous fermentation | |
| *S. cerevisiae flor* | Fernández-Espinar et al. (2000) | Identification of collection strains (from the *flor* of Jerez wines) | |
| Non-*Saccharomyces* yeasts; *S. cerevisiae* | Pramateftaki et al. (2000) | Ecological study in natural fermentation | Other techniques used: mtDNA; δ elements |
| Non-*Saccharomyces* yeasts; *S. cerevisiae* | Torija et al. (2001) | Population dynamics in a natural fermentation | Other techniques used: mtDNA |
| Non-*Saccharomyces* yeasts; *S. cerevisiae* | Esteve-Zarzoso et al. (2001) | Population study in Jerez wines | Other techniques used: mtDNA; karyotyping |
| Non-*Saccharomyces* yeasts; *S. cerevisiae* | Sabate et al. (2002) | Identification of species associated with vines and wineries | Other techniques used: mtDNA |
| Non-*Saccharomyces* yeasts | Ganga and Martínez (2004) | Ecological study of controlled fermentations | |
| *S. cerevisiae* | Capello et al. (2004) | Ecological study in natural fermentations | Other techniques used: mtDNA; δ elements |
| Non-*Saccharomyces* yeasts; *S. cerevisiae* | Rodríguez et al. (2004) | Ecological study in natural fermentation (analysis of β-glucosidase activity) | Other techniques used: mtDNA, karyotyping |
| Non-*Saccharomyces* yeasts; *S. cerevisiae* | González et al. (2007) | Ecological study in natural fermentations | Other techniques used: mtDNA |
| Non-*Saccharomyces* yeasts; *S. cerevisiae* | Lópes et al. (2007) | Analysis of culture establishment (effect on non-*Saccharomyces flora*) | Other techniques used: mtDNA |
| Non-*Saccharomyces* yeasts | Romancino et al. (2008) | Ecological study in natural fermentation | Other techniques used: restriction analysis of 26S rRNA |
| Non-*Saccharomyces* yeasts | Zott et al. (2008) | Population dynamics | Other techniques used: sequencing of 5.8S-ITS rDNA |
| Non-*Saccharomyces* yeasts; *S. cerevisiae* | Tofalo et al. (2009) | Population dynamics in a natural fermentation | Other techniques used: sequencing of D1/D2; RAPD (strain level) |

ITS = internal transcribed spacer; mtDNA = mitochondrial DNA; RAPD = random amplification of polymorphic DNA; rDNA = ribosomal DNA.

**TABLE 5.3**   Studies Using Hybridization to Characterize *Saccharomyces cerevisiae* Strains

| Species studied | Reference | Application | Observations |
|---|---|---|---|
| *S. cerevisiae* | Degré et al. (1989) | Characterization of commercial strains | Other techniques used: karyotyping |
| *S. cerevisiae* | Querol et al. (1992a) | Comparative study of characterization techniques | Development of a new method for analysis of mtDNA<br>Other techniques used: karyotyping |
| *S. cerevisiae* | Lieckfeldt et al. (1993) | Development of a PCR technique for characterization | Validation of the PCR technique by comparison with hybridization |
| *S. cerevisiae* | Lavallée et al. (1994) | Quality control in the production of commercial yeasts | Other techniques used: δ elements |
| *S. cerevisiae flor* | Ibeas and Jiménez (1996) | Analysis of chromosome polymorphism | |
| *S. cerevisiae* | Nadal et al. (1999) | Analysis of chromosome polymorphism in sparkling wines | |

mtDNA = mitochondrial DNA.

*Saccharomyces*, particularly *S. cerevisiae*, as a result of their importance in the winemaking process. Some of these studies have concluded that a combination of techniques is required for definitive characterization of individual strains (Baleiras Couto et al., 1996; Fernández-Espinar et al., 2001; Pramateftaki et al., 2000).

## 2. METHODS FOR SPECIES IDENTIFICATION

### 2.1. Methods Based on Analysis of Ribosomal DNA (rDNA)

Ribosomal genes (5.8S, 18S, and 26S) are grouped in tandem to form transcription units that are repeated 100 to 200 times throughout the genome. Each transcription unit contains another two regions, the internal transcribed spacer (ITS) and the external transcribed spacer (ETS), both of which are transcribed but not processed. The coding regions are separated by intergenic spacers (IGSs), also known as non-transcribed spacers (NTSs). Although the 5S

gene is not part of the transcription unit, in yeasts it is located adjacent to it. The sequence conservation and concerted evolution of the 5.8S, 18S, and 26S ribosomal genes and the ITS and NTS spacers means that the similarity between repeated transcription units within a given species is greater than between units from different species. This sequence similarity within species—which arises through mechanisms such as unequal crossing-over and gene conversion (Li, 1997)—makes these ribosomal DNA (rDNA) regions powerful tools with which to identify species and establish phylogenetic relationships between them (Kurztman & Robnett, 1998).

Various methods have been developed to identify yeast species based on information contained within these regions, as described below.

#### 2.1.1. Sequencing of Ribosomal DNA (rDNA)

Yeast species can be identified by comparison of nucleotide sequences from rDNA regions. The two most commonly used regions are the

**TABLE 5.4**    Studies Using Electrophoretic Separation of Chromosomes or Electrophoretic Karyotyping for Strain Characterization

| Species studied | Reference | Application | Observations |
|---|---|---|---|
| *Saccharomyces cerevisiae* | Blondin and Vezinhet (1988) | Characterization of commercial strains | |
| *S. cerevisiae* | Degré et al. (1989) | Characterization of commercial strains | Other techniques used: hybridization of total DNA |
| *S. cerevisiae* | Vezinhet et al. (1990) | Characterization of commercial strains | Other techniques used: mtDNA |
| *S. cerevisiae* | Yamamoto et al. (1991) | Characterization of commercial and collection strains | |
| *S. cerevisiae* | Bidenne et al. (1992) | Chromosomal polymorphism | |
| *S. cerevisiae* | Querol et al. (1992a) | Comparative study of characterization techniques | Development of a new method for analysis of mtDNA. Other techniques used: hybridization |
| *S. cerevisiae* | Frezier and Dubourdieu (1991) | Ecological study in natural fermentation | |
| *S. cerevisiae* | van der Westhuizen and Pretorius (1992) | Genetic improvement | |
| Non-*Saccharomyces* yeasts; *S. cerevisiae* | Schütz and Gafner (1993) | Population dynamics in controlled and natural fermentations | Strain and species level |
| *S. cerevisiae* | Grando et al. (1994) | Comparison of molecular techniques | Other techniques used: RAPD |
| *S. cerevisiae* | Schütz and Gafner (1994) | Population dynamics in natural fermentations | |
| *S. cerevisiae* flor | Martínez et al. (1995) | Population dynamics during aging of Jerez wines | Other techniques used: mtDNA |
| *S. cerevisiae* | Versavaud et al. (1995) | Ecological study and analysis of geographical distribution in natural fermentations | Other techniques used: mtDNA; δ elements |
| *S. cerevisiae* | Briones et al. (1996) | Ecological study of controlled fermentation | |
| *S. cerevisiae* | Nadal et al. (1996) | Ecological study in natural fermentation | Other techniques used: mtDNA |
| *S. cerevisiae* | Egli et al. (1998) | Population dynamics of natural and controlled fermentations | Other techniques used: δ elements |
| *S. cerevisiae* flor | Mesa et al. (1999) | Ecological study during aging of Jerez wines | Other techniques used: mtDNA |

*(Continued)*

**TABLE 5.4** Studies Using Electrophoretic Separation of Chromosomes or Electrophoretic Karyotyping for Strain Characterization—*cont'd*

| Species studied | Reference | Application | Observations |
|---|---|---|---|
| *S. cerevisiae*; *S. cerevisiae flor* | Esteve-Zarzoso et al. (2001) | Population dynamics in Jerez wines | Other techniques used: 5.8S-ITS (species level) mtDNA |
| *S. cerevisiae* | Fernández-Espinar et al. (2001) | Authentication of commercial yeasts | Other techniques used: mtDNA; δ elements |
| *S. bayanus* | Naumov et al. (2002) | Identification in Hungarian wines | Species level |
| *S. cerevisiae* | Martínez et al. (2004) | Geographic origin of native isolates | Other techniques used: mtDNA |
| *S. cerevisiae* | Martínez et al. (2004) | Geographic origin of native isolates | Other techniques used: mtDNA |
| Non-*Saccharomyces* yeasts; *S. cerevisiae* | Rodríguez et al. (2004) | Ecological study in natural fermentation (analysis of β-glucosidase activity) | Other techniques used: 5.8S-ITS (species level); mtDNA |
| *S. cerevisiae* | Schuller et al. (2004) | Characterization of commercial strains | Other techniques used: mtDNA; δ elements; microsatellites |

ITS = internal transcribed spacer; mtDNA = mitochondrial DNA; RAPD = random amplification of polymorphic DNA.

D1 and D2 regions at the 5′ end of the genes encoding the 26S (Kurtzman & Robnett, 1998) and 18S (James et al., 1997) ribosomal subunits. The availability of sequences in DNA databases, particularly for the D1/D2 region of the 26S gene, makes this technique particularly useful for assigning an unknown yeast to a specific species when the homology of the sequences is greater than 99% (Kurtzman & Robnett, 1998). Sequence comparison with the DNA databases is performed using the program WU-BLAST2, available from http://www.ebi.ac.uk/Blas2/index.html.

The use of direct sequencing of the regions of interest by polymerase chain reaction (PCR) has been combined with technology for automated sequencing to make this a relatively rapid technique. In this process, the target region is amplified by PCR from a total DNA template. The PCR products are purified using commercial kits to remove the primers and excess deoxynucleotides that interfere with the sequencing reaction. During automated sequencing, four fluorescent dyes are used to label the bases (A, G, C, and T). Dye incorporation is carried out by a second round of PCR amplification using the same primers. Fine capillaries are then used to separate the labeled DNA fragments according to size. Laser excitation of the dyes results in emission of a signal at a specific wavelength, and software can be used to transform the signals into peaks, with each color corresponding to a nucleotide. This process is rapid and allows approximately 600 nucleotides to be read in 2 or 3 h, depending on the model of the sequencer.

Given the technological advances that have been made and the widespread availability of sequencing data via the Internet, sequencing has become an extremely useful tool that complements the other molecular techniques described in this chapter.

**TABLE 5.5**   Studies Using Restriction Analysis of Mitochondrial DNA for Strain Characterization

| Species studied | Reference | Application | Observations |
|---|---|---|---|
| *Saccharomyces cerevisiae* | Vezinhet et al. (1990) | Characterization of commercial strains | Other techniques used: karyotyping |
| *S. cerevisiae* | Querol et al. (1992a) | Comparative study of characterization techniques | Development of a new method for analysis of mtDNA. Other techniques used: karyotyping; hybridization |
| *S. cerevisiae* | Querol et al. (1992b) | Monitoring establishment of yeasts in controlled fermentations | |
| *S. cerevisiae* | Querol et al. (1994) | Population dynamics in a natural fermentation | |
| | Martínez et al. (1995) | Population dynamics during aging of Jerez wines | Other techniques used: karyotyping |
| *S. cerevisiae; S. cerevisiae flor* | Versavaud et al. (1995) | Ecological study and analysis of geographical distribution in natural fermentations | Other techniques used: δ elements; karyotyping |
| *S. cerevisiae* | Guillamón et al. (1996) | Geographical study | Other techniques used: karyotyping |
| *S. cerevisiae* | Nadal et al. (1996) | Ecological study in natural fermentation | Other techniques used: karyotyping |
| *S. cerevisiae* | Gutiérrez et al. (1997) | Ecological study in spontaneous and controlled fermentations | |
| *S. cerevisiae* | Constantí et al. (1998) | Population dynamics in a controlled fermentation (effect of sulfur dioxide and inoculation) | Other techniques used: 5.8S-ITS (species level) |
| *S. cerevisiae* | Sabaté et al. (1998) | Ecological study in natural fermentation | |
| *S. cerevisiae flor* | Mesa et al. (1999) | Molecular characterization | Other techniques used: karyotyping |
| *S. cerevisiae; Saccharomyces bayanus* var. *uvarum* | Torriani et al. (1999) | Ecological study in natural fermentation | Other techniques used: RAPD |
| *S. cerevisiae* | Comi et al. (2000) | Diversity and geographic distribution on grapes | |
| *S. cerevisiae* | Pramateftaki et al. (2000) | Ecological study in natural fermentation | Other techniques used: 5.8S-ITS (species level) |
| *S. cerevisiae; S. cerevisiae flor* | Esteve-Zarzoso et al. (2001) | Population dynamics in Jerez wines | Other techniques used: 5.8S-ITS (species level); karyotyping |

*(Continued)*

**TABLE 5.5** Studies Using Restriction Analysis of Mitochondrial DNA for Strain Characterization—*cont'd*

| Species studied | Reference | Application | Observations |
|---|---|---|---|
| *S. cerevisiae* | Fernández-Espinar et al. (2001) | Authentication of commercial yeasts | Other techniques used: karyotyping, δ elements |
| *S. cerevisiae* | Fernández-González et al. (2001) | Population dynamics in a controlled fermentation | Other techniques used: PCR-TTGE |
| *S. cerevisiae* | Torija et al. (2001) | Population dynamics in a natural fermentation | Other techniques used: 5.8S-ITS (species level) |
| *S. cerevisiae* | Beltrán et al. (2002) | Ecological study of controlled fermentation | Other techniques used: 5.8S-ITS (species level) |
| *S. cerevisiae* | Lópes et al. (2002) | Ecological study and analysis of geographical distribution in natural fermentations | Other techniques used: δ elements |
| *S. cerevisiae* | Sabaté et al. (2002) | Genetic diversity of strains on the vine and in the winery | Other techniques used: 5.8S-ITS (species level) |
| *S. cerevisiae* | Granchi et al. (2003) | Ecological study of a natural fermentation (effect of nitrogen) | |
| *S. cerevisiae* | Torija et al. (2003) | Population dynamics in the laboratory | |
| *S. cerevisiae* | Capello et al. (2004) | Ecological study in natural fermentations | Other techniques used: 5.8S-ITS; δ elements |
| *S. cerevisiae (flor)* | Esteve-Zarzoso et al. (2001) | Authentication and identification of natural and collection isolates from Jerez wine *flor* | |
| *S. cerevisiae* | Martínez et al. (2004) | Geographic origin of native isolates | Other techniques used: karyotyping |
| Non-*Saccharomyces* yeasts; *S. cerevisiae* | Rodríguez et al. (2004) | Ecological study in natural fermentation (analysis of β-glucosidase activity) | Other techniques used: 5.8S-ITS; karyotyping |
| *S. cerevisiae* | Schuller et al. (2004) | Characterization of commercial yeasts | Other techniques used: δ elements; karyotyping; microsatellites |
| *S. cerevisiae* | Lópes et al. (2006) | Biodiversity study | |
| Non-*Saccharomyces* yeasts; *S. cerevisiae* | González et al. (2007) | Ecological study in natural fermentations | Other techniques used: 5.8S-ITS |
| Non-*Saccharomyces* yeasts; *S. cerevisiae* | Lópes et al. (2007) | Analysis of yeast establishment | Other techniques used: 5.8S-ITS |

mtDNA = mitochondrial DNA; ITS = internal transcribed spacer; PCR-TTGE = polymerase chain reaction-temperature gradient gel electrophoresis; RAPD = random amplification of polymorphic DNA; rDNA = ribosomal DNA.

## 2.1.2. Restriction Analysis of Ribosomal DNA (rDNA)

In an effort to develop techniques for use in industrial applications, other simpler methods have been designed based on PCR amplification of rDNA regions followed by restriction analysis of the amplified products. The principles of PCR are described in Section 3.4. Various studies have used small quantities of isolated yeast colonies added to the PCR reaction directly without prior purification of DNA. In this approach, an initial incubation step at 95°C for 15 min is included in the amplification protocol to release the DNA into the reaction mixture. Thus, by removing the requirement for DNA purification, the time required is reduced substantially. The amplification products are visualized following electrophoresis in 1.4% agarose gels. Differently sized amplification products correspond to different species; when the amplicons are of the same size, however, they do not always correspond to the same species, and digestion of these fragments with restriction enzymes is required for definitive identification. Digestion of PCR products is performed directly in the reaction mixture without prior purification. The fragments generated are separated by electrophoresis in 3% agarose gels and their size is assessed by comparison with appropriate DNA markers. This technique is both uncomplicated and reproducible. Dlauchy et al. (1999) used this method to amplify the 18S ribosomal gene and the ITS1 intergenic region from 128 species associated mainly with foodstuffs, wine, beer, and soft drinks using the primers NS1 (5'-GTA GTC ATA TGC TTG TCT C-3') and its2 (5'-GCT GCG TTC TTC ATC GAT GC-3') and digesting the PCR products with AluI, HaeIII, MspI, and RsaI. This method was later used by Redzepovic et al. (2002). Another very useful rDNA region that can be used to differentiate between species is that containing the 5.8S gene and the adjacent intergenic regions ITS1

and ITS2, which are amplified using the primers its1 (5'-TCC GTA GGT GAA CCT GCG G-3') and its4 (5'-TCC TCC GCT TAT TGA TAT GC-3'), described by White et al. (1990). This technique was used by Guillamón et al. (1998) for the rapid identification of wine yeasts, and was later extended to 191 yeasts associated with foodstuffs and beverages (de Llanos et al., 2004; Esteve-Zarzoso et al., 1999; Fernández-Espinar et al., 2000). The amplified fragments and restriction profiles for these species with HaeIII, HinfI, CfoI, and DdeI are available online at http://yeast-id.com/. The technique has been used in numerous studies for the identification of wine yeasts (see Table 5.2).

Restriction analysis of other rDNA regions has also been used to identify other yeast species, particularly those belonging to the Saccharomyces sensu stricto complex. This is the case for the NTS region (Baleiras Couto et al., 1996; Capece et al., 2003; Caruso et al., 2002; Nguyen & Gaillardin, 1997; Pulvirenti et al., 2000), the 18S gene (Capece et al., 2003), and various domains of the 26S gene (Baleiras Couto et al., 1996, 2005; Romancino et al., 2008; Smole-Mozina et al., 1997; van Keulen et al., 2003). However, the absence of a sequence database for these regions means that their use cannot be generalized for the identification of yeasts.

## 2.2. Real-time Polymerase Chain Reaction (PCR)

Real-time PCR was developed in 1996 and since then its use has increased almost exponentially across a range of applications (Wilhelm & Pingoud, 2003). In this technique, the appearance of the amplification products is monitored during each PCR cycle. It is based on the detection and quantification of a signal generated by a fluorescent donor dye. The signal is in direct proportion to the quantity of PCR product in the reaction. The process is carried out in a thermocycler coupled to a detector that can acquire

**TABLE 5.6** Studies Using Random Amplification of Polymorphic DNA for Strain Characterization

| Species studied | Reference | Application | Observations |
|---|---|---|---|
| *Saccharomyces cerevisiae; Zygosaccharomyces* species | Baleiras Couto et al. (1994) | Identification of spoilage yeasts in alcoholic beverages | Species level |
| *S. cerevisiae* | Grando et al. (1994) | Comparison of molecular techniques | Other techniques used: karyotyping |
| *S. cerevisiae* | Quesada and Cenis (1995) | Identification | Genus, species, and strain level |
| *S. cerevisiae* | Baleiras Couto et al. (1996) | Characterization of wine spoilage strains | Other techniques used: 5.8S-ITS (species level); microsatellites |
| *S. cerevisiae; Saccharomyces bayanus* var. *uvarum* | Torriani et al. (1999) | Ecological study in natural fermentation | Other techniques used: mtDNA |
| *S. cerevisiae* | Echeverrigaray et al. (2000) | Characterization of commercial strains | |
| *S. cerevisiae* | Pérez et al. (2001a) | Comparative study of molecular techniques | Other techniques used: microsatellites; CAPS |
| Non-*Saccharomyces* yeasts; *S. cerevisiae* | Lopandic et al. (2008) | Population study of natural fermentation | Species level. Other techniques used: sequencing of D1/D2; AFLP (*Saccharomyces* strains) |
| *S. cerevisiae* | Urso et al. (2008) | Analysis of establishment | Other techniques used: PCR-DGGE (biodiversity study) |
| Non-*Saccharomyces* yeasts; *S. cerevisiae* | Tofalo et al. (2009) | Population dynamics in a natural fermentation | Other techniques used: sequencing of D1/D2 (species level); 5.8S-ITS (species level) |

AFLP = amplified fragment length polymorphism; CAPS = cleaved amplified polymorphic sequence (digestion of RAPD products); ITS = internal transcribed spacer; mtDNA = mitochondrial DNA; PCR-DGGE = polymerase chain reaction-denaturing gel gradient electrophoresis.

and quantify the signal emitted by the donor in each sample at the end of each cycle. The data obtained are represented as an amplification curve with the point at which the intensity of the signal from the donor becomes greater than the background noise indicated. This is known as the threshold cycle ($C_t$) and it is inversely proportional to the number of copies of the target sequence in the sample (DNA or cells). Consequently, it can be used to assess the starting quantity of target DNA with a high degree of accuracy over a wide range of concentrations.

The fluorescent signal may be derived from intercalating agents or probes. The intercalating agent SYBR green binds to double-stranded DNA, leading to an increase in fluorescence with increasing amounts of PCR product. Three types of probe can be used: hydrolysis probes, hairpin probes, and hybridization probes. The most widely used hydrolysis probe is the Taqman probe, which has both donor and acceptor fluorochromes. When both fluorochromes are bound to the probe, the donor does not emit a signal. When the probe is bound to a sequence

**TABLE 5.7**    Strain Characterization Studies Using Polymerase Chain Reaction (PCR) Amplification of Variable Regions of the Genome (Microsatellites and Minisatellites)

| Species studied | Reference | Application | Observations |
|---|---|---|---|
| *Saccharomyces cerevisiae* | Lieckfeldt et al. (1993) | Development of a characterization technique | Validation by comparison with hybridization techniques. Visualization in agarose gels |
| *S. cerevisiae* | Baleiras Couto et al. (1996) | Characterization of wine spoilage strains | Other techniques used: 5.8S-ITS; RAPD |
| *S. cerevisiae* | Gallego et al. (1998) | Development of a characterization technique | Visualization in an automatic sequencer |
| *S. cerevisiae* | Hennequin et al. (2001) | Development of a characterization technique (wine, beer, baking, and clinical isolates) | Visualization using radioactivity following acrylamide gel electrophoresis |
| *S. cerevisiae* | Pérez et al. (2001a) | Comparative study of molecular techniques | Other techniques used: microsatellites; CAPS |
| *S. cerevisiae* | Pérez et al. (2001b) | Development of a characterization technique | Visualization in an automatic sequencer |
| *S. cerevisiae* | González Techera et al. (2001) | Development of a characterization technique (commercial strains) | Visualization by silver staining of acrylamide gels |
| *Kloeckera apiculata; S. cerevisiae* | Caruso et al. (2002) | Ecological study in natural fermentation | Microsatellites (based on Baleiras Couto et al., 1996). Other techniques used: amplification and restriction of NTS region |
| Non-*Saccharomyces* yeasts | Capece et al. (2003) | Analysis of effectiveness of the technique in non-*Saccharomyces* strains | Microsatellites (based on Baleiras Couto et al., 1996. Visualization in agarose gels.). Other techniques used: amplification and restriction of 18S gene; amplification and restriction of NTS region |
| *S. cerevisiae* | Howell et al. (2004) | Monitoring of laboratory fermentations | Based on the protocol of González Techera et al. (2001) |
| *S. cerevisiae* | Marinangeli et al. (2004) | Development of a characterization technique | Visualization on agarose gels |
| *S. cerevisiae* | Shuller et al. (2004) | Characterization of commercial strains | Microsatellites (based on Pérez et al., 2001b). Other techniques used: mtDNA, δ elements, karyotyping |
| *S. cerevisiae* | Ayoub et al. (2006) | Biodiversity of natural wine isolates | |

*(Continued)*

**TABLE 5.7**  Strain Characterization Studies Using Polymerase Chain Reaction (PCR) Amplification of Variable Regions of the Genome (Microsatellites and Minisatellites)—*cont'd*

| Species studied | Reference | Application | Observations |
|---|---|---|---|
| *S. bayanus* var. *uvarum* | Masneuf-Pomarède et al. (2007) | Characterization of natural wine isolates | |
| *S. cerevisiae* | Richards et al. (2009) | Characterization of commercial strains and natural isolates | Creation of a database on 246 genotypes |

AFLP = amplified fragment length polymorphism; CAPS = cleaved amplified polymorphic sequence (digestion of RAPD products); ITS = internal transcribed spacer; mtDNA = mitochondrial DNA; NTS = non-transcribed spacer; RAPD = random amplification of polymorphic DNA.

of interest during the PCR reaction, the exonuclease activity of *Taq* polymerase activates the donor fluorochrome in the rest of the probe, leading to emission of a fluorescent signal. This signal is monitored as it accumulates during successive PCR cycles. Hairpin probes (Molecular Beacons, Scorpions) contain inverted tandem repeats (ITRs) at their 5′ and 3′ ends. In the absence of the target sequence, this design allows them to form a stem-loop structure through sequence complementarity between the two ITR regions. When the probe is bound to the target DNA sequence, the separation of the fluorochromes results in fluorescence. Finally, hybridization probes consist of two probes, a donor and an acceptor, both designed to bind to the amplified region and each labeled with a fluorophore. Resonance energy transfer only occurs when both probes are bound to the target DNA in close proximity. The choice of which of these fluorescence systems to use is influenced by the advantages and disadvantages that they each present. For instance, SYBR green is most appropriate if a simple, cheap, and easy-to-use system is required. However, during the PCR reaction it can bind primer dimers and other nonspecific products and lead to overestimation of the concentration of target DNA. The need for greater specificity calls for the use of a system involving probes.

Real-time PCR has a number of advantages over other identification techniques. It is highly specific and sensitive, quantitative, and does not require additional analyses such as electrophoresis following PCR. The lack of requirement for additional procedures and the shorter reaction times and amplification cycles make real-time PCR a very rapid technique. This is particularly useful for routine analysis and applications requiring corrective measures. Nevertheless, despite all the advantages offered by this type of system, designing the probes and primers is very demanding, since it is this that determines the specificity and sensitivity of the method. Software is available to help in the design of appropriate primers and probes for use in real-time PCR. These are normally designed based on sequence data for genes or genomic regions that have demonstrated effectiveness for the establishment of phylogenetic relationships between yeast species. These sequences also have the advantage of being easily available via the Internet. Specifically, they correspond to the ITS (James et al., 1996) and D1/D2 (Kurztman & Robnett, 1998) rDNA regions, the mitochondrial gene COX2 (Belloch et al., 2000; Kurztman & Robnett, 2003), and the nuclear gene actin (Daniel & Meyer, 2003). These have been applied in real-time PCR systems developed for the detection and quantification of total yeasts in wine (Hierro et al., 2006a) and for the monitoring of populations of *Saccharomyces* species and *Hanseniaspora* species during alcoholic fermentation (Hierro et al., 2007). Occasionally, the differences in the nucleotide sequence between some species are insufficient to allow design of primers, and molecular

**TABLE 5.8**   Studies Using Polymerase Chain Reaction (PCR) Amplification of δ Elements for *Saccharomyces cerevisiae* Strain Characterization

| Species studied | Reference | Application | Observations |
|---|---|---|---|
| *S. cerevisiae* | Ness et al. (1993) | Development of a novel method for characterization of wine strains | Specific for *S. cerevisiae* strains |
| *S. cerevisiae* | Lavallée et al. (1994) | Quality control in the production of commercial yeasts | Other techniques used: hybridization of total DNA |
| *S. cerevisiae* | Versavaud et al. (1995) | Ecological study and analysis of geographical distribution in natural fermentations | Other techniques used: mtDNA; karyotyping |
| *S. cerevisiae* | Egli et al. (1998) | Population dynamics in spontaneous and controlled fermentations | Other techniques used: karyotyping |
| *S. cerevisiae* | Pramateftaki et al. (2000) | Ecological study in natural fermentation | Other techniques used: mtDNA |
| *S. cerevisiae* | Fernández-Espinar et al. (2001) | Authentication of commercial yeasts | Other techniques used: mtDNA; karyotyping |
| *S. cerevisiae* | Lópes et al. (2002) | Ecology and geographical distribution study in spontaneous fermentation | Other techniques used: mtDNA |
| *S. cerevisiae* | Legras and Karst (2003) | Characterization of commercial yeasts | Optimization of the method described by Ness et al. (1993) |
| *S. cerevisiae* | Capello et al. (2004) | Ecological study in natural fermentations | Other techniques used: mtDNA; 5.8S-ITS (species level) |
| *S. cerevisiae* | Ciani et al. (2004) | Analysis of the origin of strains responsible for spontaneous fermentation | |
| *S. cerevisiae* | Schuller et al. (2004) | Characterization of commercial yeasts | Other techniques used: mtDNA; karyotyping; microsatellites |
| *S. cerevisiae* | Le Jeune et al. (2006) | Population dynamics in spontaneous fermentations | |

ITS = internal transcribed spacer; mtDNA = mitochondrial DNA.

markers with greater variability are needed. An approach of this type was used by Martorell et al. (2005) and more recently by Salinas et al. (2009) using the random amplification of polymorphic DNA (RAPD) technique to design specific probes for species of *S. cerevisiae*. Cloning and sequencing of a band obtained by RAPD allowed specific primers for the species to be designed.

## 2.3. Polymerase Chain Reaction (PCR)-denaturing Gradient Gel Electrophoresis (DGGE)

Recently, a genetic fingerprinting technique based on PCR amplification and denaturing gradient gel electrophoresis (DGGE) has been introduced into microbial ecology (Muyzer et al., 1993). This technique allows DNA fragments of the same length to be separated on the basis of sequence differences. Separation of DNA amplicons is based on the decreased electrophoretic mobility of a partially melted double-stranded DNA molecule in polyacrylamide gels containing a linear gradient of denaturing agents (a mixture of urea and formamide). DNA migration is retarded when the DNA strands dissociate at a specific concentration of denaturing agent. Complete strand separation is prevented by the presence of a high-melting-point domain, created by DNA amplification using particular groups of universal primers. A sequence containing guanines (G) and cytosines (C) is added to the 5' end of one of the PCR primers, coamplified, and thus introduced into the amplified DNA fragments.

A related technique is temperature gradient gel electrophoresis (TGGE), which is based on a linear temperature gradient for separation of DNA molecules. DNA bands in DGGE and TGGE profiles can be visualized using ethidium bromide or a more recent alternative, SYBR Green I. PCR fragments can be extracted from the gel and used in sequencing reactions for species identification.

Although the use of DGGE and TGGE in microbial ecology is still in its infancy, preliminary results are encouraging (Muyzer & Smalla, 1998). The methods have only recently been used, however, for yeast identification in wine fermentations (Andorrà et al., 2008; Cocolin et al., 2000; di Maro et al., 2007; Prakitchaiwattana et al., 2004; Renouf et al., 2007; Stringini et al., 2009; Urso et al., 2008).

# 3. METHODS TO DIFFERENTIATE BETWEEN SACCHAROMYCES CEREVISIAE STRAINS

## 3.1. Hybridization Techniques

A large proportion of the *S. cerevisiae* genome is not transcribed or translated and, therefore, does not contribute to the phenotype of the yeast. Although mutations in these noncoding DNA regions do not affect phenotype, they can eliminate or create restriction sites. These variations in restriction sites can be detected by hybridization of DNA probes corresponding to the affected regions. In this technique, the restriction fragments obtained by digestion of the DNA are separated on agarose gels and transferred to nylon or nitrocellulose membranes by Southern blotting prior to hybridization with the probes (Sambrook et al., 1989. A similar technique can be applied to chromosomes separated by pulsed-field gel electrophoresis (PFGE). The probes are labeled radioactively with $^{32}$P or non-radioactively with digoxigenin or biotin. The usefulness of this technique for the characterization of *S. cerevisiae* strains or strains from other yeast species has been demonstrated using probes against *PFK2*, *PY30*, and *PDC1*, which encode glycolytic enzymes (Seehaus et al., 1985); *TRP1* and *TRP3*, which code for enzymes involved in amino acid synthesis (Braus et al., 1985; Pedersen, 1983, 1985, 1986a, 1986b); and repetitive DNA regions such as the retrotransposons *Ty1* and *Ty2* (Walmsley et al., 1989). However, few studies have applied the technique in wine yeasts (see Table 5.3). Other repetitive DNA regions (Degré et al., 1989; Ibeas & Jiménez, 1996; Lavallée et al., 1994; Lieckfeldt et al., 1993; Nadal et al., 1999) and genes encoding metabolic proteins such as *Ura3* and *TRP1* (Querol et al., 1992a) have also been used as probes.

## 3.2. Pulsed-field Gel Electrophoresis (PFGE) of Chromosomes

In PFGE, the alternating application of two transverse electric fields means that the chromosomes are continually forced to change the direction of their migration. As a result, large fragments of DNA are no longer detained in the agarose gel matrix and can be separated. The yeasts are grown in liquid medium and then combined with molten agarose and placed in small molds. The cells are then lysed in situ and the released DNA is immobilized in the agarose matrix. The blocks are inserted into agarose gels, which are then exposed to electric fields. The parameters that determine the resolution of the bands are the electric-field switching intervals, the agarose concentration, the temperature, and the angle between the electric fields.

This method of karyotype analysis has been demonstrated to be highly efficient for the differentiation of *S. cerevisiae* strains. The polymorphism revealed is the result of the addition or elimination of long fragments of DNA in homologous chromosomes during the evolution of the yeast genome (Casaregola et al., 1998; Keogh et al., 1998; Wolfe & Shields, 1997). Numerous studies that have used karyotype analysis to characterize wine strains of *S. cerevisiae* (see Table 5.4) have shown that it is a powerful technique for differentiating between these strains.

## 3.3. Restriction Analysis of Mitochondrial DNA

The mitochondrial DNA (mtDNA) of *S. cerevisiae* is a small, highly variable molecule of between 60 and 80 kb. The high degree of polymorphism revealed by restriction analysis of mtDNA has made it one of the most commonly applied techniques for the characterization of wine strains of this species (see Table 5.5).

Several methods have been developed to isolate yeast mtDNA (Aiglé et al., 1984; Gargouri, 1989; Querol & Barrio, 1990). However, the use of cesium chloride gradients and ultracentrifugation make many of them inappropriate for industrial applications. To circumvent these difficulties, Querol et al. (1992a) developed an approach to mtDNA analysis that does not require either technique. This simplified protocol relies on the composition of A-T and G-C base pairs in the yeast mtDNA,

which is 75% AT-rich but nevertheless contains some 200 GT-rich regions (Gray, 1989). GCAT-type enzymes do not recognize either GC- or AT-rich regions in digestions of total DNA. Consequently, given the small number of restriction sites in the mtDNA and the large number of sites in the nuclear DNA, the latter is digested into small fragments and the mtDNA bands can be clearly visualized over the background shadow of the digested nuclear DNA. Not all enzymes reveal the same degree of polymorphism, and digestion patterns are highly species-dependent. In the specific case of *S. cerevisiae*, the most appropriate enzymes to differentiate between strains are *Hin*fI and *Hae*III (Guillamón et al., 1994). López et al. (2001) simplified this method with a modified protocol that reduced the time required from 77 to 25 h. This rapid technique enables a greater number of strains to be analyzed in a shorter amount of time and is ideal for industrial applications due to its speed, safety, and low cost, and because it does not require sophisticated equipment or highly trained personnel.

## 3.4. Polymerase Chain Reaction (PCR)-based Methods

The quickest molecular techniques that have been used to differentiate strains of wine yeast are those based on PCR (Saiki et al., 1985, 1988). Some PCR-based techniques have been developed to detect DNA polymorphisms without the use of restriction enzymes. The techniques most frequently used to differentiate between yeast strains are RAPD and microsatellite analysis. Other techniques such as the amplification of δ sequences and intron splice sites have been developed specifically to differentiate between wine strains of the species *S. cerevisiae*.

All of these techniques use oligonucleotide primers, which bind to target sequences on each strand of the yeast DNA. The sequence of the primers varies according to the technique, as discussed below. Amplification is carried out with a thermostable DNA polymerase, and the protocol comprises a variable number of cycles (generally between 25 and 45) that always include denaturation of the DNA followed by hybridization and extension. The result is amplification of the DNA, doubling the quantity of target DNA in each cycle. The amplification conditions, especially the hybridization temperature, also differ. The amplification products are visualized in 1.4% agarose gels, and the strain-specific nature of the profiles allows differentiation between strains. Below we discuss each of these techniques in detail.

### 3.4.1. Random Amplification of Polymorphic DNA (RAPD)

The RAPD technique (Williams et al., 1990) is characterized by the use of a single short primer (around 10 nucleotides) that has a random sequence. The RAPD-PCR reaction is carried out at a low hybridization temperature (37°C). Thus, the pairings between the oligonucleotide and the DNA are determined by the short and random sequence of the primer and favored by the low hybridization temperature, leading to the amplification of a range of DNA fragments distributed throughout the genome. The result is a pattern of amplified products of different molecular weight that can be characteristic of the species or of different strains or isolates within the same species (Bruns et al., 1991; Paffetti et al., 1995).

The main advantage of RAPD is that no prior sequence information is required in order to design a primer. Furthermore, because the technique allows analysis of variability throughout the entire genome, it reveals more polymorphism than techniques that analyze specific regions. However, the low hybridization temperature means that the amplification profiles are unstable and difficult to reproduce, and multiple reactions are required for each sample using DNA from different extractions as the template. Only the bands present in all

of the reactions can be considered. Given that the results obtained with several oligonucleotides must be combined to achieve good resolution, the technique is not appropriate for routine industrial application. Consequently, it has not been used extensively for the characterization of *S. cerevisiae* strains (see Table 5.6) and is more widely applied in taxonomic studies (Fernández-Espinar et al., 2003; Molnar et al., 1995).

### 3.4.2. Polymerase Chain Reaction (PCR) Analysis of Repetitive Genomic DNA (Microsatellites and Minisatellites)

The extensive variability of repetitive regions of genomic DNA makes them suitable targets for the molecular identification of yeast strains. These motifs, known as microsatellites and minisatellites, vary substantially in length and are present as tandem repeats distributed randomly throughout the genome. Microsatellites are usually shorter than 10 base pairs whereas minisatellites are between 10 and 100 base pairs in length. The variability found in these regions can be demonstrated by PCR amplification using specific oligonucleotides, such as $(GTG)_5$, $(GAG)_5$, $(GACA)_4$, or M13. The capacity of these oligonucleotides to reveal polymorphism among strains of *S. cerevisiae* has been demonstrated previously by Lieckfeldt et al. (1993) using the hybridization techniques described in Section 3.1. The same authors were the first to use these sequences as primers in the PCR reaction, and they demonstrated the usefulness of the technique for strain characterization. The technique was later used by other authors (see Table 5.7), and Baleiras Couto et al. (1996) used it successfully to characterize spoilage strains of *S. cerevisiae* in alcoholic beverages. The amplified products obtained are approximately 700 to 3500 base pairs long and can therefore be visualized using agarose gels.

Recently, protocols similar to those used in paternity testing and assessment of ancestry in humans have been developed. Here, sequence data from *S. cerevisiae* databases is assessed to identify repetitive regions. Sequences containing microsatellite motifs are then used to design primers. This technique was applied for the first time in wine strains of *S. cerevisiae* by Gallego et al. (1998), although only four strains were analyzed. Subsequently, González Techera et al. (2001) and Pérez et al. (2001b) designed new primers to differentiate *S. cerevisiae* wine strains. The primers designed by those groups have recently been used to characterize commercial wine strains of *S. cerevisiae*. Howell et al. (2004) used the primers designed by González Techera et al. (2001) to monitor commercial strains of *S. cerevisiae* during laboratory-scale fermentations and Schuller et al. (2004) used the method described by Pérez et al. (2001b) to characterize 23 commercial wine strains. The study by Schuller et al. (2004) showed that the resolution with this technique is comparable to that obtained with δ elements and restriction analysis of mtDNA. Recently, various authors have proposed useful methods for the identification of *S. cerevisiae* based on PCR amplification of polymorphic regions of the genome using combinations of more than two primers in a single PCR reaction (Richards et al., 2009; Vaudano & García-Moruno, 2008). The system proposed by Richards et al. (2009) is of particular interest since the authors have generated a database containing 246 genotypes including 78 commercial wine strains along with other natural isolates from various different regions of the world. Clinical isolates of *S. cerevisiae* have also been characterized using this method (Hennequin et al., 2001).

Amplification products are usually visualized in acrylamide gels, although automatic sequencers can also be used. Consequently, the technique is of little use in routine applications despite its high resolution and reproducibility.

Marinangeli et al. (2004) observed that some genes encoding cell-wall proteins from *S. cerevisiae* contain variable numbers of microsatellites that lead to strain variation in gene size. Those authors developed a method for the characterization of *S. cerevisiae* wine strains based on the amplification of these genes. In this technique, the high annealing temperatures used (60–65°C) in the PCR reaction ensure stable and reproducible amplification profiles.

### 3.4.3. Amplification of δ Sequences

Delta sequences are 0.3 kb elements that flank *Ty1* retrotransposons (Cameron et al., 1979). Around 100 copies of the δ element are present in the yeast genome as part of *Ty1* retrotransposons or as isolated elements. However, the δ sequences are concentrated in genomic regions adjacent to the transfer RNA genes (Eigel & Feldmann, 1982). The number and localization of these elements display a degree of intraspecific variability that Ness et al. (1993) exploited to develop specific primers ($\delta_1$ and $\delta_2$) for the differentiation of *S. cerevisiae* strains. They demonstrated that the δ elements were stable enough for this technique to be used on an industrial scale, and this was later confirmed by other groups (see Table 5.8). Comparison with other high-resolution techniques, such as restriction analysis of mtDNA and chromosome electrophoresis, has shown that analysis of δ elements can reveal extensive variability among *S. cerevisiae* isolates (Fernández-Espinar et al., 2001; Pramateftaki et al., 2000).

Recently, Legras and Karst (2003) optimized the technique by designing two new primers ($\delta_{12}$ and $\delta_{21}$) that are located very close to $\delta_1$ and $\delta_2$. The use of $\delta_{12}$ and $\delta_{21}$ or of $\delta_{12}$ and $\delta_2$ reveals greater polymorphism, which is reflected by the appearance of a larger number of bands. Consequently, the new primers are able to differentiate more strains, with 53 commercial strains unequivocally differentiated in their study. Schuller et al. (2004) confirmed this finding by showing that the combined use of $\delta_2$ and $\delta_{12}$ identified twice as many strains as the set of primers designed by Ness et al. (1993).

An important drawback of this technique is the influence of DNA concentration on the profile obtained, as shown by Fernández-Espinar et al. (2001) and discussed later by Schuller et al. (2004). Although this problem is avoided by standardizing the concentration of DNA, comparison of results between laboratories is complicated. Another problem is the appearance of "ghost" bands due to the low annealing temperature (42°C) used during the amplification reaction. Recently, Ciani et al. (2004) used an annealing temperature of 55°C to characterize wine strains of *S. cerevisiae*. Although this resulted in much more stable amplification profiles, fewer bands were obtained.

## 3.5. Amplified Fragment Length Polymorphism (AFLP)

Although amplified fragment length polymorphism (AFLP) is fundamentally based on PCR amplification, we will consider it in a separate section, owing to its complex methodology involving the use of other techniques.

AFLP involves the restriction of genomic DNA followed by the binding of adapters to the fragments obtained and their selective amplification by PCR. The adapter sequence and restriction sites are collectively used as the targets for the primers during PCR amplification. The fragments are separated in DNA sequencing gels and visualized by autoradiography or automated sequencing (Vos et al., 1995). As in the case of RAPD, no prior sequence information is required in order to design primers. Furthermore, the technique is easily reproduced and yields extensive information. However, although AFLP is a useful technique to discriminate between yeast strains (de Barros Lopes et al., 1999), it is very laborious, it requires automatic sequencers (which are not appropriate for routine industrial applications), and the data produced are difficult to interpret. Although the technique has been very widely used to study bacteria, plants, and animals, fewer studies have addressed its use in yeasts (Boekhout, 2001; Borst et al., 2003; Dassanayake

& Samaranayake, 2003; Theelen et al., 2001; Trilles et al., 2003). The technique has nevertheless been used for the characterization of different species of wine yeast (Azumi & Goto-Yamamoto, 2001; Curtin et al., 2007; de Barros Lopes et al., 1999; Flores Berrios et al., 2005; Lopandic et al., 2008).

## 4. APPLICATIONS

In this section we will discuss industrial applications of identifying species and strains of wine yeast.

### 4.1. Analysis of Variation in Yeast Populations During Natural Fermentation: Wine Ecology

The microbiological fermentation that is used to produce wine involves the growth of a series of microbial populations that have a direct influence on the final product. It is therefore of interest to characterize the microbial ecology of this process in an effort to control fermentation and, ultimately, the final quality of the wine. This requires techniques that can differentiate between species. The molecular techniques that have been most widely used for the differentiation of wine yeasts present in natural fermentations include electrophoretic karyotyping (Nadal et al., 1996; Schütz & Gafner, 1993); restriction analysis of the 5.8S-ITS region (Granchi et al., 1999; Pramateftaki et al., 2000; Rodríguez et al., 2004; Torija et al., 2001); restriction analysis of other rDNA regions (van Keulen et al., 2003); and a combination of techniques, such as RAPD and mtDNA restriction analysis (Torriani et al., 1999) or repetitive intergenic consensus PCR and PCR of intron splice sites combined with restriction fragment length polymorphism (RFLP) and sequence analysis of the 5.8S-ITS and D1/D2 rDNA regions (Hierro et al., 2006b).

The results of these studies have revealed microbial diversity not only between wine-growing regions but also from year to year in the same winery. Furthermore, the use of these techniques has allowed the identification of yeasts that have not been described previously in studies using conventional identification methods. Sabaté et al. (2002) undertook a study of the yeasts present on the vine and grape, in the winery, and during fermentation in the Spanish Priorat appellation using restriction analysis of the 5.8S-ITS rDNA region. They found that the soil of the vineyards contained strains of *Hanseniaspora uvarum*, a species that had previously been described as associated with the grapes and particularly as present during the initial phases of fermentation. They also isolated species of the genus *Cryptococcus* (*Cryptococcus uniguttulatum*, *Cryptococcus laurentii*, and *Cryptococcus ateren*) in the soil and on the stems and leaves of the vines, while other species, such as *Aureobasdium pullulans*, which are typically associated with soil environments, were found in the must. Studies undertaken in the Jerez region in the south of Spain by Esteve-Zarzoso et al. (2001) also identified soil species (*Issatchenkia terricola*) in the initial phases of fermentation. Those authors also reported that the *flor* of biologically aged wines from the Jerez region contained not only *S. cerevisiae* but also *Candida cantarrelli* (in 91.6% of samples) and *Dekkera bruxellensis* (which in this case did not cause wine spoilage). Finally, this technique has been used to show that strains of *Pichia guilliermondii* produce 4-ethylphenol, which is responsible for the aroma of stables in wines (Dias et al., 2003). Prior to this study, only *D. bruxellensis* and *Dekkera anomala* were known to produce this compound.

Studies have also been undertaken to analyze molecular variation in natural populations of *S. cerevisiae* using restriction analysis of mtDNA (Granchi et al., 2003; Gutiérrez et al., 1997; Pramateftaki et al., 2000; Querol et al., 1994; Sabaté et al., 1998; Torija et al., 2001), electrophoretic karyotyping (Frezier & Dubourdieu, 1992; Schütz & Gafner, 1993, 1994), microsatellite analysis (Caruso et al., 2002; Howell et al., 2004), fluorescent in situ hybridization (FISH) (Xufre et al., 2006), or a combination of techniques (Egli et al., 1998; Lópes et al., 2002; Nadal et al., 1996; Rodríguez et al., 2004; Torriani et al., 1999; Versavaud et al., 1995). These studies have shown that different strains of *S. cerevisiae* dominate fermentation in different appellations. Determining which strains dominate in each region is essential in order to select appropriate strains for use in starter cultures, since the use of autochthonous strains to achieve controlled fermentations will help to maintain the sensory characteristics typical of the region.

## 4.2. Analysis of Population Variation in Inoculated Fermentations: Monitoring Establishment

Variability of the yeast flora in musts can be reduced by addition of a microbial inoculum year after year. This inoculum normalizes the initial flora and gives rise to a homogeneous fermentation irrespective of vintage. Over the course of fermentation, the winery should check that the inoculated yeast displaces the existing flora and dominates fermentation as a result of its numerical superiority. In order to monitor the establishment of an inoculated strain, it is necessary to be able to differentiate it from the other yeasts present over the course of fermentation. This task is complicated by the fact that the inoculated yeast belongs to the same species as most of the other yeasts present in the must, namely *S. cerevisiae*. The main techniques that have been used to analyze the ecology of controlled fermentations are electrophoretic karyotyping (Briones et al., 1996), restriction analysis of mtDNA (Beltrán et al., 2002; Constantí et al., 1998; Gutiérrez et al., 1997), or a combination of different techniques (Egli et al., 1998; Esteve-Zarzoso et al., 2001; Fernández-González et al., 2001; Martínez et al., 1995; Mesa et al., 1999). Studies have also addressed the effect of inoculating a commercial strain of

*S. cerevisiae* on the population of non-*Saccharomyces* yeasts in the must (Beltrán et al., 2002; Constantí et al., 1998; Ganga & Martínez, 2004; Urso et al., 2008).

Although it is usually assumed that when a yeast starter culture is used growth of the autochthonous yeasts is suppressed, some studies have shown that these yeasts can still participate in fermentation (Querol et al., 1992b; Schütz & Gafner, 1993, 1994), and in some cases commercial yeasts only account for 50% of the population (Esteve-Zarzoso et al., 2000). Consequently, it is important to develop simple methods for routine analysis of commercial yeasts in industrial fermentations. Querol et al. (1992b) used restriction analysis of mtDNA to analyze the population dynamics of *Saccharomyces* yeast strains during wine fermentation involving inoculation with an industrial strain. López et al. (2002) developed a method for monitoring establishment of inoculated yeasts based on PCR amplification of variable regions in the mtDNA. The method was based on variability in the number and position of introns in the *COX2* gene between strains of *S. cerevisiae*. This method is particularly useful since it can be used to assess whether the inoculated yeast has become established within just 8 h, thus allowing wineries to initiate corrective measures to prevent stuck fermentation.

## 4.3. Characterization of Commercial Yeasts

Dried wine yeasts were developed in the 1950s when laboratories in Canada (Adams, 1954) and the United States (Castor, 1953) independently carried out selection of wine strains that were subsequently used in directed fermentations. More than 100 different strains are currently marketed, mainly by six companies. Molecular characterization of commercial yeast strains is necessary for two reasons. Firstly, it is needed for quality-control purposes to confirm that the obtained yeast is the one that

was originally selected and not a contaminant, and, secondly, to detect fraud.

Given that most active dried yeasts belong to the species *S. cerevisiae*, the techniques used must be able to differentiate clearly between strains. Most of the techniques described in Section 3 are useful for this purpose, as was recently shown by Schuller et al. (2004) in a comparative study of 23 commercial strains by electrophoretic karyotyping, restriction analysis of mtDNA, amplification of δ elements, and microsatellite analysis. Electrophoretic karyotyping (Blondin & Vezinhet, 1998; Yamamoto et al., 1991), amplification of δ elements (Legras & Karst, 2003; Ness et al., 1993), and microsatellite analysis (González Techera et al., 2001) had previously been used for this purpose. Other studies have been reported in which more than one technique was used to characterize commercial isolates: mtDNA analysis and karyotyping (Schuller et al., 2004; Vezinhet et al., 1990), amplification of δ elements and DNA fingerprinting (Lavallée et al., 1994), and karyotyping with hybridization (Degré et al., 1989). In fact, Fernández-Espinar et al. (2001) showed that definitive characterization of commercial strains requires a combination of various molecular techniques. The techniques applied in that study were restriction analysis of mtDNA with *Hin*fI, electrophoretic karyotyping, and PCR amplification of genomic δ elements. One of the most interesting findings reported by Fernández-Espinar et al. (2001) was the large number of errors or fraudulent practices by companies that produce commercial yeasts. Commercial strains have also been characterized by Echeverrigaray et al. (2000) using the RAPD technique and by Manzano et al. (2006) using TGGE-PCR and restriction analysis. De Barros Lopes et al. (1996) have developed a technique based on amplification of introns for the characterization of commercial strains. However, this technique has not been applied subsequently by other authors, possibly as a result of the complexity of the profiles generated.

## 4.4. Identification of New Species and Hybrids Involved in Wine Fermentation

Clearly, only certain yeasts, selected to produce desirable organoleptic properties following fermentation of the must, are appropriate for use in starter cultures. Most of the yeast strains used to date belong to the species *S. cerevisiae* and, to a lesser extent, *S. bayanus*. These species have clearly different metabolic characteristics, and, as a result, the choice of which one to use in the fermentation depends on the desired outcome (Giudici et al., 1995; Naumov et al., 1993, 2000; Torriani et al., 1999). *S. bayanus*, for example, is cryotolerant and therefore used in fermentations carried out at low temperatures. It would be of particular interest to identify other species with new enologic properties that are able to complete fermentation. Recently, in a study involving restriction analysis of the 18S-ITS1 rDNA region, *S. paradoxus* was isolated in Croatian wines, where it was found to predominate during fermentation (Redzepovic et al., 2002). As discussed below, the presence in alcoholic fermentations of natural hybrids resulting from crosses between different species of the *S. cerevisiae sensu stricto* complex suggests that some species normally associated with natural environments may be present in fermentations, as in the case of *S. paradoxus*. For instance, *Saccharomyces kudriavzevii* was identified as one of the parent species of the hybrid cider strain CID1, along with *S. cerevisiae* and *Saccharomyces uvarum* (Groth et al., 1999), and other hybrid strains that predominate in wine fermentations from central European regions (González et al., 2006; Lopandic et al., 2007).

In addition to identifying these and other, as yet unidentified, species in winemaking contexts, it would be of interest to identify hybrids that are better adapted than their parent strains to those winemaking conditions in which they have arisen. The presence of these hybrid species is common in brewing. For instance, *Saccharomyces pastorianus* is the result of a cross between *S. cerevisiae* and *Saccharomyces monacensis* (currently included in the *S. bayanus* taxon) (Hansen & Kielland-Brandt, 1994; Yamagishi & Ogata, 1999) and the hybrid *S. bayanus*-type strain derived from *S. uvarum* (wine yeasts included in the *S. bayanus* taxon) and *S. cerevisiae* (Nguyen & Gaillardin, 1997; Nguyen et al., 2000). Although less frequently, hybrids have also been identified in cider and wine, as in the case of the cider strain CID1 mentioned earlier and the wine strain S6U (Masneuf et al., 1998).

Molecular techniques such as electrophoretic karyotyping, AFLP, and RAPD can reveal hybrid character by analysis of the fraction of bands shared between a hybrid strain and each of the parental strains (Azumi & Goto-Yamamoto, 2001; de Barros Lopes et al., 2002; Fernández-Espinar et al., 2003; Masneuf et al., 1998). These techniques are time-consuming and laborious. In contrast, the method of restriction analysis and sequencing of the nuclear gene *Met2* developed by Masneuf et al. (1996) has been used successfully in combination with analysis of mitochondrial genes (*ATP8*, *ATP9*, or *SSU*) for the identification of double and triple hybrids (Groth et al., 1999; Masneuf et al., 1998). Other more recent approaches have involved the amplification of a nuclear DNA region (YBR033w) without the need to resort to restriction analysis of the amplicons (Torriani et al., 2004), and restriction analysis of five nuclear genes (*CAT8*, *CYR1*, *GSY1*, *MET6*, and *OPY1*) from different chromosomes and the 5.8S-ITS rDNA region alongside sequence analysis of the mitochondrial gene *COX2* (González et al., 2006).

## 4.5. Detection of Wine Spoilage Yeasts

Wine is a highly appropriate culture medium for the growth of a large number of microorganisms, in part due to its richness in organic acids, amino acids, residual sugars, growth factors, and mineral salts. The main negative effects of yeasts in wine are the generation of undesirable aromas and flavors during winemaking and the formation of biofilms or turbidity, or the production of gas, during storage.

Various species of wine spoilage yeast have been described. The typical contaminating species found during winemaking belong to the genera *Pichia* and *Candida* or the species *Saccharomycodes ludwigii*. *Zygosaccharomyces bailii*, *S. cerevisiae*, and *S. ludwigii* are the principal contaminants during bottling, whereas species belonging to the genera *Dekkera/Brettanomyces* are found during barrel aging. The high sugar content of sweet and sparkling wines favors the growth of *Zygosaccharomyces* species, particularly *Z. bailii*. Few studies have addressed the identification and molecular characterization of these species, and those that have done so used standard techniques such as analysis of mtDNA or microsatellites, restriction analysis of rDNA, or RAPD, alone or in combination (Baleiras Couto et al., 1994, 1996; Cocolin et al., 2004). Less common techniques have also been used, such as those based on restriction analysis combined with PFGE (Miot-Sertier & Lonvaud-Funel, 2007; Oelofse et al., 2009) or FISH (Röder et al., 2007; Stender et al., 2001).

There are no legal limits in terms of the number of yeasts permitted in wine, but recommendations do exist. The International Organization of Vine and Wine (OIV) recommends a maximum of $10^2–10^5$ colony-forming units (CFU) per mL. Wineries apply their own criteria regarding acceptable levels of contamination, and these are much stricter than the OIV recommendations. For instance, they recommend no more than 1 CFU/mL in sweet wines (Loureiro & Malfeito-Ferreira, 2003).

Clearly, techniques for the detection of spoilage yeasts are essential. These techniques must be very sensitive and allow quantification of the number of microorganisms present. They must also be rapid to allow the application of corrective measures on the production line prior to release of the products onto the market. Ibeas et al. (1996) have developed a system for detection of species of the genera *Dekkera/Bretanomyces* based on two consecutive PCR reactions (nested PCR). Using this system, they were able to detect contaminations of less than 10 cells in samples of Jerez wine. Currently, real-time or quantitative PCR represents a good alternative technique with which to resolve these types of problem, since it is both rapid and highly sensitive. Systems of this type have been developed by Phister and Mills (2003) and Delaherche et al. (2004) for the detection and quantification of *D. bruxellensis* strains. Recently, Hayashi et al. (2007) developed a set of primers for the detection and identification of *Brettanomyces/Dekkera* species using the ITS rDNA region. The technique employed a novel loop-mediated isothermal amplification method, which appears to be more specific, sensitive, and straightforward than standard PCR techniques. The use of such systems for other species of wine spoilage yeast would be of particular interest for wineries in order to avoid spoilage during wine storage prior to sale.

## Acknowledgments

Research by our group is funded by the Spanish Interministerial Commission on Science and Technology (Ref. AGL2006-12703-CO2 and AGL2009-12673-CO2) and by the Regional Government of Valencia (Generalitat Valenciana; Ref. PROMETEO/2009/019).

## References

Adams, A. M. (1954). A simple continuous propagator for yeast. *Rept. Ontario Hort. Expt. Sta., and Products Lab* 102–103.

Aiglé, M., Erbs, D., & Moll, M. (1984). Some molecular structures in the genome of larger brewing yeast. *Am. Soc. Brew. Chem., 42*, 1–7.

Andorrà, I., Landi, S., Mas, A., Guillamón, J. M., & Esteve-Zarzoso, B. (2008). Effect of oenological pratices on microbial populations using culture-independent techniques. *Food Microbiol., 25*, 849–856.

Ayoub, M. J., Legras, J. L., Saliba, R., & Gaillardin, C. (2006). Application of multi locus sequences typing to the analysis of the biodiversity of indigenous *Saccharomyces cerevisiae* wine yeasts from Lebanon. *J. Appl. Microbiol., 100,* 699–711.

Azumi, M., & Goto-Yamamoto, N. (2001). AFLP analysis of type strains and laboratory and industrial strains of *Saccharomyces* sensu stricto and its application to phenetic clustering. *Yeast, 18,* 1145–1154.

Baleiras Couto, M. M., Eijsma, B., Hofstra, H., Huis in't Veld, J. H. H., & van der Vossen, J. M. B. M. (1996). Evaluation of molecular typing techniques to assign genetic diversity among strains of *Saccharomyces cerevisiae. Appl. Environ. Microbiol., 62,* 41–46.

Baleiras Couto, M. M., Reizinho, R. G., & Duarte, F. L. (2005). Partial 26S rDNA restriction analysis as a tool to characterise non-Saccharomyces yeasts present during red wine fermentations. *Int. J. Food Microbiol., 102,* 49–56.

Baleiras Couto, M. M., van der Vossen, J. M., Hofstra, H., & Huis in't Veld, J. H. (1994). RAPD analysis: A rapid technique for differentiation of spoilage yeasts. *Int. J. Food Microbiol., 24,* 249–260.

Barnett, J. A., Payne, R. W., & Yarrow, I. J. (1990). *Yeast: Characterization and identification* (2nd ed.). London, UK: Cambridge University Press.

Belloch, C., Querol, A., Garcia, M. D., & Barrio, E. (2000). Phylogeny of the genus *Kluyveromyces* inferred from the mitochondrial cytochrome-c oxidase II gene. *Int. J. Syst. Evol. Microbiol., 50,* 405–416.

Beltrán, G., Torija, M. J., Novo, M., Ferrer, N., Poblet, M., Guillamón, J. M., et al. (2002). Analysis of yeast populations during alcoholic fermentation: Six year follow-up study. *Syst. Appl. Microbiol., 25,* 287–293.

Bidenne, C., Blondin, B., Dequin, S., & Vezinhet, F. (1992). Analysis of the chromosomal DNA polymorphism of wine strains of *Saccharomyces cerevisiae. Curr. Genet., 22,* 1–7.

Blondin, B., & Vezinhet, F. (1988). Identification de souches de levures oenologiques par leurs caryotypes obtenus en électrophorèse en champ pulsé. *Rev. Fr. Oenol., 115,* 7–11.

Boekhout, T., Theelen, B., Diaz, M., Fell, J. W., Hop, W. C., Abeln, E. C., et al. (2001). Hybrid genotypes in the pathogenic yeast *Cryptococcus neoformans. Microbiology, 147,* 891–907.

Borst, A., Theelen, B., Reinders, E., Boekhout, T., Fluit, A. C., & Savelkoul, P. H. (2003). Use of amplified fragment length polymorphism analysis to identify medically important *Candida* spp., including C. dubliniensis. *J. Clin. Microbiol., 41,* 1357–1362.

Braus, G., Furter, R., Prantl, F., Niederberger, P., & Hütter, R. (1985). Arrangement of genes TRP1 and TRP3 of *Saccharomyces cerevisiae* strains. *Arch. Microbiol., 142,* 383–388.

Briones, A. I., Ubeda, J., & Grando, M. S. (1996). Differentiation of Saccharomyces cerevisiae strains isolated from

fermenting musts according to their karyotype patterns. *Int. J. Food Microbiol., 28,* 369–377.

Bruns, T. D., White, T. J., & Taylor, J. W. (1991). Fungal molecular systematics. *Annu. Rev. Ecol. Syst., 22,* 524–564.

Bureau, G., Brun, O., Vigues, A., Maujean, A., Vesselle, G., & Feuillat, A. (1982). Etude d'une microflore levurienne champenoise. *Connaiss. Vigne Vin., 16,* 15–32.

Cameron, J. R., Loh, E. Y., & Davis, R. W. (1979). Evidence for transposition of dispersed repetitive DNA families in yeast. *Cell, 16,* 739–751.

Capece, A., Salzano, G., & Romano, P. (2003). Molecular typing techniques as a tool to differentiate non-*Saccharomyces* wine species. *Int. J. Food Microbiol., 84,* 33–39.

Cappello, M. S., Bleve, G., Grieco, F., Dellaglio, F., & Zacheo, G. (2004). Characterization of *Saccharomyces cerevisiae* strains isolated from must of grape grown in experimental vineyard. *J. Appl. Microbiol., 97,* 1274–1280.

Caruso, M., Capece, A., Salzano, G., & Romano, P. (2002). Typing of *Saccharomyces cerevisiae* and *Kloeckera apiculata* strains from Aglianico wine. *Lett. Appl. Microbiol., 34,* 323–328.

Casaregola, S., Nguyen, H. V., Lepingle, A., Brignon, P., Gnedre, F., & Guillardin, C. (1998). A family of laboratory strains of *Saccharomyces cerevisiae* carry rearrangements involving chromosomes I and III. *Yeast, 14,* 551–561.

Castor, J. G. (1953). Experimental development of compressed yeast fermentation starters. *Wines & Vines, 34,* 8–27.

Ciani, M., Mannazzu, I., Marinangeli, P., Clementi, F., & Martini, A. (2004). Contribution of winery-resident *Saccharomyces cerevisiae* strains to spontaneous grape must fermentation. *Anton. Leeuw., 85,* 159–164.

Cocolin, L., Bisson, L. F., & Mills, D. A. (2000). Direct profiling of the yeast dynamics in wine fermentations. *FEMS Microbiol. Lett., 189,* 81–87.

Cocolin, L., Rantsiou, K., Iacumin, L., Zironi, R., & Comi, G. (2004). Molecular detection and identification of *Brettanomyces/Dekkera bruxellensis* and *Brettanomyces/Dekkera anomalus* in spoiled wines. *Appl. Environ. Microbiol., 70,* 1347–1355.

Comi, G., Maifreni, M., Manzano, M., Lagazio, C., & Cocolin, L. (2000). Mitochondrial DNA restriction enzyme analysis and evaluation of the enological characteristics of Saccharomyces cerevisiae strains isolated from grapes of the wine-producing area of Collio Italy. *Int. J. Food Microbiol., 58,* 117–121.

Constantí, M., Reguant, C., Poblet, M., Zamora, F., Mas, A., & Guillamón, J. M. (1998). Molecular analysis of yeast population dynamics: Effect of sulphur dioxide and inoculum on must fermentation. Int. *J. Food Microbiol., 41,* 169–175.

Cottrell, M., Kovk, J. K. F., Lategam, P. M., & Britz, T. Z. (1986). Long chain fatty acid composition as an aid in the classification of the genus. *Saccharomyces. Syst. Appl. Microbiol.*, *8*, 166–168.

Curtin, C. D., Bellon, J. R., Henschke, P. A., Godden, P. W., & de Barros Lopes, M. A. (2007). Genetic diversity of *Dekkera bruxellensis* yeasts isolated from Australian wineries. *FEMS Yeast Res.*, *7*, 471–481.

Daniel, H. M., & Meyer, W. (2003). Evaluation of ribosomal RNA and actin gene sequences for the identification of ascomycetous yeasts. *Int. J. Food Microbiol.*, *86*, 61–78.

Dassanayake, R. S., & Samaranayake, L. P. (2003). Amplification-based nucleic acid scanning techniques to assess genetic polymorphism in *Candida. Crit Rev Microbiol.*, *29*, 1–24.

de Barros Lopes, M., Bellon, J. R., Shirley, N. J., & Gaute, P. R. (2002). Evidence for multiple interspecific hybridization in *Saccharomyces sensu stricto* species. *FEMS Yeast Res.*, *1*, 323–331.

de Barros Lopes, M., Rainieri, S., Henschkle, P. A., & Longridge, P. (1999). AFLP fingerprinting for analysis of yeast genetic variation. *Int. J. Syst. Bacteriol.*, *49*, 915–924.

de Barros Lopes, M., Soden, A., Henschke, P. A., & Langridge, P. (1996). PCR differentiation of commercial strains using intron splice site primers. *Appl. Environ. Microbiol.*, *62*, 4514–4520.

Degré, R., Thomas, D. Y., Ash, J., Mailhiot, K., Morin, A., & Dubord, C. (1989). Wine yeasts strain identification. *Am. J. Enol. Vitic.*, *40*, 309–315.

Delaherche, A., Claisse, O., & Lonvaud-Funel, A. (2004). Detection and quantification of *Brettanomyces bruxellensis* and ropy *Pediococcus damnosus* strains in wine by real-time polymerase chain reaction. *J. Appl. Microbiol.*, *97*, 910–915.

de Llanos, R., Fernández-Espinar, M. T., & Querol, A. (2004). Identification of species of the genus *Candida* by RFLP analysis of the 5.8S rRNA gene and the two ribosomal internal transcribed spacers. *Anton. Leeuw.*, *85*, 175–185.

Dias, L., Dias, S., Sancho, T., Stender, T., Querol, A., Malfeito-Ferreiro, M., et al. (2003). Identification of yeasts isolated from wine-related environments and capable of producing 4-ethylphenol. *Food Microbiol.*, *20*, 567–574.

di Maro, E., Ercolini, D., & Coppola, S. (2007). Yeast dynamics spontaneous wine fermentation of the Catalanesca grape. *Int. J. Food Microbiol.*, *117*, 201–210.

Dlauchy, D., Tornai-Lehoczki, J., & Gábor, P. (1999). Restriction enzyme analysis of PCR amplified rDNA as a taxonomic tool in yeast identification. *System. Appl. Microbiol.*, *22*, 445–453.

Duarte, F. L., Pais, C., Spencer-Martins, I., & Leao, C. (1999). Distinctive electrophoretic isoenzyme profiles in *Saccharomyces sensu stricto. Int. J. Syst. Bacteriol*, *49*, 1907–1913.

Echeverrigaray, S., Paese-Toresan, S., & Carrau, J. L. (2000). RAPD marker polymorphism among comercial winery yeast strains. *World J. Microbiol. Biotechnol.*, *16*, 143–146.

Egli, C. M., Edinger, W. D., Mitrakul, C. M., & Henick-Kling, T. (1998). Dynamics of indigenous and inoculated yeast populations and their effect on the sensory character of Riesling and Chardonnay wines. *J. Appl. Microbiol.*, *85*, 779–789.

Eigel, A., & Feldmann, H. (1982). Ty1 and delta elements occur adjacent to several tRNA genes in yeast. *EMBO J*, *1*, 1245–1250.

Esteve-Zarzoso, B., Belloch, C., Uruburu, F., & Querol, A. (1999). Identification of yeast by RFLP analysis of the 5.8S rRNA gene and the two ribosomal internal transcribed spacers. *Int. J. Syst. Bacteriol.*, *49*, 329–337.

Esteve-Zarzoso, B., Gostíncar, A., Bobet, R., Uruburu, F., & Querol, A. (2000). Selection and molecular characterisation of wine yeasts isolated from the "El Penedés" area (Spain). *Food Microbiol.*, *17*, 553–562.

Esteve-Zarzoso, B., Peris-Toran, M. J., García-Maiquez, E., Uruburu, F., & Querol, A. (2001). Yeast population dynamics during the fermentation and biological aging of sherry wines. *Appl. Environ. Microbiol.*, *67*, 2056–2061.

Fernández-Espinar, M. T., Barrio, E., & Querol, A. (2003). Genetic variability among species of the Saccharomyces sensu stricto. *Yeast*, *20*, 1213–1226.

Fernández-González, M., Espinosa, J. C., Ubeda, J. F., & Briones, A. I. (2001). Yeast present during wine fermentation: Comparative analysis of conventional plating and PCR-TTGE. *Syst. Appl. Microbiol.*, *24*, 634–638.

Fernández-Espinar, M. T., Esteve-Zarzoso, B., Querol, A., & Barrio, E. (2000). RFLP analysis of the ribosomal internal transcribed spacers and the 5.8S rRNA gene region of the genus *Saccharomyces*: A fast method for species identification and the differentiation of flor yeasts. *Anton. Leeuw.*, *78*, 87–97.

Fernández-Espinar, M. T., López, V., Ramón, D., Bartra, E., & Querol, A. (2001). Study of the authenticity of commercial wine yeast strains by molecular techniques. *Int. J. Food Microbiol.*, *70*, 1–10.

Fleet, G., & Heard, G. (1993). Yeast growth during fermentation. In G. Fleet (Ed.), *Wine microbiology and biotechnology* (pp. 27–57). Chur, Switzerland: Harwood Academic Publishers.

Flores Berrios, E. P., Alba González, J. F., Arrizon Gaviño, J. P., Romano, P., Capece, A., & Gschaedler Mathis, A. (2005). The uses of AFLP for detecting DNA polymorphism, genotype identification and genetic diversity between yeasts isolated from Mexican agave-distilled beverages and from grape musts. *Lett. Appl. Microbiol.*, *41*, 147–152.

Frezier, V., & Dubourdieu, D. (1992). Ecology of yeast *Saccharomyces cerevisiae* during spontaneous fermentation in a Bordeaux winery. *Am. J. Enol. Vitic., 43,* 375–380.

Gallego, F. J., Pérez, M. A., Martínez, I., & Hidalgo, P. (1998). Microsatellites obtained from database sequences are useful to characterize Saccharomyces cerevisiae. *Am. J. Enol. Vitic., 49,* 350–351.

Ganga, M. A., & Martínez, C. (2004). Effect of wine yeast monoculture practice on the biodiversity of non-*Saccharomyces* yeasts. *J. Appl. Microbiol., 96,* 76–83.

Gargouri, A. (1989). A rapid and simple method for the extraction of yeast mitochondrial DNA. *Curr. Genet., 15,* 235–237.

Giudici, P., & Pulvirenti, A. (2002). Molecular methods for identification of wine yeasts. In M. Ciani (Ed.), *Biodiversity and biotechnology of wine yeasts* (pp. 35–52). Kerala, India: Research Signpost.

Giudici, P., Zambonelli, C., Passarelli, P., & Castellari, L. (1995). Improvement of wine composition with cryotolerant *Saccharomyces* strains. *Am. J. Enol. Vitic., 46,* 143–147.

Golden, D. A., Beuchat, L. R., & Hitchcock, H. L. (1994). Changes in fatty acid composition of Zygosaccharomyces rouxii as influenced by solutes, potassium sorbate and incubation temperature. *Int. J. Food Microbiol., 21,* 293–303.

González, S. S., Barrio, E., Gafner, J., & Querol, A. (2006). Natural hybrids from *Saccharomyces cerevisiae, Saccharomyces bayanus* and *Saccharomyces kudriavzevii* in wine fermentations. *FEMS Yeast Res, 6,* 1221–1234.

González, S. S., Barrio, E., & Querol, A. (2007). Molecular identification and characterization of wine yeasts isolated from Tenerife (Canary Island, Spain). *J. Appl. Microbiol., 102,* 1018–1025.

González Techera, A., Jubany, S., Carrau, F. M., & Gaggero, C. (2001). Differentiation of industrial wine yeast strains using microsatellite markers. *Lett. Appl. Microbiol., 33,* 71–75.

Granchi, L., Bosco, M., Messini, A., & Vincenzini, M. (1999). Rapid detection and quantification of yeast species during spontaneous wine fermentation by PCR-RFLP analysis of the rDNA ITS region. *J. Appl. Microbiol., 87,* 949–956.

Granchi, L., Ganucci, D., Viti, C., Giovannetti, L., & Vincenzini, M. (2003). *Saccharomyces cerevisiae* biodiversity in spontaneous commercial fermentations of grape musts with "inadequate" assimilable-nitrogen content. *Lett. Appl. Microbiol., 36,* 54–58.

Grando, M. S., Ubeda, J., & Briones, A. I. (1994). RAPD analysis of wine *Saccharomyces* strains differentiated by pulsed field gel electrophoresis. *Biotechnol. Tech., 8,* 557–560.

Gray, M. W. (1989). Origin and evolution of mitochondrial DNA. *Annu. Rev. Cell. Biol., 5,* 25–50.

Groth, C., Hansen, J., & Piskur, J. (1999). A natural chimeric yeast containing genetic material from three species. *Int. J. Syst. Bacteriol., 49,* 1933–1938.

Guillamón, J. M., Barrio, E., Huerta, T., & Querol, A. (1994). Rapid characterization of four species of the *Saccharomyces sensu stricto* complex according to mitochondrial DNA patterns. *Int. J. Bacteriol., 44,* 708–714.

Guillamón, J. M., Barrio, E., & Querol, A. (1996). Characterization of wine yeast strains of the *Saccharomyces* genus on the basis of molecular markers: Relationships between genetic distance and geographic or ecological origin. *System. Appl. Microbiol., 19,* 122–132.

Guillamón, J. M., Sabate, J., Barrio, E., Cano, J., & Querol, A. (1998). Rapid identification of wine yeast species based on RFLP analysis of the ribosomal internal transcribed spacer ITS region. *Arch. Microbiol., 169,* 387–392.

Gutiérrez, A. R., López, R., Santamaría, M. P., & Sevilla, M. J. (1997). Ecology of inoculated and fermentations in Rioja Spain musts, examined by mitochondrial DNA restriction analysis. *Int. J. Food Microbiol., 20,* 241–245.

Hansen, J., & Kielland-Brandt, M. C. (1994). *Saccharomyces carlsbergensis* contains two functional MET2 alleles similar to homologues from *S. cerevisiae* and S. monacensis. *Gene, 140,* 33–40.

Hayashi, N., Arai, R., Tada, S., Taguchi, H., & Ogawa, Y. (2007). Detection and identification of *Brettanomyces/Dekkera* sp. yeasts with a loop-mediated isothermal amplification method. *Food Microbiol., 24,* 778–785.

Hennequin, C., Thierry, A., Richard, G. F., Lecointre, G., Nguyen, H. V., Gaillardin, C., et al. (2001). Microsatellite typing as a new tool for identification of *Saccharomyces cerevisiae* strains. *J. Clin. Microbiol., 39,* 551–559.

Hierro, N., Esteve-Zarzoso, B., González, A., Mas, A., & Guillamón, J. M. (2006a). Real-time quantitative PCR (QPCR) and reverse transcription-QPCR for detection and enumeration of total yeasts in wine. *Appl. Environ. Microbiol., 72,* 7148–7155.

Hierro, N., Esteve-Zarzoso, B., Mas, A., & Guillamón, J. M. (2007). Monitoring of *Saccharomyces* and *Hanseniaspora* populations during alcoholic fermentation by real-time quantitative PCR. *FEMS Yeast Res., 7,* 1340–1349.

Hierro, N., González, A., Mas, A., & Guillamón, J. M. (2006b). Diversity and evolution of non-*Saccharomyces* yeast populations during wine fermentation: Effect of grape ripeness and cold maceration. *FEMS Yeast Res, 6,* 102–111.

Howell, K. S., Bartowsky, E. J., Fleet, G. H., & Henschke, P. A. (2004). Microsatellite PCR profiling of *Saccharomyces cerevisiae* strains during wine fermentation. *Lett. Appl. Microbiol., 38,* 315–320.

Ibeas, J. L., & Jiménez, J. (1996). Genomic complexity and chromosomal rearrangements in wine-laboratory yeast hybrids. *Curr. Genet., 30*, 410–416.

Ibeas, J. I., Lozano, I., Perdigones, F., & Jiménez, J. (1996). Detection of *Dekkera-Brettanomyces* strains in sherry by a nested PCR method. *Appl. Environ. Microbiol., 62*, 998–1003.

James, S. A., Cai, J., Roberts, I. N., & Collins, M. D. (1997). Phylogenetic analysis of the genus *Saccharomyces* based on 18S rRNA gene sequences: Description of *Saccharomyces kunashirensis* sp. nov., and *Saccharomyces martiniae* sp. nov. *Int. J. Syst. Bacteriol., 47*, 453–460.

James, S. A., Collins, M. D., & Roberts, I. N. (1996). Use of an rRNA internal transcribed spacer region to distinguish phylogenetically closely related species of the genera *Zygosaccharomyces* and Torulaspora. *Int. J. Syst. Bacteriol., 46*, 189–194.

Keogh, R. S., Seoighe, C., & Wolfe, K. H. (1998). Evolution of gene order and chromosomal number in *Saccharomyces, Kluyveromyces* and related fungi. *Yeast, 14*, 443–457.

Kreger-van Rij, N. J. W. (1984). *The yeast, a taxonomic study*. Amsterdam, the Netherlands: Elsevier Science B.V.

Kurtzman, C. P., & Phaff, H. J. (1987). In A. H. Rose, & J. S. Harrison (Eds.), *The yeast. Molecular taxonomy, Vol. 1* (pp. 63–94). London, UK: Academic Press.

Kurtzman, C. P., & Robnett, C. J. (1998). Identification and phylogeny of ascomycetous yeasts from analysis of nuclear large subunit 26S ribosomal DNA partial sequences. *Anton. Leeuw., 73*, 331–371.

Kurtzman, C. P., & Robnett, C. J. (2003). Phylogenetic relationships among yeasts of the 'Saccharomyces complex' determined from multigene sequence analyses. *FEMS Yeast Res., 3*(4), 417–432.

Lavallée, F., Salvas, Y., Lamy, S., Thomas, D. Y., Degré, R., & Dulau, L. (1994). PCR and DNA fingerprinting used as quality control in the production of wine yeast strains. *Am. J. Enol. Vitic., 45*, 86–91.

Legras, J.-L., & Karst, F. (2003). Optimisation of interdelta for *Saccharomyces cerevisiae* strain characterization. *FEMS Microbiol. Lett., 221*, 249–255.

Le Jeune, C., Erny, C., Demuyter, C., & Lollier, M. (2006). Evolution of the population of *Saccharomyces cerevisiae* from grape to wine in a spontaneous fermentation. *Food Microbiol., 23*, 709–716.

Li, W. H. (1997). *Molecular evolution*. Sunderland, MA: Sinauer Associates.

Lieckfieldt, E., Meyer, W., & Börn, T. (1993). Rapid identification and differentiation of yeasts by DNA and PCR fingerprinting. *J. Basic Microbiol., 33*, 413–426.

Longo, E., Cansado, J., Agrelo, D., & Villa, T. (1991). Effect of climatic conditions on yeast diversity in grape musts from northwest Spain. *Am. J. Enol. Vitic., 42*, 141–144.

Lopandic, K., Gangl, H., Wallner, E., Tscheik, G., Leitner, G., Querol, A., et al. (2007). Genetically different wine yeasts isolated from Austrian vine-growing regions influence wine aroma differently and contain putative hybrids between *Saccharomyces cerevisiae* and Saccharomyces kudriavzevii. *FEMS Yeast Res., 7*, 953–965.

Lopandic, K., Tiefenbrunner, W., Gangl, H., Mandl, K., Berger, S., Leitner, G., et al. (2008). Molecular profiling of yeasts isolated during spontaneous fermentations of Austrian wines. *FEMS Yeast Res., 8*, 1063–1075.

Lópes, C. A., Lavalle, T. L., Querol, A., & Caballero, A. C. (2006). Combined use of killer biotype and mtDNA-RFLP patterns in a Patagonian wine *Saccharomyces cerevisiae* diversity study. *Anton. Leeuw., 89*, 147–156.

Lópes, C. A., Rodríguez, M. E., Sangorrín, M., Querol, A., & Caballero, A. C. (2007). Patagonia wines: Implantation of an indigenous strain of *Saccharomyces cerevisiae* in fermentations conducted in traditional and modern cellars. *J. Ind. Microbiol. Biotechnol., 34*, 139–149.

Lópes, C. A., van Broock, M., Querol, A., & Caballero, A. C. (2002). *Saccharomyces cerevisiae* wine yeast populations in a cold region in Argentinean Patagonia. A study at different fermentation scales. *J. Appl. Microbiol., 93*, 608–615.

López, V., Fernández-Espinar, M. T., Barrio, E., Ramón, D., & Querol, A. (2002). A new PCR-based method for monitoring inoculated wine fermentations. *Int. J. Food Microbiol., 81*, 63–71.

López, V., Querol, A., Ramón, D., & Fernández-Espinar, M. T. (2001). A simplified procedure to analyse mtDNA from industrial yeasts. *Int. J. Food Microbiol., 68*, 75–81.

Loureiro, V., & Malfeito-Ferreira, M. (2003). Spoilage yeasts in the wine industry. *Int. J. Food Microbiol., 86*, 23–50.

Manzano, M., Medrala, D., Giusto, C., Bartolomeoli, I., Urso, R., & Comi, G. (2006). Classical and molecular analysis to characterize commercial dry yeasts used in wine fermentations. *J. Appl. Microbiol., 100*, 599–607.

Marinangeli, P., Angelozzi, D., Ciani, M., Clementi, F., & Mannazzu, I. (2004). Minisatellites in *Saccharomyces cerevisiae* genes encoding cell wall proteins: A new way towards wine strain characterisation. *FEMS Yeast Res, 4*, 427–435.

Martínez, C., Gac, S., Lavin, A., & Ganga, M. (2004). Genomic characterization of *Saccharomyces cerevisiae* strains isolated from wine-producing areas of South America. *J. Appl. Microbiol., 96*, 1161–1168.

Martínez, J., Millán, C., & Ortega, J. M. (1989). Growth of natural flora during the fermentation of inoculated musts from "Pedro Ximenez" grapes. *S. Afr. J. Enol. Vitic., 10*, 31–35.

Martínez, P., Codon, A. C., Pérez, L., & Benitez, T. (1995). Physiological and molecular characterization of flor yeasts: Polymorphism of flor yeast populations. *Yeast, 11*, 1399—1411.

Martini, A., & Vaughan-Martini, A. (1990). Grape must fermentation: Past and Present. In J. F. T. Spencer, & D. M. Spencer (Eds.), *Yeast technology* (pp. 105—123). Berlin, Germany: Springer-Verlag.

Martorell, P., Querol, A., & Fernández-Espinar, M. T. (2005). Rapid identification and enumeration of *Saccharomyces cerevisiae* cells in wine by real-time PCR. *Appl. Environ. Microbiol., 71*, 6823—6830.

Masneuf, I., Aigle, M., & Dubourdieu, D. (1996). Development of a polymerase chain reaction/restriction fragment length polymorphism method for *Saccharomyces cerevisiae* and *Saccharomyces bayanus* identification in enology. *FEMS Microbiol. Lett., 138*, 239—244.

Masneuf, I., Hansen, J., Groth, C., Piskur, J., & Dubourdieu, D. (1998). New hybrids between *Saccharomyces* sensu stricto yeast species found among wine and cider production strains. *Appl. Environ. Microbiol., 64*, 3887—3892.

Masneuf-Pomarède, I., Le Jeune, C., Durrens, P., Lollier, M., Aigle, M., & Dubourdieu, D. (2007). Molecular typing of wine yeast strains *Saccharomyces bayanus* var. *uvarum* using microsatellite markers. *Syst. Appl. Microbiol., 30*, 75—82.

Mesa, J. J., Infante, J. J., Rebordinos, L., & Cantoral, J. M. (1999). Characterization of yeasts involved in the biological ageing of Sherry wines. *Lebensm.-Wiss. u.-Technol., 32*, 114—120.

Miot-Sertier, C., & Lonvaud-Funel, A. (2007). Development of a molecular method for the typing of *Brettanomyces bruxellensis* (*Dekkera bruxellensis*) at the strain level. *J. Appl. Microbiol., 102*, 555—562.

Molnar, O., Messner, R., Prillinger, H., Stahl, U., & Slavikova, E. (1995). Genotypic identification of *Saccharomyces* species using random amplified polymorphic DNA analysis. *System. Appl. Microbiol., 18*, 136—145.

Moreira da Silva, M., Malfeito-Ferreira, M., Aubyn, S., & Loureiro, V. (1994). Long chain fatty acid composition as criterion for yeast distinction in the brewing industry. *J. Inst. Brew., 100*, 17—22.

Muyzer, G., de Waal, E. C., & Uitterlinden, A. G. (1993). Profiling of complex microbial populations by denaturing gradient gel electrophoresis analysis of polymerase chain reaction-amplified genes encoding for 16S rRNA. *Appl. Environ. Microbiol., 59*, 695—700.

Muyzer, G., & Smalla, K. (1998). Application of denaturing gradient gel electrophoresis (DGGE) and temperature gradient gel electrophoresis (TGGE) in microbial ecology. *Anton. Leeuw., 73*, 127—114.

Nadal, D., Carro, D., Fernandez-Larrea, J., & Piña, B. (1999). Analysis and dynamics of the chromosomal complements of wild sparkling-wine yeast strains. *Appl. Environ. Microbiol., 65*, 1944—1950.

Nadal, D., Colomer, B., & Piña, B. (1996). Molecular polymorphism distribution in phenotypically distinct populations of wine yeast strains. *Appl. Environ. Microbiol., 62*, 1944—1950.

Naumov, G. I., Masneuf, I., Naumova, E. S., Aigle, M., & Dubourdieu, D. (2000). Association of *Saccharomyces bayanus* var. *uvarum* with some French wines: Genetic analysis of yeast populations. *Res. Microbiol., 151*, 683—691.

Naumov, G. I., Naumova, E. S., Antunovics, Z., & Sipiczki, M. (2002). *Saccharomyces bayanus* var. *uvarum* in Tokaj wine-making of Slovakia and Hungary. *Appl. Microbiol. Biotechnol., 59*, 727—730.

Naumov, G. I., Naumova, E. S., & Gaillardin, C. (1993). Genetic and karyotypic identification of wines *Saccharomyces bayanus* yeast isolated in France and Italy. *Syst. Appl. Microbiol., 16*, 272—215.

Ness, F., Lavallée, F., Dubordieu, D., Aigle, M., & Dulau, L. (1993). Identification of yeast strains using the polymerase chain reaction. *J. Sci. Food Agric., 62*, 89—94.

Nguyen, H. V., & Gaillardin, C. A. (1997). Two subgroups within the *Saccharomyces bayanus* species evidenced by PCR amplification and restriction polymorphism of the Non-transcribed spacer 2 in the ribosomal DNA unit. *Syst. Appl. Microbiol., 20*, 286—294.

Nguyen, H. V., Lepingle, A., & Gaillardin, C. A. (2000). Molecular typing demonstrates homogeneity of *Saccharomyces uvarum* strains and reveals the existence of hybrids between *S. uvarum* and *S. cerevisiae*, including the *S. bayanus* type strain CBS 380. *Syst. Appl. Microbiol., 23*, 71—85.

Oelofse, A., Lonvaud-Funel, A., & du Toit, M. (2009). Molecular identification of *Brettanomyces bruxellensis* strains isolated from red wines and volatile phenol production. *Food Microbiol., 26*, 377—385.

Paffetti, C., Barberió, C., Casalone, E., Cavalieri, D., Fani, R., Fia, G., et al. (1995). DNA fingerprinting by random amplified polymorphic DNA and restriction fragment length polymorphism is useful for yeast typing. *Res. Microbiol., 146*, 587—594.

Pedersen, M. B. (1983). DNA sequence polymorphisms in the genus *Saccharomyces*. I. Comparison of the HIS4 and ribosomal RNA genes in larger strains, ALE strain and various species. *Carlsberg Res. Commun., 48*, 485—503.

Pedersen, M. B. (1985). DNA sequence polymorphisms in the genus *Saccharomyces*. II. Analysis of the genes RDN1, HIS4, LEU2 and Ty transposable elements in Calrsberg, Tubog and 22 Babarian brewing strains. *Carlsberg Res. Commun., 50*, 263—272.

Pedersen, M. B. (1986a). DNA sequence polymorphisms in the genus *Saccharomyces*. III. Restriction endonuclease fragment patterns of chromosomal regions in brewing strains. *Carlsberg Res. Commun., 51*, 163—183.

Pedersen, M. B. (1986b). DNA sequence polymorphisms in the genus *Saccharomyces*. IV. Homologous chromosomes III of *Saccharomyces bayanus, S. carlsbergensis, and S. uvarum. Carlsberg Res. Commun., 51*, 185–202.

Pérez, M. A., Gallego, F. J., & Hidalgo, P. (2001a). Evaluation of molecular techniques for the genetic characterization of *Saccharomyces cerevisiae* strains. *FEMS Microbiol. Lett., 205*, 375–378.

Pérez, M. A., Gallego, F. J., Martínez, I., & Hidalgo, P. (2001b). Detection, distribution and selection of microsatellites SSRs in the gemome of the yeast *Saccharomyces cerevisiae* as molecular markers. *Lett. Appl. Microbiol., 33*, 461–466.

Phister, T. G., & Mills, D. A. (2003). Real-time PCR assay for detection and enumeration of *Dekkera bruxellensis* in wine. *Appl. Environ. Microbiol., 690*, 7430–7434.

Prakitchaiwattana, C. J., Fleet, G. H., & Heard, G. M. (2004). Application and evaluation of denaturing gradient gel electrophoresis to analyse the yeast ecology of wine grapes. *FEMS Yeast Res., 4*, 865–877.

Pramateftaki, P. V., Lanaridis, P., & Typas, M. A. (2000). Molecular identification of wine yeasts at species or strains level: A case study with strains from two vine-growing areas of Greece. *J. Appl. Microbiol., 89*, 236–248.

Price, C. W., Fuson, G. B., & Phaff, H. J. (1978). Genome comparison in yeast systematic: Delimitation of species within the genera *Schwanniomyces, Saccharomyces, Debaryomyces* and *Pichia. Microbiol. Rev., 42*, 161–193.

Pulvirenti, A., Nguyen, H., Caggia, C., Giudici, P., Rainieri, S., & Zambonelli, C. (2000). Saccharomyces uvarum, a proper species within *Saccharomyces sensu stricto. FEMS Microbiol. Lett., 192*, 191–196.

Querol, A., & Barrio, E. (1990). A rapid and simple method for the preparation of yeast mitochondrial DNA. *Nucleic Acids Res., 18*, 1657.

Querol, A., Barrio, E., Huerta, T., & Ramón, D. (1992b). Molecular monitoring of wine fermentations conducted by active dry yeast strains. *Appl. Environ. Microbiol., 58*, 2948–2953.

Querol, A., Barrio, E., & Ramón, D. (1992a). A comparative study of different methods of yeast strain characterization. *Syst. Appl. Microbiol., 15*, 439–446.

Querol, A., Barrio, E., & Ramón, D. (1994). Population dynamics of natural Saccharomyces strains during wine fermentation. *Int. J. Food Microbiol., 21*, 315–323.

Querol, A., Jiménez, M., & Huerta, T. (1990). Microbiological and enological parameters during fermentation of musts from poor and normal grape-harvest in the region of Alicante Spain. *J. Food Sci., 55*, 1603–1606.

Quesada, M. P., & Cenis, J. L. (1995). Use of random amplified polymorphic DNA RAPD-PCR in the characterization of wine yeasts. *Am. J. Enol. Vitic., 46*, 204–208.

Redzepovic, S., Orlic, S., Sikora, S., Majdak, A., & Pretorius, I. S. (2002). Identification and characterization of *Saccharomyces cerevisiae* and *Saccharomyces paradoxus* strains isolated from Croatian vineyards. *Lett. Appl. Microbiol., 35*, 305–310.

Renouf, V., Claisse, O., & Lonvaud-Funel, A. (2007). Inventory and monitoring of wine microbial consortia. *Appl. Microbiol. Biotechnol., 75*, 149–164.

Richards, K. D., Goddard, M. R., & Gardner, R. C. (2009). A database profile genotypes for *Saccharomyces cerevisiae. Anton. Leeuw., 96*, 355–359.

Röder, C., König, H., & Fröhlich, J. (2007). Species-specific identification of *Dekkera/Brettanomyces* yeasts by fluorescently labeled DNA probes targeting the 26S rRNA. *FEMS Yeast Res., 7*, 1013–1026.

Rodríguez, M. E., Lopes, C. A., van Broock, M., Valles, S., Ramon, D., & Caballero, A. C. (2004). Screening and typing of Patagonia wine yeasts for glycosidase activies. *J. Appl. Microbiol., 96*, 84–95.

Romancino, D. P., di Maio, S., Muriella, R., & Oliva, D. (2008). Analysis of non- *Saccharomyces* yeast populations isolated from grape musts from Sicily (Italy). *J. Appl. Microbiol., 105*, 2248–2254.

Rosini, G., Federichi, F., & Martini, A. (1982). Yeast flora of grape berries during ripening. *Microbiol. Ecol., 8*, 83–89.

Sabaté, J., Cano, J., Esteve-Zarzoso, B., & Guillamón, J. M. (2002). Isolation and identification of yeasts associated with vineyard and winery by RFLP analysis of ribosomal genes and mitochondrial DNA. *Microbiol. Res., 157*, 267–274.

Sabaté, J., Cano, J., Querol, A., & Guillamón, J. M. (1998). Diversity of *Saccharomyces* strains in wine fermentations: Analysis for two consecutive years. *Lett. Appl. Microbiol., 26*, 452–455.

Saiki, R. K., Gelfand, D. H., Stoffel, S., Schaf, S. J., Higuchi, R., Hoprn, G. T., et al. (1988). Primer-directed enzymatic amplification of DNA with a thermostable DNA polymerase. *Science, 239*, 487–491.

Saiki, R., Sharf, S., Faloona, F., Mullis, K. B., Horn, G. T., & Erlich, H. A. (1985). Enzymatic amplification of beta globin genomic sequences and restriction gene analysis for diagnostic of sickle-cell anaemia. *Science, 230*, 1350–1354.

Salinas, F., Garrido, D., Ganga, A., Veliz, G., & Martínez, C. (2009). Taqman real-time PCR for the detection and enumeration of *Saccharomyces cerevisiae* in wine. *Food Microbiol., 26*, 328–332.

Sambrook, J., Fritsch, E. F., & Maniatis, T. (1989). *Molecular cloning: A laboratory manual* (2nd ed.). Cold Spring Harbor, NY: Cold Spring Harbor Laboratory.

Scheda, R., & Yarrow, D. (1966). The instability of physiological properties used as criteria in the taxonomy of yeast. *Arch. Microbiol., 55*, 209–225.

Scheda, R., & Yarrow, D. (1968). Variations in the fermentative pattern of some *Saccharomyces* species. *Arch. Microbiol., 61*, 310–316.

Schuller, D., Valero, E., Dequin, S., & Casal, M. (2004). Survey of molecular methods for the typing of wine yeast strains. *FEMS Microbiol. Lett., 231*, 19–26.

Schütz, M., & Gafner, J. (1993). Analysis of yeast diversity during spontaneous and induced alcoholic fermentations. *J. Appl. Bacteriol., 75*, 551–558.

Schütz, M., & Gafner, J. (1994). Dynamics of the yeast strain population during spontaneous alcoholic fermentation determined by CHEF gel electrophoresis. *Lett. Appl. Microbiol., 19*, 253–257.

Seehaus, T., Rodizio, R., Heinisch, J., Aguilera, H. D., & Zimmerman, F. K. (1985). Specific gene probes as tools in yeast taxonomy. *Curr. Genet., 10*, 103–110.

Smole-Mozina, S., Dlauchy, D., Deak, T., & Raspor, P. (1997). Identification of *Saccharomyces* sensu stricto and *Torulaspora* yeasts by PCR ribotyping. *Lett. Appl. Microbiol., 24*, 311–315.

Stender, H., Kurtzman, C., Hyldig-Nielsen, J. J., Sørensen, D., Broomer, A., Oliveira, K., et al. (2001). Identification of *Dekkera bruxellensis (Brettanomyces)* from wine by fluorescente in situ hybridization using peptide nucleic acid probes. *Appl. Environ. Microbiol., 67*, 938–941.

Stringini, M., Comitini, F., Taccari, M., & Ciani, M. (2009). Yeast diversity during tapping and fermentation of palm wine from Cameroon. *Food Microbiol., 26*, 415–420.

Theelen, B., Silvestri, M., Gueho, E., van Belkum, A., & Boekhout, T. (2001). Identification and typing of Malassezia yeasts using amplified fragment length polymorphism (AFLP), random amplified polymorphic DNA (RAPD) and denaturing gradient gel electrophoresis (DGGE). *FEMS Yeast Res., 1*, 79–86.

Tofalo, R., Chaves-López, C., di Fabio, F., Schirone, M., Felis, G. E., Torriani, S., et al. (2009). Molecular identification and osmotolerant profile of wine yeasts that ferment a high sugar grape must. *Int. J. Food Microbiol., 130*, 179–187.

Torija, M. J., Rozes, N., Poblet, M., Guillamón, J. M., & Mas, A. (2001). Yeast population dynamics in spontaneous fermentations: Comparison between two different wine-producing areas over a period of three years. *Anton. Leeuw., 79*, 345–352.

Torija, M. J., Rozes, N., Poblet, M., Guillamón, J. M., & Mas, A. (2003). Effects of fermentation temperature on the strain population Saccharomyces cerevisiae. *Int. J. Food Microbiol., 80*, 47–53.

Torriani, S., Zapparoli, G., Malocrinò, P., Suzzi, G., & Dellaglio, F. (2004). Rapid identification and differentiation of *Saccharomyces cerevisiae, Saccharomyces bayanus* and their hybrids by multiplex PCR. *Lett. Appl. Microbiol., 38*, 239–244.

Torriani, S., Zapparoli, G., & Suzzi, G. (1999). Genetic and phenotypic diversity of *Saccharomyces sensu stricto* strains isolated from Amarone wine. Diversity of *Saccharomyces* strains from Amarone wine. *Anton. Leeuw., 75*, 207–215.

Tredoux, H. G., Kock, J. L. F., Lategan, P. M., & Muller, H. B. (1987). A rapid identification technique to differentiate between *S. cerevisiae* strains and other yeast species in the winery industry. *Am. J. Enol. Vitic., 38*, 161–164.

Trilles, L., Lazera, M., Wanke, B., Theelen, B., & Boekhout, T. (2003). Genetic characterization of environmental isolates of the *Cryptococcus neoformans* species complex from Brazil. *Med Mycol., 41*, 383–390.

Urso, R., Rantsiou, K., Dolci, P., Rolle, L., Comi, G., & Cocollin, L. (2008). Yeast biodiversity and dynamics during sweet wine production as determined by molecular methods. *FESM Yeast Res., 8*, 1053–1062.

Vacanneyt, B. P., Hennebert, G., & Kersters, K. (1991). Differentiation of yeast species based on electrophoretic whole-cell protein patterns. *Syst. Appl. Microbiol., 14*, 23–32.

Vaduano, E., & García-Moruno, E. (2008). Discrimination of *Saccharomyces cerevisiae* wine strains using microsatellite multiplex PCR and band pattern analysis. *Food Microbiol., 25*, 56–64.

van der Westhuizen, T. J., & Pretorius, I. S. (1992). The value of electrophoretic fingerprinting and karyotyping wine yeast breeding programmes. *Anton. Leeuw., 61*, 249–257.

van Keulen, H., Lindmark, D. G., Zeman, K. E., & Gerlosky, W. (2003). Yeasts present during spontaneous fermentation of Lake Erie Chardonnay, Pinot Gris and Riesling. *Anton. Leeuw., 83*, 149–154.

van Vuuren, H. J., & van der Meer, L. (1987). Fingerprinting of yeast by protein electrophoresis. *Am. J. Enol. Vitic., 38*, 49–53.

Versavaud, A., Courcoux, P., Roulland, C., Dulau, L., & Hallet, J. N. (1995). Genetic diversity and geographical distribution of wild *Saccharomyces cerevisiae* strains from the wine-producing area of Charentes, France. *Appl. Environ. Microbiol., 61*, 3521–3529.

Vezinhet, F., Blondin, B., & Hallet, J.-N. (1990). Chromosomal DNA patterns and mitochondrial DNA polymorphism as tools for identification of enological strains of Saccharomyces cerevisiae. *Appl. Microbiol. Biotechnol., 32*, 568–571.

Vos, P., Hogers, R., & Bleeker, M. (1995). AFLP: A new technique for DNA fingerprinting. *Nucleic Acids Res, 23*, 4407–4414.

Walmsley, R. M., Wikinson, B. M., & Kong, T. H. (1989). Genetic fingerprinting for yeast. *Biotechnol., 7*, 1168–1170.

White, T. J., Bruns, T., Lee, E., & Taylor, J. (1990). Amplification and direct sequencing of fungal ribosomal RNA genes for phylogenetics. In M. A. Innis, D. H. Gelfand,

J. J. Sninsky, & T. J. White (Eds.), *PCR protocols: A guide to methods and applications* (pp. 315–322). San Diego, CA: Academic Press.

Wilhelm, J., & Pingoud, A. (2003). Real-time polymerase chain reaction. *Review. Chem. Bio. Chem., 4*, 1120–1128.

Williams, J. G. K., Kubelik, A. R., Livak, K. J., Rafalski, J. A., & Tingey, S. V. (1990). DNA-polymorphism amplified by arbitrary primers are useful as genetic markers. *Nucleic Acids Res., 18*, 6531–6535.

Wolfe, K. H., & Shields, D. C. (1997). Molecular evidence for an ancient duplication of the entire yeast genome. *Nature, 387*, 708–713.

Xufre, A., Albergaria, H., Inácio, J., Spencer-Martins, I., & Gírio, F. (2006). Application of fluorescence in situ hybridisation (FISH) to the analysis of yeast populations dynamics in winery and laboratory grape must fermentations. *Int. J. Food Microbiol., 108*, 376–384.

Yamagishi, H., & Ogata, T. (1999). Chromosomal structures of bottom fermenting yeasts. *Syst. Appl. Microbiol., 22*, 341–353.

Yamamoto, N., Yamamoto, N., Amemiya, H., Yokomori, Y., Shimizu, K., & Totsuka, A. (1991). Electrophoretic karyotypes of wine yeasts. *Am. J. Enol. Vitic., 42*, 358–363.

Zott, K., Miot-Sertier, C., Claisse, O., Lonvaud-Funel, A., & Masneuf-Pomarede, I. (2008). Dynamics and diversity of non-Saccharomyces yeasts during the early stages in winemaking. *Int. J. Food Microbiol., 125*, 197–203.

# Genomic and Proteomic Analysis of Wine Yeasts

## José E. Pérez-Ortín, José García-Martínez

Departamento de Bioquímica & Biología Molecular & Laboratorio de Chips de DNA, SCSIE, Universitat de València, Spain

OUTLINE

## 1. INTRODUCTION

The selection of suitable microorganisms for use in industrial processes is a key issue in food biotechnology. One of the key challenges in this area is to improve the properties of starter cultures, such as the ability to establish reproducible growth. Many of the programs aimed at enhancing the properties of industrial microorganisms, however, are restrained by a lack of sufficient knowledge regarding the metabolic and regulatory processes occurring within the cells. These shortcomings may, however, be short-lived, considering the continuous advances being made in functional genomics and proteomics. Studies in these areas will help, for example, to identify the effects of genetic alterations on final products, generate desirable pleiotropic effects through mutations in regulatory genes, predict stress responses in the different environments to which microorganisms are exposed, and identify genomic variations associated with adaptation to the particular conditions of winemaking.

This chapter focuses exclusively on the yeast species *Saccharomyces cerevisiae*. In addition to being the main microorganism involved in wine fermentation, it has been used as a model organism in molecular biology for many years (Miklos & Rubin, 1996) and is the only wine yeast species for which abundant genomic and proteomic information is available. It was the first eukaryote to have its complete genome sequenced (Goffeau et al., 1997), and, since then, numerous functional analysis projects have uncovered enormous amounts of information on the biology of this microorganism (Dujon, 1998). It can safely be said that *S. cerevisiae* is currently the best understood of all eukaryotic organisms. Most of the techniques currently used in functional genomics and proteomics were initially developed in this yeast. DNA chip, or microarray, technology, for example, was primarily developed using *S. cerevisiae* (DeRisi et al., 1997; Schena et al., 1995; Wodicka et al., 1997), and all the latest advances in this field have also been tested using this yeast (see Section 4). Vast amounts of data have thus been compiled on gene expression in *S. cerevisiae*. Indeed, the information on *S. cerevisiae* far exceeds that available for any other prokaryotic or eukaryotic organisms. As a result, it has been possible to propose global models for genetic and metabolic regulation (Gasch et al., 2000).

The fact that *S. cerevisiae* was the first microorganism to be widely used in the development of genome technology allowed other phylogenetically related yeasts to be analyzed subsequently in global sequencing projects, and the use of comparative genomics has since led to important conclusions regarding gene functionality (Butler et al., 2009; Cliften et al., 2003; Kellis

et al., 2003; Liti et al., 2009; Souciet et al., 2000). DNA microarray analysis is a very useful tool for comparing genomes from different strains of *S. cerevisiae*, including wine strains (Carro et al., 2003; Hauser et al., 2001; Schacheter et al., 2009) and similar species.

*S. cerevisiae* has also been used in the development of the more recent field of proteomics. Proteomic studies have generated vast amounts of data on protein expression profiles and variability in laboratory strains of *S. cerevisiae* (Washburn et al., 2001), and these have recently been extended to include wine strains (Rossignol et al., 2009; Trabalzini et al., 2003; Zuzuárregui et al., 2006). Important advances have also been made in metabolomics, a new field in which *S. cerevisiae* is practically the only eukaryote to have been studied to date (Raamsdonk et al., 2001; Rossouw et al., 2008). The integration of different types of "omic" data into predictive models has provided the basis for new research strategies in systems biology (Borneman et al., 2007; Pizarro et al., 2007).

Most of the information that has been gathered in all of the above areas is related to laboratory strains of *S. cerevisiae*, although more recent studies have been extended to other strains (particularly wine strains) and industrial processes (Bisson et al., 2007). Knowledge generated from the analysis of laboratory strains may be helpful in understanding the results of studies conducted with wine strains during industrial fermentation, and it is extremely simple to apply techniques used with laboratory strains to their industrial counterparts. This chapter will therefore also look at the methods used and results obtained for non-wine strains of *S. cerevisiae*.

## 2. GENOMIC CHARACTERISTICS OF WINE YEASTS

The history of wine yeasts is as old as the earliest civilizations in the Mediterranean region, with the first references to winemaking dating back to 7400 years ago. Reports of wine production were limited to this geographical area for many centuries, until the practice was spread to other parts of the world with suitable climate conditions, as Europe embarked on its conquest of other continents in the fifteenth century (reviewed in Mortimer, 2000 and Pretorius, 2000). Must fermentation was considered to occur spontaneously until 1863, when Louis Pasteur discovered that yeasts were responsible for the process. Although numerous yeasts and bacteria contribute to must fermentation (see Chapters 2−6 and 9), the principle microorganisms responsible for this biotransformation belong to the genus *Saccharomyces*, principally *S. cerevisiae*. This is why *S. cerevisiae* is often referred to as *the* wine yeast (Pretorius, 2000).

The origin of *S. cerevisiae* has been much debated. While some authors are of the opinion that it is naturally present in fruit (Mortimer & Polsinelli, 1999), others believe that its origin is more recent and that it is the result of hybridization with other natural species and subsequent natural selection in artificial environments (Martini, 1993). This second hypothesis is supported by the fact that *S. cerevisiae* is found only in areas close to human activity. According to this theory, all the modern isolates of *S. cerevisiae* would have been transported by insects from the winery back to the vineyards (Naumov, 1996). While this debate is central to determining the true origin of the *S. cerevisiae* genome, what is known for certain is that the genomic constitution of this species has been molded by the severe fermentation-related stresses to which it has been exposed throughout the centuries. Proof of this are the genomic differences between primary and secondary fermentation wine strains and between brewing strains and bread-making strains, whose genotypes have been unknowingly selected over hundreds of years with the continual improvements made to these biotechnological processes. Another

important point is that all of today's laboratory strains are derived from natural isolates. The best-documented case is that of the most popular yeast among molecular biologists: the S288c strain, which was derived from a heterothallic (ho), diploid strain isolated in a rotten fig in California in 1938 (Mortimer & Johnston, 1986). It is very likely that the strain had been transported from a winery by insects.

Most laboratory strains of *S. cerevisiae* are ho, haploid or diploid, and have a set of 16 fixed-length chromosomes (see Figure 6.1). The majority of wine strains, in contrast, are diploid, aneuploid, or polyploid (Bakalinsky & Snow, 1990; Codón et al., 1995). They are also homothallic (HO), variably heterozygous (Barre et al., 1993; Butler et al., 2009; Carreto et al., 2008; Codón et al., 1995), and characterized by a high level of polymorphism in chromosome length (Bidenne et al., 1992; Rachidi et al., 1999). Many strains are trisomic or tetrasomic for certain chromosomes (Guijo et al., 1997; Bakalinsky & Snow, 1990). The above characteristics have numerous practical implications, including highly variable sporulation capacity (0−75%) (Bakalinsky & Snow, 1990; Barre et al., 1993; Mortimer et al., 1994) and spore viability (0−98%) (Barre et al., 1993; Codón et al., 1995; Mortimer et al., 1994). The ability of *S. cerevisiae* to alter its genome is enhanced by the existence of mitotic and meiotic cycles. Genome ploidy and plasticity provide wine yeasts with certain advantages that facilitate their adaptation to changing external environments and perhaps also increase the dosage of genes that have an important role in fermentation (Bakalinsky & Snow, 1990; Salmon, 1997). This genomic plasticity, however, is not restricted to *S. cerevisiae* and even allows stable hybridization with closely related species. Several natural strains, such as S6U and CD1, for example, are hybrids of *S. cerevisiae* and *Saccharomyces bayanus*. S6U is an allotetraploid (Naumov et al., 2000), which probably explains its stability despite having two distinct

**(a)**

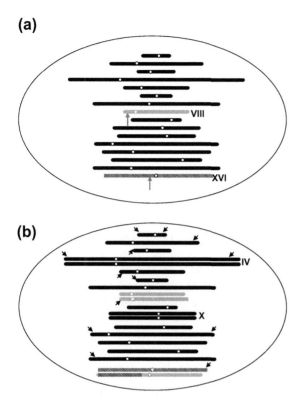

**(b)**

FIGURE 6.1 The genome of the reference *Saccharomyces cerevisiae* laboratory strain has 16 chromosomes whose lengths are shown to scale (a). The centromeres are shown as white dots. The haploid genome is shown in this figure. Diploid strains have two, probably identical, copies of each chromosome. Many variants of this reference genome have been found in wine strains. The T73 strain (b), for example, isolated in musts from the Alicante appellation (Querol et al., 1992) has, at least, the following variations: (1) a reciprocal translocation between chromosomes VIII and XVI, which generates two variants of each chromosome in T73 (Pérez-Ortín et al., 2002a) (the site of the translocation is shown by grey arrows) (a); (2) an additional, presumably identical, copy of chromosomes IV and X (Pérez-Ortín, unpublished results); (3) many variations in the copy number of genes from subtelomeric families, shown by arrowheads (b) (García-Martínez & Pérez-Ortín, unpublished results); and (4) markedly fewer copies of Ty transposons (Hauser et al., 2001). The T73 genome shown probably has two copies of each chromosome except for chromosomes IV and X. For simplicity, we have shown just a single copy of chromosomes with two copies. For chromosomes with three copies, we show the name and just two copies. We have included the two copies of chromosomes VIII and XVI to show the translocation between these chromosomes.

genomes. The same has been observed with brewing strains (Kielland-Brandt et al., 1995). The formation of interspecific hybrids between members of the *Saccharomyces sensu stricto* group appears to be one of the adaptive mechanisms employed by industrial yeasts (Belloch et al., 2009; Querol et al., 2003). This genome plasticity, which is inherent in wine strains, is not a desirable property in model organisms used in genetic studies, and laboratory strains used for such purposes are selected precisely for their lack of plasticity. Laboratory strains are also capable of adapting to changing environmental conditions, normally via point mutations (Ferea et al., 1999), although in certain circumstances large regions or entire chromosomes may also be modified (Hughes et al., 2000b).

Wine strains, unlike laboratory strains, are capable of chromosomal rearrangement during mitosis (Longo & Vézinhet, 1993). In an experiment by Puig et al. (2000), *URA3* was replaced with an exogenous marker gene, *KanMX*, in the natural wine strain T73 and used to monitor genetic variation in a series of consecutive must fermentations. The authors found that *URA3* homozygotes appeared at a rate of $2 \times 10^{-5}$ per generation in a process they attributed to mitotic recombination or gene conversion. Phenotypically, the Ura$^-$ cells were at a selective disadvantage to the Ura$^+$ cells (heterozygotes [*URA3/ura3*] and homozygotes [*URA3/URA3*]). Chromosomal changes were also detected in some cells. Because of their strong tendency towards genomic changes, wine strains do not display the same genetic uniformity as that used to define laboratory strains (Pretorius, 2000; Snow, 1983). This problem is further compounded by the HO nature of these strains. Haploid cells produced by sporulation can change their mating type and conjugate to form new diploid cells. The frequent use of such mechanisms during vinification would lead to the generation of multiple genome combinations and very rapid changes. This particular evolutionary mechanism has been termed "genome renewal" (Mortimer et al., 1994; Mortimer, 2000). The proponents of this theory suggest that this renewal would give rise to highly homozygous strains and eliminate deleterious mutations by natural selection. Natural strains are known, however, to be typically aneuploid (Bakalinsky & Snow, 1990; Guijo et al., 1997) and heterozygous for many loci (Barre et al., 1993; Kunkee & Bisson, 1993), and such properties are inconsistent with the genome renewal hypothesis (Puig et al., 2000). While the possible influence of meiotic changes cannot be entirely ruled out, there are other mechanisms that might explain the natural variation observed in wine strains. For instance, translocations mediated by Ty transposons (Rachidi et al., 1999), mitotic crossing-over (Aguilera et al., 2000), and gene conversion have all been described as mechanisms capable of causing the most rapid adaptive changes (Puig et al., 2000).

The practice of inoculating must with pure wine yeast cultures to improve the quality and homogeneity of wines produced from one year to the next dates back to the 1970s (Pretorius, 2000). Pure cultures have been obtained from natural strains in wine-producing countries around the world. In the first half of the twentieth century, these strains were selected and modified by more or less empirical methods. The selection techniques were improved in later years, however, with the emergence of classical genetic tools (reviewed in Pretorius, 2000). The end of the twentieth century brought genetic engineering methods that opened up a world of possibilities and further improved the quality of the selection methods used (see Chapter 8). The plasticity of the wine strain genome, however, poses a new challenge, as there is a risk of genetically engineered changes becoming unstable with successive generations. Mutations or insertions in a single locus, for example, could eventually be eliminated by gene conversion, homologous recombination, or even perhaps

by meiosis and conjugation. Consequently, all the homologous loci in a particular strain (two or more, depending on the case) must be manipulated in an identical fashion to ensure the phenotypic stability of the strain (Puig et al., 1998, 2000).

## 3. COMPARATIVE GENOMICS AND THE ORIGIN OF THE *S. CEREVISIAE* GENOME

Although the origin of *S. cerevisiae* is unknown, that of its genome can be investigated by comparing genomes from natural strains of this species with those from other more-or-less-related species. A better understanding of the origin and evolution of the *S. cerevisiae* genome will have a positive impact in numerous areas. It will greatly improve our knowledge of the origin of the species and the ways in which it has adapted to industrial processes over the years, and also shed light on the mechanisms underlying the evolution of its genome, and, by extension, that of other eukaryotic organisms.

Comparative genomics studies in yeasts have been performed by partial or complete sequencing followed by bioinformatic comparison of sequence data and chromosomal organization of genes. The first complete genome sequence for *S. cerevisiae* was published for a laboratory strain in 1997 (Goffeau et al., 1997). The corresponding sequences for natural wine strains were made available about 12 years later (Borneman et al., 2008; Novo et al., 2009). Today, full genome sequences are available for several dozen *S. cerevisiae* strains, including laboratory, wine, and other strains (Liti et al., 2009; Schacherer et al., 2009).

In 1997, it was suggested that the *S. cerevisiae* genome was the result of an ancient duplication, dating back approximately $10^8$ years, of an ancestral genome followed by the elimination of duplicated genes and the acquisition of new functions for other genes (Wolfe & Shields, 1997). This theory would explain the genetic redundancy detected in this species. *S. cerevisiae* has 2458 genes from 722 families containing between two and 108 members (Herrero et al., 2003). Part of the redundancy would be due to ancestral duplication and part to smaller duplications that took place later (Llorente et al., 2000). The existence of large numbers of gene families is a common feature of hemiascomycetous yeasts. In a comparative genomic study of these yeasts, Malpertuy et al. (2000) found a substantial number of genes that do not exist in other organisms. The genes, which are specific to ascomycetes, seem to have evolved more rapidly and are perhaps responsible for the biological differences that characterize this group of yeasts. When this ancient duplication actually took place in *S. cerevisiae* is a subject of debate. Langkjaer et al. (2003) postulated that it was before the divergence of *Saccharomyces* and *Kluyveromyces* but other authors have suggested that it was later (Fares & Wolfe, 2003). In a study of collinearity (synteny) between different hemiascomycete species, Llorente et al. (2000) proposed that the primary evolutionary mechanism (apart from global genome duplication) was the duplication of small regions (the length of a few genes) of the genome followed by specialization or gene loss. In related species, such as *S. cerevisiae* and *S. bayanus*, the duplication sites tend to be located close to copies of Ty transposons or in subtelomeric regions where families of repeated genes are concentrated (Fischer et al., 2001). A genomic comparison of *S. cerevisiae*, *Saccharomyces paradoxus*, *S. bayanus*, and *Saccharomyces mikatae* found the greatest variability in subtelomeric regions, particularly in terms of repeated gene families (Kellis et al., 2003). These regions range in size from 7 to 52 kb and their function might be to facilitate rapid changes via duplication and translocation. While these mechanisms have played a part in the evolution of the *Saccharomyces* genus, they have also had a much more recent role in facilitating adaptation to specific industrial processes. Indeed, various subtelomeric gene families are of immense importance to the biology of these yeast strains. Based on the results of a comparative genomic study of multiple wine and non-wine strains, Carreto et al. (2008) proposed that the diversity observed in the strains analyzed was mainly the result of Ty element insertions and subtelomeric recombination. The fact that the subtelomeric regions of different chromosomes contain many members of gene families involved in hexose transport (Bargues et al., 1996), use of natural carbon sources such as sucrose (Carlson et al., 1989), maltose (Chow et al., 1989), and melibiose (Naumova et al., 1997), flocculation (Teunissen & Steensma, 1995), and resistance to the toxicity of molasses in which industrial yeasts are cultured (Ness & Aigle, 1995) suggests that these regions might act as reservoirs of variability for rapid adaptations to the changing environments to which industrial yeasts are exposed. This mechanism may indeed still be very active in certain strains such as Cava strains, in which high rates of subtelomeric variability have been detected (Carro et al., 2003; Carro & Piña, 2001). Small and large duplications and translocations may also have contributed to speciation due to reproductive isolation in the *Saccharomyces* genus (Delneri et al., 2003; Fischer et al., 2000, 2001). There may be other cases where the selection of one particular chromosomal rearrangement rather than another is random. Nonetheless, it is reasonable to think that many of the combinations produced by the different genomic rearrangement mechanisms discussed above have been selected because they provide the organism with a particular selective advantage. Our group found a case in which reciprocal translocation between chromosomes VIII and XVI gave rise to a new, more efficient promoter for the sulfite resistance gene *SSU1* (Pérez-Ortín et al., 2002a). As sulfite has been used as a treatment in vineyards, wineries, and wines for thousands of years, resistance to this substance was probably selected by wine strains as a useful survival mechanism. In an extensive study of translocation between various wine and non-wine strains, our group found that the reciprocal translocation between chromosomes VIII and XVI was present in some but not all of the wine strains, but was absent from all the non-wine strains, providing evidence that this translocation is associated with the use of sulfite in winemaking (Pérez-Ortín et al., 2002a). In that study, we also detected a close phylogenetic relationship between wine strains from geographically distant countries such as South Africa, France, Japan, Spain, and the United States, suggesting that strains that had originated in Europe were spread to other parts of the world with the expansion of winemaking.

The recent development of high-resolution genome mapping techniques such as mass sequencing and tiling array analysis (see Section 4) has permitted the genomic sequencing of several dozen *S. cerevisiae* strains and the formulation of hypotheses regarding the origin of this species and that of other strains used for biotechnological purposes (brewing, bread making, sake production) and pathogenic strains isolated in immunosuppressed patients (Liti et al., 2009; Schacherer et al., 2009). Single nucleotide polymorphism (SNP) analysis has shown that the genomes of different strains of *S. cerevisiae* tend to represent a mosaic generated by recombination between lineages with different geographical and/or ecological origins (Liti et al., 2009). What seems clear is that this species has been domesticated on various separate occasions, at least once in the case of wine fermentation and another time in the case of sake fermentation (Liti et al., 2009). Today's strains would thus be derivatives and combinations of those initial domesticated strains. Pathogenic strains, however, seem to have arisen on multiple occasions from wild and domesticated strains opportunistically adapted to the new

ecosystem of human tissues (Schacherer et al., 2009).

Another interesting point worth noting is the discovery of hybrid wine yeasts derived from *S. cerevisiae* and other *Saccharomyces* species. It has been known for some time that certain lager brewing strains have genomes derived from more than one species (Rainieri et al., 2006). These strains are partial allotetraploids that arose from a natural hybridization event between *S. cerevisiae* and a yeast similar to *S. bayanus* (Nakao et al., 2009; Rainieri et al., 2006). More recently, however, there have also been descriptions of wine strains with a genome containing chromosomes from more than one species and wine yeast hybrids of *S. bayanus* and *Saccharomyces kudriavzevii* (González et al., 2006). Genomic analysis showed that all the hybrids arose from a single hybridization event. The resulting genome would then have evolved through successive chromosome rearrangements resulting in the generation of hybrid chromosomes and the loss of several chromosome copies (mostly corresponding to *S. kudriavzevii*). Such rearrangements affected not only sequences of transposons (as in the cases described above) but also other conserved regions such as ribosomal DNA (rDNA) and protein-encoding genes (Belloch et al., 2009). The study of these hybrids is of practical interest because they might have useful properties for biotechnological applications. It is known, for example, that *S. bayanus* var. *uvarum* is responsible for the fermentation of must at low temperatures and the production of large quantities of glycerol and β-phenylethanol (Solieri et al., 2008). In an attempt to obtain yeast strains with improved winemaking properties, Solieri et al. (2008) constructed artificial hybrids between *S. cerevisiae* and *Saccharomyces uvarum* by spore conjugation and found that the hybrids contained mitochondria from only one of the two species and that the fermentative properties of the hybrid depended on these mitochondria.

## 4. THE USE OF *S. CEREVISIAE* AS A MODEL ORGANISM FOR THE DEVELOPMENT OF DNA MICROARRAY TECHNOLOGY

There are a number of reasons why many of the technologies used in the field of genomics were developed using *S. cerevisiae*, but the main one is probably that it was the first organism to be analyzed in a genomic sequencing project that generated numerous functional genomics studies even before the full sequence was published (Goffeau et al., 1997). The fact that *S. cerevisiae* has been used as a model organism for genetics and molecular biology since the 1940s has given rise to an enormous number of very powerful tools for these types of analysis. As a result of these developments, our knowledge of the genetics and biology of this yeast is unparalleled. The only other organism that has been so thoroughly investigated is perhaps *Escherichia coli*. Even before the emergence of DNA microarray technology, *S. cerevisiae* was used in the development of numerous methods for the global analysis of gene expression such as Serial Analysis of Gene Expression (SAGE) technology, which was used to perform the first analysis of the entire messenger RNA (mRNA) complement (baptized transcriptome) of a cell (Velculescu et al., 1997). As with many other technologies, SAGE was later used to analyze other organisms with great success (Velculescu et al., 2000). While SAGE is an extremely powerful tool, capable of accurately quantifying the number of copies of mRNA present in a cell, it has largely been replaced by DNA microarray analysis, which is a much simpler and less costly technology. In recent years, however, the development of high-throughput sequencing techniques (also developed using *S. cerevisiae*) has led to a renewed interest in tag-sequencing technologies. RNA-seq, for example, has been successfully used to characterize the transcriptome of *S. cerevisiae* with considerable improvements over previous techniques in terms of sensitivity, transcript quantification, and, to some degree, resolution (Nagalakshmi et al., 2008).

DNA microarrays have been widely used to investigate many aspects of *S. cerevisiae* metabolism (Figure 6.2). The technology has other uses, however. Apart from providing valuable information on metabolic activity in different conditions and mutants, it has also been used to investigate the effects of many drugs and toxic products on gene expression and to analyze genomic variations in *S. cerevisiae* and related species. All of these uses have also been applied to wine yeast strains.

### 4.1. Metabolic Studies

Given the vast information already available on yeast regulatory pathways, global expression studies should be able to provide sufficient data to allow individual genes to be linked to one or more phenotypes or metabolic pathways. It should also theoretically be possible to determine the components of each of these pathways, to provide, for the first time, a global view of a eukaryotic cell. The first global gene expression study, performed by Pat Brown's group, used DNA microarray analysis to study gene expression in *S. cerevisiae* during growth in glucose and during the shift from fermentative to respiratory growth (DeRisi et al., 1997). The study has already become a classic in its field and has been cited over 2500 times (as of August 2009). Similar studies have analyzed other processes or situations that involve metabolic changes. Transcriptional changes in *S. cerevisiae*, for example, have been analyzed in the change from a fermentable to a nonfermentable carbon source (Kuhn et al., 2001), in aerobic compared to anaerobic conditions in a continuous-culture study (ter Linde et al., 1999), in the lag phase prior to active culture growth (Brejning et al., 2003), during sporulation (Chu et al., 1998), and during the cell cycle (Cho et al., 1998). Another major research focus is the functional

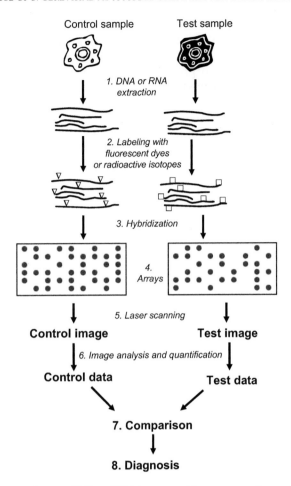

**FIGURE 6.2**   DNA microarray analysis. 1. RNA or DNA is extracted from a test and a control sample using conventional methods. 2. In the case of microarrays on glass slides, the probes are labeled with fluorescent dyes (using a different fluorophore for the test and control sample). The probes used in macroarrays on nylon filters are labeled with radioactive isotopes. 3. Just one hybridization step is used in glass-slide microarrays as these involve the use of a single array with both the test and control samples mixed together prior to hybridization. Two hybridization steps are required for nylon-filter macroarrays. These steps are preferably performed on the same filter but they need to be sequential as the probes are labeled with radioactive isotopes. 4. Following hybridization, the arrays are washed to allow detection of the different hybridization signals. 5. The hybridization images are captured using a laser scanner. This is done directly using two different lasers (one for each fluorophore) in the case of microarray analysis. In macroarray analysis, however, latent images are generated on special screens and later scanned by laser. 6. The readings generate an image for each sample. The intensity is then quantified using special software that generates hybridization intensity data that allows comparison of the samples. 7. Statistically significant differences are analyzed using purpose-designed programs. 8. The final stage involves the formulation of corresponding hypotheses and conclusions.

analysis of transcription factors via overexpression or analysis of null or conditional mutants (Carmel-Harel et al., 2001; DeRisi et al., 1997; Holstege et al., 1998).

Clusters of genes that display identical or similar expression patterns under the different conditions studied have been used to identify the functions of individual genes based on the assumption that coregulated genes must be involved in the same metabolic pathways. The most common way to conduct a study of this type is to use clustering algorithms to group genes by expression profiles (reviewed in Hughes & Shoemaker, 2001 and Brazma & Vilo, 2000) in order to identify groups that have putative functional relationships. Another way is to search for transcription-factor-binding sites in gene promoters. Two types of study have been used for this purpose: in silico comparison of promoter sequences (Brazma et al., 1998; Bussemaker et al., 2000; Hampson et al., 2000; Roth et al., 1998) and in vivo studies of genome-wide transcription-factor-binding sites using a technique called Chip-ChIP, which is a combination of DNA microarray analysis (Chip) and chromatin immunoprecipitation (ChIP).

## 4.2. Effects of Drugs and Other External Factors

DNA microarray technology can be used to measure, in a single experiment, an organism's global transcriptional response to treatment with an external factor such as a drug or environmental agent (Gasch et al., 2000; Hughes et al., 2000a; Jelinsky et al., 2000; Jelinsky & Samson, 1999). Because the response of genes to experimental conditions is a dynamic process characterized by multiple interactions, analyzing responses to external agents can reveal functional relationships within or between metabolic pathways. Such techniques have been used to analyze, for example, the transcriptional response to inhibition of translation or amino acid biosynthesis, or to compounds with antifungal activity (Bammert & Fosel, 2000; Hardwick et al., 1999; Jia et al., 2000). Molecular targets of specific drugs can also be identified by comparing expression profiles induced by a particular drug with those induced in mutants for specific genes (Hughes et al., 2000a). Similar results can be achieved by

inducing haploinsufficiency, which consists of studying growth deficiencies caused by the loss of one of the two gene copies in a diploid cell. To perform a systematic, comparative study, it is necessary to have a full collection of a diploid strain in which each gene has been deleted and replaced with a specific sequence tag (Winzeler et al., 1999). In these studies, the full collection of approximately 6000 strains with single deletions is grown together under particular conditions (such as the presence of a drug) and strains that exhibit delayed growth compared to wild-type strains indicate genes that are necessary for resistance to certain drugs or culture conditions (Giaever et al., 1999, 2002). This technique can uncover subtle growth differences that would otherwise remain undetected. Up to 6000 strains can be compared simultaneously thanks to the sequence tags present in each strain, which enable an accurate count to be made of the cells in a strain at any moment using special DNA microarrays containing probes for each sequence tag. These studies open new perspectives not only for pharmacogenomics but also for the study of the effect on wine yeasts of toxic substances such as alcohol, pesticides, and treatments such as copper and sulfite. Although most of the studies to date have been conducted using standard laboratory strains, the results can be easily extrapolated to industrial strains.

The fermentation of sugars by wine yeasts is followed by rapid growth and carbon dioxide production, which can be interrupted with the depletion of carbon or nitrogen sources or the appearance of growth inhibitors (reviewed in Pretorius, 2000). An improved understanding of the metabolic changes that occur in the shift from one carbon source to another (DeRisi et al., 1997; ter Linde et al., 1999) and of metabolic signal transduction pathways (Hardwick et al., 1999; Ogawa et al., 2000) will contribute to improving the technical aspects of fermentation processes in wineries and help to prevent stuck fermentations.

In an extensive large-scale experiment that analyzed response to many of the stress conditions to which yeasts are exposed, Gasch et al. (2000) found that the transcriptional response to almost all of the stress factors tested was practically identical across a large group of genes. The authors termed this the "environmental stress response" (ESR). The experiment provided a basis for further tests with wine strains exposed to general or specific stresses associated with wine fermentation. Indeed, the first experiments of this type have already been performed (see below). The ultimate aim of such studies is to identify the most suitable strains for the various fermentation conditions found in different wineries and wines. Similar experiments involving a limited group of genes have also been performed (Ivorra et al., 1999). A more detailed discussion is given in Chapter 2 of this book. On investigating the effect of ethanol on laboratory yeast strains by DNA microarray analysis, Alexandre et al. (2001) concluded that cells used ionic homeostasis, heat protection, and antioxidant defense, in addition to previously described mechanisms, to respond to stress. In a study of the effect of copper excess and deficiency on laboratory strains, also using DNA microarrays, Gross et al. (2000) found that a small number of genes were differentially expressed and that some of these were involved in the iron uptake system. This finding suggests that the copper and iron uptake systems might be related. Because copper is commonly used to inhibit bacterial and fungal growth in wines, wine yeast strains must be able to endure elevated copper concentrations and it would be useful to determine how they have achieved this capacity.

## 4.3. Use of DNA Microarrays in the Analysis of Wine Yeasts

S. cerevisiae was also the first microorganism in which genomic tools such as DNA microarray analysis were used to analyze natural and industrial strains. Since this yeast plays a key role in winemaking and has an enormous influence on the final product, it is important to understand the molecular events underlying fermentation and the influence of the winery and vintage, and of the physical, biological, and chemical properties of the must, on this process. Such an understanding would be greatly enhanced by analysis of the gene expression profiles of these yeasts in different growth conditions. Before the emergence of DNA microarray technology, the expression profiles of only a small number of genes at a time could be analyzed in wine yeasts (see Chapter 2). Most of the DNA microarray experiments described so far in this chapter, however, have analyzed laboratory strains, which are incapable of wine fermentation.

Various strategies have been employed in studies using DNA microarrays to analyze expression profiles in wine yeasts. Two studies have been conducted using laboratory media and culture conditions (Cavalieri et al., 2000; Hauser et al., 2001), whereas others have used synthetic musts that reproduce the conditions found in a natural environment but provide the means to accurately determine and reproduce the composition of the must (Backhus et al., 2001; Rossignol et al., 2003). Another strategy has involved the use of grape juice medium sterilized by filtration (Marks et al., 2003; Mendes-Ferreira et al., 2007a, 2007b).

The use of standard laboratory conditions has the advantage of allowing comparison of data from wine strains with those from the more extensively studied laboratory strains. The enormous amounts of information available on reference strains can thus be used to undertake a much more in-depth investigation of the metabolic pathways and molecular mechanisms underlying wine yeast fermentation. Cavalieri et al. (2000), for example, detected at least two-fold variability in global expression levels for 6% of the genome between progeny of a natural wine strain isolate. Their findings indicate that wine strains are highly heterozygous. Because most of the metabolic differences segregated as a suite of traits, the authors concluded that they were the result of changes in a small number of regulatory genes. One specific example would be the genes involved in the biosynthesis of amino acids. There have also been descriptions of other phenotypes caused by changes in structural rather than regulatory genes, explaining why these changes are not associated with other phenotypes. Examples include the *YHB1* gene (Hauser et al., 2001), genes involved in resistance to sulfite (Pérez-Ortín et al., 2002a) and copper, and the filigreed phenotype (Cavalieri et al., 2000).

It is important to conduct experiments in real-life conditions as, although laboratory culture conditions greatly facilitate analysis, they do not fully reproduce the conditions found in natural environments. Given the variability of natural musts, one option is to use synthetic musts, which mimic natural conditions but can be easily reproduced in different laboratories. In a study of this type, using macroarrays and various wine strains with different fermentative capacity, Zuzuárregui and del Olmo (2004) found that the expression levels of certain stress-response genes were similar across the strains. They also found that the mRNA levels of many of these genes remained very high in the strains with weaker fermentative capacity. Their results demonstrated that it is possible to establish a correlation between stress resistance and fermentation capacity.

The amount of available nitrogen is considered to be one of the main limiting factors for yeast growth in musts (reviewed in Pretorius, 2000). Studies performed with wine yeasts have generally found high expression levels for genes linked to amino acid and purine biosynthesis (Backhus et al., 2001; Cavalieri et al., 2000; Hauser et al., 2001), which are indicative of high growth rates. Activation of the methionine biosynthesis pathway and alterations in sulfate and nitrogen assimilation are known markers for metabolic phenotype as they are connected with cell-cycle progression (Patton et al., 2000). The effect of nitrogen availability on the growth of wine yeasts has been analyzed in two recent studies. One of these compared global gene expression profiles in synthetic media containing high and low concentrations of arginine (a source of nitrogen) (Backhus et al., 2001), whereas the other compared expression profiles in a Riesling must with normal concentrations of nitrogen and another to which diammonium phosphate (DAP) was added during the late fermentation phase, when yeast growth is no longer active (Marks et al., 2003). In the first study, it was found that nitrogen limitation induced genes that would normally be repressed by the high concentrations of glucose in the must. This suggests that, in the growth conditions that characterize the fermentation of must containing high concentrations of sugars and nitrogen, the use of glucose might be diverted, at least partly, to a respiratory metabolism (Backhus et al., 2001). This effect would be similar to what is known as the Pasteur effect, which is the inhibition of fermentation in the presence of oxygen. Although this effect has been reported to be irrelevant for yeast in laboratory growth conditions (Lagunas, 1986), it might occur in the fermentation of musts with low levels of nitrogen, and, accordingly, cause sluggish or stuck fermentations. Indeed, it is standard practice in wineries to add DAP in such cases. A study by Marks et al. (2003) found that the addition of DAP affected the expression of 350 genes. The 185 genes that were found to be downregulated encoded small-molecule transporters and nitrogen catabolic enzymes, including enzymes involved in the synthesis of urea, which is a precursor of ethyl carbamate. The other 165 genes affected were all upregulated. These included genes involved in the biosynthesis of amino acids, purines, and ribosomal proteins (suggesting a more active metabolism despite an absence of cell proliferation) and assimilation of inorganic sulfate (necessary for the elimination of hydrogen sulfide). The results of the study by Marks et al. provide a possible explanation for why the addition of DAP reduces the production of ethyl carbamate and hydrogen sulfide, two undesirable components in wines. They are also consistent with results from a study that analyzed samples taken at different time points during fermentation of a synthetic must with a relatively low level of nitrogen (300 mg/L). The authors reported that the gene expression pattern observed could be explained by entry into the stationary phase (cell proliferation arrest) in response to nitrogen depletion; they also reported that the process was regulated by the *TOR* pathway (Rossignol et al., 2003).

A more comprehensive and realistic study of transcriptional response in *S. cerevisiae* to different nitrogen concentrations during alcoholic fermentation was published more recently (Mendes-Ferreira et al., 2007a, 2007b). The authors, using real grape must, compared 11 samples from different time points of a series of control vinifications, nitrogen-limiting fermentations, and fermentations to which DAP was added. They found alterations in approximately 70% of the yeast transcriptome in at least one of the fermentation stages and also showed a clear association between these changes and nitrogen concentrations. In agreement with earlier findings published by Backhus et al. (2001), their results indicated that early response to nitrogen limitation involved the induction of genes associated with respiratory metabolism and a subsequent general decrease in the levels of genes associated with catabolism. Curiously, they also found a slight increase in the expression level of genes encoding ribosomal proteins and involved in ribosome biogenesis during nitrogen depletion. In total, 36 genes were found to be overexpressed when nitrogen levels were low or absent compared to when DAP was added. These signature genes might be useful for predicting nitrogen deficiency and detecting

sluggish or stuck fermentations (Mendes-Ferreira et al., 2007b). The study also demonstrated that the main transcriptional effect of adding nitrogen was an upregulation in genes involved in glycolysis, thiamine metabolism, and energy pathways (Mendes-Ferreira et al., 2007a), findings that are similar to those reported by Marks et al. (2003) following DAP addition. A study performed by Jiménez-Martí and del Olmo (2008) showed that the effect of nitrogen refeeding depended on the source of nitrogen used, as they detected differences in gene expression reprogramming depending on whether ammonia or amino acids were added. The addition of ammonia resulted in higher levels of genes involved in amino acid biosynthesis, whereas that of amino acids directly prepared cells for protein biosynthesis.

Global gene response has also been analyzed in low-temperature winemaking conditions, which are widely considered to improve the sensory quality of wine. In experiments carried out at 13 and 25°C, Beltrán et al. (2006) observed that the lower temperature induced cold stress-response genes at the initial stage of fermentation and increased levels of genes involved in cell cycle, growth control, and maintenance in the middle and late stages of fermentation. Furthermore, several genes involved in mitochondrial short-chain fatty acid synthesis were found to be overexpressed at 13°C compared to 25°C. These transcriptional changes were correlated with higher cell viability, improved ethanol tolerance, and increased production of short-chain fatty acids and associated esters.

The natural environment of *S. cerevisiae* has shaped the evolution of this organism's metabolism to allow it to exploit the anaerobic conditions and high ethanol levels that characterize fermentation and to tolerate high levels of certain compounds that are common during alcoholic fermentation. All these situations, however, are causes of stress for *S. cerevisiae* and are reflected in the yeast's gene expression pattern, even though the organism is capable of responding effectively to these stresses. As has already been discussed, differential expression of certain stress-response genes has been detected in wine yeasts. The expression levels of genes involved in oxidative metabolism, for example, are low (Backhus et al., 2001). The results of the fermentation monitoring study conducted by Rossignol et al. (2003) indicate that anaerobic stress is a characteristic of wine fermentation and that the absence of ergosterol synthesis, one of the main growth-limiting factors for yeasts in musts with low oxygen and high ethanol levels (see Pretorius, 2000), is due to the continuous decrease in the expression levels of genes involved in ergosterol biosynthesis.

Ethanol stress is another major pressure that *S. cerevisiae* has to deal with during vinification. Ethanol tolerance is not fully understood (Pretorius, 2000) but it is known to partly depend on alterations in the plasma membrane. Genes encoding enzymes involved in the synthesis of fatty acids, phospholipids, and ergosterol are highly expressed (Backhus et al., 2001) in *S. cerevisiae* yeasts but decrease towards the stationary phase (Rossignol et al., 2003). Using microarray analysis to identify target genes and analyze ethanol sensitivity in knockout strains, Hirasawa et al. (2007) found that the biosynthesis of tryptophan can confer ethanol tolerance. Ethanol stress, however, does not appear to be the main pressure in vinification. The greatest effect on gene expression is produced upon entry into the stationary phase (Rossignol et al., 2003). The changes in gene expression seen in this phase, however, appear to differ from those observed under laboratory conditions (Gasch et al., 2000).

In a comprehensive study of the transition from the exponential to the stationary phase in wine fermentation, Marks et al. (2008) discovered 223 genes that were dramatically induced at various points during fermentation. They called this the "fermentation stress response" (FSR). The most interesting point was that the FSR was found to overlap only partially with the ESR (Gasch et al., 2000). Interestingly, 62% of the FSR genes were novel, suggesting that the stress conditions in wine fermentation were rather different from those observed in laboratory conditions. Also of interest was the fact that respiratory and gluconeogenesis genes were expressed even in high glucose concentrations and that ethanol accumulation, at least in the experiment by Gasch et al., was the main reason for entry into the stationary phase.

Because compounds such as copper sulfate and sodium bisulfate have been used for many years to inhibit fungal and bacterial growth on vines and grapes and in wines, wine strains might very well respond more efficiently than other strains to these stresses thanks to the overexpression of certain detoxifying genes. Indeed, wine strains have been found to overexpress genes involved in the transport of sulfur (*SUL1-2*) and sulfite (*SSU1*) (Cavalieri et al., 2000; Hauser et al., 2001). It can be concluded that the pressures to which wine strains have been exposed over thousands of years have led to the selection of strains that are better adapted to the fermentation conditions found in wineries. Strains that have developed resistance to treatments such as copper sulfate and sodium bisulfate are a good example of this adaptation.

Finally, two studies have analyzed the genomic response in a commercial wine yeast strain to rehydration and adaptation to osmotic stress at the beginning of vinification. In the first study, rehydration was carried out in a complete glucose medium to identify events related to re-establishment of fermentation (Rossignol et al., 2006). The authors reported substantial transcriptional changes. The expression profile observed in the dried yeasts was characteristic of cells grown under respiratory conditions and exposed to nitrogen and carbon starvation and considerable stress during rehydration. Furthermore, many genes involved in biosynthetic pathways (transcription or protein synthesis) were coordinately induced while those subject to glucose repression were downregulated. While expression of general stress-response genes was repressed during rehydration, despite the high sugar levels, that of acid-stress genes was induced, probably in response to the accumulation of organic acids. In the second study, rehydration was carried out in water to separate this process from adaptation to osmotic pressure (Novo et al., 2007). The results of the study showed that rehydration for an additional hour (following an initial period of 30 min) did not induce any relevant changes in global gene expression. The incubation of rehydrated cells in a medium containing fermentable carbon sources activates genes involved in the fermentation pathway, the nonoxidative branch of the pentose phosphate pathway, ribosomal biogenesis, and protein synthesis.

Erasmus et al. (2003) analyzed yeast response to high sugar concentrations by inoculating rehydrated wine yeast in Riesling grape juice containing equimolar amounts of glucose and fructose to a final concentration of 40% (wt/vol) and comparing global gene expression with that observed in yeasts inoculated in the same must containing 22% sugar. Although the sugar concentration used is not generally found in winemaking conditions, some of the results coincided with those reported by Rossignol et al. (2003), with sugar stress resulting in the apparent upregulation of glycolytic and pentose phosphate pathway genes and structural genes involved in the formation of acetic acid from acetaldehyde and succinic acid from glutamate and the downregulation of genes involved in the de novo biosynthesis of purines, pyrimidines, histidine, and lysine. The authors also reported considerable changes in the expression levels of stress-response genes. These changes affected, among others, genes involved in the production of the compatible osmolyte glycerol (*GPD1*) and genes encoding the heat shock proteins HSP104/12/26/30/42/78/82 and SSA3/4.

Gene expression profiling of industrial strains may also help to uncover as-yet-unknown functions of numerous genes in the *S. cerevisiae* genome, as these genes might only have relevant functions in industrial fermentation conditions. For instance, 130 genes from various subtelomeric families of unknown function (*PAU, AAD, COS*) have been found to be induced during wine fermentation (Rossignol et al., 2003), indicating that they probably have an important role in this process. It should also be noted that 28% of the FSR genes detected in the experiment by Marks et al. (2008) described above had an unknown function.

## 4.4. Genomic Studies

DNA microarray analysis is also a promising tool for the study of wine strain genomes. This technology forms the basis for various types of study in this area, including Affymetrix oligonucleotide microarray analyses. These microarrays consist of a very large number of short oligonucleotide sequences derived from the reference *S. cerevisiae* laboratory strain S288c. The oligonucleotides represent all the open reading frames (ORFs) distributed throughout the yeast genome. In this method, hybridization is highly dependent on the identity of the sequence, and a single nucleotide change will alter the hybridization signal. Thus, the signals produced by a particular strain can be compared with those from a reference strain to identify sequence changes, including SNPs. The method has been successfully used to study polymorphisms in various strains (Primig et al., 2000; Winzeler et al., 1998). Affymetrix also manufactures tiling arrays, another type of oligonucleotide microarray system that covers the entire sequence of the yeast genome. Tiling arrays are used for transcriptome mapping and to identify transcripts that do not correspond to annotated genes (Royce et al., 2005). These arrays have also been used for detailed genomic analysis. As described in Section 3,

Schacherer et al. (2009) used this method to resequence 63 yeast strains, including 14 wine strains.

There also exist tiling arrays with long oligonucleotides (manufactured by Agilent, for example) and arrays containing probes spotted at a lower density than that seen in tiling arrays (oligonucleotides over 60 bases long or double-strand fragments of 300 or more bases). These tools, however, are not suitable for detecting isolated sequence variations. Microarrays consisting of long oligonucleotides or double-strand fragments are, however, useful for genomic comparisons designed to identify increases or decreases in the number of copies of a particular gene or chromosomal region. The first study of this type was conducted by Hughes et al. (2000b) using laboratory strains. A similar study by Infante et al. (2003) that analyzed *S. cerevisiae* flor yeast strains found that two natural strains had differences in the copy number of 38% of their genes, which illustrates the enormous genomic variability that characterizes yeasts of this type. In many cases, the differences were in regions flanked by Ty transposons and other regions with a high recombination rate, which would explain the amplification or deletion events observed. The authors suggested that such regions were the site of double-strand breaks responsible for free ends capable of recombination with short homologous regions (10−18 base pairs). A similar mechanism has been described for the *SSU1* gene region in wine strains (see Section 4.3). In the case of *flor* yeast strains, the continuous presence of acetaldehyde and ethanol in the medium would increase the frequency of double-strand breaks, conferring a selective advantage on strains that have adapted to this hostile environment.

DNA macroarray analysis has also been used to study gross gene expression profiles in the T73 wine strain (Pérez-Ortín et al., 2002b). The study revealed numerous copy-number variations for genes from subtelomeric families and a number of other genes such as the copper resistance gene *CUP1*. Curiously, *CUP1* has a deletion in the genomic region of the wine strain (Pérez-Ortín et al., 2002b), which reduces its expression levels (Hauser et al., 2001). The study by Hauser et al. found that the number of Ty transposons (Ty1, Ty2, Ty3, and Ty4) was greatly reduced in the T73 wine strain compared to the S288c laboratory strain. This finding was consistent with less-complete previously published results (Jordan & McDonald, 1999), with later results (Carreto et al., 2008), and with results for brewing strains (Codón et al., 1998) and suggests that the colonization of the genome of laboratory strains by these molecular parasites may be recent. The strong selective pressure exerted on wine strains might have prevented the excessive accumulation of sequences of this type (Jordan & McDonald, 1999).

The flexibility of DNA chip technology means that purpose-designed arrays can be created for specific studies. In a study of chromosomal rearrangements in Cava strains (secondary fermentation), Carro et al. (2003) used specially designed and constructed macroarrays containing 14 chromosome I probes and hybridized them with DNA from chromosome I isolated from various Cava strains with length variations in this chromosome. Their results indicated the existence of a subtelomeric region that tends to be deleted in the right arm of chromosome I of this highly variable strain.

# 5. PROTEOMIC ANALYSIS OF WINE STRAINS

DNA microarray technology allows the expression of all the genes in a particular organism (the transcriptome) to be analyzed. Global analyses can thus be used to assess the effects of physical, chemical, and biological agents, and even specific mutations, on gene expression. Nonetheless, analysis of mRNA levels is not sufficient for a complete description of biological systems. This also requires accurate

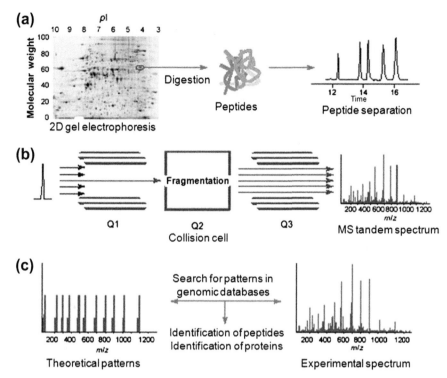

**FIGURE 6.3** Standard proteomic analysis by two-dimensional (2D) gel electrophoresis and mass spectrometry (MS). The method consists of three fully integrated steps. In the first step, the proteins are separated on 2D gels, stained, and then individual spots isolated. The protein spots are then digested with trypsin and the resulting peptides are separated by high-performance liquid chromatography (HPLC). In the second step, each eluted peptide is ionized by electrospray ionization. It then enters the mass spectrometer through the first quadrupole mass filter (Q1) and is fragmented in a collision cell (Q2). The resulting spectrum is recorded (Q3). In the third step, the tandem MS spectrum of a selected ionized peptide contains sufficient specific sequencing information to identify the peptide and its associated protein. m/z = mass to charge ratio.

measurement of the expression and activity of the corresponding proteins (the proteome). Furthermore, even though expression levels of different mRNA species and the proteins they encode are correlated, this correlation is not perfect for all genes (Futcher et al., 1999; Ideker et al., 2001). Of even greater importance, however, is the level of correlation between changes in mRNA and protein levels. While changes in the proteome and transcriptome generally occur in parallel (homodirectional changes), the multiple effects caused by post-transcriptional regulation justify the need for proteomic studies (Griffin et al., 2002; Ideker et al., 2001). Thus, proteomics, which is the analysis of the full complement of proteins expressed by a genome (Pennington et al., 1997; see Figure 6.3), is considered to be the best tool for obtaining a quantitative description of the state of a biological system. In other words, proteome analysis provides a better picture of an organism's phenotype than does the analysis of mRNA levels.

While there are vast amounts of genomic data available for yeasts (including sequence and gene expression data obtained by DNA microarray analysis), the yeast proteome is still largely undefined (Fey et al., 1997). This is particularly true for yeasts of industrial and biotechnological interest, as most of the studies to date have analyzed laboratory strains (Link et al., 1999; Washburn et al., 2001). The first comparative study in this area, performed using three haploid strains derived from laboratory strains, led the authors to conclude that differences in protein expression level and post-translational modifications influenced the molecular and biochemical characteristics of cells and were possibly responsible for the different mutant phenotypes observed in these strains (Rogowska-Wrzesinska et al., 2001).

Several studies have analyzed the effect of environmental stresses on proteome-level responses in laboratory strains. These studies are similar to those conducted in the area of genomics analyzing the influence of environmental factors on global gene expression in laboratory strains. One such proteomic study analyzed oxidative stress caused by hydrogen peroxide (Godon et al., 1998) leading to the expression of batteries of genes referred to by the authors as "stimulons." The expression of 115 proteins with different functional roles was observed. These included proteins linked to antioxidant activity, heat shock response, and protease activity. The expression of 52 proteins, including metabolic enzymes and proteins involved in translation, was repressed. In another study of S. cerevisiae, sorbic acid was found to produce slightly different and less drastic effects, although it did reveal expression of stress-response proteins (mainly linked to oxidative stress) and several molecular chaperones (Hsp12, 26, 42, and some isoforms of Hsp70) (de Nobel et al., 2001). Analysis of mRNA levels following the induction of sorbic acid stress showed that these were poorly correlated with protein abundance.

In another proteome analysis, the addition of cadmium $(Cd^{+2})$ induced expression of 54 proteins and repressed that of a further 43 (Vido et al., 2001). Of these, nine enzymes involved in the sulfur amino acid biosynthesis pathway and glutathione (GSH) synthesis were strongly induced, as were proteins with antioxidant activity. Although $Cd^{+2}$ is not an active redox ion, it can cause oxidative stress and lipid peroxidation and also affect cellular thiol redox balance. These data suggest that the two cellular thiol redox systems—GSH and thioredoxin—are essential protection mechanisms against cadmium stress, a theory later corroborated by Fauchon et al. (2002), who related cadmium stress with sulfur metabolism. As GSH is essential for the detoxification of cadmium, when exposed to this substance, cells convert most sulfur into GSH. The cells change their proteome to reduce the production of sulfur-rich proteins to permit optimal GSH turnover and ensure optimal levels of this essential compound. It has been estimated that this change allows for a 30% reduction in sulfur amino acid incorporation into proteins, which would enable a considerable increase in GSH production and thus ensure cell survival. This is a clear example of the important role of proteome plasticity in yeast cell adaptation to adverse conditions and agents.

Little information is available on the proteomic profiles of industrial yeasts as most of the studies in this area have been carried out using laboratory strains. In two studies involving the analysis and identification of over 200 proteins, Joubert et al. (2000, 2001) concluded that the K11 brewing strain was a hybrid of S. cerevisiae and Saccharomyces pastorianus (S. bayanus). Their work also led them to postulate that the physiological properties required by top-fermenting (ale) strains (flocculation and fermentation at low temperatures) might have been acquired by hybridization. Their reasoning was based on the fact that, unlike bottom-fermenting (lager) strains, which are all hybrids, top-fermenting strains are not hybrids and are very closely related to S. cerevisiae laboratory strains. The two types of brewing strain also have very different physiological properties.

Trabalzini et al. (2003) studied the proteomic response in a wine strain of S. cerevisiae (k310) isolated during spontaneous wine fermentation. Wine strains are exposed to numerous hostile conditions during fermentation. Unlike other studies, which have analyzed isolated effects of environmental stress on yeasts, the study by Trabalzini et al. investigated physiological response to fermentation stress; in particular, depletion of the main carbon source and glucose, and increasing ethanol levels. They found that specific proteins, which differed from those observed for other *S. cerevisiae* strains (such as those used in bread making), were either induced or repressed in response to these physiological stresses. The proteomic response also involved the induction of intracellular proteolysis, which appeared to be directed towards certain classes of protein. The main inference from this study is that the proteomic response to fermentation stress in a wine strain of *S. cerevisiae* is largely directed at mitigating the effects of increasing ethanol levels. Ethanol stress has been associated with both oxidative damage (due to an increased production of free radicals) and cytotoxic effects (due to acetaldehyde production). Ethanol also induces the expression of heat shock proteins and proteins involved in trehalose metabolism, whose purpose is to stabilize membranes and proteins and suppress protein aggregation. It is extremely important to further investigate proteomic responses in fermentation yeasts as a good wine strain must be capable of overcoming the hostile conditions it is faced with in industrial processes. Additionally, the cell changes that occur in *S. cerevisiae* during fermentation (autoproteolysis) and aging (autolysis) are responsible for the organoleptic properties of wine. Accordingly, the amount of nitrogen in autolysates together with free amino acid concentrations, which differ greatly depending on the yeast strain, can have a considerable influence on the flavor, composition, and quality of the final product (Martínez-Rodríguez et al., 2001a, 2001b). Proteolytic enzymes might be involved in the turnover of nitrogenous compounds before and during autolysis in winemaking conditions. It has also been proposed that yeasts might use amino acids not only as sources of nitrogen but also to restore the redox balance in critical environmental conditions (Mauricio et al., 2001).

Two recent studies have compared the transcriptome and proteome of wine yeasts. In the first of these, Zuzuárregui et al. (2006) compared two wine strains with different fermentative capacities and found that one of the strains was incapable of completing fermentation. Although the transcriptome and proteome analyses revealed specific differences, they both indicated that the strain with fermentation difficulty had defects, namely excess proton uptake (a sign of ethanol intolerance) and increased oxidative damage due to elevated levels of acetaldehyde. In the second study, Rossignol et al. (2009) compared proteomic changes in a wine strain between the exponential growth phase and the stationary phase during wine fermentation. They found major changes in the abundance of proteins related to glycolysis, ethanol production, and amino acid metabolism. The most interesting finding was that these changes were very poorly correlated with previously observed transcriptional changes (Rossignol et al., 2003), which suggests that post-transcriptional regulatory mechanisms are very important in the late stages of wine fermentation. A recent study involving laboratory strains and laboratory culture conditions with various nutrient deficiencies indicated that the response to nitrogen depletion was fundamentally controlled at a translational and not a transcriptional level (Kolkman et al., 2006).

The importance of gaining a comprehensive understanding of proteomic response in fermentation yeasts is thus clear: it will greatly contribute to improving the organoleptic properties associated with high-quality wines.

# 6. OTHER GLOBAL STUDIES

One of the aims of large-scale studies is to provide a global view of living systems. Genomics, for example, focuses on the full genome to help understand the relevance of individual genes, while transcriptomics and proteomics analyze the link between physiological changes and changes in transcript and protein expression levels with respect to total RNA or protein expression levels. Most of the large-scale functional studies conducted to date have been based on transcriptomic and proteomic analyses. A more recent "omic" approach, metabolomics, aims to characterize the physiological state of a cell by determining the concentration of all of the small molecules that comprise the metabolism and identifying metabolic pathways and fluxes. This approach may provide the best and most direct measurement of an organism's physiological activity and bring us a little closer to a true approximation of its phenotype since, as stated by Delneri et al. (2001), "mRNA molecules are not functional entities within the cell, but simply transmitters of the instructions for synthesising proteins… proteins and metabolites [in contrast] represent true functional entities within cells" (p. 87). Furthermore, the use of metabolomic data in the systematic analysis of gene function has the added advantage that there are considerably fewer metabolites than genes or gene products. Nevertheless, unlike proteins, metabolites are not directly related to genes.

Metabolomic studies have emerged in an attempt to assign functions to genes on the basis of metabolic analyses. The primary aim is to discover biochemical reactions catalyzed by enzymes encoded by genes of unknown function (Martzen et al., 1999). The difficulty with such an approach is that it assigns mechanisms rather than biological functions.

An alternative approach would be to study changes in the metabolome induced by the deletion or overexpression of a specific gene and to then assign functions by comparing the changes induced with those observed in similar manipulations of known genes. Such an approach, referred to as metabolic footprinting, was used by Raamsdonk et al. (2001) in *S. cerevisiae*. Measuring concentrations of specific metabolites in a cell, however, is a very costly process. The approach used by Raamsdonk et al. was extended in a subsequent study by the same

group (Allen et al., 2003) to permit large-scale analyses by optimizing the experimental conditions and surmounting the technical difficulty of measuring intracellular metabolites, which have a rapid turnover and need to be separated from the extracellular space. The optimization of mass spectrometry has allowed the analysis of extracellular metabolites in spent culture medium.

It is also possible to study and define specific metabolic pathways by integrating and incorporating data obtained using the technologies discussed in this chapter into biological models to predict cell behavior that can then be tested experimentally. Ideker et al. (2001), for example, used a combined genomic and proteomic approach to elucidate the galactose utilization metabolic pathway. They followed a typical strategy used in systems biology. The steps they described are summarized in the following points: (1) definition of all the genes in the pathway of interest; (2) perturbation of each pathway component through a series of genetic or environmental manipulations and quantification of global cellular response; (3) integration of the observed mRNA and protein responses with the current, pathway-specific model; and (4) formulation of new hypotheses to explain observations not predicted by the model. Although metabolomics is a relatively new field, a study by Eglinton et al. (2002), using metabolomic analysis of mutant laboratory strains, showed how genetic modification affects the production of several secondary metabolites of fermentation including acids (such as acetic acid), esters, aldehydes, and higher alcohols. Many of these metabolites make an important contribution to the flavor and aroma of the wine. A recent study by Rossouw et al. (2008) investigating the relationship between the transcriptomes of five wine strains and the aroma profile produced during fermentation found that the expression levels of five genes were related to differences in aroma. They then constructed wine strains overexpressing these genes and found that the changes in the exo-metabolome corresponded to the predicted changes.

# 7. FUTURE DIRECTIONS

The use of genomic and proteomic methods to study wine yeasts is still in its infancy. Although the results achieved so far have begun to provide molecular explanations to problems related to wine yeast physiology, we are still far from the level of detail available for laboratory strains. It is important to discover what makes wine strains capable of must fermentation in circumstances in which the much-better-known laboratory strains are not. Laboratory strains of *S. cerevisiae* are indeed nothing more than simplified genomic derivatives of natural strains. Deciphering the genome of wine strains is also interesting from a basic scientific perspective. Gaining a deeper understanding of the genome, transcriptome, and proteome of wine yeasts and integrating this information into mathematical models capable of predicting physiological changes will allow carefully constructed improvements in the characteristics of these strains and the biotechnological processes in which they participate.

Although we have been making wine for over 7000 years, we only very recently discovered, thanks to Louis Pasteur, that *S. cerevisiae* was the main driving force behind the process. Since then, this yeast has been the focus of much basic and applied research. Nowadays, the in-depth information that large-scale studies can provide on the full complement of macro-molecules found in this microorganism will help us to fully understand its physiology and elucidate the manner in which it makes wine.

# References

Aguilera, A., Chávez, S., & Malagón, F. (2000). Mitotic recombination in yeast: Elements controlling its incidence. *Yeast, 16*, 731–754.

Alexandre, H., Ansanay-Galeote, V., Dequin, S., & Blondin, B. (2001). Global gene expression during short-term ethanol stress in *Saccharomyces cerevisiae*. *FEBS Lett., 498*, 98–103.

Allen, J., Davey, H. M., Broadhurst, D., Heald, J. K., Rowland, J. J., Oliver, S. G., et al. (2003). High-throughput classification of yeast mutants for functional genomics using metabolic footprinting. *Nat. Biotech., 21*, 692–696.

Backhus, L. E., DeRisi, J., Brown, P. O., & Bisson, L. F. (2001). Functional genomic analysis of a commercial wine yeast strain of *Saccharomyces cerevisiae* under differing nitrogen conditions. *FESM Yeast Res., 1*, 1–15.

Bakalinsky, A. T., & Snow, R. (1990). The chromosomal constitution of wine strains of *Saccharomyces cerevisiae*. *Yeast, 6*, 367–382.

Bammert, G. F., & Fostel, J. M. (2000). Genome-wide expression patterns in *Saccharomyces cerevisiae*: Comparison of drug treatments and genetic alterations affecting biosynthesis of ergosterol. *Antimicrob. Agents Chemother, 44*, 1255–1265.

Bargues, M., Salom, D., Gómez, A., Paricio, N., Pérez-Alonso, M., & Pérez-Ortín, J. E. (1996). Sequencing analysis of a 4.1 kb subtelomeric region from yeast chromosome IV identifies *HXT15*, a new member of the hexose transporter family. *Yeast, 12*, 1005–1011.

Barre, P., Vézinhet, F., Dequin, S., & Blondin, B. (1993). Genetic improvement of wine yeast. In G. H. Fleet (Ed.), *Wine Microbiology and Biotechnology* (pp. 421–447). UK: Harwood Academic, Reading.

Belloch, C., Pérez-Torrado, R., González, S. S., Pérez-Ortín, J. E., García-Martínez, J., Querol, A., et al. (2009). Chimeric genomes of natural hybrids of *Saccharomyces cerevisiae* and *Saccharomyces kudriavzevii*. *Appl. Environ. Microbiol., 75*, 2534–2544.

Beltrán, G., Novo, M., Leberre, V., Sokol, S., Labourdette, D., Guillamón, J. M., et al. (2006). Integration of transcriptomic and metabolic analyses for understanding the global responses to low temperature winemaking fermentations. *FESM Yeast Res., 6*, 1167–1183.

Bidenne, C., Blondin, B., Dequin, S., & Vézinhet, F. (1992). Analysis of the chromosomal DNA polymorphism of wine strains of. *Saccharomyces cerevisiae. Curr. Genet., 22*, 1–7.

Bisson, L. F., Karpel, J. E., Ramakrishnan, V., & Joseph, L. (2007). Functional genomics of wine yeast *Saccharomyces cerevisiae*. *Adv. Food Nutr. Res., 53*, 65–121.

Borneman, A. R., Forgan, A. H., Pretorius, I. S., & Chambers, P. J. (2008). Comparative genome analysis of a *Saccharomyces cerevisiae* wine strain. *FESM Yeast Res., 8*, 1185–1195.

Borneman, A. R., Chambers, P. J., & Pretorius, I. S. (2007). Yeast systems biology: Modelling the winemaker's art. *Trends Biotechnol., 25*, 349–355.

Brazma, A., Jonassen, I., Vilo, J., & Ukkonen, E. (1998). Predicting gene regulatory elements in silico on a genomic scale. *Genome Res., 8*, 1202–1215.

Brazma, A., & Vilo, J. (2000). Gene expression data analysis. *FEBS Lett., 480*, 17–24.

Brejning, J., Jespersen, L., & Arneborg, N. (2003). Genome-wide transcriptional changes during the lag phase of *Saccharomyces cerevisiae*. *Arch. Microbiol., 179*, 278–294.

Bussemaker, H. J., Li, H., & Siggia, E. D. (2000). Building a dictionary for genomes: Identification of presumptive regulatory sites by statistical analysis. *Proc. Natl. Acad. Sci. USA, 97*, 10 096–10 100.

Butler, G., Rasmussen, M. D., Lin, M. F., Santos, M. A., Sakthikumar, S., Munro, C. A., et al. (2009). Evolution of pathogenicity and sexual reproduction in eight *Candida* genomes. *Nature, 459*, 657–662.

Carlson, M., Osmond, B. C., & Botstein, D. (1989). *SUC* genes of Yeast: A dispersed gene family. *Cold Spring Harbor Symp. Quant. Biol., 45*, 799–803.

Carmel-Harel, O., Stearman, R., Gasch, A. P., Botstein, D., Brown, P. O., & Storz, G. (2001). Role of thioredoxin reductase in the Yap1p-dependent response to oxidative stress in *Saccharomyces cerevisiae*. *Mol. Microbiol., 39*, 595–605.

Carreto, L., Eiriz, M. F., Gomes, A. C., Pereira, P. M., Schuller, D., & Santos, M. A. (2008). Comparative genomics of wild type yeast strains unveils important genome diversity. *BMC Genomics, 9*, 524.

Carro, D., García-Martínez, J., Pérez-Ortín, J. E., & Piña, B. (2003). Structural characterization of chromosome I size variants from a natural yeast strain. *Yeast, 30*, 171–183.

Carro, D., & Piña, B. (2001). Genetic analysis of the karyotype instability in natural wine yeast strains. *Yeast, 18*, 1457–1470.

Cavalieri, D., Townsend, J. P., & Hartl, D. L. (2000). Manifold anomalies in gene expression in a vineyard isolate of *Saccharomyces cerevisiae* revealed by DNA microarray analysis. *Proc. Natl. Acad. Sci. USA, 97*, 12369–12374.

Cho, R. J., Fromont-Racine, M., Wodicka, L., Feierbach, B., Stearns, T., Legrain, P., et al. (1998). Parallel analysis of genetic selections using whole genome oligonucleotide arrays. *Proc. Natl. Acad. Sci. USA, 95*, 3752–3757.

Chow, T. H. C., Sollitti, P., & Marmur, J. (1989). Structure of the multigene family of *MAL* loci in *Saccaromyces*. *Mol. Gen. Genet., 217*, 60–69.

Chu, S., DeRisi, J., Eisen, M., Mulholland, J., Botstein, D., Brown, P. O., et al. (1998). The transcriptional program of sporulation in budding yeast. *Science, 282*, 699–705.

Cliften, P., Sudarsanam, P., Desikan, A., Fulton, L., Fulton, B., Majors, J., et al. (2003). Finding functional features in *Saccharomyces* genomes by phylogenetic footprinting. *Science, 301*, 71–76.

Codón, A. C., Benítez, T., & Korhola, M. (1998). Chromosomal polymorphism and adaptation to specific industrial environments of *Saccharomyces* strains. *Appl. Microbiol. Biotechnol., 49*, 154–163.

Codón, A. C., Gasent-Ramírez, J. M., & Benítez, T. (1995). Factors which affect the frequency of sporulation and tetrad formation in *Saccharomyces cerevisiae* baker's yeasts. *Appl. Environ. Microbiol., 61*, 630–638.

de Nobel, H., Lawrie, L., Brul, S., Klis, F., Davis, M., Alloush, H., et al. (2001). Parallel and comparative analysis of the proteome and transcriptome of sorbic acid-stressed *Saccharomyces cerevisiae*. *Yeast, 18*, 1328–1413.

Delneri, D., Brancia, F. L., & Oliver, S. G. (2001). Towards a truly integrative biology through the functional genomics of yeast. *Curr. Opin. Biotech., 12*, 87–91.

Delneri, D., Colson, I., Grammenoudi, S., Roberts, I. N., Louis, E. J., & Oliver, S. G. (2003). Engineering evolution to study speciation in yeasts. *Nature, 422*, 68–72.

DeRisi, J. L., Iyer, V. R., & Brown, P. O. (1997). Exploring the metabolic and genetic control of gene expression on a genomic scale. *Science, 24*, 680–686.

Dujon, B. (1998). European Functional Analysis Network (EUROFAN) and the functional analysis of the *Saccharomyces cerevisiae* genome. *Electrophoresis, 19*, 617–624.

Eglinton, J. M., Heinrich, A. J., Pollnitz, A. P., Langridge, P., Henschke, P. A., & de Barros Lopes, M. (2002). Decreasing acetic acid accumulation by a glycerol overproducing strain of *Saccharomyces cerevisiae* by deleting the *ALD6* aldehyde dehydrogenase gene. *Yeast, 19*, 295–301.

Erasmus, D. J., van der Merwe, G. K., & van Vuuren, H. J. J. (2003). Genome-wide expression analyses: Metabolic adaptation of *Saccharomyces cerevisiae* to high sugar stress. *FESM Yeast Res., 3*, 375–399.

Fares, M. A., & Wolfe, K. (2003). Accelerated evolution of some gene copies after genome duplication in *Saccharomyces*. *Yeast, 20*, S277.

Fauchon, M., Lagniel, G., Aude, J. C., Lombardia, L., Soularue, P., Petat, C., et al. (2002). Sulfur sparing in the yeast proteome in response to sulfur demand. *Mol. Cell, 9*, 713–723.

Ferea, T. L., Botstein, D., Brown, P. O., & Rosenzweig, R. F. (1999). Systematic changes in gene expression patterns following adaptive evolution in yeast. *Proc. Natl. Acad. Sci. USA, 96*, 9721–9726.

Fey, S. J., Nawrocki, A., Larsen, M. R., Gorg, A., Roepstorff, P., Skews, G. N., et al. (1997). Proteome studies of *Saccharomyces cerevisiae*: A methodological outline. *Electrophoresis, 18*, 1361–1372.

Fischer, G., James, S. A., Roberts, I. N., Oliver, S. G., & Louis, E. J. (2000). Chromosomal evolution in *Saccharomyces*. *Nature, 405*, 451–454.

Fischer, G., Neuveglise, C., Durrens, P., Gaillardin, C., & Dujon, B. (2001). Evolution of gene order in the genomes of two related yeast species. *Genome Res., 11,* 2009–2019.

Futcher, B., Latter, G. I., Monardo, P., McLaughlin, C. S., & Garrels, J. I. (1999). A sampling of the yeast proteome. *Mol. Cell. Biol., 19,* 7357–7368.

Gasch, A. P., Spellman, P. T., Kao, C. M., Carmel-Harel, O., Eisen, M. B., Storz, G., et al. (2000). Genomic expression programs in the response of yeast cells to environmental changes. *Mol. Biol. Cell., 11,* 4241–4257.

Giaever, G., Chu, A. M., Ni, L., Connelly, C., Riles, L., Véronneau, S., et al. (2002). Functional profiling of the *Saccharomyces cerevisiae* genome. *Nature, 418,* 387–391.

Giaever, G., Shoemaker, D. D., Jones, T. W., Liang, H., Winzeler, E. A., Astromoff, A., et al. (1999). Genomic profiling of drug sensitivities via induced haploinsufficiency. *Nat. Genet., 21,* 278–283.

Godon, C., Lagniel, G., Lee, J., Buhler, J. M., Kieffer, S., Perrot, M., et al. (1998). The $H_2O_2$ stimulon in *Saccharomyces cerevisiae. J. Biol. Chem., 273,* 22 480–22 489.

Goffeau, A., et al. (1997). The yeast genome directory. *Nature, 387*(Suppl. 1), 1–105.

González, S. S., Barrio, E., Gafner, J., & Querol, A. (2006). Natural hybrids from *Saccharomyces cerevisiae, Saccharomyces bayanus,* and *Saccharomyces kudriavzevii* in wine fermentations. *FESM Yeast Res., 6,* 1221–1234.

Griffin, T. J., Gygi, S. P., Ideker, T., Rist, B., Eng, J., Hood, L., et al. (2002). Complementary profiling of gene expression at the transcriptome and proteome levels in *Saccharomyces cerevisiae. Mol. Cell Proteomics, 1,* 323–333.

Gross, C., Kelleher, M., Iyer, V. R., Brown, P. O., & Winge, D. R. (2000). Identification of the copper regulon in *Saccharomyces cerevisiae* by DNA microarrays. *J. Biol. Chem., 275,* 32 310–32 316.

Guijo, S., Mauricio, J. C., Salmon, J. M., & Ortega, J. M. (1997). Determination of the relative ploidy in different *Saccharomyces cerevisiae* strains used for fermentation and "flor" film ageing of dry sherry-type wines. *Yeast, 13,* 101–117.

Hampson, S., Baldi, P., Kibler, D., & Sandmeyer, S. B. (2000). Analysis of yeast's ORF upstream regions by parallel processing, microarrays, and computational methods. *Proc. Int. Conf. Intell. Syst. Mol. Biol., 8,* 190–201.

Hardwick, J. S., Kuruvilla, F. G., Tong, J. K., Shamji, A. F., & Schreiber, S. L. (1999). Rapamycin-modulated transcription defines the subset of nutrient-sensitive signaling pathways directly controlled by the Tor proteins. *Proc. Natl. Acad. Sci. USA, 96,* 14 866–14 870.

Hauser, N. C., Fellenberg, K., Gil, R., Bastuck, S., Hoheisel, J. D., & Pérez-Ortín, J. E. (2001). Whole genome analysis of a wine yeast strain. *Comp. Funct. Genom., 2,* 69–79.

Herrero, E., de la Torre, M. A., & Valentín, E. (2003). Comparative genomics of yeast species: New insights into their biology. *Int. Microbiol., 6,* 183–190.

Hirasawa, T., Yoshikawa, K., Nakakura, Y., Nagahisa, K., Furusawa, C., Katakura, Y., et al. (2007). Identification of target genes conferring ethanol stress tolerance to *Saccharomyces cerevisiae* based on DNA microarray data analysis. *J. Biotechnol., 131,* 34–44.

Holstege, F. C., Jennings, E. G., Wyrick, J. J., Lee, T. I., Hengartner, C. J., Green, M. R., et al. (1998). Dissecting the regulatory circuitry of a eukaryotic genome. *Cell, 95,* 717–728.

Hughes, T. R., Marton, M. J., Jones, A. R., Roberts, C. J., Stoughton, R., Armour, C. D., et al. (2000a). Functional discovery via a compendium of expression profiles. *Cell, 102,* 109–126.

Hughes, T. R., Roberts, C. J., Dai, H., Jones, A. R., Meyer, M. R., Slade, D., et al. (2000b). Widespread aneuploidy revealed by DNA microarray expression profiling. *Nat. Genet., 25,* 333–337.

Hughes, T. R., & Shoemaker, D. D. (2001). DNA microarrays for expression profiling. *Curr. Opin. Chem. Biol., 5,* 21–25.

Ideker, T., Thorsson, V., Ranish, J. A., Christmas, R., Buhler, J., Eng, J. K., et al. (2001). Integrated genomic and proteomic analyses of a systematically perturbed metabolic network. *Science, 292,* 929–934.

Infante, J. J., Dombek, K. M., Rebordinos, L., Cantoral, J. M., & Young, E. T. (2003). Genome-wide amplifications caused by chromosomal rearrangements play a major role in the adaptive evolution of natural yeast. *Genetics, 165,* 1745–1759.

Ivorra, C., Pérez-Ortín, J. E., & del Olmo, M. (1999). An inverse correlation between stress resistance and stuck fermentations in wine yeasts. A molecular study. *Biotechnol. Bioeng., 64,* 698–708.

Jelinsky, S. A., Estep, P., Church, G. M., & Samson, L. D. (2000). Regulatory networks revealed by transcriptional profiling of damaged *Saccharomyces cerevisiae* cells: Rpn4 links base excision repair with proteasomes. *Mol. Cell. Biol., 20,* 8157–8167.

Jelinsky, S. A., & Samson, L. D. (1999). Global response of *Saccharomyces cerevisiae* to an alkylating agent. *Proc. Natl. Acad. Sci. USA, 96,* 1486–1491.

Jia, M. H., Larossa, R. A., Lee, J. M., Rafalski, A., Derose, E., Gonye, G., et al. (2000). Global expression profiling of yeast treated with an inhibitor of amino acid biosynthesis, sulfometuron methyl. *Physiol. Genomics, 3,* 83–92.

Jiménez-Martí, E., & del Olmo, M. (2008). Addition of ammonia or amino acids to a nitrogen-depleted medium affects gene expression patterns in yeast cells during alcoholic fermentation. *FESM Yeast Res., 8,* 245–256.

Jordan, I. K., & McDonald, F. (1999). Tempo and mode of Ty evolution in *Saccharomyces cerevisieae. Genetics, 151,* 1341–1351.

Joubert, R., Brignon, P., Lehmann, C., Monribot, C., Gendre, F., & Boucherie, H. (2000). Two-dimensional gel analysis of the proteome of lager brewing yeasts. *Yeast, 16*, 511–522.

Joubert, R., Strub, J. M., Zugmeyer, S., Kobi, D., Carte, N., van Dorsselaer, A., et al. (2001). Identification by mass spectrometry of two-dimensional gel electrophoresis-separated proteins extracted from lager brewing yeast. *Electrophoresis, 22*, 2269–2282.

Kellis, M., Patterson, N., Endrizzi, M., Birren, B., & Lander, E. S. (2003). Sequencing and comparison of yeast species to identify genes and regulatory elements. *Nature, 423*, 241–254.

Kielland-Brandt, M. C., Nilsson-Tillgren, T., Gjermnsen, C., Holmberg, S., & Pedersen, M. B. (1995). Genetics of brewing yeasts. In A. Wheals, A. Rose & J. Harrison (Eds.), *The yeasts* (pp. 223–254). London, UK: Academic Press.

Kolkman, A., Daran-Lapujade, P., Fullaondo, A., Olsthoorn, M. M., Pronk, J. T., Slijper, M., et al. (2006). Proteome analysis of yeast response to various nutrient limitations. *Mol. Syst. Biol., 2*, 2006, 0026.

Kuhn, K. M., DeRisi, J. L., Brown, P. O., & Sarnow, P. (2001). Global and specific translational regulation in the genomic response of *Saccharomyces cerevisiae* to a rapid transfer from a fermentable to a nonfermentable carbon source. *Mol. Cell. Biol., 21*, 916–927.

Kunkee, R. E., & Bisson, L. F. (1993). Wine-making yeasts. In A. H. Rose & J. S. Harrison (Eds.), *The yeasts: Yeast Technology* (pp. 69–125). London, UK: Academic Press.

Lagunas, R. (1986). Misconceptions about the energy metabolism of. *Saccharomyces cerevisiae. Yeast, 2*, 221–228.

Langkjaer, R. B., Cliften, P. F., Johnston, M., & Piskur, J. (2003). Yeast genome duplication was followed by asynchronous differentiation of duplicated genes. *Nature, 421*, 848–852.

Link, A. J., Eng, J., Schieltz, D. M., Carmack, E., Mize, G. J., Morris, D. R., et al. (1999). Direct analysis of protein complexes using mass spectrometry. *Nat. Biotech., 17*, 676–682.

Liti, G., Carter, D. M., Moses, A. M., Warringer, J., Parts, L., James, S. A., et al. (2009). Population genomics of domestic and wild yeasts. *Nature, 458*, 337–341.

Llorente, B., Durrens, P., Malpertuy, A., Aigle, M., Artiguenave, F., Blandin, G., et al. (2000). Genomic exploration of the hemiascomycetous yeasts: 20. Evolution of gene redundancy compared to *Saccharomyces cerevisiae. FEBS Lett., 487*, 122–133.

Longo, E., & Vézinhet, F. (1993). Chromosomal rearrangements during vegetative growth of a wild strain of *Saccharomyces cerevisiae. Appl. Environ. Microbiol., 59*, 322–326.

Malpertuy, A., Tekaia, F., Casarégola, S., Aigle, M., Artiguenave, F., Blandin, G., et al. (2000). Genomic exploration of hemiascomycetous yeasts: 19. Ascomycetes-specific genes. *FEBS Lett., 487*, 113–121.

Marks, V. D., Ho Sui, S. J., Erasmus, D., van der Merwe, G. K., Brumm, J., Wasserman, W. W., et al. (2008). Dynamics of the yeast transcriptome during wine fermentation reveals a novel fermentation stress response. *FESM Yeast Res., 8*, 35–52.

Marks, V. M., van der Merwe, G. K., & van Vuuren, H. J. J. (2003). Transcriptional profiling of wine yeast in fermenting grape juice: Regulatory effect of diammonium phosphate. *FESM Yeast Res., 3*, 269–287.

Martínez-Rodríguez, A. J., Carrascosa, A. V., & Polo, M. C. (2001b). Release of nitrogen compounds to the extracellular medium by three strains of Saccharomyces cerevisiae during induced autolysis in a model wine system. *Int. J. Food Microbiol., 68*, 155–160.

Martínez-Rodríguez, A. J., Polo, M. C., & Carrascosa, A. V. (2001a). Structural and ultrastructural changes in yeast cells during autolysis in a model wine system and in sparkling wines. *Trends Biochem. Sci., 15*, 305–309.

Martini, A. (1993). The origin and domestication of the wine yeast *Saccharomyces cerevisiae. J. Wine Res., 4*, 165–176.

Martzen, M. R., McCraith, S. M., Spinelli, S. L., Torres, F. M., Fields, S., Grayhack, E. J., et al. (1999). A biochemical genomics approach for identifying genes by the activity of their products. *Science, 286*, 1153–1155.

Mauricio, J. C., Valero, E., Millan, C., & Ortega, J. M. (2001). Changes in nitrogen compounds in must and wine during fermentation and biological aging by flor yeast. *J. Agric. Food Chem., 49*, 3310–3315.

Mendes-Ferreira, A., del Olmo, M., García-Martínez, J., Jiménez-Martí, E., Leão, C., Mendes-Faia, A., et al. (2007b). Signature genes for predicting nitrogen deficiency during alcoholic fermentation. *Appl. Environm. Microbiol., 73*, 5363–5369.

Mendes-Ferreira, A., del Olmo, M., García-Martínez, J., Jiménez-Martí, E., Mendes-Faia, A., Pérez-Ortín, J. E., et al. (2007a). Transcriptional response of *Saccharomyces cerevisiae* to different nitrogen concentrations during alcoholic fermentation. *Appl. Environm. Microbiol., 73*, 3049–3060.

Miklos, G. L. G., & Rubin, G. M. (1996). The role of the genome project in determining gene function: Insight for model organisms. *Cell, 86*, 521–529.

Mortimer, R. K. (2000). Evolution and variation of the yeast (*Saccharomyces*) genome. *Genome Res., 10*, 403–409.

Mortimer, R. K., & Johnston, J. R. (1986). Genealogy of principal strains of the yeast genetic stock center. *Genetics, 113*, 35–43.

Mortimer, R. K., & Polsinelli, M. (1999). On the origins of wine yeast. *Res. Microbiol., 150*, 199–204.

Mortimer, R. K., Romano, P., Suzzi, G., & Polsinelli, M. (1994). Genome renewal: A new phenomenon revealed from a genetic study of 43 strains of *Saccharomyces cerevisiae* derived from natural fermentation of grape musts. *Yeast, 10,* 1543–1552.

Nagalakshmi, U., Wang, Z., Waern, K., Shou, C., Raha, D., Gerstein, M., et al. (2008). The transcriptional landscape of the yeast genome defined by RNA sequencing. *Science, 320,* 1344–1349.

Nakao, Y., Kanamori, T., Itoh, T., Kodama, Y., Rainieri, S., Nakamura, N., et al. (2009). Genome sequence of the lager brewing yeast, an interspecies hybrid. *DNA Res., 16,* 115–129.

Naumov, G. I. (1996). Genetic identification of biological species in the *Saccharomyces* sensu stricto complex. *J. Ind. Microbiol., 17,* 295–302.

Naumov, G. I., Naumova, E. S., Masneuf, I., Aigle, M., Kondratieva, V. I., & Dubourdieu, D. (2000). Natural polyploidization of some cultured yeast *Saccharomyces sensu stricto*: Auto- and allotetraploidy. *Syst. Appl. Microbiol., 23,* 429–442.

Naumova, E. S., Turakainen, H., Naumov, G. I., & Korhola, M. (1997). Superfamily of alpha-galactosidase *MEL* genes of the *Saccharomyces sensu stricto* species complex. *Mol. Gen. Genet., 253,* 111–117.

Ness, F., & Aigle, M. (1995). *RTM1* a member of a new family of telomeric genes in yeast. *Genetics, 113,* 35–43.

Novo, M., Beltran, G., Rozes, N., Guillamon, J. M., Sokol, S., Leberre, V., et al. (2007). Early transcriptional response of wine yeast after rehydration: Osmotic shock and metabolic activation. *FESM Yeast Res., 7,* 304–316.

Novo, M., Bigey, F., Beyne, E., Galeote, V., Gavory, F., Mallet, S., et al. (2009). Eukaryote-to-eukaryote gene transfer events revealed by the genome sequence of the wine yeast *Saccharomyces cerevisiae* EC1118. *Proc. Natl. Acad. Sci. USA., 106,* 16 333–16 338.

Ogawa, N., DeRisi, J., & Brown, P. O. (2000). New components of a system for phosphate accumulation and polyphosphate metabolism in *Saccharomyces cerevisiae* revealed by genomic expression analysis. *Mol. Biol. Cell., 11,* 4309–4321.

Patton, E. E., Peyraud, C., Rouillon, A., Surdin-Kerjan, Y., Tyers, M., & Thomas, D. (2000). SCF(Met30)-mediated control of the transcriptional activator Met4 is required for the G(1)-S transition. *EMBO J., 19,* 1613–1624.

Pennington, S. R., Wilking, M. R., Hochstrasser, D. F., & Dunn, M. J. (1997). Proteome analysis: From protein characterization to biological function. *Trends Cell Biol., 7,* 168–173.

Pérez-Ortín, J. E., García-Martínez, J., & Alberola, T. (2002b). DNA chips for yeast biotechnology. The case of wine yeasts. *J. Biotechnol., 98,* 227–241.

Pérez-Ortín, J. E., Querol, A., Puig, S., & Barrio, E. (2002a). Molecular characterization of a chromosomal rearrangement involved in the adaptive evolution of yeast strains. *Genome Res., 12,* 1533–1539.

Pizarro, F., Vargas, F. A., & Agosin, E. (2007). A systems biology perspective of wine fermentations. *Yeast, 24,* 977–991.

Pretorius, I. S. (2000). Tailoring wine yeast for the new millennium: Novel approaches to the ancient art of winemaking. *Yeast, 16,* 675–729.

Primig, M., Williams, R. M., Winzeler, E. A., Tevzadze, G. G., Conway, A. R., Hwang, S. Y., et al. (2000). The core meiotic transcriptome in budding yeasts. *Nat. Genet., 26,* 415–423.

Puig, S., Querol, A., Barrio, E., & Pérez-Ortín, J. E. (2000). Mitotic recombination and genetic changes in *Saccharomyces cerevisiae* during wine fermentation. *Appl. Environ. Microbiol., 66,* 2057–2061.

Puig, S., Ramón, D., & Pérez-Ortín, J. E. (1998). An optimized method to obtain stable food-safe recombinant wine yeast strains. *J. Agr. Food Chem., 46,* 1689–1693.

Querol, A., Barrio, E., Huerta, T., & Ramón, D. (1992). Molecular monitoring of wine fermentations conducted by active dry yeast strains. *Appl. Environ. Microbiol., 58,* 2948–2953.

Querol, A., Belloch, C., Fernández-Espinar, M. T., & Barrio, E. (2003). Molecular evolution in yeast of biotechnological interest. *Int. Microbiol., 6,* 201–205.

Raamsdonk, L. M., Teusink, B., Broadhurst, D., Zhang, N., Hayes, A., Walsh, M. C., et al. (2001). A functional genomics strategy that uses metabolome data to reveal the phenotype of silent mutations. *Nat. Biotech., 19,* 45–50.

Rachidi, N., Barre, P., & Blondin, B. (1999). Multiple Ty-mediated chromosomal translocations lead to karyotype changes in a wine strain of *Saccharomyces cerevisiae*. *Mol. Gen. Genet., 261,* 841–850.

Rainieri, S., Kodama, Y., Kaneko, Y., Mikata, K., Nakao, Y., & Ashikari, T. (2006). Pure and mixed genetic lines of *Saccharomyces bayanus* and *Saccharomyces pastorianus* and their contribution to the lager brewing strain genome. *Appl. Environ. Microbiol., 72,* 3968–3974.

Rogowska-Wrzesinska, A., Larsen, P. M., Blomberg, A., Görg, A., Roepstorff, P., Norberk, J., et al. (2001). Comparison of the proteomes of three yeast wild type strains: CEN.PK2, FY1679 and W303. *Comp. Funct. Genom, 2,* 207–225.

Rossignol, T., Dulau, L., Julien, A., & Blondin, B. (2003). Genome-wide monitoring of wine yeast gene expression during alcoholic fermentation. *Yeast, 20,* 1369–1385.

Rossignol, T., Kobi, D., Jacquet-Gutfreund, L., & Blondin, B. (2009). The proteome of a wine yeast strain during fermentation, correlation with the transcriptome. *J. Appl. Microbiol., 107,* 47–55.

Rossignol, T., Postaire, O., Storaï, J., & Blondin, B. (2006). Analysis of the genomic response of a wine yeast to rehydration and inoculation. *Appl. Microbiol. Biotechnol., 71*, 699–712.

Rossouw, D., Naes, T., & Bauer, F. F. (2008). Linking gene regulation and the exo-metabolome: A comparative transcriptomics approach to identify genes that impact on the production of volatile aroma compounds in yeast. *BMC Genomics, 9*, 530.

Roth, F. P., Hughes, J. D., Estep, P. W., & Church, G. M. (1998). Finding DNA regulatory motifs within unaligned non-coding sequences clustered by whole-genome mRNA quantitation. *Nat. Biotech., 16*, 939–945.

Royce, T. E., Rozowsky, J. S., Bertone, P., Samanta, M., Stolc, V., Weissman, S., et al. (2005). Issues in the analysis of oligonucleotide tiling microarrays for transcript mapping. *Trends Genet., 21*, 466–475.

Salmon, J.-M. (1997). Enological fermentation kinetics of an isogenic ploidy series derived form an industrial *Saccharomyces cerevisiae* strain. *J. Ferment. Bioeng, 83*, 253–260.

Schacherer, J., Shapiro, J. A., Ruderfer, D. M., & Kruglyak, L. (2009). Comprehensive polymorphism survey elucidates population structure of *Saccharomyces cerevisiae*. *Nature, 458*, 342–345.

Schena, M., Shalon, D., Davis, R. W., & Brown, P. O. (1995). Quantitative monitoring of gene expression patterns with a complementary DNA microarray. *Science, 270*, 467–470.

Snow, R. (1983). Genetic improvement of wine yeast. In J. F. T. Spencer, D. M. Spencer & A. R. V. Smith (Eds.), *Yeast genetics — Fundamental and applied aspects* (pp. 439–459). New York, NY: Springer-Verlag.

Solieri, L., Antúnez, O., Pérez-Ortín, J. E., Barrio, E., & Giudici, P. (2008). Mitochondrial inheritance and fermentative: Oxidative balance in hybrids between *Saccharomyces cerevisiae* and *Saccharomyces uvarum*. *Yeast, 25*, 485–500.

Souciet, J., Aigle, M., Artiguenave, F., Blandin, G., Bolotin-Fukuhara, M., Bon, E., et al. (2000). Genomic exploration of the hemiascomycetous yeasts: 1. A set of yeast species for molecular evolution studies. *FEBS Lett., 487*, 3–12.

ter Linde, J. J., Liangm, H., Davism, R. W., Steensma, H. Y., van Dijken, J. P., & Pronk, J. T. (1999). Genome-wide transcriptional analysis of aerobic and anaerobic chemostat cultures of *Saccharomyces cerevisiae*. *J. Bacteriol., 181*, 7409–7413.

Teunissen, A. W., & Steensma, H. Y. (1995). The dominant flocculation genes of *Saccharomyces cerevisiae* constitute a new subtelomeric gene family. *Yeast, 11*, 1001–1013.

Trabalzini, L., Paffetti, A., Scaloni, A., Talamo, F., Ferro, E., Coratza, G., et al. (2003). Proteomic response to physiological fermentation stresses in a wild-type wine strain of *Saccharomyces cerevisiae*. *Biochem. J., 370*, 35–46.

Velculescu, V. E., Vogelstein, B., & Kinzler, K. W. (2000). Uncharted transcriptomes with SAGE. *Trends Genet., 16*, 423–425.

Velculescu, V. E., Zhang, L., Zhou, W., Vogelstein, J., Basrai, M. A., Bassett, D. E., et al. (1997). Characterization of the yeast transcriptome. *Cell, 88*, 243–251.

Vido, K., Spector, D., Lagniel, G., Lopez, S., Toledano, M. B., & Labarre, J. (2001). A proteome analysis of the cadmium response in *Saccharomyces cerevisiae*. *J. Biol. Chem., 276*, 8469–8474.

Washburn, M. P., Wolters, D., & Yates, J. R. (2001). Large-scale analysis of the yeast proteome by multidimensional protein identification technology. *Nat. Biotech., 19*, 242–247.

Winzeler, E. A., Richards, D. R., Conway, A. R., Goldstein, A. L., Kalman, S., McCullough, M. J., et al. (1998). Direct allelic variation scanning of the yeast genome. *Science, 281*, 1194–1197.

Winzeler, E. A., Shoemaker, D. D., Astromoff, A., Liang, H., Anderson, K., Andre, B., et al. (1999). Functional characterization of the *S. cerevisiae* genome by gene deletion and parallel analysis. *Science, 285*, 901–906.

Wodicka, L., Dong, H., Mittmann, M., Ho, M. H., & Lockhart, D. J. (1997). Genome-wide expression monitoring in *Saccharomyces cerevisiae*. *Nat. Biotechnol., 15*, 1359–1367.

Wolfe, K. H., & Shields, D. C. (1997). Molecular evidence for an ancient duplication of the entire yeast genome. *Nature, 387*, 708–713.

Zuzuárregui, A., & del Olmo, M. L. (2004). Expression of stress response genes in wine strains with different fermentative behavior. *FESM Yeast Res., 4*, 699–710.

Zuzuárregui, A., Monteoliva, L., Gil, C., & del Olmo, M. (2006). Transcriptomic and proteomic approach for understanding the molecular basis of adaptation of *Saccharomyces cerevisiae* to wine fermentation. *Appl. Environ. Microbiol., 72*, 836–847.

# Improvement of Wine Yeasts by Genetic Engineering

*Daniel Ramón* [1], *Ramón González* [2]

[1] Biópolis SL, Valencia, Spain and [2] Instituto de Ciencias de la Vid y del Vino (CSIC-UR-CAR), Logroño, Spain

OUTLINE

## 1. WINE, BIOTECHNOLOGY, AND GENETICS

### 1.1. Wine and Classical Genetics

To a scientist, biotechnology is simply the use of a living organism or any of its parts to produce a commercial product. Fermenting a microscopic fungus such as *Penicillium chrysogenum* in order to produce cephalosporin or penicillin, for instance, is a form of biotechnology, since the metabolites synthesized by a living organism are subsequently sold as pharmaceutical products. When the living organism is used to produce a food or beverage, the process is referred to as food biotechnology. Since at least two different living organisms are required to make wine (the vines that produce the grapes and the yeasts that are responsible for fermentation of the grape must), there is more than one justification for classifying wine as a product of food biotechnology. This is the scientist's view, however, and it is quite different from that held by the consumer, particularly in the European Union (EU). In the minds of most consumers, food biotechnology is the use of genetics to produce foods and beverages, or, put more crudely, it is about putting genes in your soup.

Even if we apply this incorrect definition of biotechnology, however, such activities have been linked with wine production for thousands of years, starting with the application of genetics to the cultivation of vines. The first archeological evidence of vine cultivation dates back to 7000 BC in Mesopotamia. From this point onwards, multiple references can be found to the production of wine in the Egyptian and Phoenician cultures, and, later, of course, to the spectacular winemaking practices of the Greek and Roman civilizations. Since then, wine has been produced in many regions of the world using different vine cultivars, all belonging to the genus *Vitis*. These plants are classified into two subgenera—*Euvitis* and *Muscadinia*—that each have specific morphological characteristics as a result of their genomic differences. Most of the species currently used in viticulture belong to the subgenus *Euvitis*, specifically *Vitis vinifera*, which is the most widely cultivated species (Antcliff, 1992). The domestication of this plant took place around 5000 years ago in the region now occupied by parts of Azerbaijan, Georgia, the north of Iraq, and the northeast of Turkey, although some authors have suggested that an independent domestication event occurred in Spain (Núñez & Walker, 1989; Stevenson, 1985). These domestication events are unlikely to have been particularly complicated, since the evolutionary ancestor of this plant displayed a series of characteristics—such as the natural ability to climb, minimal requirements for water and minerals, and a high propagative capacity—that facilitated cultivation without significant effort. However, as some experts have suggested, the most important adaptations were undoubtedly the switch to functional bisexuality and the increase in the size of the fruit (Carbonneau, 1983; Vivier & Pretorius, 2002). All of these phenotypic changes occurred when growers empirically employed natural selection of spontaneous mutations; in other words, they were generated by the application of a genetic technique. Improvement programs have continued to be implemented through the use of deliberate cross-pollination in an effort to achieve greater resistance to pests and environmental stresses. This genetic history and the effect of human migration on the spread of vine cultivation have together resulted in almost nine million hectares currently planted with vines worldwide and more than 24 000 different extant cultivars, of which 5000 belong to the species *V. vinifera*.

Genetics has also been used empirically for many years to improve wine yeast, a popular but incorrect term used to refer to the microbial species *Saccharomyces cerevisiae*. There are currently hundreds of strains of the species that over centuries have adapted through a process of mutation and selection to the different ecological niches of each must and winegrowing region. Whereas laboratory strains of *S. cerevisiae* have 16 chromosomes and a genome containing around 12 megabases of DNA, industrial wine strains tend to be diploid, aneuploid, or occasionally polyploid (Snow, 1983). As a result, the genetic origin of existing wine strains has generated much debate among research groups. While some suggest that *S. cerevisiae* is a natural organism present on fruits (Mortimer & Polsinelli,

1999), there is increasing evidence that many strains are the result of hybridization between natural strains of *Saccharomyces* species that have been spontaneously selected in winemaking environments (Querol & Bond, 2009). What seems clear is that the genome of existing wine yeasts has arisen as a result of powerful selection pressure over millions of generations (Querol et al., 2003). In addition, in the last 20 years efforts have been made to modify the genome of wine strains by mutagenesis and selection, hybridization, cytoduction, and protoplast fusion. In the vast majority of cases, these approaches have been unsatisfactory from an industrial point of view, although they have increased our understanding of the genome of these yeast strains (Barre et al., 1993; González et al., 2003; González-Ramos et al., 2009; Hammond, 1996; Pretorius, 2000; Pretorius & van der Westhuizen, 1991; Quirós et al., 2010; Rainieri & Pretorius, 2000).

## 1.2. Wine and Genetic Engineering

Clearly, then, both vines and wine yeasts have been subject to genetic improvement. From a scientific point of view, we can say that no vine cultivars or strains of wine yeast are free of genetic modification. Nevertheless, since these changes have been introduced through the use of classical genetics, they appear not to be linked with biotechnology in the minds of consumers. For some years now, genetic engineering techniques have been available to allow specific genes to be isolated, modified in the laboratory, and reintroduced into the original organism or a different one to produce so-called transgenic or genetically modified organisms (GMOs). It is these targeted molecular changes that consumers think of as real biotechnology. Enology and viticulture have not been exempt from these developments, and genetic engineering has begun to be applied to both vines and wine yeasts, although to varying extents and with differing outcomes.

Genetic engineering relies on the ability to isolate genes from a genome and to then introduce them into cells through the use of genetic transformation techniques. The *S. cerevisiae* genome has been completely sequenced thanks to the efforts of an international, publicly funded research project (Goffeau et al., 1997). The primary structure has been determined for the first 6000 genes that make up the genome and projects have been initiated to determine their function. As a result, the molecular make-up of *S. cerevisiae* is better understood than that of any other eukaryotic organism. In recent years, genomes have been decoded not only for the principal laboratory strain but also for strains of interest in industry and medicine, or for research into the mechanisms of evolution. The genomes of two industrial wine yeasts have been fully sequenced and annotated (Borneman et al., 2008; Novo et al., 2009). Others have been sequenced with a lower coverage for the purpose of evolutionary studies (Liti et al., 2009). In addition to their relevance for the understanding of how strains have adapted to winemaking environments during evolution, these sequence data have enormous potential for use in biotechnology.

It can be reasoned that any gene introduced into a yeast that is then inoculated in a winery fermenter, if expressed during fermentation, will lead to accumulation of the protein that it encodes as vinification advances and will therefore introduce the technological activity of interest. Extensive biochemical and physiological data have been accumulated over many years in studies of the growth of laboratory strains of *S. cerevisiae* in defined media and under controlled laboratory conditions. Furthermore, many of the genes associated with the metabolic generation of physicochemical or organoleptic properties that are relevant to winemaking have been cloned and sequenced. Since effective methods are available for the transformation of *S. cerevisiae*, including most of the wine strains studied to date, it has been possible to use all of

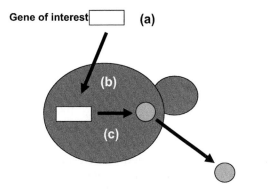

**Phenotype of interest**

FIGURE 7.1  Schematic diagram showing the approach used to obtain improved industrial wine yeasts through genetic engineering. Firstly, biochemical information is required and cloned genes must be available for the gene products implicated in the process of interest (a). Secondly, an effective genetic transformation system is required in order to introduce the gene into the chosen wine yeast (b). Finally, regulatory sequences must be introduced to allow expression of the cloned gene to be activated at a specific point (c).

this information to investigate the generation of improved wine yeasts through the use of genetic engineering (Cebollero et al., 2007; Dequin, 2001; Pretorius, 2000; Pretorius & Bauer, 2002; Querol & Ramón, 1996). This process has also been reliant on the use of molecular techniques to confirm the establishment and eventual dominance of inoculated wine yeasts during fermentation (Querol et al., 1992).

Genetic and metabolic engineering strategies for the improvement of wine yeasts require three tools (see Figure 7.1):

1) Cloned genes with known biochemical and genetic links to a given enological problem
2) An effective genetic transformation system that will allow the genes to be introduced into an industrial wine yeast
3) Regulatory sequences known as promoters that allow expression of the genes of interest at the appropriate point during vinification

In the following sections, we will address each of these elements.

# 2. SYSTEMS FOR THE GENETIC TRANSFORMATION OF WINE YEASTS

## 2.1. Transformation Procedures

Unlike other microorganisms such as *Bacillus subtilis* and *Haemophilus influenzae*, *S. cerevisiae* does not undergo natural genetic transformation. In order for DNA to be introduced into the cells of this organism, reach the nucleus, and become stably expressed generation after generation, cells must be made competent for the entry of exogenous DNA. These artificial genetic transformation techniques allow millions of transformants to be obtained with just a microgram of exogenous DNA in laboratory strains of *S. cerevisiae* and are now employed routinely by thousands of scientists around the world.

Four different techniques are commonly used. The first is based on the use of protoplasts or spheroplasts, which are obtained by treating the yeast cells with a mixture of enzymes to digest their cell wall and allow exogenous DNA to enter (Hinnen et al., 1978). Since one of the main functions of the cell wall is to maintain osmotic balance, protoplasts must be maintained in a medium that is equilibrated with the intracellular osmotic pressure. When the protoplasts are placed in contact with a suspension of exogenous DNA, treatment with polyethylene glycol (PEG) will induce DNA entry into the cell. After a period of incubation, the protoplasts are transferred to fresh medium containing an osmotic stabilizer and they regenerate the cell wall to yield a normal cell containing the exogenous DNA. The second method also involves creation of artificial permeability, which in this case is achieved by treating intact yeast cells and DNA with high concentrations of lithium salts, normally lithium acetate, and PEG (Gietz et al., 1992). The third method involves the use of a device known as an electroporator, which subjects intact cells or protoplasts to short, high-voltage electrical discharges (Delorme, 1989). This treatment opens small pores in the surface of the cell or protoplast through which exogenous DNA can penetrate and reach the interior of the cell. Finally, although its use is very limited, the biolistic technique used for transformation of plant cells has been employed by some authors for yeast transformation (Armaleo et al., 1990). Here, yeast cultures are bombarded with tungsten microparticles coated with exogenous DNA in an attempt to introduce the DNA into the cytoplasm and nucleus of the cell. In the vast majority of cases, transformation of industrial wine yeasts has been carried out using the lithium salts protocol, although protoplasts and electroporation have also been used (see Table 7.1).

## 2.2. Selection Markers

To achieve transformation, the gene of interest is usually inserted into a plasmid to form a transformation vector. In order to select cells that have acquired the exogenous DNA, genes that function as selection markers, usually by conferring resistance to a drug or by complementing auxotrophies, are introduced into the vector. Since the marker gene is included in the same plasmid, detection indicates the presence of the gene of interest. In the case of markers conferring drug resistance, genes are used to generate resistance to certain antibiotics, herbicides, or amino acid analogs that are toxic for yeast, such as canavanine, chloramphenicol, cycloheximide, diuron, geneticin, hygromicin, kanamycin, methotrexate, sulfometuron, p-fluorophenylalanine, phleomycin, and zeocin. Various resistance mechanisms can be involved, the most common being the inactivation of the corresponding antibiotic. Alternatively, the resistance gene may be a mutant version or allele of a gene encoding the primary target of the selective agent (for instance in the

**TABLE 7.1** Transformation Methods Used in the Generation of Transgenic Wine Yeasts

| Transformation method | Reference |
| --- | --- |
| Electroporation | Salek et al. (1990) |
| | Pérez-González et al. (1993) |
| | van Rensburg et al. (2005) |
| | Coulon et al. (2006) |
| | Husnik et al. (2006) |
| Protoplasts | Lee and Hassan (1988) |
| | Pérez-González et al. (1993) |
| Lithium salts | Lee and Hassan (1988) |
| | Boone et al. (1990) |
| | Petering et al. (1991) |
| | Laing and Pretorius (1993) |
| | Pérez-González et al. (1993) |
| | Dequin and Barre (1994) |
| | van Rensburg et al. (1994) |
| | González-Candelas et al. (1995) |
| | Ansanay et al. (1996) |
| | Sánchez-Torres et al. (1996) |
| | Michnick et al. (1997) |
| | Puig et al. (1998) |
| | Sánchez-Torres et al. (1998) |
| | Dequin et al. (1999) |
| | Ganga et al. (1999) |
| | Remize et al. (1999) |
| | González-Candelas et al. (2000) |
| | Vilanova et al. (2000) |
| | Pérez-Torrado et al. (2002) |
| | Manzanares et al. (2003) |
| | Walker et al. (2003) |
| | Fernández-González et al. (2005) |
| | Cambon et al. (2006) |
| | van Rensburg et al. (2007) |
| | Swiegers et al. (2007) |
| | González-Ramos et al. (2008) |
| | Herrero et al. (2008) |
| | González-Ramos et al. (2009) |
| | Ehsani et al. (2009) |

case of cycloheximide, sulfometuron, or p-fluorophenylalanine resistance). As shown in Figure 7.2a, the cells that take up the exogenous DNA containing the resistance gene are easily differentiated from those that do not by growing them in medium containing the

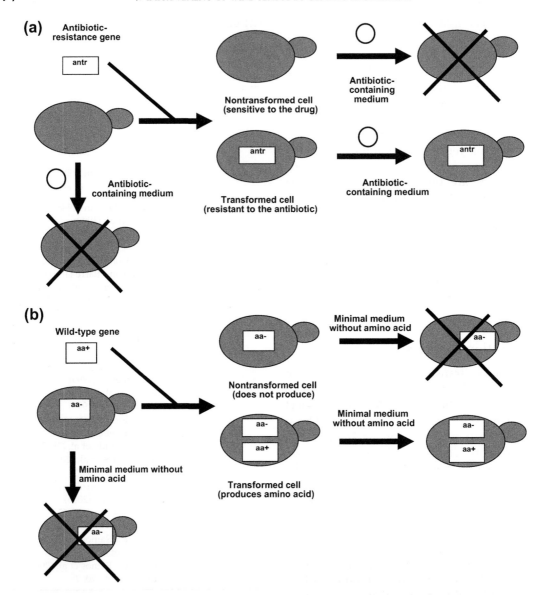

**FIGURE 7.2** Schematic diagram showing the genetic selection of transformed wine yeasts. Selection based on acquisition of antibiotic-resistance genes (a). The recipient cell cannot grow in medium containing an antibiotic to which it is sensitive (white circle). The gene used for selection encodes a protein that inactivates the antibiotic. Consequently, the transformed cells that contain the gene will grow in selective medium containing the drug while the nontransformed cells will die. Selection by auxotrophic complementation (b). The system is based on the use of cells carrying a mutation in the pathway used to synthesize an essential cell metabolite, in this case an amino acid. The cell is therefore auxotrophic for that nutritional requirement and cannot grow in minimal medium without addition of the metabolite. The gene used for selection is a wild-type version of the mutated gene in the receiving cell. As a result, the mutation is complemented in the transformed cells, which will be converted into prototrophs. The transformed cells can therefore grow in minimal medium without addition of nutrients whereas untransformed cells will not grow in this selective medium.

antibiotic. In the case of auxotrophic complementation, the strains used carry mutations that prevent the synthesis of a particular amino acid, nitrogenous base, or vitamin. For instance, mutations have been used in the *URA3* gene, which encodes an enzyme in the synthesis pathway for the nucleotide uracil, or *TRP1*, which codes for an enzyme involved in the synthesis of the amino acid tryptophan. These mutants are unable to grow in minimal medium without the addition of uracil or tryptophan, respectively. Selection involves inclusion of the wild-type allele in the transformation vector so that the receiving cells repair the molecular lesion and are able to grow in minimal medium without the addition of supplements (see Figure 7.2b).

In the case of wine yeasts, selection by auxotrophic complementation is not straightforward since, as indicated previously, many, if not all, industrial strains are polyploid and have more than one copy of each gene. As a result, each copy must be mutated, with the corresponding difficulties that this entails. In contrast, the use of antibiotic-resistance genes does not present methodological difficulties but may not receive consumer support. This public rejection rests on the possibility that the marker gene might be transferred to gut bacteria by conjugation or natural transformation and lead to the generation of antibiotic-resistant pathogenic bacterial flora. However, although such a view is supported by certain environmental non-government organizations, it has no scientific basis and, according to the World Health Organization (WHO), evidence suggests that no such risk exists (World Health Organization, 1993). Nevertheless, as a result of public pressure, the use of antibiotic-resistance genes in foodstuffs has been prohibited in the EU since 2005.

Since many of the transgenic wine yeasts that have been produced carry antibiotic-resistance genes as selection markers (see Table 7.2), systems have been developed to eliminate the

**TABLE 7.2** Selection Markers Used to Generate Transgenic Wine Yeasts

| Selection method | Reference |
|---|---|
| *URA3* complementation | Dequin and Barre (1994) |
| | Ansanay et al. (1996) |
| | Pérez-Torrado et al. (2002) |
| | González-Ramos et al. (2008) |
| | Herrero et al. (2008) |
| | González-Ramos et al. (2009) |
| *LEU2* complementation | Vilanova et al. (2000) |
| Cycloheximide resistance | Pérez-González et al. (1993) |
| | González-Candelas et al. (1995) |
| | Sánchez-Torres et al. (1996) |
| | Sánchez-Torres et al. (1998) |
| | Ganga et al. (1999) |
| | González-Candelas et al. (2000) |
| | Manzanares et al. (2003) |
| Killer factor resistance | Lee and Hassan (1988) |
| | Boone et al. (1990) |
| | Salek et al. (1990) |
| Phleomycin resistance | Remize et al. (1999) |
| | Cambon et al. (2006) |
| | Coulon et al. (2006) |
| | Husnik et al. (2006) |
| G418 resistance | Laing and Pretorius (1993) |
| | van Rensburg et al. (1994) |
| | Puig et al. (1998) |
| | Dequin et al. (1999) |
| | Walker et al. (2003) |
| | Cambon et al. (2006) |
| | van Rensburg et al. (2007) |
| | González-Ramos et al. (2008) |
| | Ehsani et al. (2009) |
| | González-Ramos et al. (2009) |
| Sulfometuron resistance | Petering et al. (1991) |
| | Fernández-González et al. (2005) |
| | van Rensburg et al. (2005) |
| | Swiegers et al. (2007) |
| | van Rensburg et al. (2007) |
| p-fluorophenylalanine resistance | González-Ramos et al. (2008) |
| | González-Ramos et al. (2009) |

gene following selection (Puig et al., 1998; Walker et al., 2003). Instead of plasmid vectors, these systems use linear DNA fragments containing the gene of interest and the selection gene. They also contain fragments of DNA corresponding to the *S. cerevisiae* genome that allow them to be targeted to specific loci on a chromosome of the wine yeast. Once the transgenic wine yeast has been selected, the resistance gene is excised to produce a final transgenic yeast that carries only the gene of interest at a defined position in the genome. However, in most cases, elimination of the antibiotic-resistance gene leaves a trace or scar in the genome in the form of a short fragment of exogenous DNA. To prevent this occurring, genes belonging to the same yeast can be used as resistance markers. These are generally mutant alleles of the protein targets of antibiotics such as cycloheximide, p-fluorophenylalanine, or sulfometuron. In order for this technique to be effective, however, the transformation methods must prevent carryover of DNA derived from the vectors used to generate the DNA construct. Alternatively, it is possible to use a cotransformation strategy (see Figure 7.3). In this case, two types of DNA molecule are used in the transformation. One carries the antibiotic-resistance gene in a plasmid that replicates in the yeast nucleus but is unstable in the absence of

selective pressure. The other construct carries the gene of interest and is designed to integrate in the yeast genome. This strategy takes advantage of the lack of specificity that is characteristic of microbial transformation techniques. In other words, when a cell is competent, it is likely to take up both types of DNA present in the transformation medium. Once the transformants have been selected using the resistance marker, integration of the construct at the expected site is confirmed. Then, growth of the yeast in the absence of selective pressure for the marker allows it to be eliminated along with all associated bacterial sequences. This strategy has been used to generate two yeasts that have been approved by the United States Food and Drug Administration (FDA) and Health Canada for use in foodstuffs (Coulon et al., 2006; Husnik et al., 2006). These types of yeast are referred to by the acronym GRAS (generally recognized as safe). Unlike in classical genetics, where little is known about the changes produced by mutation or crossing, here the nature of the genetic modification is understood in detail, with complete sequence information available for both the inserted fragment and the integration site.

A different situation occurs in transgenic wine yeasts in which improvements are generated not by expression of a new gene (or a change in the pattern or level of expression for the yeast's own gene) but rather by elimination of a gene. This involves integration of a marker gene in a specific locus, normally replacing the gene of interest. The difficulty of this approach lies in the diploid or aneuploid character of industrial strains, since two or more copies of the same gene must be eliminated in order to generate an improvement in the characteristics of the strain. A recent example involved the production of yeasts with enhanced release of mannoproteins through the elimination of genes involved in cell wall biogenesis (González-Ramos et al., 2008, 2009).

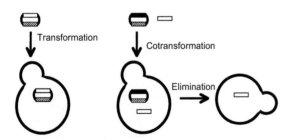

FIGURE 7.3 Schematic representation of conventional transformation and cotransformation systems. In conventional transformation, the selection marker (hatched rectangle) and the gene or construct of interest (white rectangle) form part of the same molecule. In cotransformation, these two elements are present in different molecules, allowing easy elimination (curing) of the marker and other exogenous sequences without leaving scars or traces in the genome.

## 3. REGULATION OF GENE EXPRESSION: PROMOTERS OF INTEREST IN BIOTECHNOLOGY

To produce a transgenic wine yeast, it is not sufficient to simply introduce a gene of interest into the genome of the yeast. For a variety of reasons, it is also important to control the expression of the exogenous gene. In many cases, overexpression can be toxic for the transformed yeast. At other times, it may be necessary to express the transgenic protein at a specific point during vinification. For instance, if the intention is to enhance aroma, it is necessary to express the gene towards the end of vinification, when the yeast has entered the stationary phase. This is because aromatic compounds are volatile and, if the gene is expressed from the beginning, much of the aromatic product will be lost. Nevertheless, there are also situations in which it is appropriate to maintain constant expression. Each of these cases requires the use of different regulatory sequences known as promoters, derived from inducible, delayed, or constitutively expressed genes, accordingly.

A great deal of information is available on the molecular control of gene expression in laboratory strains of *S. cerevisiae* grown under controlled conditions. This information continues to be expanded through the use of DNA microarrays (Pérez-Ortín et al., 2002). In a pioneering study, Puig et al. (1996) showed that genes such as *HSP104*, *POT1*, and *SSA3*, which are expressed during the late stationary phase in laboratory yeast strains, were expressed early during vinification. In the same study, those authors found that the promoter of the gene *ACT1* is able to drive expression during late phases of vinification. Subsequent, more detailed studies have shown that gene expression patterns during vinification depend on the strain of wine yeast analyzed (Carrasco et al., 2001; Riou et al., 1997). That

said, the expression of some genes, such as *HSP12*, can be considered a marker of resistance to the stress imposed by vinification (Ivorra et al., 1999). It has been shown that the promoter for the *TDH2/3* gene, which encodes the glyceraldehyde-3-phosphate dehydrogenase enzyme, drives the highest levels of expression during vinification and that some genes such as *SPI1* appear to display enhanced expression during the stationary phase (Puig & Pérez-Ortín, 2000).

More promoters have become available for use in controlling transgene expression in wine yeasts as a consequence of DNA microarray analysis of transcription profiles during vinification (Backus et al., 2001; Beltrán et al., 2006; Jiménez-Martí & del Olmo, 2008; Marks et al., 2008; Mendes-Ferreira et al., 2007; Rossignol et al., 2003, 2006; Zuzuarregui et al., 2006). To date, however, few promoters have been used and most drive constitutive expression (see Table 7.3).

## 4. TRANSGENIC WINE YEASTS

The tools described earlier have been used to produce transgenic wine yeasts that have been produced to display improved metabolic characteristics or enhanced production of compounds with interesting organoleptic or nutritional properties (see Table 7.4). Most of these yeasts have been developed in the laboratories of the Australian Wine Research Institute at the University of Adelaide (http://www.awri.com.au/), the Institute for Wine Research at the University of Stellenbosch in South Africa (http://academic.sun.ac.za/wine_biotechnology/), the Institut de Produits de la Vigne del Institut Nacional de la Recherche Agronomique (INRA) in Montpellier, France (http://www.montpellier.inra.fr/), the Instituto de Agroquímica y Tecnología de Alimentos del Consejo Superior de Investigaciones Científicas in Valencia, Spain (http://www.iata.csic.es/), the Instituto de Fermentaciones Industriales del Consejo Superior de

**TABLE 7.3**    Promoters Used in the Generation of Transgenic Wine Yeasts

| Gene[1] | Protein | Reference |
|---|---|---|
| ACT1 | Actin | Pérez-González et al. (1993)<br>González-Candelas et al. (1995)<br>Sánchez-Torres et al. (1996)<br>Sánchez-Torres et al. (1998)<br>Ganga et al. (1999)<br>González-Candelas et al. (2000)<br>Manzanares et al. (2003) |
| ADH1 | Aldehyde dehydrogenase | Petering et al. (1991)<br>Dequin and Barre (1994)<br>Ansanay et al. (1996)<br>Dequin et al. (1999)<br>Cambon et al. (2006)<br>van Rensburg et al. (2007) |
| PGK1 | Phosphoglycerate kinase | Vilanova et al. (2000)<br>Fernández-González et al. (2005)<br>Coulon et al. (2006)<br>Husnik et al. (2006)<br>Swiegers et al. (2007) |
| TDH3 | Glyceraldehyde-3-phosphate dehydrogenase | Herrero et al. (2008)<br>Ehsani et al. (2009) |

[1]*Gene from which the promoter is derived.*
Examples are shown only for cases in which the regulatory sequences differ from those normally associated with the gene used in the transformation experiment. Examples involving introduction of multiple copies of a gene belonging to the transformed yeast are not shown.

**TABLE 7.4**    Phenotypes Improved by Genetic Engineering in Wine Yeasts

| Modified phenotype | Reference |
|---|---|
| Elevated glycogen to increase resistance to nutritional stress | Pérez-Torrado et al. (2002) |
| Synthesis of pectinases to improve filtration | González-Candelas et al. (1995)<br>Vilanova et al. (2000)<br>Fernández-González et al. (2005)<br>van Rensburg et al. (2007) |
| Synthesis of killer factor | Lee and Hassan (1988)<br>Boone et al. (1990)<br>Salek et al. (1990) |
| Synthesis of pediocin as an antimicrobial agent | Schoeman et al. (1999) |
| Synthesis of glucose oxidase as an antimicrobial agent | Malherbe et al. (2003) |
| Increased acidity | Dequin and Barre (1994)<br>Dequin et al. (1999) |
| Reduced acidity | Bony et al. (1997)<br>Volschenk et al. (1997a)[1]<br>Volschenk et al. (1997b)[1]<br>Husnik et al. (2006) |

*(Continued)*

**TABLE 7.4**   Phenotypes Improved by Genetic Engineering in Wine Yeasts—*cont'd*

| Modified phenotype | Reference |
| --- | --- |
| Synthesis of glucose oxidase and reduced ethanol levels | Malherbe et al. (2003) |
| Increased glycerol and reduced ethanol | Michnick et al. (1997)<br>Remize et al. (1999)<br>de Barros Lopes et al. (2000)<br>Eglinton et al. (2002)<br>Cambon et al. (2006)<br>Ehsani et al. (2009) |
| Synthesis of β-(1,4)-endoglucanase to improve aroma | Pérez-González et al. (1993) |
| Synthesis of α-L-arabinofuranosidase to improve aroma | Sánchez-Torres et al. (1996) |
| Synthesis of β-glucosidase to improve aroma | Sánchez-Torres et al. (1998)<br>van Rensburg et al. (2005) |
| Synthesis of β-(1,4)-endoxylanase to improve aroma | Ganga et al. (1999) |
| Synthesis of alcohol acetyltransferase to improve aroma | Lilly et al. (2000)<br>Verstrepen et al. (2003) |
| Synthesis of α-L-rhamnosidase to improve aroma | Manzanares et al. (2003) |
| Synthesis of phenol decarboxylase to improve aroma | Smit et al. (2003)[1] |
| Increased resveratrol levels | González-Candelas et al. (2000)<br>Becker et al. (2003)[1] |
| Synthesis of β-glucuronidase as a marker | Petering et al. (1991) |
| Reduction of ethyl carbamate levels | Coulon et al. (2006) |
| Increased mannoprotein content | González-Ramos et al. (2008)<br>González-Ramos et al. (2009) |
| Synthesis of monoterpenes to improve aroma | Herrero et al. (2008) |
| Synthesis of tryptophanase to improve aroma | Swiegers et al. (2007) |

[1]*Modification analyzed only in laboratory strains.*

Investigaciones Científicas in Madrid, Spain (http://www.ifi.csic.es/), and the Wine Research Center at the University of British Columbia in Canada (http://www.landfood.ubc.ca/wine/index.html).

## 4.1. Improvements Affecting the Industrial Winemaking Process

During the industrial production of active dried yeast, cells are grown under aerobic conditions until a considerable biomass is obtained. Before the cells are dried and packed, feeding with the carbon source is stopped and the yeast cells are forced to switch to respiratory metabolism and consume the ethanol produced during fermentation. The dried cells are then rehydrated before being added to the fermentation tank. However, the prior processes of glucose starvation and dehydration stress the cells and reduce their viability and fermentative capacity (discussed in more detail in Chapter 11). In an effort to alleviate these problems, transgenic wine yeasts have been produced that overexpress the

*GSY2* gene (Pérez-Torrado et al., 2002). This gene encodes the enzyme glycogen synthase, and its overexpression leads to increased accumulation of glycogen, a cell metabolite linked to the stress response. As a result, cell viability is increased.

It is desirable for vinification to occur in the shortest time possible without negatively influencing the organoleptic properties of the wine obtained. In an effort to achieve this, transgenic wine yeasts that overexpress the glycolytic enzyme glyceraldehyde-3-phosphate dehydrogenase have been produced. However, although these yeasts have been reported to indirectly reduce fermentation time, they also lead to substantial changes in the volatile profile and a reduction in the alcohol content of the wine (Remize et al., 1999).

During the winemaking process, pectinolytic enzymes are usually introduced to enhance the extraction of the must and improve clarification and filtration of the wine. This practice increases the cost of the process and risks the introduction of contaminating activities that affect the aroma or color of the wine, since pure enzyme preparations are not commercially available. To solve this problem, transgenic wine yeasts have been produced that carry a gene, *pelA* from the filamentous fungus *Fusarium solani*, which encodes a pectate lyase (González-Candelas et al., 1995). This transgenic yeast secretes the pectinolytic enzymes into the must during fermentation. Vilanova et al. (2000) adopted a different approach involving the *S. cerevisiae* polygalacturonase *PGU1*. Since expression of *PGU1* is weak or absent in many strains, the authors engineered a wine strain that constitutively expressed the gene.

Finally, it is worth mentioning the production of various transgenic wine strains that are able to eliminate microorganisms that interfere with the fermentation process or lead to microbial spoilage. Wine strains have been produced that contain extra copies of genes coding for killer factors or other genes encoding different killer factors from those already carried by the strain (Boone et al., 1990; Lee & Hassan, 1988; Salek et al., 1990). More recently, the gene from *Pediococcus acidilactici* that encodes the bacteriocin pediocin A has been expressed in laboratory strains of *S. cerevisiae* (Schoeman et al., 1999). Pediocin A is active against, among others, *Listeria monocytogenes* and is effectively secreted by the transgenic yeast. According to Schoeman et al., these developments may make it possible to produce strains of baker's, wine, or brewer's yeast with biocontrol activity. Similarly, a gene from the filamentous fungus *Aspergillus niger*, which encodes the enzyme glucose oxidase, when introduced into laboratory strains results in inhibition of the growth of spoilage yeasts such as *Acetobacter aceti* and *Gluconobacter oxidans* when cocultured with the transgenic yeast (Malherbe et al., 2003). This effect is mediated by enhanced production of hydrogen peroxide.

## 4.2. Improvements Affecting the Physicochemical Properties of Wine

In hot winegrowing regions, it is common to obtain musts with low acidity and unbalanced organoleptic properties. A metabolic engineering strategy to increase the acidity of these wines has been developed based on the introduction of a lactic acid fermentation pathway into wine yeasts (Dequin & Barre, 1994; Dequin et al., 1999). This was achieved by expressing a gene from the lactic acid bacteria *Lactobacillus casei*, which encodes L(+)-lactate dehydrogenase and leads to the conversion of 20% of the glucose into lactate during fermentation. The resulting pH reduction of 0.3, however, was also associated with a slightly reduced ethanol content.

In contrast, wines from cooler regions have an excess of malic acid, which must be corrected by the addition of lactic acid bacteria such as *Oenococcus oeni* to produce malolactic fermentation and deacidification. Experiments have been performed in both laboratory and wine strains to coexpress the *Schizosaccharomyces pombe* gene *mae1*, which encodes malate permease, with a malic enzyme from the lactic acid bacteria *Lactococcus lactis* (Bony et al., 1997; Volschek et al., 1997a). Under those conditions, the transgenic *S. cerevisiae* strain was able to transport malic acid into the cell and transform it into lactate in a reaction catalyzed by the malic enzyme, thus reducing the acidity of the wine. In an alternative approach, the *mae1* and *mae2* genes from *S. pombe* were coexpressed in *S. cerevisiae* (Volschenk et al., 1997b). The *mae2* gene encodes the *S. pombe* malic enzyme and coexpression led to the effective degradation of 8 g of malate per liter of must, compared with 4.5 g/L degraded by earlier transgenic yeast strains. A recent advance has been the production by cotransformation of the commercial yeast strain ML01, which coexpresses the malate permease of *S. pombe* with the malolactic enzyme of *O. oeni* (Husnik et al., 2006).

Genetic engineering has also been used successfully to produce low-alcohol wines. Strategies have involved expressing the *A. niger* gene encoding the glucose oxidase enzyme (Malherbe et al., 2003), enhancing expression of the glyceraldehyde-3-phosphate dehydrogenase-encoding genes *GPD1* or *GPD2* (de Barros Lopes et al., 2000; Michnick et al., 1997; Remize et al., 1999), or deleting the *FPS1* gene encoding the glycerol transporter (Eglinton et al., 2002). However, many of these studies have only been performed in laboratory strains. Furthermore, reducing the ethanol concentration implies increasing that of glycerol and leads to a reduced generation of biomass and changes in the production of volatile compounds such as acetate, acetoin, 2,3-butanediol, or succinate. Its use thus needs to be carefully reviewed before implementation (Remize et al., 1999). Efforts have continued to be made in recent years to redirect the carbon flux of *S. cerevisiae* towards pathways that are more compatible with wine quality through metabolic engineering

and manipulation of cofactors (Cambon et al., 2006; Ehsani et al., 2009).

Chemical stability is the main physicochemical property of concern to enologists, particularly during the production of white wines. The principal considerations are protein and tartrate precipitation. The first involves the precipitation of complexes containing mainly grape-derived proteins that are unstable in the presence of ethanol, especially over prolonged periods at insufficiently cool temperatures. Tartrate precipitation involves the formation of potassium bitartrate crystals in bottled wine. It has been reported that, in both cases, the problem is reduced by the presence of higher concentrations of mannoproteins (Dupin et al., 2000; Feuillat et al., 1998). Transgenic wine strains have therefore been developed in which the secretion of mannoproteins during fermentation is increased by eliminating all copies (two or three) of certain genes linked to cell wall biosynthesis (González-Ramos et al., 2008, 2009).

## 4.3. Improvement of Organoleptic, Nutritional, and Safety-related Properties

The fruity aroma of certain wines has been enhanced through the use of genetically engineered wine yeasts. This aroma depends mainly on the presence of certain terpenes—particularly geraniol, nerol, and linalool—that are found in the must in two fractions: a free fraction that produces aroma by virtue of being volatile and a bound fraction, associated via diglycosidic bonds with fragments of cell wall from the grape berries. The bound fraction does not contribute to wine aroma unless the glycosidic bonds are cleaved by glycosidases. This process is favored by the prior activity of cellulases and hemicellulases (Villanueva et al., 2000). In order to perform this enzymatic treatment, genes from filamentous fungi and aerobic yeast that code for these enzymes have been expressed in wine yeasts, with the result that concentrations of one or more aromatic terpenes were increased (Ganga, 1999; Manzanares, 2003; Pérez-González et al., 1993; Sánchez-Torres et al., 1996, 1998). Transgenic wine yeasts have also been produced that are able to produce terpenes directly as a product of their metabolism (Herrero et al., 2008).

Alternative strategies have sought to express the *PADC* gene from *B. subtilis* and the *PDC* gene from the lactic acid bacteria *Lactobacillus plantarum*, genes that encode decarboxylases for phenolic compounds (Smit et al., 2003). Microvinification experiments using these transgenic wine yeasts showed that phenolic compounds present in the must were converted into 4-vinyl and 4-ethyl derivatives, which are volatile and contribute to an enhanced aroma. A more complicated strategy involved increasing the levels of alcohol acetyltransferase by overexpression of *ATF1* in wine yeasts in an attempt to induce overproduction of volatile aromatic esters (Lilly et al., 2000). Various genes encode enzymes with this activity and comparative deletion studies have been performed in laboratory strains (Verstrepen et al., 2003). These experiments have shown that *ATF1* has a role in the synthesis of volatile esters.

There is an important contribution to the quality of traditional-method sparkling wines by products derived from yeast autolysis (see Chapter 2). Given that this is a slow process, transgenic wine yeasts have been produced that undergo more rapid autolysis as a result of the manipulation of genes related to autophagy (Cebollero et al., 2009).

Finally, a *Candida molischiana* gene encoding a β-glucosidase has been expressed in a wine yeast. This enzyme can mobilize resveratrol conjugates, compounds thought to be linked to the so-called "French paradox" due to their anticholesterolemic and antitumoral effects in vitro. The transgenic yeast that has been developed produces wines with a higher resveratrol content than that produced with conventional yeasts (González-Candelas et al., 2000). Alternative strategies have been developed to increase resveratrol production in laboratory strains by coexpressing a gene from the poplar tree encoding a coenzyme A (CoA) ligase with the resveratrol synthase gene from grape (Becker et al., 2003). It should be remembered, however, that the effect of resveratrol has only been analyzed in vitro and that wine is an alcoholic beverage and therefore not a suitable vehicle for the administration of active pharmaceutical ingredients.

One of the main food safety concerns associated with wine production is the formation of ethyl carbamate. In fact, there are legal limits for this compound that are applicable for commercial wine import and export. It is a potential carcinogen mainly derived from a spontaneous reaction between ethanol and urea formed by the yeast as a result of arginine metabolism. Kitamoto et al. (1991) approached this problem in a sake-producing yeast strain by eliminating both copies of the gene *CAR1*, which encodes an enzyme that catalyzes the first step in *S. cerevisiae* arginine metabolism. More recently, Coulon et al. (2006) used cotransformation to produce a yeast that constitutively expresses the urease-encoding gene *Dur1,2*. In some experiments, the use of this yeast has led to a 90% reduction in the ethyl carbamate content of the wines produced when compared with those produced using the parent strain. The commercial name of this strain is EMLo1.

There are undoubtedly many more developments underway in genetic engineering laboratories that will come to light in the next few years. The advances made to date suggest various options for the industrial application of these new technologies. However, the use of transgenic approaches will have to be considered within the appropriate legal framework and it will be necessary to convince consumers of the benefits that are offered. These considerations will be discussed in the following sections.

# 5. LEGISLATION AFFECTING THE USE OF GENETIC ENGINEERING IN THE WINE INDUSTRY

## 5.1. European Legislation

Wine yeasts that have been modified by genetic engineering are transgenic organisms and their industrial use would thus lead to the production of transgenic wines. With the exception of a small production in Moldavia a few years ago (I.S. Pretorius, personal communication), there have been no commercial references to the production or sale of transgenic wines. Sales of beer labeled as transgenic have recently begun, however, in Sweden, although the beer is not produced with a transgenic brewer's yeast but rather with a conventional yeast and a barley wort to which fragments of transgenic maize have been added as a nitrogen source (Editorial, 2004). What would be the implications of deciding to market a wine produced with a transgenic yeast strain? The response to this question will vary according to the legislation on transgenic food and beverages in the country in question. Below we discuss the European model, which affects most of the producing countries.

If approval is requested for marketing in the EU, or more specifically in a member state such as Spain, the evaluation process would fall under legislation regulating research and development, environmental release, patentability, marketing, and labeling of transgenic foods produced by genetic engineering. Currently, following a recent modification, the heart of that legislation is EU Directive 2001/18/EC on the deliberate release into the environment of GMOs, which has been developed and adapted in various documents, particularly Regulation (EC) 1829/2003 on genetically modified food and feed and Regulation (EC) 1830/2003 on the traceability and labeling of GMOs and the traceability of food and feed products produced from GMOs.

The most notable element of this legislation refers to the marketing of transgenic products, which must be considered on a case-by-case basis, with obligatory evaluation by panels of scientific experts. It is usually a long process that involves submission of a dossier by the company seeking marketing authorization to demonstrate the lack of adverse effects on human health or the environment associated with the product for which authorization is sought. The minimum requirement in terms of health assessment is an analysis of the nutritional composition of the transgenic product compared with the corresponding conventional product, an assessment of allergenicity, and a preclinical study of toxicity in laboratory animals. In the first health assessment of a transgenic wine yeast, performed in a strain that expresses a xylanase from the filamentous fungus *A. nidulans* (Picó et al., 1999), it was seen that no additional risk was associated with ingestion of the transgenic product compared to its conventional counterpart. Assessment of environmental impact must be carried out by controlled environmental release of the GMO. These processes tend to take an average of 5 years and require significant financial investment. The results are assessed by a panel of scientific experts.

In Europe, evaluation of transgenic foods is undertaken by the panel on genetically modified (GM) foods of the European Food Safety Authority (EFSA). This panel is composed of scientific experts in various fields related to food safety evaluation and molecular biology. Based on their evaluation, a final decision is made on whether to accept or reject the proposal. In addition, it is obligatory for the public to be informed of all authorizations and for the European Parliament to be consulted. Finally, the Council of the EU can approve or reject by majority vote the commission's proposal to authorize a transgenic product.

In this new directive, member states must guarantee labeling and monitoring throughout all phases of marketing, and initial approval is limited to a period of 10 years. Following marketing authorization, there is an obligatory monitoring period during which possible long-term effects can be assessed, particularly in relation to environmental impact.

## 5.2. Labeling

Regulation 1830/2003 on traceability and labeling of GMOs and the traceability of food and feed products produced from GMOs is the current point of reference for legislation on labeling. The regulation states that anyone in the EU who plants transgenic grape varieties must inform all of their clients in writing and store a copy of this communication for at least 5 years. The same applies to anyone who markets a transgenic yeast in the EU. In addition, the recently approved Regulation 65/2004 established a system for creating and assigning unique identifiers for GMOs.

According to these regulations, in the EU, a wine would be considered transgenic and thus require labeling as such when it is prepared from a transgenic grape variety or wine yeast. In contrast, if an enzyme obtained from a GMO is added to the must or wine, it need not be labeled as genetically modified so long as the enzyme is not active in the bottled wine. This decision, for which there is no scientific justification, has been questioned by some scientists (Ramón et al., 2004).

## 5.3. The Situation in Other Countries

Many consider the extensive legislation affecting the marketing and labeling of GM foods in the EU to be a clear example of over-legislation. Political pressure, particularly from certain environmental groups and multinational food companies, has had more influence on the preparation of these directives and regulations than have common sense and consumer interests (Ramón et al., 1998). This has given rise to a complex situation that has affected and will continue to affect all links in the food production chain, from the farmer to the consumer, via the food processing and distribution industry.

The situation in other non-EU countries is less complex. In the United States, the FDA made a public declaration in 1992 affirming that it was not necessary to develop specific legislation for the commercialization of GM foods. In their view, the legislation applicable to the marketing of foods obtained using conventional genetic methods was sufficient, as it required detailed analysis of potential adverse effects on hygiene, health, and the environment (Kessler et al., 1992). Currently, the FDA requires prior evaluation of genetic modifications to be used in food or feeds but does not require specific labeling when the food is sufficiently similar in its nutritional composition to the conventional alternatives. There is a clear distinction between the American model, which assesses the final product irrespective of the method used to obtain it, and the European model, in which both elements are considered.

In other countries, such as Argentina, Australia, Canada, and Japan, the system is more similar to the American model. Some other countries still have very loose legislation or are discussing how to proceed while continuing to market GM foods. These differences could lead to havens of permissiveness that ultimately undermine efforts to guarantee the lack of additional risk associated with the commercialization of these products, a possibility that is clearly undesirable.

## 6. THE FUTURE

Is there a market for wines produced by fermentation with a transgenic yeast? The answer to this question is not straightforward. It will depend on the country in which the wine is produced, the country in which it is sold, and, of course, the benefits offered by the transgenic modification.

For many years, there was a clear lack of interest in these new technologies shown by the International Organization of Wine and Vine (OIV). Partly for this reason, the United States has abandoned this organization. Many of the countries that are most active in wine biotechnology research, including the United States and Canada, are not members of the OIV. A number of them have undertaken genome sequencing projects for grape varieties or projects to develop transgenic vines or yeasts that are financed by public and private funds. However, the situation in the EU, and in particular the major wine-producing countries such as France and Spain, is quite different. There are leading groups on the development of transgenic yeasts as well as some groups involved in sequencing the genome of grape varieties, but there is no interest in these groundbreaking technologies from the wineries themselves. Some Latin American countries such as Argentina and Chile have begun to employ these new technologies. Their future direction is more likely to follow that of Australia, the United States, and South Africa than that of the EU.

It is currently unimaginable for transgenic wines to be sold in the EU. This is different from the situation in the United States, where it is unlikely that specific labeling would be required for the internal market. One can reasonably assume that the response of consumers to transgenic wines will reflect their feelings towards GM foods in general. If this is the case, we can define a region of strong rejection in the EU represented by Austria, France, Germany, and some Scandinavian countries, and a more receptive region comprising Spain and Portugal.

It remains unlikely that genetic engineering will be applied to wines from famous appellations such as Rioja or Bordeaux; it is more likely to be used in those wines for which improvements in value for money are key determinants of sales. These wines account for a much higher percentage of annual sales than the more famous wines and they also have problems that might be resolved through the use of genetic engineering. We should not forget, though, that wine is a very traditional product with a strong geographic influence. It is much more than an alcoholic beverage—it represents part of the identity of many countries and regions. In products of this type with a long tradition, it is much more complicated for new technologies to be introduced and, when they are, they only succeed when they offer something of genuine interest to the consumer. It is therefore fair to predict that transgenic wines produced from vine cultivars modified to be resistant to pests are likely to fail, as the benefit is for the grower rather than the consumer. In contrast, a transgenic wine in which improvements have been made in aroma or color without affecting price could have a greater likelihood of acceptance. A transgenic wine containing elevated concentrations of a compound offering nutritional or health benefits might have an even greater chance of success despite the fact that, as an alcoholic beverage, it should never be considered an appropriate vehicle for active ingredients designed to improve consumer health. As mentioned earlier, some such examples already exist.

In the United States and Canada, where administrative procedures are apparently less costly and in the absence of pressure from organizations such as the OIV, two transgenic wine yeasts have already been approved by the relevant authorities (Coulon et al., 2006; Husnik et al., 2006). Furthermore, these countries do not require specific labeling for transgenic products of this kind. To obtain approval, it was essential to provide scientific evidence of functional equivalence (apart from those characteristics that have been intentionally modified) between the original yeast and the corresponding transgenic strain. This entailed the use of cotransformation systems and the removal of unnecessary sequences from the final construct, along with the use of tools to analyze the transcriptome and provide additional data to demonstrate equivalence (Coulon et al., 2006; Husnik et al., 2006). From a commercial point of view and in terms of public opinion, these yeasts differ substantially. One aims to resolve a problem that is partly technological and partly affects the sensory properties of the final product (malolactic fermentation) and for which nonrecombinant options are available, such as conventional malolactic fermentation involving O. oeni. The other addresses a problem for which there is no easily implemented conventional option and that can also affect the health of the consumer. It is reasonable to assume that wines labeled as transgenic are more likely to be accepted in the latter case than in the former.

There is still some way to go. It will certainly be necessary to generate recombinant yeast strains that are as "clean" as possible, to undertake extensive assessment of health and environmental impacts, and to explain clearly to consumers exactly how and why the modifications have been made. It will also be important to learn the lessons provided by the industrial use and commercialization of the yeasts mentioned above. This is not going to be an easy task, but it will undoubtedly be easier for our colleagues in Australia, North America, and South Africa. With time and patience, we will discover whether or not it was worth the effort.

# References

Ansanay, V., Dequin, S., Camarasa, C., Schaeffer, V., Grivet, J. P., Blondin, B., et al. (1996). Malolactic fermentation by engineered Saccharomyces cerevisiae as compared with engineered Schizosaccharomyces pombe. Yeast, 12, 215–225.

Antcliff, A. J. (1992). Taxonomy, the grapevine as a member of the plant kingdom. In B. G. Coombe & P. R. Dry (Eds.), Viticulture, Vol. 1 (pp. 107–118). Underdale, Australia: Winetitles.

Armaleo, D., Ye, G. N., Klein, T. M., Shark, K. B., Sanford, J. C., & Johnston, S. A. (1990). Biolistic nuclear transformation of Saccharomyces cerevisiae and other fungi. Curr. Genet., 17, 97–103.

Backhus, L. E., DeRisi, J., Brown, P. O., & Bisson, L. F. (2001). Functional genomic analysis of a commercial wine strain of Saccharomyces cerevisiae under differing nitrogen conditions. FEMS Yeast Res., 1, 111–125.

Barre, P., Vézinhet, F., Dequin, S., & Blondin, B. (1993). Genetic improvement of wine yeast. In G. H. Fleet (Ed.), Wine microbiology and biotechnology (pp. 421–447). Singapore: Harwood Academic Publishers.

Becker, J. V. W., Gareth, O. A., van der Merwe, M. J., Lambrechts, M. G., Vivier, M. A., & Pretorius, I. P. (2003). Metabolic engineering of Saccharomyces cereviseae for the synthesis of the wine-related antioxidant resveratrol. FEMS Yeast Res., 4, 79–85.

Beltrán, G., Novo, M., Leberre, V., Sokol, S., Labourdette, D., Guillamon, J. M., et al. (2006). Integration of transcriptomic and metabolic analyses for understanding the global responses of low-temperature winemaking fermentations. FEMS Yeast Res., 6, 1167–1183.

Bony, M., Bidart, F., Camarasa, C., Ansanay, V., Dulau, L., Barre, P., et al. (1997). Metabolic analysis of S. cerevisiae strains engineered for malolactic fermentation. FEBS Lett., 410, 452–456.

Boone, C., Sdicu, A. M., Wagner, J., Degré, R., Sanchez, C., & Bussey, H. (1990). Integration of the yeast K1 killer toxin gene into the genome of marked wine yeasts and its effect on vinification. Am. J. Enol. Vitic., 41, 37–42.

Borneman, A. R., Forgan, A. H., Pretorius, I. S., & Chambers, P. J. (2008). Comparative genome analysis of a Saccharomyces cerevisiae wine strain. FEMS Yeast Res., 8, 1185–1195.

Cambon, B., Monteil, V., Remize, F., Camarasa, C., & Dequin, S. (2006). Effects of GPD1 overexpression in Saccharomyces cerevisiae comercial wine yeast strains lacking ALD6 genes. Appl. Environ. Microbiol., 72, 4688–4694.

Carbonneau, A. (1983). Stèrilitès male et female dans le genre Vitis. II. Consequènces en gènètique et sèlection. Agronomie, 3, 645–649.

Carrasco, P., Querol, A., & del Olmo, M. (2001). Analysis of the stress resistance of comercial wine yeast strains. Arch. Microbiol., 174, 450–457.

Cebollero, E., Gonzalez-Ramos, D., & Gonzalez, R. (2009). Construction of a recombinant autolytic wine yeast strain overexpressing the csc1–1 allele. Biotechnol. Prog., 25, 1598–1604.

Cebollero, E., González-Ramos, D., Tabera, L., & González, R. (2007). Transgenic wine yeast technology comes of age: Is it time for transgenic wine? Biotechnol. Lett., 29, 191–200.

Coulon, J., Husnik, J. I., Inglis, D. L., van der Merwe, G. K., Lovaud, A., Erasmus, D. J., & van Vuuren, H. J. J. (2006). Metabolic engineering of *Saccharomyces cerevisiae* to minimize the production of ethyl carbamate in wine. *Am. J. Enol. Vitic., 57*, 113–124.

de Barros Lopes, M., Rehman, A. U., Gockowiak, H., Heinrich, A. J., Langridge, P., & Henschke, P. A. (2000). Fermentation properties of a wine yeast overexpressing the Saccharomyces cerevisiae glycerol 3-phosphate dehydrogenase gene (GPD2). *Aust. J. Grape Wine Res., 6*, 208–215.

Delorme, E. (1989). Transformation of *Saccharomyces cerevisiae* by electroporation. *Appl. Environ. Microbiol., 55*, 2242–2246.

Dequin, S. (2001). The potential of genetic engineering for improving brewing, wine-making and baking yeasts. *Appl. Microbiol. Biotechnol., 56*, 577–588.

Dequin, S., Baptista, E., & Barre, P. (1999). Acidification of grape musts by *Saccharomyces cerevisiae* wine yeast strains genetically engineered to produce lactic acid. *Am. J. Enol. Vitic., 50*, 45–50.

Dequin, S., & Barre, P. (1994). Mixed lactic acid-alcoholic fermentation by *Saccharomyces cerevisiae* expressing the *Lactobacillus casei* L(+)-LDH. *Biotechnology, 12*, 173–177.

Dupin, I. V., McKinnon, B. M., Ryan, C., Boulay, M., Markides, A. J., Jones, G. P., et al. (2000). *Saccharomyces cerevisiae* mannoproteins that protect wine from protein haze: Their release during fermentation and lees contact and a proposal for their mechanism of action. *J. Agric. Food Chem., 48*, 3098–3105.

Editorial. (2004). With just a hint of transgenes... *New Scientist, 2433*, 6.

Eglinton, J. M., Heinrich, A. J., Pollnitz, A. P., Langridge, P., Henschke, P. A., & de Barros Lopes, M. (2002). Decreasing acetic acid accumulation by a glycerol over-producing strain of *Saccharomyces cerevisiae* by deleting the *ALD6* aldehyde dehydrogenase gene. *Yeast, 19*, 295–301.

Ehsani, M., Fernández, M. R., Biosca, J. A., Julien, A., & Dequin, S. (2009). Engineering of 2, 3-butanediol dehydrogenase to reduce acetoin formation by glycerol-overproducing, low-alcohol *Saccharomyces cerevisiae*. *Appl. Environ. Microbiol., 75*, 3196–3205.

Fernández-González, M., Ubeda, J. F., Cordero-Otero, R. R., Gururajan, V. T., & Briones, A. I. (2005). Engineering of an oenological Saccharomyces cerevisiae strain with pectinolytic activity and its effect on wine. *Int. J. Food Microbiol., 102*, 173–183.

Feuillat, M., Charpentier, C., & Nguyen van Long., T. (1998). Yeast's mannoproteins: A new possible enological adjuvant. *Bull. O.I.V., 71*, 945–967.

Ganga, M. A., Piñaga, F., Vallés, S., Ramón, D., & Querol, A. (1999). Aroma improving in microvinification processes by the use of a recombinant wine yeast strain expressing the Aspergillus nidulans xlnA gene. *Int. J. Food Microbiol., 47*, 171–178.

Gietz, D., St. Jean, A., Woods, R. A., & Schiestl, R. H. (1992). Improved method for high efficiency transformation of intact yeast cells. *Nucl. Acids Res., 20*, 1425.

Goffeau, A., et al. (1997). The yeast genome directory. *Nature, 387*(suppl. 1), 1–105.

González, R., Martínez-Rodríguez, A., & Carrascosa, A. (2003). Yeast autolytic mutants potentially useful for sparkling wine production. *Int. J. Food Microbiol., 84*, 21–26.

González-Candelas, L., Cortell, A., & Ramón, D. (1995). Construction of a recombinant wine yeast strain expressing a fungal pectate lyase gene. *FEMS Microbiol. Lett., 126*, 263–270.

González-Candelas, L., Gil, J. V., Lamuela-Raventós, R., & Ramón, D. (2000). The use of transgenic yeasts expressing a gene encoding a glycosyl-hydrolase as a tool to increase resveratrol content in wine. *Int. J. Food Microbiol., 59*, 179–183.

González-Ramos, D., Cebollero, E., & González, R. (2008). A recombinant *Saccharomyces cerevisiae* strain over-producing mannoproteins stabilizes wine against protein haze. *Appl. Environ. Microbiol., 74*, 5533–5540.

González-Ramos, D., Quirós, M., & González, R. (2009). Three different targets for the genetic modification of wine yeast strains resulting in improved effectiveness of bentonite fining. *J. Agric. Food Chem., 57*, 8373–8378.

Hammond, J. R. M. (1996). Yeast genetics. In F. G. Priest & I. Campbell (Eds.), *Brewing microbiology* (pp. 45–82). London, UK: Chapman and Hall.

Herrero, O., Ramon, D., & Orejas, M. (2008). Engineering the *Saccharomyces cerevisiae* isoprenoid pathway for de novo production of aromatic monoterpenes in wine. *Metabolic Engin., 10*, 78–86.

Hinnen, A., Hicks, J. B., & Fink, G. R. (1978). Transformation of yeast. *Proc. Natl. Acad. Sci. USA, 75*, 1929–1933.

Husnik, J. I., Volschenk, H., Bauer, J., Colavizza, D., Luo, Z., & van Vuuren, H. J. J. (2006). Metabolic engineering of malolactic wine yeast. *Metabolic Engin., 8*, 315–323.

Ivorra, C., Pérez-Ortin, J. E., & del Olmo, M. (1999). An inverse correlation between stress resistance and stuck fermentations in wine yeasts, a molecular study. *Biotechnol. Bioeng., 64*, 698–708.

Jimenez-Marti, E., & del Olmo, M. L. (2008). Addition of ammonia or amino acids to a nitrogen-depleted medium affects gene expression patterns in yeast cells during alcoholic fermentation. *FEMS Yeast Res., 8*, 245–256.

Kessler, D. A., Taylor, M. R., Maryanski, J. H., Flamm, E. L., & Kahl, L. S. (1992). The safety of foods developed by biotechnology. *Science, 256*, 1747–1749, 1832.

Kitamoto, K., Oda, K., Gomi, K., & Takahashi, K. (1991). Genetic engineering of a sake yeast producing no urea by successive disruption of arginase gene. *Appl. Environ. Microbiol., 57*, 301–306.

Laing, E., & Pretorius, I. S. (1993). Co-expression of an *Erwinia chrysanthemi* pectate lyase-encoding gene (*pelE*) and an *E. carotovora* polygalacturonase-encoding gene (*peh1*) in *Saccharomyces cerevisiae. Appl. Microbiol. Biotechnol., 39*, 181–188.

Lee, F. J. S., & Hassan, L. H. M. (1988). Stability and expression of a plasmid-containing killer toxin cDNA in batch and chemostat cultures of *Saccharomyces cerevisiae. Biotechnol. Bioengin., 31*, 783–789.

Lilly, M., Lambrechts, M. G., & Pretorius, I. S. (2000). Effect of increased yeast alcohol acetyltransferase activity on flavor profiles of wine and distillates. *Appl. Environ. Microbiol., 66*, 744–753.

Liti, G., Carter, D. M., Moses, A. M., Warringer, J., Parts, L., James, S. A., et al. (2009). Population genomics of domestic and wild yeasts. *Nature, 458*, 337–341.

Malherbe, D. F., du Toit, M., Cordero, R. R., van Rensburg, P., & Pretorius, I. S. (2003). Expression of the *Aspergillus niger* glucose oxidase gene in *Saccharomyces cerevisiae* and its potential applications in wine production. *Appl. Microbiol. Biotechnol., 61*, 502–511.

Manzanares, P., Orejas, M., Gil, J. V., de Graaff, L., Visser, J., & Ramón, D. (2003). Construction of a genetically modified wine yeast strains expressing the *Aspergillus aculeatus rhaA* gene, encoding an α-L-rhamnosidase of enological interest. *Appl. Environ. Microbiol., 69*, 7558–7562.

Marks, V. D., Sui, S. J. H., Erasmus, D., van der Merwe, G. K., Brumm., J., Wasserman, W. W., et al. (2008). Dynamics of the yeast transcriptome during wine fermentation reveals a novel fermentation stress response. *FEMS Yeast Res., 8*, 35–52.

Mendes-Ferreira, A., del Olmo, M., Garcia-Martinez, J., Jimenez-Marti, E., Mendes-Faia, A., Perez-Ortin, J. E., et al. (2007). Transcriptional response of *Saccharomyces cerevisiae* to different nitrogen concentrations during alcoholic fermentation. *Appl. Environ. Microbiol., 73*, 3049–3060.

Michnick, S., Roustan, J. L., Remize, F., Barre, P., & Dequin, S. (1997). Modulation of glycerol and ethanol yields during alcoholic fermentation in *Saccharomyces cerevisiae* strains overexpressed or disrupted for *GPD1* encoding glycerol 3-phosphate dehydrigenase. *Yeast, 13*, 783–793.

Mortimer, R. K., & Polsinelli, M. (1999). On the origins of wine yeast. *Res. Microbiol., 150*, 199–204.

Novo, M., Bigey, F., Beyne, E., Galeote, V., Gavory, F., Mallet, S., et al. (2009). Eukaryote-to-eukaryote gene transfer events revealed by the genome sequence of the wine yeast *Saccharomyces cerevisiae* EC1118. *Proc. Natl. Acad. Sci. USA, 106*, 16 333–16 338.

Núñez, D. R., & Walker, M. J. (1989). A review of paleobotanical findings of early *Vitis* in the Mediterranean and of the origins of cultivated grapevines, with special reference to prehistoric exploitations in the western Mediterranean. *Rev. Paleobot. Palynol., 61*, 205–237.

Pérez-González, J. A., González, R., Querol, A., Sendra, J., & Ramón, D. (1993). Construction of a recombinant wine yeast strain expressing a β-(1, 4)-endoglucanase activity and its use in microvinification processes. *Appl. Environ. Microbiol., 59*, 2801–2806.

Pérez-Ortín, J. E., García Martinez, J., & Alberola, T. M. (2002). DNA chips for yeast biotechnology, the case of wine yeasts. *J. Biotechnol., 98*, 227–241.

Pérez-Torrado, R., Gimeno-Alcañiz, J. V., & Matallana, E. (2002). Wine yeast strains engineered for glycogen overproduction display enhanced viability under glucose deprivation conditions. *Appl. Environ. Microbiol., 68*, 3339–3344.

Petering, J. E., Henschke, P. A., & Langridge, P. (1991). The *Escherichia coli* β-glucuronidase gene as a marker for *Saccharomyces* yeast strain identification. *Am. J. Enol. Vitic., 42*, 6–12.

Picó, Y., Fernández, M., Rodríguez, R., Almudéver, J., Mañes, J., Font, G., Marín, R., Carda, C., Manzanares, P., & Ramón, D. (1999). Toxicological assesment of xylanase $X_{22}$ expressed by transgenic wine yeast. *J. Agric. Food Chem., 47*, 1597–1602.

Pretorius, I. S. (2000). Tailoring wine yeast for the new millennium, novel approaches to the ancient art of winemaking. *Yeast, 16*, 675–729.

Pretorius, I. S., & Bauer, F. F. (2002). Meeting the consumer challenge through genetically customized wine-yeast strains. *Trends Biotechnol., 20*, 426–432.

Pretorius, I. S., & van der Westhuizen, T. J. (1991). The impact of yeast genetics and recombinant DNA technology on the wine industry, a review. *S. Afr. J. Enol. Vitic., 12*, 3–31.

Puig, S., & Pérez-Ortín, J. E. (2000). Expression levels and patterns of glycolytic yeast genes during wine fermentation. *System. Appl. Microbiol., 23*, 300–303.

Puig, S., Querol, A., Ramón, D., & Pérez-Ortín, J. E. (1996). Evaluation of the use of phase-specific gene promoters for the expression of enological enzymes in an industrial wine yeast strain. *Biotechnol. Lett., 18*, 887–892.

Puig, S., Ramón, D., & Pérez Ortín, J. E. (1998). Optimized method to obtain stable food-safe recombinant wine yeast strains. *J. Agric. Food Chem., 46*, 1689–1693.

Querol, A., Barrio, E., Huerta, T., & Ramón, D. (1992). Molecular monitoring of wine fermentations conducted by dry yeast strains. *Appl. Environ. Microbiol., 58*, 2948–2953.

Querol, A., & Bond, U. (2009). The complex and dynamic genomes of industrial yeasts. *FEMS Microbiol. Lett., 293*, 1–10.

Querol, A., Fernández-Espinar, M. T., del Olmo, M., & Barrio, E. (2003). Adaptative evolution of wine yeast. *Int. J. Food Microbiol., 86*, 3–10.

Querol, A., & Ramón, D. (1996). The application of molecular techniques in wine microbiology. *Trends Food Sci. Technol., 7*, 73–78.

Quirós, M., Gonzalez-Ramos, D., Tabera, L., & González, R. (2010). A new methodology to obtain wine yeast strains overproducing mannoproteins. *Int. J. Food Microbiol., 139*, 9–14.

Rainieri, S., & Pretorius, I. S. (2000). Selection and improvement of wine yeasts. *Ann. Microbiol., 50*, 15–31.

Ramón, D., Calvo, M. D., & Peris, J. (1998). New regulation for labelling of genetically modified foods, a solution or a problem? *Nature Biotechnol., 16*, 889.

Ramón, D., MacCabe, A. P., & Gil, J. V. (2004). Questions linger over European GM food regulations. *Nature Biotechnol., 22*, 149.

Remize, F., Roustan, J. L., Sablayrolles, J. M., Barre, P., & Dequin, S. (1999). Glycerol overproduction by engineered *Saccharomyces cerevisiae* wine yeast strains leads to substantial changes in by-product formation and to a stimulation of fermentation rate in stationary phase. *Appl. Environ. Microbiol., 65*, 143–149.

Riou, C., Nicaud, J. M., Barre, P., & Gaillardin, C. (1997). Stationary-phase gene expression in *Saccharomyces cerevisiae* during wine fermentation. *Yeast, 13*, 903–915.

Rossignol, T., Dulau, L., Julien, A., & Blondin, B. (2003). Genome-wide monitoring of wine yeast gene expression during alcoholic fermentation. *Yeast, 20*, 1369–1385.

Rossignol, T., Postaire, O., Storai, J., & Blondin, B. (2006). Analysis of the genomic response of a wine yeast to rehydration and inoculation. *Appl. Microbiol. Biotechnol., 71*, 699–712.

Salek, A., Schnettler, R., & Zimmermann, U. (1990). Transmission of killer activity into laboratory and industrial strains of *Saccharomyces cerevisiae* by electroinjection. *FEMS Microbiol. Lett., 70*, 67–72.

Sánchez-Torres, P., González-Candelas, L., & Ramón, D. (1996). Expression in a wine yeast strain of the *Aspergillus niger abfB* gene. *FEMS Microbiol. Lett., 145*, 189–194.

Sánchez-Torres, P., González-Candelas, L., & Ramón, D. (1998). Heterologous expression of a *Candida molischiana* anthocyanin-β-glucosidase in a wine yeast strain. *J. Agric. Food Chem., 46*, 354–360.

Schoeman, H., Vivier, M. A., du Toit, M., Dicks, L., & Pretorius, I. S. (1999). The development of bactericidal yeast strains by expressing the *Pediococcus acidilactici* pediocin gene (*pedA*) in *Saccharomyces cerevisiae. Yeast, 15*, 647–656.

Smit, A., Cordero, R. R., Lambrechts, M. G., Pretorius, I. S., & van Rensburg, P. (2003). Enhancing voletile phenol concentrations in wine by expressing various phenolic acid decarboxylase genes in *Saccharomyces cerevisiae. J. Agric. Food Chem., 51*, 4909–4915.

Snow, R. (1983). Genetic improvement of wine yeast. In J. F. T. Spencer, D. M. Spencer & A. R. W. Smith (Eds.), *Yeast genetics, fundamental and applied aspects* (pp. 439–459). New York, NY: Springer-Verlag.

Stevenson, A. C. (1985). Studies in the vegetational history of S.W. Spain. II. Palynogical investigations at Laguna de los Madres, Spain. *J. Biogeogr., 12*, 293–314.

Swiegers, J. H., Capone, D. L., Pardon, K. H., Elsey, G. M., Sefton, M. A., Francis, I. L., et al. (2007). Engineering volatile thiol release in *Saccharomyces cerevisiae* for improved wine aroma. *Yeast, 7*, 561–574.

van Rensburg, P., Stidwell, T., Lambrechts, M. G., Otero, R. C., & Pretorius, I. S. (2005). Development and assessment of a recombinant *Saccharomyces cerevisiae* wine yeast producing two aroma-enhancing beta-glucosidases encoded by the *Sacharomycopsis fibuligera* BGL and BGL genes. *Ann. Microbiol., 55*, 33–42.

van Rensburg, P., Strauss, M. L. A., Lambrechts, M. G., Otero, R. R. C., & Pretorius, I. S. (2007). The heterologous expression of polysaccharidase-encoding genes with oenological relevance in *Saccharomyces cerevisiae. J. Appl. Microbiol., 103*, 2248–2257.

van Rensburg, P., van Zyl, W. H., & Pretorius, I. S. (1994). Expression of the *Butyrivibrio fibrisolvens* endo-β-1, 4-glucanase gene together with the Erwinia pectate liase and polygalacturonase genes in *Saccharomyces cerevisiae. Curr. Genet., 27*, 17–22.

Verstrepen, K. J., van Laere, S. D. M., Vanderhaegen, B. M. P., Derdelinckx, G., Dufour, J. P., Pretorius, I. P., et al. (2003). Expression levels of the yeast alcohol acetyltransferase genes *ATF1*, *Lg-ATF1*, and *ATF2* control the formation of a broad range of volatile esters. *Appl. Environ. Microbiol., 69*, 5228–5237.

Vilanova, M., Blanco, P., Cortés, S., Castro, M., Villa, T. G., & Sieiro, C. (2000). Use of a *PGU1* recombinant *Saccharomyces cerevisiae* strain in oenological fermentations. *J. Appl. Microbiol., 89*, 876–883.

Villanueva, A., Ramón, D., Vallés, S., Lluch, M. A., & MacCabe, A. P. (2000). Heterologous expression in *Aspergillus nidulans* of a *Trichoderma longibrachiatum* endoglucanase of oenological relevance. *J. Agric. Food Chem., 48*, 951–957.

Vivier, M. A., & Pretorius, I. S. (2002). Genetically tailored grapevines for the wine industry. *Trends Biotechnol., 20*, 472–478.

Volschenk, H., Viljoen, M., Grobler, J., Bauer, F., Lonvaud, A., Denayrolles, M., et al. (1997a). Malolactic fermentation in grape musts by a genetically engineered strain of *Saccharomyces cerevisiae*. *Am. J. Enol. Vitic., 48*, 193—197.

Volschenk, H., Viljoen, M., Grobler, J., Petzold, B., Bauer, F., Subden, R. E., et al. (1997b). Engineering pathways for malate degradation in *Saccharomyces cerevisiae*. *Nature Biotechnol., 15*, 253—257.

Walker, M. E., Gardner, J. M., Vystavelova, A., McBryde, C., de Barros Lopes, M., & Jiranek, V. (2003). Application of the reusable, *KanMX* selectable marker to industrial yeast, construction and evaluation of heterothallic wine strains of *Saccharomyces cerevisiae*, possesing minimal foreign DNA sequences. *FEMS Yeast Res., 4*, 339—347.

World Health Organization (WHO). (1993). *Health aspects of marker genes in genetically modified plants*. Geneva, Switzerland: WHO.

Zuzuarregui, A., Monteoliva, L., Gil, C., & del Olmo, M. L. (2006). Transcriptomic and proteomic approach for understanding the molecular basis of adaptation of *Saccharomyces cerevisiae* to wine fermentation. *Appl. Environ. Microbiol., 72*, 836—847.

# Lactic Acid Bacteria

*Rosario Muñoz*[1]*, M. Victoria Moreno-Arribas*[2]*,*
*Blanca de las Rivas*[1]

[1] Instituto de Ciencia y Tecnología de los Alimentos y Nutrición (ICTAN, CSIC), Madrid, Spain and
[2] Instituto de Investigación en Ciencias de la Alimentación (CIAL, CSIC-UAM), Madrid, Spain

## OUTLINE

# 1. GENERAL CHARACTERISTICS OF LACTIC ACID BACTERIA

Lactic fermentation is a bacterial process that takes place during the production of numerous food products. It provides the final products with characteristic aromas and textures and plays a crucial role in food safety and hygiene. Among the bacteria responsible for lactic fermentation are lactic acid bacteria, which display high morphological and physiological diversity. The term lactic acid bacteria emerged at the beginning of the twentieth century to describe a heterogeneous group of bacteria that are currently defined as spherical (cocci) or rod-shaped (bacilli), gram-positive, catalase-negative, immobile, nonsporulating, anaerobic, aerotolerant, and producers of lactic acid (the main metabolite generated during the fermentation of sugars by these bacteria).

In winemaking, lactic acid bacteria are doubly important as they can both enhance and diminish the quality of wine. They are responsible for malolactic fermentation but they can also cause changes that adversely affect the organoleptic properties of the final product. During malolactic fermentation, the concentration of lactic acid bacteria reaches approximately $10^7$ colony-forming units (CFU)/mL, which gives an indication of their importance in winemaking.

In 1886, Louis Pasteur demonstrated that microbial growth was a common feature of all fermentation processes. Different types of fermentation were defined according to the predominant organic products present at the end of the process, and each was associated with a specific type of microorganism. Pasteur was also the first to demonstrate the presence of lactic acid bacteria in wines. These bacteria

were subsequently linked to the wine defect known as *tourné* or tartaric spoilage. Müller-Thurgau (1891) and Koch (1900) later attributed the presence of lactic acid bacteria to a reduction in the acidity of wines and shortly afterwards, in 1901, Seifert reported that these bacteria were capable of degrading malic acid. More recent studies, particularly from the 1970s onwards, confirmed the importance of malolactic fermentation in reducing acidity (essential in red wines) and ensuring the microbiological stability of the final product by preventing the onset of fermentation after bottling.

Another important and particularly relevant line of research is the study of the negative effects that lactic acid bacteria have on the quality and composition of wine. The emergence of molecular tools based on DNA analysis has provided greater insights into many known alterations and also helped to uncover new ones.

Finally, advances in recent years have led to a spectacular improvement in our understanding of the physiology, metabolism, and genetics of the lactic acid bacteria involved in winemaking. Thanks to the wealth of information now available, winemakers are better positioned to control the activity of these bacteria and to analyze and exploit their impact on the quality of wine from a broader, multidisciplinary perspective.

This chapter will review the main aspects of malolactic fermentation and the growth of lactic acid bacteria in wine. Following a short review of the basic, practical principles underlying the metabolism of lactic acid bacteria during malolactic fermentation, we will review key studies that have analyzed the effect of malolactic fermentation on the organoleptic properties of wine, examine certain properties of these bacteria that are of importance to winemaking, and, finally, discuss the main defects that the metabolism of these bacteria can cause in wine.

## 2. IDENTIFYING LACTIC ACID BACTERIA

Only a few species are capable of growing in grape must and wine because of the hostile conditions that they encounter: mainly low pH, a lack of nutrients, and the presence of ethanol. The main species of lactic acid bacteria that can survive in this environment are shown in Table 8.1. The species *Leuconostoc oenos* was considered to form part of the *Leuconostoc* genus until as recently as 1995, when analysis of the 16S ribosomal DNA (rDNA) sequence showed that it was different from the other members of the genus. This led to the creation of a new genus, *Oenococcus*, which includes just two species, the malolactic *Oenococcus oeni* (Dicks et al., 1995) and the nonmalolactic *Oenococcus kitaharae* (Endo & Okada, 2006). Of all

**TABLE 8.1**  Lactic Acid Bacteria in Wine

| Genus | Metabolism of sugars | Species |
| --- | --- | --- |
| *Pediococcus* | Homofermentative | *P. damnosus* |
| | | *P. parvulus* |
| | | *P. pentosaceus* |
| *Leuconostoc* | Heterofermentative | *Leu. mesenteroides* |
| *Oenococcus* | Heterofermentative | *O. oeni* |
| *Lactobacillus* | Homofermentative | *L. mali* |
| | Facultatively heterofermentative | *L. casei* |
| | | *L. plantarum* |
| | Heterofermentative | *L. brevis* |
| | | *L. buchneri* |
| | | *L. fermentum* |
| | | *L. fructivorans* |
| | | *L. hilgardii* |

the bacteria that perform malolactic fermentation, *O. oeni* has the greatest capacity to grow in acidic pH and in the presence of 10% (vol/vol) ethanol (Versari et al., 1999).

To identify bacteria in fermented foods and beverages such as wine, it is first necessary to isolate them through inoculation in suitable growth media. The most common medium used to isolate lactic acid bacteria is de Man Rogosa Sharpe (MRS) medium. Wibowo et al. (1985) recommended the addition of grape juice, tomato, cysteine, malic acid, and several sugars to this medium. Cycloheximide (100 mg/L) and pimaricin (50 mg/L) are also needed to inhibit the growth of yeast and fungi. The presence of carbon dioxide, in turn, favors the growth of bacteria. The colonies are left to grow until sufficient biomass for performing all the tests required is obtained. Terrade et al. (2009) recently described a chemically defined medium that satisfies the nutritional requirements and favors the growth of lactic acid bacteria from wine.

For many years, the standard, classical methods for identifying and classifying bacteria were based on phenotypic characteristics. Important advances, however, have been made in recent years thanks to the continuing developments in the field of molecular biology. Molecular tools can now be used to reveal the genetic diversity of a particular species and to study populations of microorganisms in wine without the need for prior isolation and culture.

### 2.1. Classical Identification Methods

It is standard practice to perform Gram staining and catalase tests following the isolation and purification of wine bacteria in a suitable solid medium. Bacteria that are found to be both gram-positive and catalase-negative are classified as lactic acid bacteria. This initial classification can be confirmed by growing the corresponding isolates in a liquid medium containing hexoses (glucose/fructose) but not malic acid and then testing for the production of lactic acid using a suitable method such as paper chromatography.

Additional tests are required to identify isolates at the genus level. The first step involves observing the bacteria under a microscope to determine whether they are spherical or rod-shaped. Another test involves the identification of how lactic acid is produced. This is generally performed by observing the production of gas from hexoses. The formation of gas indicates that the bacteria are heterofermentative (capable of producing multiple products from the fermentation of carbohydrates) while the absence of gas indicates that they are homofermentative (able to produce only lactic acid).

This basic information can be used to make a putative identification of the genus. *Oenococcus* and *Leuconostoc* species are heterofermentative, spherical, have single cells that form pairs or short chains, and often have an elongated, lenticular shape, giving them the appearance of short bacilli. Members of the genus *Pediococcus* are also spherical and occur in pairs or tetrads; cells arranged singly or in chains are uncommon. Members of the genus *Lactobacillus*, in contrast, are homofermentative or heterofermentative; they are rod-like and arranged singly or in pairs or chains. Some of these morphologies are shown in Figure 8.1.

Other tests used to confirm genus include analysis of the lactic acid isomer formed from hexoses and detection of the hydrolysis of arginine to ammonium or the production of mannitol from fructose (Pilone et al., 1991).

These conventional characterization tests are still used in routine practice but they can occasionally give rise to ambiguous results, particularly when assessing fermentable carbohydrates. Even in optimal conditions, these carbohydrates can slowly change the color of the medium, making it difficult to distinguish between positive and negative results. Furthermore, certain lactic acid bacteria often contain plasmids encoding enzymes involved in important biochemical pathways. Because of their instability, particularly in the absence of selective pressure, certain tests that normally yield positive results can give negative results. Characters that depend on phages can also cause similar problems.

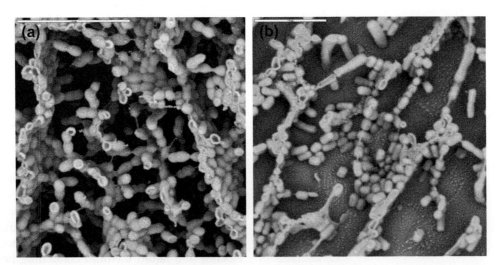

**FIGURE 8.1** Micrographs of *Oenococcus oeni* (a) and *Lactobacillus brevis* (b) taken using low-temperature scanning electron microscopy. Scales of 5 μm are shown at the top left of both images. Images provided by Dr. A.V. Carrascosa.

## 2.2. Molecular Identification Methods

It is often difficult to distinguish between lactic acid bacteria on the basis of physiological and biochemical criteria as most have very similar nutritional and growth requirements in environmental conditions (Vandamme et al., 1996). In recent years, however, molecular biology has increasingly been used to circumvent these difficulties. A wide variety of molecular techniques have been used to characterize lactic acid bacteria from wine. Examples are techniques based on restriction fragment length polymorphisms (RFLPs) (Zapparoli et al., 2000), pulsed-field gel electrophoresis (PFGE) (Gindreau et al., 1997), DNA—DNA hybridization (Dicks et al., 1995; Lonvaud-Funel et al., 1991; Sato et al., 2001), specific DNA hybridization probes (Lonvaud-Funel et al., 1991; Sohier et al., 1999), polymerase chain reaction (PCR) (Groisillier & Lonvaud-Funel, 1999; Lonvaud-Funel et al., 1993), randomly amplified polymorphic DNA (RAPD) (Zavaleta et al., 1997) (Figure 8.2), amplified fragment length polymorphisms (AFLPs) (Cappello et al., 2008), and the study of genes encoding 16S ribosomal RNA (rRNA) (Guerrini et al., 2003; Sato et al., 2001).

Amplified rDNA restriction analysis (ARDRA) has been used as a rapid, reliable method of identifying the main lactic acid bacteria involved in winemaking (Rodas et al., 2003; Ventura et al., 2000). This method, however, has certain limitations. It is not suitable, for example, for comparing *Lactobacillus plantarum* or *Lactobacillus pentosus* as these display a high level of similarity in their 16S rDNA sequence (Collins et al., 1991; Quere et al., 1997). Amplified rDNA fragment analysis via PCR followed by denaturing gradient gel electrophoresis (DGGE) has also been used to compare diversity and monitor changes in populations of lactic acid bacteria during the winemaking process (López et al., 2003).

A more recent system for identifying species of lactic acid bacteria in wine involves the use of several stages (Rodas et al., 2005). It has been concluded that RAPD and ribotyping are useful for identifying and classifying these bacteria, whereas ARDRA is useful only for identification purposes and PFGE-RFLP is useful for distinguishing between different strains of the same species (Rodas et al., 2003).

Because *O. oeni* is the main species of lactic acid bacteria associated with malolactic fermentation in wines, real-time quantitative PCR methods are currently being developed to enable the rapid detection and quantification of these bacteria in samples obtained during fermentation. The main advantage of methods of this type is that they enable rapid corrective action to be taken in order to control bacterial growth (Pinzani et al., 2004). A molecular typing method that combines RAPD and multiplex PCR has been described for characterizing different strains of *O. oeni* during winemaking and evaluating the impact of malolactic starter cultures (Reguant & Bordons, 2003).

Analysis of the population structures of *O. oeni* has yielded contradictory results. While molecular techniques such as DNA—DNA hybridization and sequencing of the genes encoding 16S

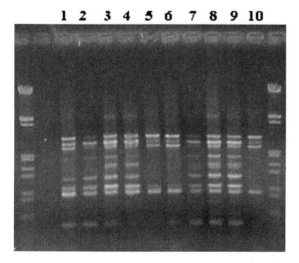

**FIGURE 8.2** Genetic diversity among 10 strains of *Oenococcus oeni* revealed by random amplification of polymorphic DNA.

and 23S rRNA and the intergenic region between 16S and 23S rDNA have shown that *O. oeni* is highly homogeneous, analysis of metabolic and physiological characteristics, such as fatty acid profile and sugar fermentation patterns, have shown quite the opposite. The results of such studies even led to a proposal to divide the species into two separate species or subspecies (Tenreiro et al., 1994). The recent use of multilocus sequence typing showed that *O. oeni* strains can be classified into well-differentiated groups and that recombination events play an important role in the genetic heterogeneity of this species (Bilhère et al., 2009; de las Rivas et al., 2004). One study identified regions of variability in the *O. oeni* genome that were the site of both recombination and gene insertion/deletion (Bon et al., 2009). This enormous variability is largely due to the loss of the DNA mismatch repair genes *mutS* and *mutL*, which may also have contributed to this species' greater adaptation to the conditions found in winemaking (Marcobal et al., 2008).

## 3. POPULATION DYNAMICS IN LACTIC ACID BACTERIA DURING WINEMAKING

Lactic acid bacteria are present throughout all stages of winemaking. They can be isolated on many surfaces and environments including vine leaves, grapes, winery equipment, and barrels. A number of recent molecular identification studies have detected new species of lactic acid bacteria in both musts (*Lactobacillus bobalius* [Mañes-Lázaro et al., 2008a] and *Lactobacillus uvarum* [Mañes-Lázaro et al., 2008b]) and wines (*Lactobacillus nagelii* [Edwards et al., 2000], *Lactobacillus vini* [Rodas et al., 2006], and *Lactobacillus oeni* [Mañes-Lázaro et al., 2009]).

The density of lactic acid bacteria in the initial phases of winemaking (the must phase and onset of alcoholic fermentation) ranges from approximately $10^3$ to $10^4$ CFU/mL. The bacteria are from a variety of mostly homofermentative

species. The most abundant are *L. plantarum*, *Lactobacillus casei*, *Lactobacillus hilgardii*, *Leuconostoc mesenteroides*, and *Pediococcus damnosus*. Less common species include *O. oeni* and *Lactobacillus brevis*. The numbers and proportions of species vary according to the ripeness and condition of the grapes at the time of harvesting. Nonetheless, as the must is fermented by the yeasts, there is a gradual reduction in the quantity and composition of the microflora. In the case of lactic acid bacteria, only those with the greatest resistance to ethanol and low pH survive this stage.

Grape must contains nutrients that favor the growth of yeasts, which rapidly proliferate and initiate alcoholic fermentation. Once this occurs, the bacterial population decreases to between approximately $10^2$ and $10^3$ CFU/mL. The addition of sulfite at concentrations typically used in prefermentation phases reduces the bacterial population but does not inhibit growth, particularly in the presence of high pH. Yeasts, which are less sensitive to the effects of sulfites, grow rapidly and initiate alcoholic fermentation, which occurs practically in the absence of lactic acid bacteria. Lactic acid bacteria have greater difficulty growing in this environment as they are less well adapted to the high sugar concentrations ($>210$ g/L) and low pH of the must (3.0–3.3). By the end of alcoholic fermentation, the density of lactic acid bacteria will have fallen to approximately $10^2$ CFU/mL. There is also a marked reduction in the variety of species present. This is logical as the metabolism of the yeasts responsible for alcoholic fermentation gradually increases ethanol levels and generates compounds that are toxic to bacteria as well as fatty acids and sulfur dioxide, thus altering the composition of the bacterial wall (Edwards et al., 1990). In normal conditions, once alcoholic fermentation is complete, there is a lag phase lasting between 10 and 15 d during which the population of lactic acid bacteria remains unchanged as their growth is inhibited by the presence of live yeasts and inhibitory substances secreted by these. Once this phase is complete, the bacteria begin to multiply until they reach a density of approximately $10^6$ CFU/mL and begin malolactic fermentation. This propagation phase is influenced mainly by pH, temperature, sulfur dioxide and ethanol levels. Growth is favored by a relatively high pH level ($>3.5$), a sulfur dioxide concentration of no more than 50 mg/L, an ethanol content of 13% (vol/vol), and a temperature of between 19 and 26°C. Other factors that favor malolactic fermentation are prolonged contact between the wine and grape skins after alcoholic fermentation and on-lees aging, as yeast autolysis generates nutrients that stimulate the growth of lactic acid bacteria (Guilloux-Benatier et al., 1993). The composition of the bacterial population changes during this phase as strains that are better equipped to resist this hostile environment are gradually selected. The first species to disappear are homofermentative lactic acid bacteria, followed by their heterofermentative counterparts and *Pediococcus* species. The dominant species at the end of alcoholic fermentation is *O. oeni*. This is the species that is best adapted to the difficult growth conditions (low pH and high ethanol content) that characterize this phase (Davis et al., 1985a; van Vuuren & Dicks, 1993), which explains why it is the primary species responsible for malolactic fermentation in most wines. Certain strains belonging to the *Pediococcus* and *Lactobacillus* genera, however, can also survive in this phase.

Malolactic fermentation starts once the bacterial population reaches a density of $10^6$ CFU/mL, following a lag phase that can last days or even months. During this stage, all the malic acid in the wine is transformed into lactic acid by the bacteria present. Malolactic fermentation usually takes between 5 d and 2 or 3 weeks, depending on the physicochemical conditions of the environment and the amount of malic acid to be transformed. This acid is thought to play an important role in stimulating bacterial growth but not in forming the biomass. The disappearance of malic acid ensures microbiological stability by inhibiting

further development of lactic acid bacteria. Aeration of the wine and light carbon dioxide pressure also favor growth. pH is one of the main factors that affects bacterial growth in wine (Wibowo et al., 1985) and each species has a different pH threshold for growth. Lactic, succinic, and tartaric acid all inhibit malolactic activity in bacteria, as do high concentrations of malic acid. Fumaric acid, in contrast, stimulates activity when present at low levels but inhibits it once it reaches a level of between 0.4 and 1.5 g/L. Fatty acids such as decanoic acid, a product of yeast metabolism, have a strong inhibitory effect on malolactic activity (Edwards & Beelman, 1987).

The lag phase often does not occur when the acidity of the grapes used to make the wine is low or when they have a high pH. In such cases, lactic acid bacteria may appear before the end of alcoholic fermentation, causing what is known as lactic taint, which is a considerable increase in the volatile acidity of the wine as the lactic acid bacteria start to metabolize sugars at the same time as malic acid.

The survival of bacteria after malolactic fermentation depends on the environment, particularly on conditions such as pH, ethanol content, and, above all, sulfur dioxide levels. In practice, lactic acid bacteria are eliminated by the addition of sulfite once malic acid transformation is complete. *O. oeni* bacteria disappear rapidly, leaving bacteria from the genera *Pediococcus* and *Lactobacillus* to dominate. In wines to which no sulfites are added, certain strains of lactic acid bacteria can alter the quality of the wine by degrading components such as citric acid, tartaric acid, and glycerol. This is particularly common in low-acid wines.

While only the free form of sulfur dioxide has antiseptic properties against yeast in wine, all forms of sulfur dioxide have activity against bacteria. The antibacterial effect of sulfur dioxide depends mainly on the pH of the wine. The levels of free sulfur dioxide required to inhibit the activity of lactic acid bacteria range from 10 to 20 mg/L for wines with a low pH and from 20 to 40 mg/L for wines with a high pH. Coccoid species (*Pediococcus*, *Oenococcus*, and *Leuconostoc*) are less resistant than *Lactobacillus* species to the effect of sulfur dioxide.

## 4. METABOLISM

While degradation of malic and citric acid by lactic acid bacteria has the greatest bearing on the final quality of the wine, these bacteria also metabolize other substrates to ensure their propagation. Of note among these substrates are sugars, tartaric acid, glycerol, and certain amino acids. The most important—and desirable—activity performed by lactic acid bacteria in wine is malic acid degradation; indeed, malolactic fermentation is recommended only when the aim is to eliminate all traces of this acid from the wine. Lactic acid bacteria are also responsible for other enzymatic activities, most of which result in changes that can detract from or even completely spoil the quality of the wine.

### 4.1. Carbohydrate Metabolism

#### 4.1.1. Monosaccharide and Disaccharide Metabolism

Grape must contains monosaccharides, disaccharides, and oligosaccharides. Monosaccharides include hexoses (glucose, fructose, galactose, and mannose) and pentoses (arabinose, xylose, ribose, and rhamnose). The most abundant sugars are glucose and fructose. Disaccharides (maltose, raffinose, and trehalose) and oligosaccharides occur at lower concentrations (Liu & Davis, 1994). It has been shown that lactic acid bacteria from wine use sugars as a source of carbon and energy (Davis et al., 1986a, 1986b; Liu et al., 1995a; Salou et al., 1994) and that they preferentially use glucose and trehalose (Liu et al., 1995a).

Homofermentative lactic acid bacteria ferment hexoses via the Embden-Meyerhof-Parnas (EMP) pathway and produce two moles of lactate and adenosine triphosphate (ATP) per mole of hexose. Heterofermentative lactic acid bacteria (*O. oeni*, *L. brevis*, *L. hilgardii*, and *Lactobacillus buchneri*) and facultative homofermentative bacteria (*L. plantarum*), in contrast, ferment hexoses and pentoses via the phosphate pentose or phosphoketolase pathway to produce one mole of lactate, ethanol, carbon dioxide, and ATP per mole of hexose. Fructose can also serve as an electron acceptor and is reduced to mannitol. Consequently, the acetyl phosphate formed during hexose fermentation is converted to acetate instead of being reduced to ethanol, thus generating an additional ATP molecule (Pilone et al., 1991; Salou et al., 1994). Heterofermentative lactic acid bacteria can also use other substances such as oxygen and pyruvate as electron acceptors, leading to the production of acetate and additional ATP.

Full-genome analysis of the *O. oeni* strain PSU-1 led to the identification of all of the genes encoding the pentose phosphate pathway as well as several sugar transporter systems (Mills et al., 2005). *O. oeni*, like other heterofermentative bacteria, is capable of converting fructose into mannitol in a reaction catalyzed by mannitol dehydrogenase. This, however, can cause problems in wines, as excessive mannitol levels can lead to high concentrations of acetic acid. Nonetheless, the gene encoding malate dehydrogenase was not found in *O. oeni* PSU-1, indicating that another dehydrogenase must be responsible for the formation of mannitol.

The fermentation of disaccharides by lactic acid bacteria from wine has not been studied in depth, and it is not clear whether these bacteria metabolize disaccharides by hydrolysis or by conversion to monosaccharides via the action of hydrolases or phosphorylases. Once the disaccharides have been released, the resulting monosaccharides enter the common sugar fermentation pathways. Although sugar transporters have been identified in the *O. oeni* PSU-1 genome, no genes linked to the transport of sucrose, lactose, maltose, or raffinose have been found (Mills et al., 2005).

### 4.1.2. Polysaccharide Metabolism

Polysaccharides can have a detrimental effect on wine, first by increasing viscosity (which requires correction by filtration) and second by altering sensory properties such as body, consistency, and roundness. Excessive concentrations of polysaccharides are thus undesirable as they cause a ropy wine, but moderate levels have a beneficial effect on both body and roundness. Enzymes with the capacity to degrade polysaccharides can improve grape must and wine by breaking down the walls of plant cells and improving the extraction of color and aroma precursors. Not many studies, however, have focused on identifying this capacity in lactic acid bacteria from wine. Guilloux-Benatier et al. (2000) demonstrated that *O. oeni* has extracellular β $(1 \rightarrow 3)$ glucanase activity in the stationary phase of growth, providing the first evidence that this species has the capacity to degrade glucan-type polysaccharides. Increases observed in glucose and fructose levels during malolactic fermentation may thus be, at least partly, due to this glucanase activity.

### 4.1.3. Polyalcohol Metabolism

Using nuclear magnetic resonance (NMR), Veiga-da-Cunha et al. (1992) confirmed the synthesis of glycerol and erythritol from glucose in *O. oeni* and reported that the erythritol-to-glycerol ratio was dependent on oxygen levels. Other researchers have also reported the production of glycerol, erythritol, and other polyalcohols by *O. oeni* and other lactic acid bacteria from wine (Firme et al., 1994; Liu et al., 1995a). According to results published by Veiga-da-Cunha et al. (1993), the pathway responsible for the production of erythritol from glucose involves the isomerization of glucose-6-phosphate followed by cleavage to produce erythrose-4-phosphate and acetyl phosphate, reduction of erythrose-4-phosphate to erythritol-4-phosphate, and finally hydrolysis of erythritol-4-phosphate to form erythritol. Nonetheless, no specific enzymes (or the genes encoding them) for the conversion of erythrose-4 and erythritol have been identified in the *O. oeni* genome (Zaunmüller et al., 2006). The formation of polyalcohols is essentially an alternative pathway for the reoxidation of NAD(P)H. Coenzyme A (CoA) deficiency appears to be responsible for the shift to the formation of erythritol, acetate, and glycerol from glucose in the absence of pantothenic acid, as evidenced by the fact that phosphotrans-acetylase and acetaldehyde dehydrogenase are limiting under conditions of pantothenic acid deficiency (Ritcher et al., 2001). Glycerol is a minor product of NAD(P)H reoxidation that is formed by the reduction of glyceraldehyde-3-phosphate to glycerol-1-phosphate followed by dephosphorylation. The *O. oeni* genome contains genes that may encode the enzymes glycerol-1-phosphate dehydrogenase and phosphatase. Because the biochemical reactions involved in the formation of glycerol and erythritol are similar, the two compounds may be synthesized by the same enzymes. Mannitol, which is one of the predominant polyalcohols in wine, is formed by reduction of fructose, as mentioned earlier.

Some lactobacilli isolated in wine have the capacity to degrade glycerol and mannitol, two of the most abundant polyalcohols found in wine. *L. brevis* and *L. buchneri* strains isolated in a spoiled wine were found to metabolize glycerol in the presence of glucose or fructose, leading to the formation of 3-hydroxypropanal (3-hydroxypropionaldehyde), which in turn is reduced to 1,3-propanediol (Schutz & Radler, 1984a, 1984b). 3-Hydroxypropionaldehyde is a precursor of acrolein, a bitter compound found in alcoholic beverages such as wine and cider.

Unlike glycerol, mannitol has been found to be used as the sole source of carbon and energy for growth by *L. plantarum* isolated in wine (Davis et al., 1988; Liu et al., 1995a). The catabolism of mannitol in *L. plantarum*, however, requires the presence of either oxygen (aerobic metabolism) or compounds such as citrate and α-keto acids that can act either directly or indirectly as electron acceptors (anaerobic metabolism) (Chen & McFeeters, 1986a, 1986b; McFeeters & Chen, 1986).

The metabolism of polyalcohols in lactic acid bacteria in wine has an important contribution in winemaking. The production of polyalcohols can influence both the sensory quality of wine (e.g., body, viscosity, and roundness) and technological processes such as filtration. The formation of acrolein from glycerol can confer a bitter taste. As far as microbiological stability is concerned, *L. plantarum* can sometimes develop after malolactic fermentation due to an increase in the pH of the wine and the production of mannitol by *O. oeni*. This mannitol may then be fermented by *L. plantarum*, resulting in high levels of lactate and a risk of spoilage.

## 4.2. Organic Acid Metabolism

Lactic acid bacteria are capable of metabolizing the main organic acids present in grape musts and wines. While they mostly act on malic and citric acid, they can also metabolize tartaric acid. Citric acid is only used in co-fermentation with hexoses, whereas malic acid and tartaric acid can be degraded without a co-substrate. Many of the strains that develop after malolactic fermentation can metabolize malic and citric acid and as a result cause a wide range of organoleptic changes. The changes linked to the degradation of malic acid have been studied in the greatest detail but, more recently, the metabolization of citric acid and its association with enhanced sensory properties have started to draw increasing attention.

### 4.2.1. Malic Acid Metabolism

Malic acid is a major acid in wines; the conversion of a dicarboxylic acid (L-malic acid) into a monocarboxylic acid (L-lactic acid) increases pH and modifies the sensory properties of wine. As ascertained by Seifert in 1901, lactic acid bacteria from wine transform L-malic acid into L-lactic acid and carbon dioxide via a direct reaction, meaning that the intermediate pyruvic acid is not formed during this conversion.

The malolactic enzyme, which was purified for the first time in *L. plantarum* (Lonvaud-Funel & Strasser de Saad, 1982), has been found in all species of lactic acid bacteria isolated in wine (Batterman & Radler, 1991; Lonvaud-Funel, 1995; Naouri et al., 1990). This enzyme is dimeric and formed by two identical 60 kDa subunits. The active form is dimeric and the monomer–dimer transition is pH-dependent (Batterman & Radler, 1991). It catalyzes a redox reaction involving NAD followed by $NADH_2$. The malolactic enzyme has two NAD-binding domains, an L-malate binding site, and an amino acid motif with a sequence that is characteristic of malic enzymes (Labarre et al., 1996).

Many studies have analyzed the biochemical characteristics of the malolactic enzyme in numerous bacterial species such as *L. casei* (Battermann & Radler, 1991), *L. plantarum* (Schüzt & Radler, 1974), *Leu. mesenteroides* (Lonvaud-Funel & Strasser de Saad, 1982), and *O. oeni* (Naouri et al., 1990). These have shown that it functions according to an ordered sequential mechanism in which the cofactors $Mn^{2+}$ and $NAD^+$ are bound before L-malate. This activity can also be induced by the substrate for the reaction, malic acid. Figure 8.3 shows a diagram of the mechanism underlying the generation of metabolic energy by lactic acid bacteria during malolactic fermentation.

The genetic locus involved in malolactic conversion (*mle*) has been identified in *O. oeni* and other lactic acid bacteria. In *O. oeni*, this

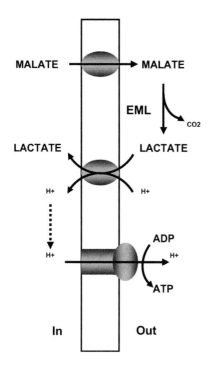

**FIGURE 8.3** Mechanism of metabolic energy production in lactic acid bacteria during malolactic fermentation.

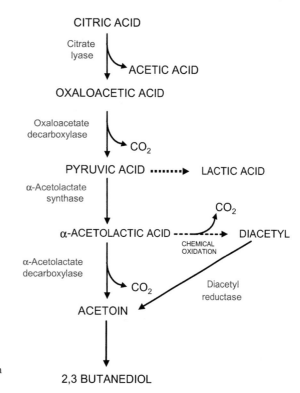

**FIGURE 8.4** Main metabolic pathway for citric acid in lactic acid bacteria.

locus contains the malolactic operon, which in turn contains three genes: *mleA*, which encodes the malolactic enzyme; *mleP*, which encodes malate-permease; and *mleR*, which possibly encodes the regulator responsible for activating the transcription of the malolactic operon (Labarre et al., 1996). This gene arrangement is conserved in other wine-related lactic acid bacteria such as *L. plantarum*, *L. brevis*, *L. casei*, *Leu. mesenteroides*, and *Pediococcus pentosaceus* (Makarova et al., 2006).

### 4.2.2. Citric Acid Metabolism

Citric acid, which is one of the acids present both on grapes and in must, is generally found at lower concentrations (0.1–1 g/L) than those of major organic acids such as tartaric acid (2–8 g/L) and malic acid (1–7 g/L). Wine lactic acid bacteria can metabolize citrate, as shown by Ramos et al. (1995) using NRM spectroscopy

and isotope labeling. Like other lactic acid bacteria, *O. oeni* does not use citrate as a sole carbon source but metabolizes it together with glucose; the resulting biomass is greater than that produced when grown in the presence of glucose alone. After being transported to the interior of the cell, citrate is converted to a mixture of lactate, acetate, diacetyl, acetoin, and 2,3-butanediol (see Figure 8.4). The bacteria break down the citrate into oxaloacetic acid in a reaction catalyzed by citrate lyase. This acid is converted by oxaloacetate decarboxylase to pyruvate, which is mostly reduced to lactate in the presence of NADH. Some pyruvate, however, is converted by acetolactate decarboxylase to acetolactic acid, giving rise to acetoin and 2,3-butanediol following decarboxylation.

The chemical oxidation of acetoin, in turn, yields diacetyl. The precursor of diacetyl (and acetoin), α-acetolactate, is also an intermediate in the biosynthesis of the amino acids valine and leucine from pyruvate.

The co-fermentation of citrate and glucose in *O. oeni* plays an important role in the physiology of these bacteria, leading to increased growth rate and biomass production, which in turn leads to increased ATP production (Ramos & Santos, 1996; Salou et al., 1994). As far as its relevance to the winemaking process is concerned, the co-metabolism of citrate and glucose increases the formation of volatile acids (acetate), which can have adverse effects on wine aroma if excessive levels are reached.

The greatest impact that citrate fermentation has on wine, however, is linked to the production of diacetyl, as this imparts a buttery aroma. Wines that undergo malolactic fermentation generally have a greater concentration of diacetyl than those that do not (Martineau et al., 1995). While moderate levels of diacetyl have a positive effect on aroma, high levels cause an unpleasant aroma, leading to spoilage (Davis et al., 1985a; Nielsen & Richelieu, 1999). The final concentration of diacetyl in wine depends on various factors including bacterial strain, wine type, sulfur dioxide, and oxygen (Martineau & Henick-Kling, 1995; Nielsen & Richelieu, 1999). Aeration, high levels of citrate and sugars, low temperature (18°C), and the elimination of yeast cells prior to malolactic fermentation all favor the production of diacetyl (Martineau et al., 1995). Sulfur dioxide, in turn, inhibits diacetyl production and therefore reduces the impact of this compound on aroma (Nielsen & Richelieu, 1999).

Analysis of the *O. oeni* genome showed the presence of the typical *cit* gene group, which includes genes that encode citrate lyase (*citDEF*), citrate lyase ligase (*citC*), oxaloacetate decarboxylase (*mae*), and the citrate transporter (*maeP* o *citP*) (Mills et al., 2005). The genome also contains genes involved in the butanediol pathway (*ilvB*, *alsD*, *butA*).

### 4.2.3. Tartaric Acid Metabolism

Certain strains of lactic acid bacteria (particularly *Lactobacillus* strains) are also capable of degrading tartaric acid, although this capacity is much less common than that of malic and citric acid metabolism. Tartaric acid is only degraded in certain conditions after the metabolism of other organic acids. The catabolism of this acid always alters wine by causing a slight reduction in fixed acidity and an increase in volatile acidity. *L. plantarum* employs a dehydratase to convert tartaric acid into oxaloacetic acid, which, in turn, is decarboxylated to pyruvate. Full genome sequencing of *L. plantarum* revealed genes encoding tartrate dehydratase (*ttdAB*), oxaloacetate/malate decarboxylase (*mae*), and pyruvate dehydrogenase. *L. brevis* and other heterofermentative bacteria use tartrate differently in that they mostly break it down into succinic acid. The metabolism of this acid varies from one strain of lactic acid bacteria to the next and also depends on environmental conditions.

The increase in volatile acidity in wines due to tartaric acid was described by Louis Pasteur and is known in winemaking as *tourné* or tartaric spoilage. It mostly affects wines from warm climates that have a pH of over 3.5 and a low level of sulfur dioxide, factors that favor the growth of certain *Lactobacillus* species. Indeed, only certain strains of *Lactobacillus* are able to degrade tartaric acid. Wines affected by tartaric spoilage generally turn cloudy, become darker, and change color. This defect also causes noticeable alterations in aroma and flavor, creating organoleptically unacceptable wines.

## 4.3. Metabolism of Phenolic Compounds

Phenolic compounds are of enormous importance in winemaking as they have both direct and indirect impacts on final quality. They are responsible not only for the color and astringency of wine but also for certain nutritional and pharmacological properties. A recent review by Rodríguez et al. (2009) analyzed the relationship between lactic acid bacteria and phenolic compounds.

In *O. oeni*, while malolactic fermentation is activated in the presence of catechin or quercetin, it is inhibited by increasing levels of *p*-coumaric acid (Reguant et al., 2000). Gallic acid delays or inhibits the formation of acetic acid, meaning that greater control of malolactic fermentation is achieved and increases in volatile acidity are prevented. In contrast, it has been demonstrated in *O. oeni* that phenolic compounds reduce the consumption of sugars and increase that of citric acid, resulting in greater concentrations of acetic acid (Rozès et al., 2003).

Several phenolic acids such as ferulic acid and *p*-coumaric are natural components of grape must and wine, and can be decarboxylated by numerous bacteria, including *L. brevis*, *L. plantarum*, and *Pediococcus* species. In *L. plantarum*, this decarboxylation is accompanied by the formation of volatile phenols (4-ethylphenol and 4-ethylguaiacol) (Cavin et al., 1993; de las Rivas et al., 2009).

Phenolic acids derived from cinnamic acid are generally esterified with tartaric acid in grape must and wines and can be released as free acids through the action of esterases. *L. plantarum* possesses two phenolic acid decarboxylases. One of these, *p*-coumaric acid decarboxylase (PCD), has been characterized (Cavin et al., 1997a) and metabolizes only *p*-coumaric, caffeic, and ferulic acids into their corresponding 4-vinyl derivatives (Cavin et al., 1997b; Rodríguez et al., 2008). In one study, the *PCD* gene was expressed in *Saccharomyces cerevisiae* to create strains capable of decarboxylating phenolic acids in wines (Smit et al., 2003). Barthelmebs et al. (2000), on knocking out this gene, found that *L. plantarum* had a second decarboxylase that was preferentially induced by ferulic acid. However, this enzyme has not yet been characterized. The same study also

reported that *L. plantarum* had an inducible phenolic acid reductase (also uncharacterized to date) that degrades phenolic acids into substituted phenylpropionic acids and converts *p*-coumaric acid into phloretic acid. It has been proposed that these inducible activities may participate in the phenolic acid stress response by reducing these acids into less toxic compounds (Gury et al., 2004).

*L. plantarum* species are the only lactic acid bacteria to have been found to have tannase activity (Vaquero et al., 2004). The biochemical characterization of this enzyme in *L. plantarum* has shown that it hydrolyzes the gallic tannins present in wine (Curiel et al., 2009). This activity, thus, is very important in winemaking because of its impact on color and turbidity.

## 4.4. Aldehyde Catabolism

Wine contains volatile aldehydes that have an important impact on the sensory quality of the final product (de Revel & Bertrand, 1993). The most abundant of these compounds is acetaldehyde, which is mostly produced by yeasts during alcoholic fermentation and can affect aging and color stability. Furthermore, acetaldehyde, hexanal, *cis*-hexen-3-al, and *trans*-hexen-2-al are all responsible for unpleasant odors.

Because aldehydes make such an important contribution to wine aroma, excess quantities must be eliminated. Sulfite has traditionally been used for this purpose but certain lactic acid bacteria, in particular *O. oeni*, offer an alternative solution as they can metabolize acetaldehyde and convert it to ethanol and acetate (Osborne et al., 2000). Lactic acid bacteria, and *O. oeni* in particular, might also be useful for metabolizing other aldehydes that can give rise to unpleasant aromas.

While certain lactic acid bacteria are also able to produce acetaldehyde, it is not yet known whether common wine strains have this capacity.

## 4.5. Glycoside Hydrolysis

When found in grapes and wine, most mono-terpenes (important aromatic compounds) and anthocyanidins (the main pigments in red grapes and red wine) are bound to sugars such as glucose (Ebeler, 2001). Glycosylated monoter-penes are not volatile and therefore do not confer aroma. Aroma is released when a glycosi-dase such as β-glucosidase hydrolyzes the sugars bound to the monoterpene and produces sugar and a volatile monoterpene. In contrast, hydrolysis of the sugar component of anthocya-nins by glycosidase enzymes known as antho-cyaninases leads to the spontaneous formation of a brown or colorless compound. In other words, anthocyaninases have a decolorizing activity.

Certain strains of O. oeni have been reported to metabolize anthocyanins and other compounds via glycosidase activity, producing compounds that have an important impact on wine aroma (Bloem et al., 2008; Boido et al., 2002; de Revel et al., 2005; d'Incecco et al., 2004; McMahon et al., 1999; Ugliano et al., 2003). Considerable variations in β-glucosidase activity have been detected between different wine and commer-cial strains of O. oeni (Barbagallo et al., 2004; Grimaldi et al., 2000, 2005), indicating that lactic acid bacteria from wine have the capacity to hydrolyze glycoconjugates that affect aroma and color.

## 4.6. Ester Synthesis and Hydrolysis

Esters are volatile compounds that are present in wine at concentrations above the perception threshold. Derived from grapes, yeast metabolism, or the esterification of alco-hols and acids during winemaking, esters make a key contribution to the aroma and there-fore the quality of wine. While esters such as ethyl acetate are responsible for the fruity aroma of wines (Ebeler, 2001), they can also have nega-tive effects at high concentrations. There is

abundant evidence that ethyl esters such as ethyl acetate, ethyl lactate, ethyl hexanoate, and ethyl octanoate are formed during malo-lactic fermentation (de Revel et al., 1999). It would therefore appear that lactic acid bacteria from wine are able to synthesize esters, but further studies are needed to identify and study the enzyme systems involved.

Esterases are involved in both the synthesis of esters and their hydrolysis in aqueous solu-tion. Davis et al. (1988) showed that the majority of O. oeni, Pediococcus, and Lactobacillus strains have esterase activity. Gaining further insight into the production and hydrolysis of esters by lactic acid bacteria in wine will improve our understanding of the impact of malolactic fermentation on wine aroma.

## 4.7. Lipid Hydrolysis

Wine contains mono-, di-, and triacylglycer-ols. The lipids found in wines are derived from grapes or released during yeast autolysis in alcoholic fermentation (Pueyo et al., 2000). These lipids can affect the flavor of wines as they form volatile fatty acids with a very low perception threshold when broken down by lipases. The fatty acids formed, additionally, can give rise to esters, ketones, and aldehydes. Volatile fatty acids are natural components of alcoholic beverages such as cider and wine (Blanco-Gomis et al., 2001) and excessive levels can have a negative organoleptic effect. There is no information on the lipolytic system in wine lactic acid bacteria. While a study by Davis et al. (1988) showed that certain strains of O. oeni displayed esterase and/or lipase activities, a later study found no lipolytic activity in 32 strains of Lactobacillus, two strains of Leuconos-toc, and three strains of Lactococcus isolated in wine (Herrero et al., 1996). Because lipases are generally extracellular or associated with whole cells, lactic acid bacteria have the potential to modify the lipid content of the musts or wines in which they grow.

## 4.8. Protein and Peptide Degradation

Wine contains proteins that can be hydrolyzed by bacterial proteases and peptidases to form peptides and amino acids that influence the flavor and stability of wine. Because lactic acid bacteria need amino acids for growth, they must necessarily have the corresponding enzyme activities to obtain the peptides and amino acids they require. Analysis of the *O. oeni* PSU-1 genome suggests that this strain possesses enzymes for the biosynthesis of eight amino acids, namely alanine, aspartic acid, asparagine, cysteine, glutamate, lysine, methionine, and threonine. The enzymes required for the synthesis of other amino acids such as isoleucine, leucine, and valine, however, are not present in the genome (Mills et al., 2005). Curiously, *O. oeni* PSU-1 lacks the ability to synthesize proline and serine—the two most abundant amino acids in grape must—yet conserves the ability to synthesize cysteine and methionine, which are only present in very low concentrations in this substrate.

Proteases and peptidases have been described in different genera and species of lactic acid bacteria commonly isolated in wine. While Davis et al. (1988) did not detect protease activity in different strains of lactic acid bacteria from wine (including *O. oeni*, *Pediococcus*, and *Lactobacillus* strains), a later study by Rollan et al. (1993) detected the production of extracellular proteases in *O. oeni*. These proteases have been partially characterized in this species (Farias et al., 1996; Rollan et al., 1995a). While lactic acid bacteria have the potential to hydrolyze proteins in wine, this ability does not seem to be very common among *O. oeni* strains (Leitao et al., 2000). Manca de Nadra et al. (1997, 1999) demonstrated that the *O. oeni* strain $X_2L$, which produces an extracellular protease, is capable of releasing peptides and amino acids during malolactic fermentation in both red and white wines. Folio et al. (2008), in turn, recently characterized EprA, an extracellular protein

with protease activity in *O. oeni*. No genes encoding a possible extracellular protease or peptidase containing a clear peptide signal were detected in the *O. oeni* PSU-1 genome (Mills et al., 2005). Ritt et al. (2009), on studying the use of peptides by *O. oeni* by analyzing the activity and biosynthesis of PepN, PepX, and PepI, found that the biosynthesis of these three peptidases depended on the peptides in the culture medium; they also reported that these peptidases, which are specific for proline-containing peptides, were important for *O. oeni* nitrogen metabolism. Finally, it has been reported that *O. oeni* has the capacity to transport and hydrolyze oligopeptides composed of two to five amino acids (Ritt et al., 2008).

## 4.9. Catabolism of Amino Acids

The first reactions that take place in the metabolism of amino acids are decarboxylation, transamination, deamination, and desulfation. The decarboxylation of amino acids leads to the formation of carbon dioxide and amines, which can have harmful health effects (see the case of biogenic amines in Section 6.1). Transamination, in turn, produces amino acids and $\alpha$-keto acids, while deamination leads to the formation of ammonia and $\alpha$-keto acids. Sulfur-containing amino acids such as methionine and cysteine produce volatile sulfur compounds via desulfation. Secondary reactions in amino acid catabolism involve the conversion of amines, $\alpha$-keto acids, and amino acids to aldehydes. The final reactions in the transformation of amino acids are the reduction of aldehydes to alcohols or their oxidation to acids.

Various studies have analyzed the metabolism of amino acids by lactic acid bacteria in wine, with particular emphasis on arginine, histidine, methionine, ornithine, and tyrosine. These metabolic processes have a major impact on the quality of wine because of the compounds they can produce (e.g., alcohols, aldehydes, and amines).

The amino acid composition of wine is complex (Lehtonen, 1996). During malolactic fermentation, for example, concentrations can increase or decrease depending on the type of amino acid involved (Davis et al., 1986a, 1986b).

Arginine is the most abundant amino acid in wine, and extensive information is available on its catabolism (Liu & Pilone, 1998). Lactic acid bacteria degrade arginine via the arginine deiminase (ADI) pathway (Liu et al., 1996). This generates energy (ATP), which favors the survival and growth of these bacteria in wine and provides them with greater viability during the stationary phase of growth under anaerobic conditions (Tonon & Lonvaud-Funel, 2000; Tonon et al., 2001). Three enzymes, which act sequentially, are involved in this pathway: ADI, ornithine transcarbamylase (OTC), and carbamate kinase (Liu et al., 1995b). The reactions they catalyze are shown in Figure 8.5a. The ADI pathway was recently characterized in several lactic acid bacteria (Arena et al., 2003; Divol et al., 2003; Tonon et al., 2001). This pathway generally involves three genes organized in an operon, arcABC. The arcA gene encodes the ADI enzyme, while the arcB and arcC genes encode OTC and carbamate kinase, respectively (see Figure 8.5b). All three genes have been detected in the majority of O. oeni,

L. brevis, L. hilgardii, L. buchneri, and P. pentosaceus strains as well as in several strains of L. plantarum and Leu. mesenteroides (Araque et al., 2009). The presence of these genes has been associated with the capacity to degrade arginine, although the amount degraded varies greatly from one strain to the next (Araque et al., 2009). Arginine degradation, for example, is influenced by the strain of lactic acid bacteria, pH, arginine concentration, and type of sugar (Granchi et al., 1998; Liu et al., 1995b; Mira de Orduña et al., 2000a, 2000b, 2001). When broken down, arginine secretes citrulline, which can subsequently be metabolized by several wine lactic acid bacteria (Liu et al., 1994; Mira de Orduña et al., 2000a).

Several strains of O. oeni can metabolize serine with the generation of ammonia (Granchi et al., 1998). The catabolism of this amino acid has not been studied in lactic acid bacteria but it is probably degraded by deamination via the action of serine dehydratase, which converts serine to ammonia and pyruvate. Pyruvate can be metabolized to formate, acetate, carbon dioxide, and ethanol or diacetyl, depending on the enzyme system present. O. oeni strains have been found to metabolize the amino acid methionine, resulting in the formation of characteristic aromas that contribute to the aromatic

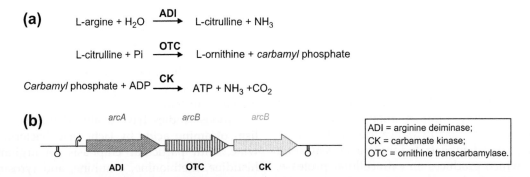

FIGURE 8.5 Degradation of arginine via the arginine deiminase pathway. Reactions catalyzed by the enzymes involved in the pathway (a). Genetic organization of the arc locus in Lactobacillus hilgardii $X_1$B (b). The arrows indicate open reading frames. The locations of a possible promoter and possible transcription termination sites are also shown. Adapted from Arena et al. (2003).

complexity of wine; examples include methanethiol, dimethyl disulfide, 3-(methylsulfanyl) propan-1-ol, and 3-(methylsulfanyl) propionic acid (Pripis-Nicolau et al., 2004). The reduction of methanethiol is the last stage in the enzymatic synthesis of methanethiol from methionine. Vallet et al. (2009) purified the alcohol dehydrogenase enzyme involved in this conversion.

Glutamic acid transport has also been described in *O. oeni*. Vasserot et al. (2003) reported that the process is energy-dependent and can be activated by the metabolism of arginine and sugars and stimulated by malic acid and acidic pH. In a study of the influence of aspartic acid on growth and malic acid and glucose metabolism in *O. oeni*, Vasserot et al. (2001) found that low concentrations (<0.3 mM) stimulated the growth of *O. oeni*, while high concentrations (>6 mM) inhibited growth and caused a reduction in the degradation of malic acid and an increase in that of glucose. It is important for winemakers to determine whether high concentrations of aspartic acid can lead to overproduction of acetic acid and reduced ethanol production in winemaking conditions.

As more studies appear in this area, we can assume that we will soon have a greater understanding of the biochemical activities involved in the metabolism of amino acids by lactic acid bacteria in wine and the corresponding impact on the sensory and health-related properties of wine.

# 5. MALOLACTIC FERMENTATION

## 5.1. Use of Malolactic Starter Cultures

The idea of using lactic acid bacteria to induce malolactic fermentation was first proposed in 1961. Studies at the time highlighted the difficulties of inducing malolactic fermentation in wine, mostly because of the poor viability of the inoculated bacteria. Later studies described reactivation protocols based on the incubation of bacteria in grape must enriched with yeast extract for 24 to 48 h followed by the inoculation of this culture as soon as possible after alcoholic fermentation (Lonvaud-Funel, 1995; Maicas, 2001; Nielsen et al., 1996). The greatest advances achieved in this area, however, came with the use of *O. oeni* starter cultures (generally lyophilized) that were ready for direct inoculation (Maicas et al., 2000). *O. oeni* is still the main species used in the many commercial starter cultures available today. Most of the cultures are prepared with single strains or a mixture of two or three strains. For more information on malolactic starter cultures, see Chapter 11.

The use of selected bacterial strains in starter cultures prevents the development of *Lactobacillus* and *Pediococcus* spoilage bacteria, which can produce high concentrations of acetic acid and affect the quality of the wine (acetic acid accounts for over 90% of total volatile acidity).

Nutrients or activators formed by inactive yeasts and substances such as casein and cellulose are often used to activate malolactic fermentation. These cultures contain amino acids and vitamins that function as growth factors for lactic acid bacteria and also absorb inhibitory substances such as sulfites and medium-chain fatty acids (Lonvaud-Funel et al., 1988).

A possible alternative to these activators is the use of *O. oeni* cells immobilized in different matrices. This strategy can increase the productivity of fermentation because of the higher packing density and the greater protection afforded to cells. Examples of different materials used as immobilization matrices in studies analyzing the use of immobilized forms of *O. oeni* to deacidify wine include alginates, polyacrylamide, wood shavings, and cellulose sponges (Crapisi et al., 1987; Maicas et al., 2001). Not all of these agents, however, have been accepted by winemakers as they imply the use of additional chemical compounds.

## 5.2. Contribution of Malolactic Fermentation to the Sensory Properties of Wine

In addition to reducing the total acidity of wine, malolactic fermentation modifies the organoleptic properties of the final product as it converts malic acid—which has a bitter flavor—to the smoother-tasting lactic acid. Although this transformation is the main reaction that occurs in malolactic fermentation, it is not the only one. Recent studies have clearly demonstrated the existence of other metabolic reactions that can have both positive and negative effects on the quality of wine.

The results of studies that have analyzed the effect of malolactic fermentation on the sensory properties of wine using gas chromatography combined with olfactometry and mass spectrometry clearly indicate that malolactic fermentation affects aroma and adds complexity to the flavor of wine (Henick-Kling, 1993; Rodríguez et al., 1990; Sauvageot & Vivier, 1997). They have also shown that the effect on flavor varies according to the strain of lactic acid bacteria and the type of wine involved.

Wine acquires a new aromatic profile following malolactic fermentation, with a decrease in varietal aromas due to the degradation or hydrolysis of the aromatic compounds in grapes and a reduction in the number and concentration of the volatile compounds produced during alcoholic fermentation. Despite the enormous influence that malolactic fermentation has on aroma, only certain changes to wine attributes that occur at this stage are related to the production or use of specific chemical compounds by lactic acid bacteria. According to Henick-Kling (1993), malolactic fermentation increases the fruity, buttery aroma of wine but reduces vegetal, grassy notes. This increase in fruitiness is possibly caused by the formation of esters by lactic acid bacteria (see Section 4.6) and the increased buttery aroma is due to the formation of diacetyl from the fermentation of citrate by lactic acid bacteria (see Section 4.2.2). The reduction in vegetative, grassy aromas, in turn, may be due to the catabolism of aldehydes by lactic acid bacteria (see Section 4.4). Malolactic fermentation also results in other changes, however, such as increased body, viscosity, and roundness due to the production of polyalcohols and polysaccharides by lactic acid bacteria (see Section 4.1).

Other characteristic aromas associated with malolactic fermentation include floral, toasty, vanilla, sweet, wood, smoky, bitter, and honey aromas (Henick-Kling, 1993; Sauvageot & Vivier, 1997). Further studies are required to link wine attributes that are altered during malolactic fermentation to the production or degradation of specific chemical compounds by lactic acid bacteria. With such information, enologists will be able to choose specific strains to obtain desired aromas and flavors.

Two studies recently undertook a metabolomic characterization of malolactic fermentation (Lee et al., 2009; Son et al., 2009). Wine contains metabolites produced during alcoholic fermentation, malolactic fermentation, and aging that have an important impact on its quality. In their study, Son et al. (2009) studied changes in metabolites in wine via NMR spectroscopy and statistical analysis and found a clear difference between wines that underwent malolactic fermentation and those that did not. Specifically, they found low levels of malate and citrate and high levels of lactate in the former. Also contributing to this differentiation were metabolites such as alanine, γ-aminobutyric acid, 2,3-butanediol, choline, glycerol, isoleucine, lactate, leucine, polyphenols, proline, succinate, and valine. Using an identical approach, Lee et al. (2009) studied the effect of different commercial strains of O. oeni on variations in metabolites during malolactic fermentation. They identified 17 primary metabolites and 65 secondary metabolites of volatile compounds. The significant differences between the wines fermented with different *O. oeni* strains were determined by secondary rather than primary metabolites, with the effects of these strains visible only in terms of the secondary metabolites. Twelve volatile compounds (2-butanol, butyl butyrate, diethyl succinate, 2-ethyl-1-hexanol, ethyl hexanoate, ethyl octanoate, 9-hexadecanoic acid, hexadecanoic acid, isoamyl alcohol, isobutyric acid, octanoic acid, and 2-phenylethanol) contributed to this differ-entiation.

# 6. ADDITIONAL CONSIDERATIONS

This section will take an in-depth look at other aspects of bacteria that are relevant to winemaking such as the formation of toxic compounds (e.g., biogenic amines and ethyl carbamate precursors) and factors that play an important role in correct malolactic fermentation such as stress resistance, presence of bacteriophages, and production of bacteriocins.

## 6.1. Formation of Biogenic Amines

Several lactic acid bacteria metabolize amino acids in grape must and wine to form ethyl carbamate precursors and biogenic amines. The impact on the quality of the wine in both cases is important as these compounds can have harmful health effects.

As discussed in Section 4.9, certain lactic acid bacteria possess decarboxylases that convert amino acids to amines and carbon dioxide. Some of these amines, known as biogenic amines, are toxic substances associated with adverse health effects (Shalaby, 1996). They are found mostly in fermented food and drinks. Wine, for example, contains as many as 25 different biogenic amines, the most abundant of which are histamine, tyramine, and putrescine, which are produced by the decarboxylation of the amino acids, histidine, tyrosine, and ornithine, respectively (Lehtonen, 1996;

Lonvaud-Funel, 2001). Because different *O. oeni* and *Lactobacillus* strains are capable of decarboxylating these amino acids, it would appear that lactic acid bacteria are responsible for the formation of biogenic amines in wine.

The formation of these harmful amines in wine, thus, probably depends on the presence of lactic acid bacteria with the necessary decarboxylation capacity. Another factor that can influence the abundance of amines in wine is the presence and concentration of precursor amino acids, which, in turn, are influenced by the composition of the must, the type of vinification, and factors such as pH and sulfur dioxide, which influence bacterial populations and activities (Lonvaud-Funel, 2001).

Lactic acid bacteria vary in terms of their capacity to produce biogenic amines from amino acids. While some studies have indicated that *O. oeni* are the main bacterial species responsible for the formation of histamine (Coton et al., 1998b; Lucas et al., 2008), others have found that histidine decarboxylase activity is an unstable property in *O. oeni* (Coton et al., 1998b), occurring only in certain strains (Lucas et al., 2008). This would explain why not all studies have found histamine-producing *O. oeni* strains (Constantini et al., 2006; Moreno-Arribas et al., 2003). Lucas et al. (2008) recently reported that *O. oeni* strains rapidly lose their ability to produce histamine because this trait is encoded on an unstable 100 kb plasmid. A similar finding was reported for the plasmid responsible for histamine synthesis in a wine strain of *L. hilgardii* (Lucas et al., 2005). Histidine decarboxylase has been purified and characterized in the *O. oeni* strain 9204 (Coton et al., 1998b; Rollan et al., 1995b). It consists of two distinct subunits, $\alpha$ and $\beta$, that are synthesized from a single polypeptide that is subsequently processed. The results of the above studies indicate that the active protein has a hexameric $(\alpha\beta)_6$ structure. This protein is a decarboxylase specific for the amino acid histidine (Coton et al., 1998a).

Certain strains of *L. brevis* appear to be responsible for the formation of tyramine in wine. Tyrosine decarboxylase has been biochemically and genetically purified and characterized in *L. brevis* IOEB 9809 (Lucas et al., 2003; Moreno-Arribas & Lonvaud-Funel, 2001). The gene encoding this enzyme forms part of an operon composed of four genes encoding a tyrosyl-tRNA synthetase, tyrosine decarboxylase, a probable tyrosine permease, and a $Na^+/K^+$ transporter (Lucas et al., 2003).

It was recently seen that *O. oeni*, the main species responsible for malolactic fermentation, may be involved in the production of putrescine in wines, with the identification of the ornithine decarboxylase gene in a putrescine-producing *O. oeni* strain isolated in wine lees (Marcobal et al., 2004). The gene encodes a 745-amino-acid protein containing conserved pyridoxal phosphate cofactor binding domains and amino acid residues involved in enzymatic activity (see Figure 8.6). This gene does not appear to be common in *O. oeni* as it was not detected in any of the 42 other *O. oeni* strains analyzed by Marcobal et al. (2004). In a later study, the same group showed that this putrescine-producing strain had acquired the genes regulating the synthesis of this compound by horizontal gene transfer from an unknown bacteria (Marcobal et al., 2006).

Most of the studies analyzing amine toxicity in humans have been performed using histamine. While data show that a healthy man can consume relatively high doses of histamine (up to 2.75 mg/kg of bodyweight), there have been reports of histamine-induced food intolerance leading to hypotension, digestive and liver disorders, migraine, and other disorders. Tyramine can also cause illness due to its vasoconstrictive properties. The presence of other amines such as putrescine and cadaverine favor the passage of histamine and tyramine into the bloodstream as these amines inhibit the activity of detoxification enzymes in the body.

Apart from the obvious health implications, biogenic amines can also have important commercial repercussions as several countries have established maximum limits for these substances in wine. In recent years, research efforts have focused on lactic acid bacteria in an effort to find rapid, reliable methods for detecting strains that synthesize these amines. The traditional microbiological method used for this purpose involved the use of a culture medium containing the precursor amino acid and a pH indicator. With the production of amine, the medium would become alkaline and the pH indicator would change color accordingly (see Figure 8.7). False negative results, however, may sometimes be obtained, as lactic acid bacteria produce large quantities of acid. One solution to this problem was the development of modified media (Bover-Cid & Holzapfel, 1999; Maijala, 1993). Another simple, rapid method for avoiding false results involves thin-layer chromatography of culture supernatants from lactic acid bacteria (García-Moruno et al., 2005).

More recently, strategies based on molecular biology techniques have been designed to detect biogenic amine-producing lactic acid bacteria (Landete et al., 2007). The most common methods are based on PCR amplification as it is rapid and has high sensitivity and specificity. The principle underlying these procedures is that all bacteria that produce biogenic amines possess the gene encoding the enzyme responsible for their formation. Other methods described include a multiplex PCR method (Marcobal et al., 2005) and a quantitative PCR method (Nannelli et al., 2008) for the detection of lactic acid bacteria that produce histamine, tyramine, and putrescine (the predominant biogenic amines in wine) (see Figure 8.8).

The aim in all cases is to design tools that facilitate the control of amine production in wine through the early detection of strains with the capacity to produce biogenic amines, thus allowing the winemaker to take the necessary corrective measures.

```
OEN  MDSEINDDSVHQTNLPHKFSTDELKIASTAHATAY-FDTNRTVVDADNSDFVDVAAVVVM  59
L30  ------------------MSSS******QE*RQ*-***D*V****VG***T**G**IA*  40


OEN  DDEK-AIINKADETK---FNIPIFIITDDSSKVDGETMSKIFHIIDWHNNYDRRLYDREI  115
L30  *Y*T-DV*DA**A**---*G**V*AV*K*AQAISADELK******LE*KF*ATVNA***  96
```

###### Binding domain
```
OEN  EAAAKKYEDGVLPPFFKALKAYVERGNIQFDCPGHQGGQYFRKSPAGREFYNFYGENIFR  175
L30  *T*VNN***SI*****S**E**S**L*************Y**H*******D*F**TV**  156
```

```
                                                      ▼▼
OEN  SDICNADVDLGDLLIHEGPAMDAEKHAARVFNADKTYFVMNGTTTSNNIAITAAVAPGDL  235
L30  A*L**:**A***********VA********Y********LG*SSNA**TVTS*L*SN***  216
```

```
                   ▼
OEN  VLFDRNNHKSVYNAALVQAGGRPVYLETSRDSYGFIGGIYSKDFDEKSIREKIAKVDPEK  295
L30  *************S**AM********Q*N*NP********DS****K***LA*****R  276
```

###### PLP binding domain       ▼▼▼▼▼                    ▼
```
OEN  AKAKRPFRLAVIQLGTYDGTIYNAKQVVERIGHLCDYILFDSAWVGYEQFIPMMKDSSPL  355
L30  **W*****************************HE**K*********E*****************RN****  336
```

```
          ▼▼                                            ▼
OEN  LLN-LGPDDPGILVTQSTHKQQAGFSQASQIHKKDSHIKGQKRYINHKQFNNAYMKFSST  414
L30  *IDD***E****I*V**V*********T**************L**CD**H***SFNL*M**  396
```

```
       ▼
OEN  SPFYPLFATLDINAKMQEGEAGKKLWHDALVTSVNARKNLLKNATMIKPFLPPVVHGKPW  474
L30  *****MY****V**A**:******R*****L*I*TIE***K*I*AGS*FR**V****N**K*  456
```

###### Specificity domain
```
OEN  QDADTEKIVSDIDYWKFEKGAKWHGFDGYADNQYFVDPNKFMLTTPGIDVETGEYEDFGI  534
L30  E*G***DMANN****R********AYE**G****Y***************NP***D******  516
```

```
OEN  PAVILANYLREHGIIPEKNDLNSILFLMTPAETQAKMDNLVTQIVKFESLVKADAPLDEV  595
L30  **T*V*****D*******S************P***N**I**LLQLQR*IEE****KV*  576
```

###### C-terminal domain
```
OEN  LPRLYSEHQDRYEGYTIKQLCQEVHDFYKNNNTKEYQKEMFLGKYFPEQAMTPYQANVEL  654
L30  **SI*AANEE**N****RE****L*********FT***RL*LREF****G*L*YE**Q*F  636
```

```
OEN  LKNNAKLVPLTDIEGLAALEGALPYPPGIFCIVPGEKWTKVAQKYFLILEESINRFPGFA  714
L30  IR*HN*****NK***EI***********V**VA****SET*V***T**QDG**NF****  696
```

```
OEN  PEIQGVYFEKE-NGKSVAYGYVYD—KSKDEEKR--  745
L30  ********KQ*-GD*V****EVYDAEVA*NDDRYNN  731
```

FIGURE 8.6 Comparison of ornithine decarboxylase gene sequences in *Oenococcus oeni* (OEN) and *Lactobacillus* 30a (L30) generated by the ClustalW sequence alignment program. The residues involved in binding to the cofactor PLP (▼) are shown in boldface in *Lactobacillus* 30a and are underlined in *O. oeni*. The dotted vertical lines indicate the separation between the different domains described for ornithine decarboxylase in *Lactobacillus* 30a. *Adapted from Marcobal et al. (2004).*

## 6.2. Formation of Ethyl Carbamate Precursors

Ethyl carbamate is a carcinogen found in fermented food and beverages such as wine (Ough, 1976). It is formed by a chemical reaction between ethanol and a precursor containing an N-carbamyl group, such as urea, citrulline, or carbamyl phosphate. The most abundant precursor is urea, which is produced by yeast during alcoholic fermentation. Citrulline and carbamyl phosphate, in turn, are produced by lactic acid bacteria during malolactic fermentation. Both

FIGURE 8.7 Detection of biogenic amine-producing bacteria in culture media. Lactic acid bacteria grown in the decarboxylation medium described by Maijala (1993). Solid medium (a). Liquid medium (b).

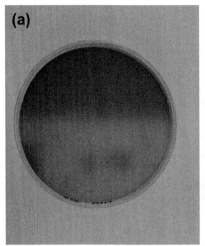

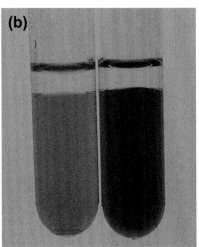

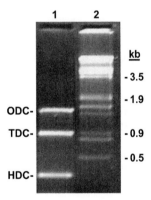

FIGURE 8.8 Amplification via multiplex polymerase chain reaction of decarboxylase genes in lactic acid bacteria. Amplification of a 1.4 kb fragment of the gene encoding ornithine decarboxylase (ODC), a 0.9 kb fragment of the gene encoding tyrosine decarboxylase (TDC), and a 0.3 kb fragment of the gene encoding histidine decarboxylase (HDC) (lane 1). The size (in kb) of some DNA fragments λ digested with EcoRI and BamHI (lane 2) is also shown. *Adapted from Marcobal et al. (2005).*

de Orduña et al., 2000a). Liu et al. (1994) reported a good correlation between citrulline excretion and the formation of ethyl carbamate precursors during the degradation of arginine by O. oeni and L. buchneri. Arginine degradation is, thus, a potential source of citrulline.

To date, there have been no reports of the excretion of carbamyl phosphate during the degradation of arginine. Carbamyl phosphate is also a pyrimidine precursor and can be synthesized by certain lactic acid bacteria from glutamine, bicarbonate, and ATP (Nicoloff et al., 2001). These compounds thus all represent new sources of ethyl carbamate precursors.

As with biogenic amines, maximum allowable limits have also been established for ethyl carbamate in different countries, reflecting the importance of keeping these levels at an absolute minimum to prevent possible health risks. Thanks to the knowledge that has been generated in the area, today's winemakers are better equipped than ever to take steps to prevent or reduce the formation of ethyl carbamate during the winemaking process. They can now implement in-process controls to monitor levels and also inoculate selected strains of yeasts and lactic acid bacteria that do not produce ethyl carbamate during alcoholic or malolactic

of these substances are metabolic intermediates in the degradation of arginine, which is one of the predominant amino acids in wine. The excretion of citrulline is very common during the degradation of arginine by wine lactic acid bacteria (Granchi et al., 1998; Mira fermentation. Araque et al. (2009) recently described a molecular method for detecting genes responsible for the synthesis of ethyl carbamate in lactic acid bacteria. It should also be noted that legislation allows the use of a special adjuvant consisting of an acid urease isolated in *Lactobacillus fermentum* (and currently sold under various trade names) in wines containing excess levels of urea. This enzyme is active at a pH of between 3 and 4 and acts by hydrolyzing urea, thus preventing the formation of ethyl carbamate without altering the chemical composition of the wine.

## 6.3. Stress Resistance

Lactic acid bacteria perform malolactic fermentation in highly adverse conditions. Strains of *L. plantarum* and *O. oeni* display the greatest resistance to the pH and ethanol levels found in this stage of the winemaking process (Alegría et al., 2004). Stress-inducing factors such as ethanol, acidic pH, phenolic compounds, sulfur dioxide, and fatty acids in wine have an inhibitory effect on growth and the duration of malolactic fermentation that has been linked to inhibition of ATPase activity (Carreté et al., 2002). The expression of the malolactic operon in *O. oeni* appears to be regulated by another factor linked to metabolic energy (Galland et al., 2003). Analysis of the complete genome of the *O. oeni* PSU-1 strain showed the presence of the full *atp* operon (*atpBEFHAGDC*), which encodes the $F_0$-$F_1$ ATPase system (Mills et al., 2005). Two proton-translocating ATPases involved in pH homeostasis in *O. oeni* have also been identified (Fortier et al., 2003).

Variations in membrane composition have also been observed when cells are exposed to stress, with a reduction in phospholipids and up to a five-fold increase in protein content. Guzzo et al. (1997) found that *O. oeni* responded to stress by synthesizing six proteins. Of these, the 18 kDa membrane-linked protein Hsp18 has been purified. Other studies with *O. oeni* have

shown that the gene encoding thioredoxin, *trxA*, is expressed under thermal and hydrogen peroxide stress (Jobin et al., 1999a) and that the homologue of *clpX*, an ATPase regulator, is also expressed under heat shock conditions and preferentially during the exponential phase of growth (Jobin et al., 1999b). In later studies, the expression levels of two proteins—the protease FtsH (Bourdineaud et al., 2003) and the transporter OmrA (Bourdineaud et al., 2004)—were found to increase in response to stress, suggesting their involvement in stress protection. In a proteomics study, Silveira et al. (2004) demonstrated an adaptive response in *O. oeni* to the presence of ethanol involving both cytoplasmic and membrane proteins (including those involved in cell-wall synthesis).

Genes previously implicated in the stress response in *O. oeni* (*clpX*, *clpLP*, *trxA*, *hsp18*, *ftsH*, *ormA*, and the operons *groESL* and *dnaK*) were also found in the fully sequenced PSU-1 strain (Mills et al., 2005). As far as oxidative stress is concerned, like all lactic acid bacteria, *O. oeni* is microaerophilic and does not possess catalase activity. It does, however, have the genes *trxA* and *trxB* and systems to eliminate reactive oxygen species (ROS) such as NADH-oxidase and NADH-peroxidase (Mills et al., 2005). No superoxide dismutase homologues were identified in the *O. oeni* PUS-1 strain (Mills et al., 2005).

The potential of other lactic acid bacteria to tolerate the hostile environment of winemaking has also been studied, leading to the identification of three cold-stress genes (*cspL*, *cspP*, and *cspC*) in *L. plantarum* (Derzelle et al., 2000, 2002, 2003). In a relatively recent study, Spano et al. (2004) cloned three genes (*hsp18.5*, *hsp19.3*, and *hsp18.55*) involved in heat-stress resistance in a wine strain of *L. plantarum* (Spano et al., 2004, 2005).

## 6.4. Bacteriophages

Bacteriophage infection of lactic acid bacteria has enormous economic repercussions in the fermented food industry. The first infection of this type to be reported in wine was detected by electron microscopy in Switzerland in 1976 (Sozzi et al., 1976). Later studies isolated bacterio-phages in wines from other geographic regions such as Australia (Davis et al., 1985b), South Africa (Nel et al., 1987), Germany (Arendt & Hammes, 1992), and France (Poblet-Icart et al., 1998). Some studies have linked difficulties associated with malolactic fermentation in certain wines to the presence of high levels of bacteriophages and interruption of malic acid metabolism (Davis et al., 1985b; Henick-Kling et al., 1986).

Poblet-Icart et al. (1998), on analyzing lysogeny of a large number of *O. oeni* wine strains, found that 45% of the strains analyzed were lysogenic. This would suggest that lysogeny is common in this species.

Some of these bacteriophages have been analyzed in an attempt to shed greater light on aspects such as morphology, protein composition, and genome size and structure. The most common bacteriophages in *O. oeni* are *Siphoviridae* species, which have a hexagonal, icosahedral head and a long, flexible, noncontractile tail. The diameter of the head ranges from 60 to 66 nm and the length of the tail from 180 to 260 nm. None of the bacterio-phages studied in *O. oeni* have a collar-whisker complex (Poblet-Icart et al., 1998).

The genome of the *O. oeni* PSU-1 strain does not contain intact, temperate bacteriophages or large fragments of a clear phagic origin (Mills et al., 2005), although the strain may act as a host for bacteriophages. Sao-José et al. (2004) identified bacteriophage integration sites in the *O. oeni* PSU-1 strain that were generally located adjacent to transfer RNA genes. Furthermore, the regions closest to these anchor sites had open reading frames (ORFs) of a phagic origin. These genes may be remnants of incomplete excision of the phage from the PSU-1 genome.

It has been shown that wine composition can affect the infective capacity of bacteriophages. Low pH levels and sulfur dioxide, for example, can inactivate these phages, preventing them from infecting sensitive bacteria (Davis et al., 1985b; Henick-Kling et al., 1986). Based on the above data, it is generally thought that infection by bacteriophages is not the main factor respon-sible for difficult malolactic fermentations. Nevertheless, it is important to bear in mind potential lysogeny when selecting *O. oeni* strains for starter cultures.

## 6.5. Bacteriocin Production

The production of bacteriocins and other anti-microbial compounds by bacterial strains of enological origin is another research area that is drawing increasing attention, particularly in terms of how to gain greater control over malo-lactic fermentation. Bacteriocins are peptide- or protein-based compounds that are ribosomally synthesized and that display antimicrobial activity against genetically related strains. They are odorless, colorless, and nontoxic.

Several wine strains are capable of pro-ducing bacteriocins, including *L. plantarum* J-23 (Rojo-Bezares et al., 2007), *L. plantarum* J-51 (Navarro et al., 2000), *P. pentosaceus* (Strasser de Saad & Manca de Nadra, 1993), *L. plantarum* LMG 2379 (Holo et al., 2001), *L. delbrueckii* subsp. *delbrueckii*, *Leu. mesenteroides* subsp. *cremoris*, and *Lactobacillus fructivorans* (Yurdugül & Bozoglu, 2002). Knoll et al. (2008) recently reported antimi-crobial activity in 8% of lactic acid bacteria, mostly *L. plantarum* species, isolated in wine. Furthermore, all the commercial malolactic fermentation starter cultures (containing *O. oeni* and *L. plantarum* strains) tested displayed activity against wine-related indicator strains, suggesting that they produce bacteriocins.

PCR analysis has been used to study genetic variability in genes involved in the synthesis of bacteriocins (*pln* genes) in *L. plantarum* strains (Sáenz et al., 2009). The *pln* locus was present in 94% of *L. plantarum* wine strains and displays considerable plasticity, with variable regions associated with the regulation of bacteriocin production.

Several of the bacteriocins produced by lactic acid bacteria isolated in grape musts and wines have been characterized. *L. plantarum* J-51, for example, produces a heat-resistant bacteriocin and has broad-spectrum antibacterial activity (Navarro et al., 2000). The bactericidal effects of pediocin N5p, produced by a strain of *P. pentosaceus* isolated in a wine in Argentina, have been studied in strains of *O. oeni*, *P. pentosaceus*, and *L. hilgardii* (Strasser de Saad & Manca de Nadra, 1993). Pediocin N5p is resistant to the physicochemical factors associated with winemaking such as pH, temperature, ethanol, and sulfur dioxide (Manca de Nadra et al., 1998; Strasser de Saad et al., 1995). This bacteriocin may also be useful in controlling the growth of spoilage bacteria during vinification. Pediocin PD-1, produced by a strain of *P. damnosus* isolated in beer (Green et al., 1997), is active against a wide range of gram-positive bacteria, including *O. oeni*. Nel et al. (2002) found that this pediocin was more efficient than either nisin or plantaricin 423 in eliminating the film formed by *O. oeni* in stainless steel tanks in wineries. The mode of action of this bacteriocin in metabolically active cells in *O. oeni* involves the cytoplasmic membrane (Bauer et al., 2005).

Studies have also investigated how to improve the efficiency of malolactic fermentation using nisin and nisin-resistant strains of *O. oeni* (Daeschel et al., 1991; Radler, 1990a, 1990b). Nisin has no effect on the organoleptic properties of wine and could therefore be used to inhibit the growth of undesirable bacteria by adding it directly or by using bacterial strains that produce it during malolactic fermentation.

# 7. INTERACTIONS BETWEEN LACTIC ACID BACTERIA AND OTHER MICROORGANISMS

The development of fungi in grapes plays a very important role in the onset of malolactic fermentation because these fungi can generate inhibitors such as organic acids or activators such as polysaccharides. Their effect varies depending on the fungi involved and their level of growth during the rotting of the grape. The most common fungi found in grapes are *Aspergillus*, *Botrytis cinerea*, *Mucor*, Penicillium, and *Rhizopus stolonifer* species. The growth of these fungi generally modifies oxalic, succinic, and fumaric acid concentrations in must and wine, reducing the viability of *O. oeni* and slowing malolactic fermentation. *B. cinerea*, in contrast, increases the degradation of malic acid by causing a shift from the fermentative metabolism of sugars to the formation of glycerol (San Romao & Lafon-Lafourcade, 1979).

*O. oeni* and *L. hilgardii* are capable of using polysaccharides synthesized by grape fungi in the absence of other assimilable organic molecules for growth. In wine, however, these fungal polysaccharides absorb long-chain fatty acids ($C_8$ and $C_{10}$), which weakens their inhibitory effect on malolactic activity.

The growth of acetic acid bacteria in grape must affects the growth and metabolism of lactic acid bacteria, with variations according to the age of the culture and the species of acetic and lactic acid bacteria present. The different species of lactic acid bacteria in wine also interact with each other. In wines with a pH greater than 3.5, for example, the growth of *Pediococcus* and *Lactobacillus* bacteria, which can reach densities of $10^8$ CFU/mL, leads to the death of *O. oeni* due to the presence of pediocins that are toxic for gram-positive species (Strasser de Saad et al., 1995).

As mentioned earlier, the growth of yeasts responsible for alcoholic fermentation in wine inhibits bacterial growth. In other cases, however, yeasts can stimulate the growth of lactic acid bacteria through the release of nutrients such as vitamins and amino acids into the medium. The type of strain that conducts alcoholic fermentation thus has a considerable influence on malolactic fermentation. Osborne and Edwards (2007) described *S. cerevisiae* strains that produced a 5.9 kDa peptide responsible for inhibiting *O. oeni* growth during malolactic fermentation. It can thus be concluded that the interactions between yeasts and lactic acid bacteria are complex and greatly depend on the strains present (Avedovech et al., 1992; Wibowo et al., 1985).

# 8. SENSORY CHANGES IN WINE DUE TO LACTIC ACID BACTERIA

In most cases, malolactic fermentation is properly controlled and the propagation of lactic acid bacteria and the biochemical reactions they participate in contribute to improving both the quality and stability of wine. There are, however, certain species and strains of lactic acid bacteria that can reduce the quality and acceptability of wine and, at times, even make it unfit for consumption (Bartowsky & Henschke, 2004; Bartowsky, 2009). Although some of these alterations, or defects, have been known for a long time (such as glycerol degradation or *piqûre lactique*), great advances have been made in our understanding of these processes with the emergence of new molecular techniques. More recent studies have described "newer" defects such as undesirable odors caused by the production of volatile phenols or aromatic heterocyclic bases, but much has still be learned about these. Of particular interest in recent years has been the analysis of alterations associated with the metabolism of amino acids in lactic acid bacteria that can have important health repercussions such as the case of ethyl carbamate and biogenic amines.

## 8.1. *Piqûre Lactique* (Lactic Taint)

*Piqûre lactique* or lactic taint is one of the most common wine flaws and is therefore among the best studied. It can occur during the production or even the storage of wine. It is typically associated with conditions that favor bacterial growth such as stuck or incomplete alcoholic fermentation. Lactic acid bacteria that appear before all the sugar in the must has been transformed into ethanol convert hexoses to acetic acid as well as to ethanol and carbon dioxide (which are also produced by yeasts). The presence of acetic acid and excessive amounts of lactic acid in the medium results in a considerable increase in volatile acidity (Lonvaud-Funel, 1999). The D-isomer of lactic acid has been associated with *piqûre lactique* in wine while the L-isomer is produced by malolactic fermentation. Most of the lactic acid bacteria associated with this flaw belong to the species *L. hilgardii* or *L. fructivorans*.

## 8.2. Glycerol Degradation and Production of Acrolein

Several lactic acid bacteria convert glycerol into 3-hydroxypropionaldehyde via the activity of glycerol dehydratase. This reaction generates acrolein. Alone, this compound is not problematic, but when it reacts with certain groups of phenolic compounds such as tannins it can cause bitter flavors. Glycerol is one of the most abundant compounds in wine and is generally found at concentrations of 5 to 8 g/L. It is one of the main products of yeast metabolism and plays a key role in wine flavor. The metabolism of glycerol thus affects the quality of wine, not only because it reduces glycerol levels but also because of the metabolic products it generates. Strains of lactic acid bacteria capable of degrading glycerol can be detected in wine using a special PCR-based molecular method (Claisse & Lonvaud-Funel, 2001).

## 8.3. Production of Extracellular Polysaccharides

Several strains of lactic acid bacteria can synthesize extracellular polysaccharides (exopolysaccharides or EPSs) from residual sugars, detracting from the quality of the wine. Such wines are characterized by abnormal viscosity.

Although this problem can occur during production, in most cases it develops gradually and appears weeks or even months after the wine has been bottled.

Different species of lactic acid bacteria can produce EPSs. Llaubères et al. (1990) found that *Pediococcus* and *Lactobacillus* strains isolated from spoiled wine and cider produced an identical EPS (a D-glucan consisting of a trisaccharide repeating unit of D-glucose attached by $(1 \rightarrow 3)$ bonds and side branches of D-glucose with $(1 \rightarrow 2)$ bonds. In *Pediococcus parvulus* 2.6 (formerly *P. damnosus* 2.6), *P. damnosus* IOEB8801, and *Lactobacillus diolivorans* G77 strains isolated in cider and wine, the production of EPSs has been associated with the presence of plasmids (Walling et al., 2001; Werning et al., 2006). The gene responsible for the production of the EPS in *O. oeni*, however, appears to be chromosomal (Dols-Lafargue et al., 2008; Werning et al., 2006). The EPS-producing *Lactobacillus collinoides* IOEB0203 and *L. hilgardii* IOEB0204 strains, however, do not appear to contain sequences similar to the glycosyltransferases responsible for the production of EPSs in the strains discussed above (Walling et al., 2005).

Molecular techniques have been developed to detect the presence of EPS-producing strains that can alter wine during production (Gindreau et al., 2001; Walling et al., 2004). Our knowledge of EPS-producing lactic acid bacteria is far from complete. An improved understanding of how these bacteria behave in wine and interact with other microorganisms together with greater knowledge of the factors involved in the synthesis of polysaccharides will help winemakers to predict their growth. Extensive filtration or heat-treatment methods are necessary to eliminate these bacteria prior to bottling. It should also be noted that these bacteria are highly tolerant of hostile conditions and sulfur dioxide, as EPSs exert a protective effect on the cell. The most important measure for preventing subsequent contamination, is, thus, rigorous cleaning of winery surfaces.

## 8.4. Production of Off-flavors

### 8.4.1. Production of Volatile Phenols

Lactic acid bacteria are responsible for a variety of off-flavors in wine, including animal-like odors attributed to excessive levels of volatile phenols. The main volatile phenols in red wines are 4-ethylphenol, 4-ethylguaiacol, 4-vinylphenol, and 4-vinyl guaiacol. The origin of ethylphenols has been a topic of debate for many years. While *Brettanomyces/Dekkera* yeasts with cinnamate decarboxylase and vinylphenol reductase activities are the main species responsible for the biosynthesis of these phenols, certain *Pediococcus* and *Lactobacillus* strains also have a role (Cavin et al., 1993; Chatonnet et al., 1995). In a recent study, de las Rivas et al. (2009) analyzed the capacity of lactic acid bacteria to produce volatile phenols in wine and described a PCR method for detecting bacteria with this potential. *L. plantarum*, *L. brevis*, and *P. pentosaceus* strains produced vinyl derivatives from hydroxycinnamic acids, but only *L. plantarum* strains produced the corresponding ethyl derivatives. *O. oeni*, *L. hilgardii*, and *Leu. mesenteroides* strains, in contrast, did not decarboxylate the hydroxycinnamic acids *p*-coumaric and ferulic acids, meaning that they are not responsible for the production of volatile phenols.

### 8.4.2. Production of Aromatic Heterocyclic Compounds

The production of undesirable aromas and flavors in wine described as "mousy" or "acetamide" has been associated with several lactic acid bacteria (Costello et al., 2001). A mousy odor or flavor is specifically attributed to the production of three volatile heterocyclic compounds: 2-ethyltetrahydropyridine, 2-acetyltetrahydopyridine, and 2-acetylpyrroline. Certain winemaking conditions such as high pH ($>3.5$) or low sulfur dioxide levels can favor the growth of the bacterial strains involved in the production of these bases

(Snowdon et al., 2006). A mousy taint can render a wine unpalatable and cannot be eliminated. This flaw has been associated with heterofermentative lactic acid bacteria (most often via the production of N-heterocycles by heterofermentative strains of *Lactobacillus* and *L. hilgardii* in particular, followed by *O. oeni* and *Pediococcus* strains) and homofermentative *Lactobacillus* species (Snowdon et al., 2006).

Very few studies have analyzed the origin of this flaw and little is known about the extent to which it affects the quality of the wine, mostly because of the complex nature of the processes involved but also because it occurs in conjunction with other defects. The presence of D-fructose, a fermentable sugar, has been associated with the production of volatile heterocyclic compounds and it has been suggested that the formation of these compounds involves ornithine and lysine metabolism in the presence of ethanol, although much remains to be discovered regarding the mechanisms underlying this process (Costello & Henschke, 2002).

## Acknowledgments

We thank the Spanish Ministry of Science and Innovation (MICINN) (grants AGL2005-00470, AGL2008-01052, Consolider INGENIO 2010 CSD2007-00063 FUN-C-FOOD), the Autonomous Community of Madrid (CAM) (S 2009/AGR-1469), Spanish Council for Scientific Research (CSIC) 2009201155, and the Spanish National Institute of Agricultural and Food Research (INIA) RM2008-00002 for the financial support provided.

## References

Alegría, E. G., López, I., Ruiz, J. I., Sáenz, J., Fernández, E., Zarazaga, M., et al. (2004). High tolerance of wild *Lactobacillus plantarum* and *Oenococcus oeni* strains to lyophilisation and stress environmental conditions of acid pH and etanol. *FEMS Microbiol. Lett., 230*, 53–61.

Araque, I., Gil, J., Carreté, R., Bordons, A., & Reguant, C. (2009). Detection of *arc* genes related with the ethyl carbamate precursors in wine lactic acid bacteria. *J. Agric. Food Chem., 57*, 1841–1847.

Arena, M. E., Manca de Nadra, M. C., & Muñoz, R. (2003). The arginine deiminase pathway in the wine lactic acid bacterium *Lactobacillus hilgardii* $X_1B$: Structural and functional study of the *arcABC* genes. *Gene, 301*, 61–66.

Arendt, E. K., & Hammes, W. P. (1992). Isolation and characterization of *Leuconostoc oenos* phages from German wines. *Appl. Microbiol. Biotechnol., 37*, 643–646.

Avedovech, R. M., McDavid, M. R., Watson, B. T., & Sandine, W. E. (1992). An evaluation of combinations of wine yeast and *Leconoctoc oenos* strains in malolactic fermentation of Chardonnay wine. *Am. J. Enol Vitic., 43*, 256–260.

Barbagallo., R. N., Spagan, G., Palmeri, R., & Torriani, S. (2004). Assessment of beta-glucosidase activity in selected wild strains of *Oenococcus oeni* for malolactic fermentation. *Enz. Microb. Technol., 34*, 292–296.

Barthelmebs, L., Diviès, C., & Cavin, J.-F. (2000). Knockout of the *p*-coumarate decarboxylase gene from *Lactobacillus plantarum* reveals the existence of two other inducible enzymatic activities involved in phenolic acid metabolism. *Appl. Environ. Microbiol., 66*, 3368–3375.

Bartowsky, E. J. (2009). Bacterial spoilage of wine and approaches to minimize it. *Lett. Appl. Microbiol., 48*, 149–156.

Bartowsky, E. J., & Henschke, P. A. (2004). The "buttery" attribute of wine-diacetyl-desirability, spoilage and beyond. *Int. J. Food Microbiol., 96*, 235–252.

Battermann, G., & Radler, F. (1991). A comparative study of malolactic enzyme and malic enzyme of different lactic acid bacteria. *Can. J. Microbiol., 37*, 211–217.

Bauer, R., Chikindas, M. L., & Dicks, L. M. T. (2005). Purification, partial amino acid sequence and mode of action of pediocin PD-1, a bacteriocin produced by *Pediococcus damnosus* NCFB 1832. *Int. J. Food. Microbiol., 101*, 17–27.

Bilhère, E., Lucas, P. M., Claisse, O., & Lonvaud-Funel, A. (2009). Multilocus sequences typing of *Oenococcus* oeni: Detection of two subpopulations shaped by intergenic recombination. *Appl. Environ. Microbiol., 75*, 1291–1300.

Blanco-Gomis, D., Mangas Alonso, J.-J., Cabrales, I. M., & Abrodo, P. A. (2001). Gas chromatographic of total fatty acids in cider. *J. Agric. Food Chem., 49*, 1260–1263.

Bloem, A., Lonvaud-Funel, A., & de Revel, G. (2008). Hydrolysis of glycosidically bound flavour compounds from oak wood by *Oenococcus oeni*. *Food Microbiol., 25*, 99–104.

Boido, E., Lloret, A., Medina, K., Carrau, F., & Dellacassa, E. (2002). Effect of β-glycosidase activity of *Oenococcus oeni* on the glycosilated flavour precursors of Tannat wine during malolactic fermentation. *J. Agric. Food Chem., 50*, 2344–2349.

Bon, E., Delaherche, A., Bilhère, E., de Daruvar, A., Lonvaud-Funel, A., & Le Marrec, C. (2009). *Oenococcus oeni* genome plasticity is associated with fitness. *Appl. Environ. Microbiol., 75*, 2079–2090.

Bourdineaud, J. P., Nehmé, B., Tesse, S., & Lonvaud-Funel, A. (2003). The *ftsH* gene of the wine bacterium *Oenococcus oeni* is involved in protection against environmental stress. *Appl. Environ. Microbiol., 69,* 2512−2520.

Bourdineaud, J. P., Nehmé, B., Tesse, S., & Lonvaud-Funel, A. (2004). A bacterial gene homologous to ABC transporters protects Oenococcus oeni from ethanol and other stress factors in wine. *Int. J. Food Microbiol., 92,* 1−14.

Bover-Cid, S., & Holzapfel, W. P. (1999). Improved screening procedure for biogenic amine production by lactic acid bacteria. *Int. J. Food Microbiol., 53,* 33−41.

Cappello, M. S., Stefani, D., Grieco, F., Logrieco, A., & Zapparoli, G. (2008). Genotyping by amplified fragment length polymorphism and malate metabolism performances of indigenous Oenococcus oeni strains isolated from Primitivo wine. *Int. J. Food Microbiol., 127,* 241−245.

Carreté, R., Vidal, M. T., Bordons, A., & Constantí, M. (2002). Inhibitory effect of sulfur dioxide and other stress compounds in wine on the ATPase activity of *Oenococcus oeni. FEMS Microbiol. Lett., 211,* 155−159.

Cavin, J.-F., Andioc, V., Etievant, P. X., & Diviès, C. (1993a). Ability of wine lactic acid bacteria to metabolize phenol carboxylic acids. *Am. J. Enol. Vitic., 44,* 76−80.

Cavin, J.-F., Barthelmebs, L., & Diviès, C. (1997a). Molecular characterization of an inducible *p*-coumaric acid decarboxylase from *Lactobacillus plantarum*: Gene cloning, transcriptional analysis, overexpression in *Escherichia coli*, purification, and characterization. *Appl. Environ. Microbiol., 63,* 1939−1944.

Cavin, J.-F., Barthelmebs, L., Guzzo, J., van Beeumen, J., Samyn, B., Travers, J.-F., et al. (1997b). Purification and characterization of an inducible *p*-coumaric acid decarboxylase from *Lactobacillus plantarum. FEMS Microbiol. Lett., 147,* 291−295.

Chatonnet, P., Dubourdieu, D., & Boidron, J. N. (1995). The influence of *Brettanomyces/Dekkera* sp. yeasts and lactic acid bacteria on the ethylphenol content of red wine. *Am. J. Enol. Vitic., 46,* 463−468.

Chen, K. H., & McFeeters, R. F. (1986a). Utilization of electron acceptors for anaerobic metabolism by *Lactobacillus plantarum*. Enzymes and intermediates in the utilization of citrate. *Food Microbiol., 3,* 83−92.

Chen, K. H., & McFeeters, R. F. (1986b). Utilization of electron acceptors for anaerobic mannitol metabolism by *Lactobacillus plantarum*. Reduction of alpha-keto acids. *Food Microbiol., 3,* 93−99.

Claisse, O., & Lonvaud-Funel, A. (2001). Détection de bacteries lactiques produisant du 3-hydroxypropionaldéhyde (précurseur d'acroléine) à partir du glycérol par tests moléculaires. *Lait, 81,* 173−181.

Collins, M. D., Rodrigues, U. M., Aguirre, M., Farrow, J. A. E., Martinez-Murcia, A., Phillips, B. A., et al. (1991). Phylogenetic analysis of the genus *Lactobacillus* and related lactic acid bacteria as determined by reverse transcriptase sequencing of 16S rRNA. *FEMS Microbiol. Lett., 123,* 241−248.

Constantini, A., Cercosimo, M., del Prete, V., & García-Moruno, E. (2006). Production of biogenic amine by lactic acid bacteria, screening by PCR, thin layer chromatography, and high-performance liquid chromatography of strains isolated from wine and must. *J. Food Prot., 69,* 391−396.

Costello, P. J., & Henschke, P. A. (2002). Mousy off-flavor of wine: Precursors and byosinthesis of the causative N-heterocycles 2-ethyltetrahydropyridine, 2-acetyltetrahydropyridine, and 2-acetyl-1-pyrroline by *Lactobacillus hilgardii* DSM 20176. *J. Agric. Food Chem., 50,* 7079−7087.

Costello, P. J., Lee, T. H., & Henschke, P. A. (2001). Ability of lactic acid bacteria to produce N-heterocycles causing mousy off-flavour in wine. *Aus. J. Grape Wine Res., 7,* 160−167.

Coton, E., Rollan, G., Bertrand, A., & Lonvaud-Funel, A. (1998b). Histamine-producing lactic acid bacteria in wines: Early detection, frequency and distribution. *Am. J. Enol. Vitic., 49,* 199−204.

Coton, E., Rollan, G. C., & Lonvaud-Funel, A. (1998a). Histidine carboxylase of *Leuconostoc oenos* 9204: Purification, kinetic properties, cloning and nucleotide sequence of the *hdc* gene. *J. Appl. Microbiol., 84,* 143−151.

Crapisi, A., Spettoli, P., Nuti, M. P., & Zamorani, A. (1987). Comparative traits of *Lactobacillus brevis, Lactobacillus fructivorans* and *Leuconostoc oenos* immobilized cells for the control of malo-lactic fermentation in wine. *J. Appl. Bacteriol., 61,* 145−151.

Curiel, J. A., Rodríguez, H., Acebrón, I., Mancheño, J. M., de las Rivas, B., & Muñoz, R. (2009). Production and physicochemical properties of recombinant *Lactobacillus plantarum* tannase. *J. Agric. Food Chem., 57,* 6224−6230.

d'Incecco, N., Bartowsky, E., Kassara, S., Lante, A., Spettoli, P., Henschke, P. (2004). Release of glycosidically bound flavour compounds of Chardonnay by *Oenococcus oeni* during malolactic fermentation. *Food Microbiol., 21,* 257−265.

Daeschel, M. A., Jung, D.-S., & Watson, B. T. (1991). Controlling wine malolactic fermentation with nisin and nisin-resistant strains of *Leuconostoc oenos. Appl. Environ. Microbiol., 57,* 601−603.

Davis, C. R., Wibowo, D., Eschenbruch, R., Lee, T. H., & Fleet, G. H. (1985a). Practical implications of malolactic fermentation. A review. *Am. J. Enol. Vitic., 36,* 290−301.

Davis, C. R., Wibowo, D., Fleet, G., & Lee, T. H. (1986a). Growth and metabolism of lactic acid bacteria during and after malolactic fermentation of wines at different pH. *Appl. Environ. Microbiol., 51*, 539–545.

Davis, C. R., Wibowo, D., Fleet, G., & Lee, T. H. (1986b). Growth and metabolism of lactic acid bacteria during fermentation and conservation of some Australian wines. *Food Technol. Austr., 38*, 35–40.

Davis, C. R., Wibowo, D., Fleet, G. H., & Lee, T. H. (1988). Properties of wine lactic acid bacteria: Their potential oenological significance. *Am. J. Enol. Vitic., 39*, 137–142.

Davis, G., Silveira, N. F. A., & Fleet, G. H. (1985b). Occurrence and properties of bacteriophages of *Leuconostoc oenos* in Australian wines. *Appl. Environ. Microbiol., 50*, 872–876.

de las Rivas, B., Marcobal, A., & Muñoz, R. (2004). Allelic diversity and population structure in *Oenococcus oeni* as determined from sequence analysis of housekeeping genes. *Appl. Environ. Microbiol., 70*, 7210–7219.

de las Rivas, B., Rodríguez, H., Curiel, J. A., Landete, J. M., & Muñoz, R. (2009). Molecular screening of wine lactic acid bacteria degrading hydroxycinnamic acids. *J. Agric. Food Chem., 57*, 490–494.

de Revel, G., & Bertrand, A. (1993). A method for the detection of carbonyl compounds in wine: Glyoxal and methylglyoxal. *J. Sci. Food Agric., 61*, 267–272.

de Revel, G., Bloem, A., Augustin, M., Lonvaud-Funel, A., & Bertrand, A. (2005). Interaction of *Oenococcus oeni* and oak wood compounds. *Food Microbiol., 22*, 569–575.

de Revel, G., Martín, N., Pripis-Nicolau, L., Lonvaud-Funel, A., & Bertrand, A. (1999). Contribution to the knowledge of malolactic fermentation influence on wine aroma. *J. Agric. Food Chem., 47*, 4003–4008.

Derzelle., S., Hallet, B., Ferrain, T., Delcour, J., & Hols, P. (2002). Cold shock induction of the *cspL* gene in *Lactobacillus plantarum* involves transcriptional regulation. *J. Bacteriol., 184*, 5518–5523.

Derzelle., S., Hallet, B., Ferrain, T., Delcour, J., & Hols, P. (2003). Improved adaptation to cold shock, stationary-phase, and freezing stresses in *Lactobacillus plantarum* overproducing cold shock proteins. *Appl. Environ. Microbiol., 69*, 4285–4290.

Derzelle., S., Hallet, B., Francis, K. P., Ferrain, T., Delcour, J., & Hols, P. (2000). Changes in *cspL, cspP,* and *cspC* mRNA abundance as a function of cold shock and growth phase in *Lactobacillus plantarum*. *J. Bacteriol., 182*, 5105–5113.

Dicks, L. M. T., Dellaglio, F., & Collins, M. D. (1995). Proposal to reclassify *Leuconostoc oenos* as *Oenococcus oeni* [corrig.] gen.nov., comb. nov. *Int. J. System. Bacteriol., 45*, 395–397.

Divol, B., Tonon, T., Morichon, S., Gindreau, E., & Lonvaud-Funel, A. (2003). Molecular characterization of *Oenococcus oeni* genes encoding proteins involved in arginine transport. *J. Appl. Microbiol., 94*, 738–746.

Dols-Lafargue, M., Lee, H. Y., Le Marrec, C., Heyraud, A., Chambat, G., & Lonvaud-Funel, A. (2008). Characterization of *gtf*, a glucosyltransferase gene in the genomes of *Pediococcus parvulus* and *Oneococcus oeni*, two bacterial species commonly found in wine. *Appl. Environ. Microbiol., 74*, 4079–4090.

Ebeler, S. E. (2001). Analytical chemistry: Unlocking the secrets of wine flavor. *Food Rev. Int., 17*, 45–64.

Edwards, C. G., & Beelman, R. B. (1987). Inhibition of the malolactic bacterium *Leuconostoc oenos* (PSU-1), by decanoic acid and subsequent removal of the inhibition by yeast ghosts. *Am. J. Enol. Vitic., 38*, 239–242.

Edwards, C. G., Beelman, R. B., Bartley, C. E., & Mc Connell, A. (1990). Production of decanoic acid and other volatile compounds and the growth of yeast and malolactic bacteria during vinification. *Am. J. Enol. Vitic., 41*, 48–56.

Edwards, C. G., Collins, M. D., Lawson, P. A., & Rodríguez, A. V. (2000). *Lactobacillus nagelii* sp. nov., and organism isolated from a partially fermented wine. *Int. J. Syst. Evol. Microbiol., 50*, 699–702.

Endo, A., & Okada, S. (2006). *Oenococcus kitaharae* sp. nov., a non-acidophilic and non-malolactic-fermenting oenococcus isolated from a composting distilled shochu residue. *Int. J. Syst. Evol. Microbiol., 56*, 2345–2348.

Farias, M. E., Rollan, G. C., & Manca de Nadra, M. C. (1996). Influence of nutritional factors on the protease production by *Leuconostoc oenos* from wine. *J. Appl. Bacteriol., 81*, 398–402.

Firme, M. P., Leitao, M. C., & Sanromao, M. V. (1994). The metabolism of sugar and malic acid by *Leuconostoc oneos*. Effect of malic acid, pH and aeration conditions. *J. Appl. Bacteriol., 76*, 173–181.

Firme, M. P., Leitao, M. C., & San Romào, M. V. (1996). The metabolism of sugar and malic acid by *Leuconostoc oenos*: Effect of malic acid, pH and aireation conditions. *J. Appl. Bacteriol., 76*, 173–181.

Folio, P., Ritt, J.-F., Alexandre, H., & Remize, F. (2008). Characterization of EprA, a major extracellular protein of Oenococcus oeni with protease activity. *Int. J. Food Microbiol., 127*, 26–31.

Fortier, L. C., Tourdot-Marechal, R., Divies, C., Lee, B. H., & Guzzo, J. (2003). Induction of *Oenococcus oeni* H⁺-ATPase activity and mRNA trasncription under acidic condictions. *FEMS Microbiol. Lett., 222*, 165–169.

Galland, D., Tourdot-Maréchal, R., Abraham, M., Chu, K. S., & Guzzo, J. (2003). Absence of malolactic activity is a characteristic of H⁺-ATPase-deficient mutants of the lactic acid bacterium *Oenococcus oeni*. *Appl. Environ. Microbiol., 69*, 1973–1979.

García-Moruno, E., Carrascosa, A. V., & Muñoz, R. (2005). A rapid and inexpensive method for the determination of biogenic amines from bacterial cultures by thin-layer chromatography. *J. Food. Prot., 68*, 625–629.

Gindreau, E., Joyeux, A., de Revel, G., Claisse, O., & Lonvaud-Funel, A. (1997). Evaluation de l'établissement des levains malolactiques au sein de la microflore bactérienne indigene. *J. Int. Sci. Vigne Vin, 31*, 197–202.

Gindreau, E., Wailling, E., & Lonvaud-Funel, A. (2001). Direct polymerase chain reaction detection of ropy *Pediococcus damnosus* strains in wine. *J. Appl. Microbiol., 90*, 535–542.

Granchi, L., Paperi, R., Rosellini, D., & Vincenzini, M. (1998). Strain variation of arginine catabolism among malolactic *Oenococcus oeni* strains of wine origin. *Ital. J. Food. Sci., 10*, 351–357.

Green, G., Dicks, L. M. T., Bruggeman, G., Vandamme, E. J., & Chikindas, M. L. (1997). Pediocin PD-1, a bactericidal antimicrobial peptide from *Pediococcus damnosus* NCFB 1832. *J. Appl. Microbiol., 83*, 13–127.

Grimaldi, A., Bartowski, E., & Jiranek, V. (2005). A survey of glycosidase activities of commercial wine strains of *Oenococcus oeni*. *Int. J. Food Microbiol., 105*, 233–244.

Grimaldi, A., McLean, H., & Jiranek, V. (2000). Identification and partial characterization of glycosidic activities of commercial strains of the lactic acid bacterium *Oenococcus oeni*. *Am. J. Enol. Vitic., 51*, 362–369.

Groisillier, A., & Lonvaud-Funel, A. (1999). Comparison of partial malolactic enzyme gene sequences for phylogenetic analysis of some lactic acid bacteria species and relationships with the malic enzyme. *Int. J. Syst. Bacteriol., 49*, 1417–1428.

Guerrini, S., Bastianini, A., Blaiotta, G., Granchi, L., Moschetti, G., Coppola, S., Romano, P., & Vincenzini, M. (2003). Phenotypic and genotypic characterization of Oenococcus oeni strains isolated from Italian wines. *Int. J. Food Microbiol., 83*, 1–14.

Guilloux-Benatier, M., Pageault, O., Man, A., & Feuillat, M. (2000). Lysis of yeast cells by *Oenococcus oeni* enzymes. *J. Ind. Microbiol. Biotech., 25*, 193–197.

Guilloux-Benatier, M., Son, H. S., Bouhier, S., & Feuillat, M. (1993). Activités enzymatiques: Glycosidases et peptidases chez *Leuconostoc oenos* au cours de la croissance bactérienne. Influence des macromolécules de levures. *Vitis, 32*, 51–57.

Gury, J., Barthelmebs, L., Tran, N. P., Diviès, C., & Cavin, J.-F. (2004). Cloning, deletion, and characterization of PadR, the transcriptional repressor of the phenolic acid decarboxylase-encoding *padA* gene of *Lactobacillus plantarum*. *Appl. Environ. Microbiol., 70*, 2146–2153.

Guzzo, J., Delmas, F., Pierre, F., Jobin, M. P., Samyn, B., van Beewmer, I., et al. (1997). A small heat-shock protein from *Leuconostoc oenos* induced by multiple stresses and during stationary growth phase. *Lett. Appl. Microbiol., 24*, 393–396.

Henick-Kling, T. (1993). Malolactic fermentation. In G. H. Fleet (Ed.), *Wine microbiology and biotechnology* (pp. 286–326). Berlin, Germany: Springer-Verlag.

Henick-Kling, T., Lee, T. H., & Nicholas, D. J. D. (1986). Characterization of the lytic activity of bacteriophages of *Leuconostoc oenos* isolated from wine. *J. Appl. Bacteriol., 61*, 525–534.

Herrero, M., Mayo, B., Gonzalez, B., & Suárez, U. E. (1996). Evaluation of technologically important traits in lactic acid bacteria isolated from spontaneosus fermentations. *J. Appl. Bacteriol., 81*, 565–570.

Holo, H., Jeknic, Z., Daeschel, M., Stevanovic, S., & Nes, I. F. (2001). Plantaricin W from *Lactobacillus plantarum* belongs to a new family of two-peptide lantibiotics. *Microbiology, 147*, 643–651.

Jobin, M. P., Garmyn, D., Divies, C., & Guzzo, J. (1999a). Expression of the *Oenococcus oeni trxA* gene is induced by hydrogen peroxide and heat shock. *Microbiology, 145*, 1245–1251.

Jobin, M. P., Garmyn, D., Divies, C., & Guzzo, J. (1999b). The *Oenococcus oeni clpX* homologue is a heat shock gene preferentially expressed in exponential growth phase. *J. Bacteriol., 181*, 6634–6641.

Knoll, C., Divol, B., & du Toit, M. (2008). Genetic screening of lactic acid bacteria of oenological origin for bacteriocin-encoding genes. *Food Microbiol., 25*, 983–991.

Koch, A. (1900). Uber die ursachem des verschwidens des saure bei garung und lagerrung des weines. *Weibau Weinhandel, 18*, 395–419.

Labarre, C., Guzzo, J., Cervin, J. F., & Divies, C. (1996). Cloning and characterization of the genes encoding the malolactic enzyme and the malate permease of *Leuconostoc oenos*. *Appl. Environ. Microbiol., 62*, 1274–1282.

Landete, J. M., de las Rivas, B., Marcobal, A., & Muñoz, R. (2007). Molecular methods for the detection of biogenic amine-producing bacteria on foods. *Int. J. Food Microbiol., 117*, 258–269.

Lee, J.-E., Hong, Y.-S., & Lee, C.-H. (2009). Characterization of fermentative behaviours of lactic acid bacteria in grape wine through $^1$H NMR- and GC-based metabolic profiling. *J. Agric. Food Chem., 57*, 4810–4817.

Lehtonen, P. (1996). Determination of amines and amino acids in wine: A review. *Am. J. Enol. Vitic., 47*, 127–133.

Leitao, M. C., Teixeira, H. C., Barreto Crespo, M. T., & San Romao, N. V. (2000). Biogenic amine ocurrence in wine. Amino acid decarboxylase and proteolytic activities expression by *Oenococcus oeni*. *J. Agric. Food Chem., 48*, 2780–2784.

Liu, S.-Q., & Davis, C. R. (1994). Analysis of wine carbohydrates using capillary gas-liquid chromatography. *Am. J. Enol. Vitic., 45*, 229–234.

Liu, S.-Q., Davis, C. R., & Brooks, J. D. (1995a). Growth and metabolism of selected lactic acid bacteria in synthetic wine. *Am. J. Enol. Vitic., 45*, 229–234.

Liu, S.-Q., & Pilone, G. J. (1998). A review: Arginine metabolism in wine lactic acid bacteria and its practical significance. *J. Appl. Microbiol., 84*, 315–327.

Liu, S.-Q., Pritchard, G. G., Hardman, M. J., & Pilone, G. J. (1994). Citrulline production and ethyl carbamate (urethane) precursor formation from arginine degradation by wine lactic acid bacteria *Leuconsotoc oenos* and *Lactobacillus buchneri*. *Am. J. Enol. Vitic., 45*, 235–242.

Liu, S.-Q., Pritchard, G. G., Hardman, M. J., & Pilone, G. J. (1995). Occurrence of arginine deiminase pathway enzymes in arginine catabolism by wine lactic acid bacteria. *Appl. Environ. Microbiol., 61*, 310–316.

Liu, S.-Q., Pritchard, G. G., Hardman, M. J., & Pilone, G. J. (1996). Arginine catabolism in wine lactic acid bacteria: Is it via the arginine deiminase pathway or the arginase-urease pathway? *J. Appl. Bacteriol., 81*, 486–492.

Llaubères, R. M., Richard, B., Lonvaud, A., & Dubourdieu, D. (1990). Structure of an exocellular β-D-glucan from *Pediococcus* sp. A wine lactic acid bacteria. *Carbohydr. Res., 203*, 103–107.

Lonvaud-Funel, A. (1995). Microbiology of the malolactic fermentation: Molecular aspects. *FEMS Microbiol Lett., 126*, 209–214.

Lonvaud-Funel, A. (1999). Lactic acid bacteria in the quality improvement and depreciation of wine. *Anton. Leeuw., 76*, 317–331.

Lonvaud-Funel, A. (2001). Biogenic amines in wine: Role of lactic acid bacteria. *FEMS Microbiol. Lett., 199*, 9–13.

Lonvaud-Funel, A., Guilloux, Y., & Joyeux, A. (1993). Isolation of a DNA probe for identification of glucan-producing *Pediococcus damnosus* in wines. *J. Appl. Bacteriol., 74*, 41–47.

Lonvaud-Funel, A., Joyeux, A., & Desens, C. (1988). The inhibition of malolactic fermentation of wines by products of yeast metabolims. *J. Food Sci. Agric., 44*, 183–191.

Lonvaud-Funel, A., Joyeux, A., & Ledoux, O. (1991). Specific enumeration of lactic acid bacteria in fermenting grape must and wine by colony hybridization with non-isotopic DNA probes. *J. Appl. Bacteriol., 71*, 501–508.

Lonvaud-Funel, A., & Strasser de Saad, A. M. (1982). Purification and properties of a malolactic enzyme from a strain of *Leuconostoc mesenteroides* isolated from grapes. *Appl. Environ. Microbiol., 43*, 357–361.

López, I., Ruiz-Larrea, F., Cocolin, L., Orr, E., Phister, T., Marshall, M., et al. (2003). Design and evaluation of PCR primers for analysis of bacterial populations in wine by denaturing gradient gel electrophoresis. *Appl. Environ. Microbiol., 69*, 6801–6807.

Lucas, P., Landete, J., Coton, M., Coton, E., & Lonvaud-Funel, A. (2003). The tyrosine decarboxylase operon of *Lactobacillus brevis* IOEB 9809: Characterization and conservation in tyramine-producing bacteria. *FEMS Microbiol. Lett., 229*, 65–71.

Lucas, P. M., Claisse, O., & Lonvaud-Funel, A. (2008). High frequency of histamine-producing bacteria in the enological environment and instability of the histidine decraboxylase production phenotype. *Appl. Environ. Microbiol., 74*, 811–817.

Lucas, P. M., Wolken, W. A. M., Claisse, O., Lolkema, J. S., & Lonvaud-Funel, A. (2005). Histamine-producing pathway encoded on an unstable plasmid in *Lactobacillus hilgardii* 0006. *Appl. Environ. Microbiol., 71*, 1417–1424.

Maicas, S. (2001). The use of alternative technologies to develop malolactic fermentation in wine. *Appl. Microbiol. Biotechnol., 56*, 35–39.

Maicas, S., Pardo, I., & Ferrer, S. (2000). The effects of freezing and freeze-drying of Oenococcus oeni upon induction of malolactic fermentation in red wine. *Int. J. Food Sci. Technol., 35*, 75–79.

Maicas, S., Pardo, I., & Ferrer, S. (2001). The potential of positively-charged cellulose sponge for malolactic fermentation of wine using *Oenococcus oeni*. *Enz. Microb. Technol., 28*, 415–419.

Maijala, R. L. (1993). Formation of histamine and tyramine by some lactic acid bacteria in MRS-broth and modified decarboxylation agar. *Lett. Appl. Microbiol., 17*, 40–43.

Makarova, K., Slesarev, A., Wolf, Y., Sorokin, A., Mirkin, B., Koonin, E., et al. (2006). Comparative genomics of the lactic acid bacteria. *Proc. Nat. Acad. Sci. USA, 103*, 15 611–15 616.

Manca de Nadra, M. C., Farias, M. E., Moreno-Arribas, M. V., Pueyo, E., & Polo, M. C. (1997). Proteolytic activity of *Leuconostoc oenos*. Effect on proteins and polypeptides from white wines. *FEMS Microbiol. Lett., 150*, 135–139.

Manca de Nadra, M. C., Farias, M. E., Moreno-Arribas, M. V., Pueyo, E., & Polo, M. C. (1999). A proteolytic effect of *Oenococcus oeni* on the nitrogenous macromolecular fraction of red wine. *FEMS Microbiol. Lett., 174*, 41–47.

Manca de Nadra, M. C., Sandino de Lamelas, D., & Strasser de Saad, M. C. (1998). Pediocin N5p from *Pediococcus pentosaceus*: Adsortion on bacterial strains. *Int. J. Food. Microbiol., 39*, 79–85.

Mañes-Lázaro, R., Ferrer, S., Rodas, A. M., Urdiain, M., & Pardo, I. (2008a). *Lactobacillus bobalius* sp. nov., a lactic acid bacterium isolated from Spanish Bobal grape must. *Int. J. Syst. Evol. Microbiol., 58*, 2699–2703.

Mañes-Lázaro, R., Ferrer, S., Rosselló-Mora, R., & Pardo, I. (2008b). *Lactobacillus uvarum* sp. nov. A new lactic acid bacterium isolated from Spanish Bobal grape must. *Syst. Appl. Microbiol., 31*, 425–433.

Mañes-Lázaro, R., Ferrer, S., Rosselló-Mora, R., & Pardo, I. (2009). *Lactobacillus oeni* sp. nov. from wine. *Int. J. Syst. Evol. Microbiol., 59*, 2010–2014.

Marcobal, A., de las Rivas, B., Moreno-Arribas, M. V., & Muñoz, R. (2004). Identification of the ornithine decarboxylase gene in the putrescine-producer *Oenococcus oeni* BIFI-83. *FEMS Microbiol. Lett., 239*, 213–220.

Marcobal, A., de las Rivas, B., Moreno-Arribas, M. V., & Muñoz, R. (2005). Multiplex PCR method for the simultaneous detection of histamine-, tyramine-, and putrescine-producing lactic acid bacteria in foods. *J. Food Prot., 68*, 874–878.

Marcobal, A., de las Rivas, B., Moreno-Arribas, M. V., & Muñoz, R. (2006). Evidence for horizontal gene transfer as origin of putrescine production in *Oenococcus oeni* RM83. *Appl. Environ. Microbiol., 72*, 7954–7958.

Marcobal, A. M., Sela, D. A., Wolf, Y. I., Makarova, K. S., & Mills, D. A. (2008). Role of hypermutability in the evolution of the genus *Oenococcus*. *J. Bacteriol., 190*, 564–570.

Martineau, B., & Henick-Kling, T. (1995). Formation and degradation of diacetyl in wine during alcoholic fermentation with *Saccharomyces cerevisiae* strain EC1118 and malolactic fermentation with *Leuconostoc oenos* strain MCW. *Am. J. Enol. Vitic., 46*, 442–448.

Martineau, B., Henich-Kling, T., & Acree, T. (1995). Reassessment of the influence of malolactic fermentation on the concentration of diacetyl in wines. *Am. J. Enol. Vitic., 46*, 385–388.

McFeeters, R. S., & Chen, K. H. (1986). Utilization of electron acceptors for anaerobic mannitol metabolism by *Lactobacillus plantarum*. Compounds which serve as electron acceptors. *Food Microbiol., 3*, 73–81.

McMahon, H., Zoecklein, B. W., Fugelsang, K., & Jasinski, Y. (1999). Quantification of glycosidase activities in selected yeasts and lactic acid bacteria. *J. Ind. Microbiol. Biotechnol., 23*, 198–203.

Mills, D. A., Rawsthorne, H., Parker, C., Tamir, D., & Makarova, K. (2005). Genomic analysis of *Oenococcus oeni* PSU-1 and its relevance to winemaking. *FEMS Microbiol. Rew., 29*, 465–475.

Mira de Orduña, R., Liu, S.-Q., Patchett, M. L., & Pilone, G. J. (2000a). Ethyl carbamate precursor citrulline formation from arginine degradation by malolactic wine lactic acid bacteria. *FEMS Microbiol. Lett., 183*, 31–35.

Mira de Orduña, R., Liu, S.-Q., Patchett, M. L., & Pilone, G. J. (2000b). Kinetics of the arginine metabolism of malolactic wine lactic acid bacteria *Lactobacillus buchneri* CUC-3 and *Oenococcus oeni* Lo11. *J. Appl. Microbiol., 89*, 547–552.

Mira de Orduña, R., Patchett, M. L., Liu, S.-Q., & Pilone, G. J. (2001). Growth, arginine metabolism of the wine lactic acid bacteria *Lactobacillus buchneri*, *Oenococcus oeni* at different pH values and arginine concentrations. *Appl. Environ. Microbiol., 67*, 1657–1662.

Moreno-Arribas, M. V., & Lonvaud-Funel, A. (2001). Purification and characterization of tyrosine decarboxylae of *Lactobacillus brevis* IOEB 9809 isolated from wine. *FEMS Microbiol. Lett., 195*, 103–107.

Moreno-Arribas, M. V., Polo, M. C., Jorganes, F., & Muñoz, R. (2003). Screening of biogenic amine production by lactic acid bacteria isolated from grape must and wine. *Int. J. Food Microbiol., 84*, 117–123.

Müller-Thurgau, H. (1981). Uber die ergebrisse, neuer untersughungen auf den gebite der meinbersitung. *Dtsch. Weinbaukong (Worms), 128*.

Nannelli, F., Claisse, O., Gindreau, E., de Revel, G., Lonvaud-Funel, A., & Lucas, P. M. (2008). Detemination of lactic acid bacteria producing biogenic amines in wine by quantitative PCR methods. *Lett. Appl. Microbiol., 47*, 594–599.

Naouri, P., Chagnaud, P., Arnaud, A., & Galzy, P. (1990). Purification and properties of a malolactic enzyme from *Leuconostoc oenos* ATCC 23278. *J. Basic Microbiol., 30*, 577–585.

Navarro, L., Zarazaga, M., Sáenz, J., Ruiz-Larrea, F., & Torres, C. (2000). Bacteriocin production by lactic acid bacteria isolated from Rioja red wines. *J. Appl. Microbiol., 88*, 41–51.

Nel, H. A., Bauer, R., Wolfaardt, G. M., & Dicks, L. M. T. (2002). Effects of bacteriocins pediocin PD-1, plantaricin 423, and nisin on biofilms of *Oenocuccus oeni* on a stainless steel surface. *Am. J. Enol. Vitic., 53*, 191–196.

Nel, L., Wingfield, B. D., van der Meer, L. J., & van Vuuren, H. J. J. (1987). Isolation and characterization of *Leuconsotoc oenos* bacteriophages from wine and sugarcane. *FEMS Microbiol. Lett., 44*, 63–67.

Nicoloff, H., Hubert, J.-C., & Bringel, F. (2001). Carbamoyl-phosphate synthetases (CPS) in lactic acid bacteria and other Gram-positive bacteria. *Lait, 81*, 151–159.

Nielsen, J. C., Prahl, C., & Lonvaud-Funel, A. (1996). Malolactic fermentation in wine by direct inoculation with freeze-dried *Leuconostoc oenos* cultures. *Am. J. Enol. Vitic., 47*, 42–48.

Nielsen, J. C., & Richelieu, M. (1999). Control of flavor development in wine during and after malolactic fermentation by *Oenococcus oeni*. *Appl. Environ. Microbiol., 65*, 740–745.

Osborne, J. P., & Edwards, C. G. (2007). Inhibition of malolactic fermentation by a peptide produced by *Saccharomyces cerevisiae* during alcoholic fermentation. *Int. J. Food Microbiol., 118*, 27–34.

Osborne, J. P., Mira de Orduña, R., Pilone, G. J., & Liu, S.-Q. (2000). Acetaldehyde metabolism by wine lactic acid bacteria. *FEMS Microbiol. Lett., 191*, 51–55.

Ough, C. S. (1976). Ethyl carbamate in fermented beverages and foods. I. Naturally occurring ethyl carbamate. *J. Agric. Food Chem., 24*, 323–328.

Pilone, G. J., Clayton, M. G., & van Duivenboden, R. J. (1991). Characterization of wine lactic acid bacteria: Single broth culture for tests of heterofermentation, mannitol from fructose, and ammonia from arginine. *Am. J. Enol. Vitic., 42*, 153–157.

Pinzani, P., Bonciani, L., Pazzagli, M., Orlando, C., Guerrini, S., & Granchi, L. (2004). Rapid detection of *Oenococcus oeni* in wine by real-time quantitative PCR. *Lett. Appl. Microbiol., 38*, 118–124.

Poblet-Icart, M., Bordons, A., & Lonvaud-Funel, A. (1998). Lysogeny of *Oenococcus oeni* (syn. *Leuconostoc oenos*) and study of their induced bacteriophages. *Curr. Microbiol., 36*, 365–369.

Pripis-Nicolau, L., de Revel, G., Bertrand, A., & Lonvaud-Funel, A. (2004). Methionine catabolism and production of volatile sulphur compounds by *Oenococcus oeni*. *J. Appl. Microbiol., 96*, 1176–1184.

Pueyo, E., Martínez-Rodríguez, A., Polo, M. C., Santa-María, G., & Bartolomé, B. (2000). Release of lipids during yeast autolysis in a model wine system. *J. Agric. Food Chem., 48*, 116–122.

Quere, F., Deschanps, A., & Urdaci, M. C. (1997). DNA probe and PCR-specific reaction for *Lactobacillus plantarum*. *J. Appl. Microbiol., 82*, 783–790.

Radler, F. (1990a). Possible use of nisin in winemaking. I. Action of nisin against lactic acid bacteria and wine yeasts in solid and liquid media. *Am. J. Enol. Vitic., 41*, 1–6.

Radler, F. (1990b). Possible use of nisin in winemaking. II. Experiments to control lactic acid bacteria in the production of wine. *Am. J. Enol. Vitic., 41*, 7–11.

Ramos, A., Lolkema, J. S., Konings, W. S., & Santos, H. (1995). Enzyme basis for pH regulation of citrate and pyruvate metabolism by *Leuconostoc oenos*, a $^{13}$C nuclear magnetic resonance study. *Appl. Environ. Microbiol., 61*, 1303–1310.

Ramos, A., & Santos, H. (1996). Citrate and sugar cofermentation in *Leuconostoc oenos*, a $^{13}$C nuclear magnetic resonance study. *Appl. Environ. Microbiol., 62*, 2577–2585.

Reguant, C., & Bordons, A. (2003). Typification of *Oenococcus oeni* strains by multiplex RAPD-PCR and study of population dynamics during malolactic fermentation. *J. Appl. Microbiol., 95*, 344–353.

Reguant, C., Bordons, A., Arola, L., & Rozès, N. (2000). Influence of phenolic compounds on the physiology of *Oenococcus oeni* from wine. *J. Appl. Microbiol., 88*, 1065–1071.

Richter, H., Vlad, D., & Unden, G. (2001). Significance of pantothenate for glucose fermentation by *Oenococcus oeni* and for suppression of the erythritol and acetate production. *Arch. Microbiol., 175*, 26–31.

Ritt, J.-F., Guilloux-Benatier, M., Guzzo, J., Alexandre, H., & Remize, F. (2008). Oligopeptide assimilation and transport by *Oenococcus oeni*. *J. Appl. Microbiol., 104*, 573–580.

Ritt, J.-F., Remize, F., Grandvalet, C., Guzzo, J., Atlan, D., & Alexandre, H. (2009). Peptidases specific for proline-containing peptides and their unusual peptide-dependent regulation in *Oenococcus oeni*. *J. Appl. Microbiol., 106*, 801–813.

Rodas, A. M., Chenoll, E., Macián, M. C., Ferrer, S., Pardo, I., & Aznar, R. (2006). *Lactobacillus vini* sp. nov., a wine lactic acid bacterium homofermentative for pentoses. *Int. J. Syst. Evol. Microbiol., 56*, 513–517.

Rodas, A. M., Ferrer, S., & Pardo, I. (2003). 16S-ARDRA, a tool for identification of lactic acid bacteria isolated from grape must and wine. *System. Appl. Microbiol., 26*, 412–422.

Rodas, A. M., Ferrer, S., & Pardo, I. (2005). Polyphasic study of wine *Lactobacillus* strains: Taxonomic implications. *Int. J. Syst. Evol. Microbiol., 55*, 197–207.

Rodríguez, S. B., Amberg, E., & Thornston, R. J. (1990). Malolactic fermentation in Chardonnay: Growth and sensory effects of commercial strains of *Leuconostoc oenos*. *J. Appl. Bacteriol., 68*, 139–144.

Rodríguez, H., Curiel, J. A., Landete, J. M., de las Rivas, B., López de Felipe, F., Gómez-Cordovés, C., et al. (2009). Food phenolics and lactic acid bacteria. *Int. J. Food Microbiol., 132*, 79–90.

Rodriguez, H., Landete, J. M., Curiel, J. A., de las Rivas, B., Mancheño, J. M., & Muñoz, R. (2008). Characterization of the *p*-coumaric acid decarboxylase from *Lactobacillus plantarum* CECT 748$^T$. *J. Agric. Food Chem., 56*, 3068–3072.

Rojo-Bezares, B., Saenz, Y., Navarro, L., Zarazaga, M., Ruíz-Larrea, F., & Torres, C. (2007). Coculture-inducible bacteriocin activity of *Lactobacillus plantarum* strain J23 isolated from grape must. *Food Microbiol., 24*, 482–491.

Rollan, G. C., Coton, C., & Lonvaud-Funel, A. (1995b). Histidine decarboxylase activity of *Leuconostoc oenos* 9204. *Food Microbiol., 12*, 455–461.

Rollan, G. C., Farias, M. E., & Manca de Nadra, M. C. (1993). Protease production by *Leuconostoc oenos* strains isolated from wine. *World J. Microbiol. Biotechnol., 9*, 587–589.

Rollan, G. C., Farias, M. E., & Manca de Nadra, M. C. (1995a). Characterization of two extracellular proteases from *Leuconostoc oenos* strains isolated from wine. *World J. Microbiol. Biotechnol., 9*, 153–155.

Rozès, N., Arola, L., & Bordons, A. (2003). Effect of phenolic compounds on the co-metabolism of citric acid and sugars by *Oenococcus oeni* from wine. *Lett. Appl. Microbiol., 36*, 337–341.

Sáenz, Y., Rojo-Bezares, B., Navarro, L., Díez, L., Somalo, S., Zaragaza, M., et al. (2009). Genetic diversity of the pln locus among oenological *Lactobacillus plantarum* strains. *Int. J. Food Microbiol., 134*, 176–183.

Salou, P., Loubiere, P., & Pareilleux, A. (1994). Growth and energetics of *Leuconostoc oenos* during cometabolism of glucose with citrate or fructose. *Appl. Environ. Microbiol.*, *60*, 1459–1466.

San Romao, M. V., & Lafon-Lafourcade, S. (1979). Premieres observations sur l'action de *Botrytis cinerea* cultivé sur moût de raisins, a légard du métabolisme des bactéries lactiques dans les mouts et les vins. *Vitis*, *18*, 155–160.

Sao-José, C., Santos, S., Nascimento, J., Brito Madurro, A. G., Parreira, R., Vieira, G., et al. (2004). Diversity in the lysis-integration region of oenophage genomes and evidence for multiple tRNA loci, as target for prophage integration in *Oenococcus oeni*. *Virology*, *325*, 95–98.

Sato, H., Yanagida, F., Shinohara, T., Suzuki, M., Suzuki, K., & Yokotsuka, K. (2001). Intraspecific diversity *Oenococcus oeni*. *FEMS Microbiol. Lett.*, *202*, 109–114.

Sauvageot, F., & Vivier, P. (1997). Effects of malolactic fermentation on sensory properties of four Burgundy wines. *Am. J. Enol. Vitic.*, *48*, 187–192.

Schultz, H., & Radler, F. (1984a). Anaerobic reduction of glycerol to propanediol-1, 3 by Lactobacillus brevis and *Lactobacillus buchneri*. *Syst. Appl. Microbiol.*, *5*, 169–178.

Schultz, H., & Radler, F. (1984b). Propanediol-1, 2-dehydratase and metabolism of glycerol of *Lactobacillus brevis*. *Arch. Microbiol.*, *139*, 366–370.

Schütz, M., & Radler, F. (1974). Das vorkommen von malatenzym und malolactac-enzym bei verschiedenen milchsaürenbakterien. *Arch. Mikrobiol.*, *96*, 329–339.

Shalaby, A. R. (1996). Significance of biogenic amines to food safety and human health. *Food Res. Int.*, *29*, 675–690.

Silveira, M. G., Baumgärtner, M., Rombouts, F. M., & Abee, T. (2004). Effect of adaptation to ethanol on cytoplasmic and membrane protein profiles of *Oenococcus oeni*. *Appl. Environ. Microbiol.*, *70*, 2748–2755.

Smit, A., Cordero Otero, R. R., Lambrechts, M. G., Pretorius, I. S., & van Rensburg, P. (2003). Enhancing volatile phenol concentrations in wine by expressing various phenolic acid decarboxylase genes in *Saccharomyces cerevisiae*. *J. Agric. Food Chem.*, *51*, 4909–4915.

Snowdon, E. M., Bowyer, M. C., Grbin, P. R., & Bowyer, P. K. (2006). Mousy off-flavor: A review. *J. Agric. Food Chem.*, *54*, 6465–6474.

Sohier, D., Coulon, J., & Lonvaud-Funel, A. (1999). Molecular identification of *Lactobacillus hilgardii* and genetic relatedness with *Lactobacillus brevis*. *Int. J. Syst. Bacteriol.*, *49*, 1075–1079.

Son, H.-S., Hwang, G.-S., Park, W.-M., Hong, Y.-S., & Lee, C.-H. (2009). Metabolomic characterization of malolactic fermentation and fermentative behaviours of wine yeasts in grape wine. *J. Agric. Food Chem.*, *57*, 4801–4809.

Sozzi, T., Maret, R., & Poulin, J. M. (1976). Mise en evidence de bactériophages dans le vin. *Experientia*, *32*, 568–569.

Spano, G., Beneduce, L., Perrotta, C., & Massa, S. (2005). Cloning and characterization of the hsp 18.55 gene, a new member of the small heat shock gene family isolated from wine *Lactobacillus plantarum*. *Res. Microbiol.*, *156*, 219–224.

Spano, G., Capozzi, V., Vernile, A., & Massa, S. (2004). Cloning, molecular characterization and expression analysis of two small heat shock genes isolated from wine *Lactobacillus plantarum*. *J. Appl. Microbiol.*, *97*, 774–782.

Strasser de Saad, A. M., & Manca de Nadra, M. C. (1993). Characterization of bacteriocin produced by *Pediococus pentosaceus* from wine. *J. Appl. Bacteriol.*, *74*, 406–410.

Strasser de Saad, A. M., Pasteris, S. E., & Manca de Nadra, M. C. (1995). Production and stability of pediocin N5p in grape juice medium. *J. Appl. Bacteriol.*, *78*, 473–476.

Tenreiro, R. M., Santos, M. A., Pavela, H., & Vieira, G. (1994). Inter-strain relationships among wine leuconostocs and their divergence from other *Leuconostoc* species, as revealed by low frequency restriction fragment analysis of genomic DNA. *J. Appl. Bacteriol.*, *77*, 271–280.

Terrade, N., Nöel, R., Couillaud, R., & de Orduña, R. M. (2009). A new chemically defined medium for wine lactic acid bacteria. *Food Res. Int.*, *42*, 363–367.

Tonon, T., Bourdineaud, J. P., & Lonvaud-Funel, A. (2001). The *arcABC* gene cluster encoding the arginine deiminase pathway of *Oenococcus oeni*, and arginine induction of a CRP-like gene. *Res. Microbiol.*, *152*, 653–661.

Tonon, T., & Lonvaud-Funel, A. (2000). Metabolism of arginine and its positive effect on growth and revival of *Oenococcus oeni*. *J. Appl. Microbiol.*, *89*, 526–531.

Ugliano, M., Genovese, A., & Moio, L. (2003). Hydrolysis of wine aroma precursors during malolactic fermentation with four commercial starter cultures of *Oenococcus oeni*. *J. Agric. Food Chem.*, *51*, 5073–5078.

Vallet, A., Santarelli, X., Lonvaud-Funel, A., de Revel, G., & Cabanne, C. (2009). Purification of an alcohol dehydrogenase involved in the conversion of methional to methionol in *Oenococcus oeni* IOEB 8406. *Appl. Microbiol. Biotechnol.*, *82*, 87–94.

van Vuuren, H. J. J., & Dicks, L. M. T. (1993). *Leuconostoc oenos*: A review. *Am. J. Enol. Vitic.*, *44*, 99–112.

Vandamme, P., Pot, B., Gillis, M., de Vos, P., Kersters, K., & Swings, J. (1996). Polyphasic taxonomy, a consensus approach to bacterial systematics. *Microbiol. Rev.*, *60*, 407–438.

Vaquero, I., Marcobal, A., & Muñoz, R. (2004). Tannase activity by lactic acid bacteria isolated from grape must and wine. *Int. J. Food Microbiol.*, *96*, 199–204.

Vasserot, Y., Dion, C., Bonnet, E., Maujean, A., & Jeandet, P. (2001). A study into the role of L-aspartic acid on the metabolism of L-malic acid and D-glucose by *Oenococcus oeni*. *J. Appl. Microbiol., 90*, 380–387.

Vasserot, Y., Dion, C., Bonnet, E., Tabary, I., Maujean, A., & Jeandet, P. (2003). Transport of glutamate in *Oenococcus oeni* 8403. *Int. J. Food Microbiol., 85*, 307–311.

Veiga-da-Cunha, M., Firme, P., San Romào, M. V., & Santos, H. (1992). Application of $^{13}$C nuclear magnetic resonance to elucidate the unexpected biosynthesis of erythritol by *Leuconostoc oenos*. *Appl. Environ. Microbiol., 58*, 2271–2279.

Veiga-da-Cunha, M., Santos, H., & van Schaftingen, E. (1993). Pathway and regulation of erythritol formation in *Leuconostoc oenos*. *J. Bacteriol., 175*, 3941–3948.

Ventura, M., Casas, I. A., Morelli, L., & Callegari, M. L. (2000). Rapid amplified ribosomal DNA restriction analysis (ARDRA) identification of *Lactobacillus spp.* Isolated from fecal and vaginal samples. *System. Appl. Microbiol., 23*, 504–509.

Versari, A., Parpinello, G. P., & Cattaneo, M. (1999). *Leuconostoc oenos* and malolactic fermentation in wine: A review. *J. Ind. Microbiol. Biotechnol., 23*, 447–455.

Walling, E., Gindreau, E., & Lonvaud-Funel, A. (2001). La biosnthèse d'exopolysaccharide par des souches de *Pediococcus damnosus* isolées du vin: Mise au point d'outils moléculaires de detection. *Lait, 81*, 289–300.

Walling, E., Gindreau, E., & Lonvaud-Funel, A. (2004). A putative glucan synthase gene *dps* detected in exopolysaccharide-producing *Pediococcus damnosus* and *Oenococcus oeni* strains isolated from wine and cider. *Int. J. Food. Microbiol., 98*, 53–62.

Walling, E., Gindreau, E., & Lonvaud-Funel, A. (2005). A putative glucan synthase gene dps detected in exopolysaccharide-producing Pediococcus damnosus and Oenococcus oeni strains isolated from wine and cider. *Int. J. Food Microbiol., 98*, 53–62.

Werning, M. L., Ibarburu, I., Dueñas, M. T., Irastorza, A., Navas, J., & López, P. (2006). *Pediococcus parvulus gtf* gene encoding the GTF glycosyltransferase and its application for specific PCR detection of β-D-Glucan-producing bacteria in foods and beverages. *J. Food Prot., 69*, 161–169.

Wibowo, D., Eschenbruch, R., Davis, C. R., Fleet, G. H., & Lee, T. H. (1985). Occurrence and growth of lactic acid bacteria in wine: A review. *Am. J. Enol. Vitic., 36*, 302–313.

Yurdugül, S., & Bozoglu, F. (2002). Studies on an inhibitor producer by lactic acid bacteria on wines on the control of malolactic fermentation. *Eur. Food Res. Technol., 215*, 38–41.

Zapparoli, G., Reguant, C., Bordons, A., Torriani, S., & Dellaglio, F. (2000). Genomic DNA fingerprinting of *Oenococcus oeni* strains by pulsed-field gel electrophoresis and randomly amplified polymorphic DNA-PCR. *Curr. Microbiol., 40*, 351–355.

Zaunmüller, T., Eichert, M., Ritcher, H., & Under, G. (2006). Variations in the energy metabolism of biotechnologically relevant heterofermentative lactic acid bacteria during growth on sugars and organic acids. *Appl. Microbiol. Biotechnol., 72*, 421–429.

Zavaleta, A. I., Martínez-Murcia, A. J., & Rodríguez-Valera, F. (1997). Intraspecific genetic diversity of *O. oeni* as derived from DNA fingerprinting and sequence analysis. *Appl. Environ. Microbiol., 63*, 1261–1267.

CHAPTER

# 9

# Acetic Acid Bacteria

*José M. Guillamón*[1], *Albert Mas*[2]

[1] Departamento de Biotecnología de los Alimentos, Instituto de Agroquímica y Tecnología de Alimentos (CSIC), Valencia, Spain, [2] Departamento de Bioquímica y Biotecnología, Facultad de Enología, Universitat Rovira i Virgili, Tarragona, Spain

## OUTLINE

*Molecular Wine Microbiology* **Doi: 10.1016/B978-0-12-375021-1.10009-8**

# 1. INTRODUCTION

Acidophilic bacteria can grow on substrates with a pH of less than 5 and are found in acidic foodstuffs such as fruit juice. They comprise two main groups: acetic acid bacteria and lactic acid bacteria. In grape must and wine, the low pH (between 3 and 4) and the presence of alcohol and/or high concentrations of sugar limit the microbial flora to just a few yeasts and bacteria. Among these, lactic acid bacteria play an important role in winemaking, since they are responsible for malolactic fermentation. In contrast, acetic acid bacteria are only linked to wine spoilage processes, mainly through the production of acetic acid, acetaldehyde, and ethyl acetate. This form of wine spoilage has been recognized since its initial description by Pasteur (1868), but winemakers still remain vigilant towards the risk of these bacteria causing an increase in volatile acidity at some point during the vinification process and producing what is widely known as "pricked" wine.

In this chapter, we will review the influence of acetic acid bacteria on winemaking and provide up-to-date information that will be of use in helping enologists to detect and control their growth and, thus, prevent spoilage of the final product. Acetic acid bacteria are found on grapes and in wine and must, and their growth depends on the phase in the winemaking process and the treatments that have been used. Although the literature available on these bacteria and their effects on winemaking is less extensive than that on the other two groups of organisms of relevance in winemaking (yeasts and lactic acid bacteria), some excellent, highly recommendable reviews have been published (Bartowsky & Henschke, 2008; Drysdale & Fleet, 1988; du Toit & Pretorius, 2002; Polo & Sánchez-Luengo, 1991).

# 2. GENERAL CHARACTERISTICS

Acetic acid bacteria are gram-negative or gram-variable, ellipsoid or cylindrical bacteria that appear under the microscope as individual cells, in pairs, in chains, or in clumps. Their size varies between 0.4 and 1 μm wide and between 0.8 and 4.5 μm long. They are clearly motile under the microscope and have polar or peritrichous flagella. They do not form endospores as resistant forms. They are aerobic and usually display respiratory metabolism with oxygen functioning as a terminal electron acceptor. Nevertheless, in unfavorable conditions (anaerobic or with low concentrations of oxygen), alternative electron acceptors can be used with a considerable associated slowing of bacterial metabolism and, therefore, growth. They are catalase-positive and oxidase-negative. The optimal temperature for growth is 25 to 30°C and the optimal pH is 5 to 6, although they can still grow well at pHs below 4 (de Ley et al., 1984). Some species produce pigments on solid growth medium and can produce different types of polysaccharides.

These bacteria are found in substrates containing sugar and/or ethanol, such as fruit juices, wine, cider, beer, and vinegar. On these substrates, bacterial metabolism involves incomplete oxidation of the sugars and alcohols and leads to accumulation of organic acids as end products. The production of acetic acid on ethanol-containing substrates accounts for the common name ascribed to these bacteria. However, these microorganisms are also able to oxidise glucose to gluconic acid, galactose to galactonic acid, and arabinose to arabonic acid. Some of these reactions are of significant interest to the winemaking industry. The traditional industrial application of acetic acid bacteria is in the production of vinegar; however, lesser-known applications include the production of cellulose and the conversion of sorbitol into sorbose.

# 3. NUTRITION AND METABOLISM

Acetic acid bacteria are obligate aerobes and, as a result, their growth is highly dependent upon the availability of molecular oxygen. Nevertheless, under conditions such as those found during winemaking (for instance alcoholic fermentation or aging), alternative terminal electron acceptors such as quinones can be used. Consequently, acetic acid bacteria can survive under the almost completely anaerobic conditions that are generally present during winemaking. Under those conditions, the bacteria may also display limited growth. This growth will be enhanced by any process involving aeration or oxygenation of the medium as a result of the increased levels of the principal electron acceptor.

The other important factor in the growth of these bacteria is the carbon source. This will determine which metabolic pathways are used and, therefore, which metabolic intermediates and end products will ultimately influence the quality of the wine.

## 3.1. Carbohydrate Metabolism

Acetic acid bacteria can metabolize various carbohydrates as carbon sources. As in other microorganisms, glucose acts as a carbon source for most strains of acetic acid bacteria. Unlike in yeasts, however, this glucose is not metabolized as part of glycolysis. Although most of the individual reactions are functional, the complete pathway is inactive as a result of the lack of the phosphofructokinase enzyme. Consequently, acetic acid bacteria must use alternative pathways in order to employ carbohydrates as sources of carbon and energy (see Figure 9.1):

1) The glucose-6-phosphate dehydrogenase system transforms a mole of glucose-6-phosphate into a mole of ribulose-5-phosphate with the formation of two moles of reduced nicotinamide adenine dinucleotide phosphate (NADPH) via three reactions. In all cases, the production of energy takes place as a consequence of the oxidation of NADPH via the respiratory chain and the production of adenosine triphosphate (ATP) by oxidative phosphorylation.

2) The Entner-Doudoroff pathway, which should be considered as an extension of the glucose-6-phosphate system, converts

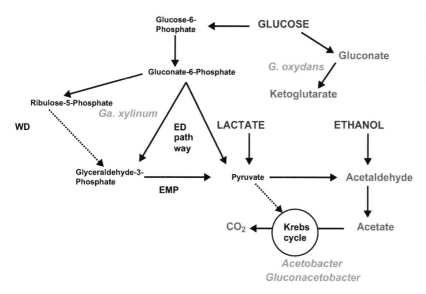

FIGURE 9.1 Schematic diagram of the metabolism of acetic acid bacteria. ED = Entner-Doudoroff; EMP = Embden-Meyerhoff-Parnas; WD = Warburg-Dickens.

glucose-6-phosphate into gluconate-6-phosphate. This is then hydrolyzed by an aldolase to form pyruvate and glyceraldehyde-3-phosphate. Both molecules are converted into acetate by decarboxylation. This reaction is characteristic of some strains of *Gluconobacter oxydans*.

3) The Warburg-Dickens pathway or hexose monophosphate cycle is the most common pathway for the metabolism of glucose and is present in all species of acetic acid bacteria. It includes the three reactions that comprise the glucose-6-phosphate dehydrogenase system and produces monosaccharides of varying size that are ultimately converted into triose phosphates, which can then be metabolized in the Embden-Meyerhoff-Parnas (EMP) pathway.

The end products of these pathways can be completely oxidized to produce carbon dioxide and water through the Krebs cycle. However, among the acetic acid bacteria only the genera *Acetobacter*, *Acidomonas*, and *Gluconacetobacter* are able to carry out this overoxidation of ethanol (Cleenwerck & de Vos, 2008). Species of the genus *Gluconobacter* do not have a functional Krebs cycle and are therefore unable to completely oxidize the molecules formed in the earlier metabolic pathways. In this genus, the lack of α-ketoglutarate and succinate dehydrogenases prevents completion of the cycle. The failure to oxidize acetic acid is common to *Saccharibacter* and *Neoasaia* species, whereas other genera display limited oxidation of this molecule. Those species that can carry out complete oxidation only do so through the Krebs cycle when all sugars or alcohols have been fully consumed. Consequently, the Krebs or tricarboxylic acid cycle will never be functional in winemaking processes, since the pathway is inhibited by glucose or fructose and ethanol (Saeki et al., 1997a).

Although acetic acid bacteria can completely degrade glucose, they characteristically display incomplete oxidation of carbon sources via one or two biochemical reactions, resulting in accumulation of intermediate metabolites. Although these partial oxidations are typical of substrates such as alcohols, they also occur with some monosaccharides such as glucose. Glucose is directly oxidized to glucono-δ-lactone, which is then oxidized to gluconic acid (see Figure 9.1). Although most acetic acid bacteria can carry out this reaction, it is especially active in *G. oxydans* and is characteristic in sugar-containing substrates such as grapes and must. Under these conditions, the accumulation of gluconic acid is typical of the growth of acetic acid bacteria. The use of glucose and, presumably, other sugars, either through the hexose monophosphate pathway or via direct oxidation to gluconic acid, depends on the pH and the glucose concentration of the medium. At pHs below 3.5 or glucose concentrations above 0.9 to 2.7 g/L, oxidation of glucose via the hexose monophosphate pathway is inhibited (Drysdale & Fleet, 1988; du Toit & Pretorius, 2002). The end result under these conditions would be the accumulation of gluconic acid in the culture medium.

## 3.2. Metabolism of Ethanol and Other Alcohols

Transformation of ethanol into acetic acid is the most well-known characteristic of acetic acid bacteria and by far the most relevant in winemaking. This transformation involves two biochemical reactions: ethanol is first transformed into acetaldehyde in a reaction catalyzed by alcohol dehydrogenase and the acetaldehyde is then transformed into acetic acid by aldehyde dehydrogenase. Both reactions involve electron transfer to molecular oxygen. The availability of other terminal electronic acceptors such as quinones (Drysdale & Fleet, 1988) may explain the survival and even limited growth of these bacteria under anaerobic conditions such as those found in bottled wine and even during alcoholic fermentation. *Acetobacter* alcohol dehydrogenase activity is more stable under winemaking conditions than that of *Gluconobacter*, which may explain the greater production of acetic acid by *Acetobacter*. Aldehyde dehydrogenase is more sensitive than alcohol dehydrogenase to the concentration of ethanol in the medium, and this could lead to greater accumulation of acetaldehyde in wines with a higher alcohol concentration (Muraoka et al., 1983). Drysdale and Fleet (1989a) also observed increasing concentrations of acetaldehyde as the concentration of oxygen dissolved in the wine diminished. The presence of excess sulfite also leads to an increase in the concentration of acetaldehyde as a consequence of the stable bond formed between these molecules, which prevents its transformation into acetic acid (Lafon-Lafourcade, 1985). Both alcohol and aldehyde dehydrogenase are located in the cell membrane with their active sites orientated outwards, and as a consequence the products of the reactions they catalyze are usually found in the medium and not inside the cell (Saeki et al., 1997b). Although a cytoplasmic form of alcohol dehydrogenase has been described, it has a much lower specific activity (Adachi et al., 1978).

Glycerol, the main byproduct of alcoholic fermentation, also acts as a carbon source for acetic acid bacteria (de Ley et al., 1984). Most of the glycerol is transformed into dihydroxyacetone (ketogenesis), although some may be used effectively for the production of biomass.

In a similar process to that seen with glycerol, this metabolic capacity of acetic acid bacteria is extended to the direct oxidation of other primary alcohols and polyalcohols, which are converted by oxidation into their respective ketones and ketoses. In this way, acetoin is produced from 2,3-butanediol or acetol from 1,2-propanediol. Some polyalcohols are converted into their corresponding sugars. Thus, fructose is produced from mannitol, sorbose from sorbitol, erythrose from erythritol, etc. Since most of the enzymes that catalyze these reactions are located in the cell membrane, a wide range of substrates accumulate in the medium and, as a result, acetic acid bacteria are particularly appropriate microorganisms for use in biotechnology (Deppenmeier et al., 2002).

## 3.3. Metabolism of Organic Acids

Although incomplete oxidation is a common metabolic characteristic of acetic acid bacteria, some substrates that are present in a state of intermediate oxidation, such as organic acids, can continue to be oxidized. Lactic acid is a good carbon source for many acetic acid bacteria and it can be oxidized through various pathways with different end products. In one of these, lactic acid is oxidized to pyruvate, which is then hydrolyzed to acetaldehyde and carbon dioxide in a reaction catalyzed by pyruvate decarboxylase. Acetaldehyde is then oxidized to form the end product, acetic acid. The activity of this pyruvate decarboxylase is dependent on the predominant substrate in the culture medium. This activity is not detected in the presence of mannitol, whereas it is maximal in the presence of lactic acid (Raj et al., 2001). An alternative to this pathway is the production of acetoin from lactic acid via an acetolactate intermediate (de Ley, 1959).

Complete oxidation of many organic acids is dependent on the availability of a functional Krebs cycle. Since *Gluconobacter* is unable to completely oxidize acetic acid, it is also unable to completely oxidize many organic acids. Acetic acid is oxidized in a reaction catalyzed by acetyl-coenzyme A (CoA) synthase, which leads to the production of actetyl-CoA (Saeki et al., 1997b). Acetyl-CoA enters the Krebs cycle and is converted into intermediate metabolites in the pathway before oxidation is completed with the generation of carbon dioxide and water. The presence of acetic acid leads to an increase in the activity of acetyl-CoA synthase, but this activity is strongly inhibited by ethanol and glucose.

## 3.4. Nitrogen Metabolism

Some species of acetic acid bacteria can fix atmospheric nitrogen. This was first described some years ago for *Gluconoacetobacter diazotrophicus* (Gillis et al., 1989), and six more species have recently been reported to display this capacity: *Gluconoacetobacter johannae, Gluconoacetobacter azotocaptans, Acetobacter peroxydans, Swaminathania salitolerans, Acetobacter nitrogenifigens,* and *Gluconoacetobacter kombuchae,* the seventh and last reported to date (Dutta & Gachhui, 2007). Other species use ammonia as a simpler nitrogen source (de Ley et al., 1984). Thus, these bacteria can synthesize all of their amino acids and nitrogen compounds from ammonia. The presence of amino acids in the growth medium can have a stimulatory or inhibitory effect on growth, depending on the amino acid. Thus, glutamate, glutamine, proline, and histidine stimulate the growth of acetic acid bacteria, whereas valine in the case of *G. oxydans* and threonine and homoserine in the case of *Acetobacter aceti* appear to inhibit growth (Belly & Claus, 1972). However, no studies have addressed the nutritional requirements of acetic acid bacteria in terms of nitrogenous compounds under winemaking conditions. A preference for certain amino acids has been observed for acetic acid bacteria during the production of vinegar (Maestre et al., 2008; Valero et al., 2005), in some cases leaving substantial amounts of ammonia in the medium. The preferential use of proline is noteworthy, as this amino acid is not used by yeast under fermentative conditions and is therefore particularly abundant in grape must and wine. The release of amino acids and other nitrogenous compounds following yeast autolysis may also be sufficient to make a significant contribution to the growth of acetic acid bacteria in wine. In fact, while additional nutrients must be added during the production of vinegars from alcohol or substrates low in amino acids (such as cider; Valero et al., 2005), this is not generally necessary for wines.

## 4. TAXONOMY

The first nomenclature for the classification of acetic acid bacteria is attributed to Peerson in 1822 with the proposal of the name *Mycoderma* for this group of microorganisms. Pasteur (1868) carried out the first systematic study of acetic fermentation. He recognized that the "vinegar mother" was a mass of live microorganisms that induced acetic acid fermentation and that this was not possible in the absence of *Mycoderma aceti.* Later, around 1879, Hansen observed that the microbial flora that converted alcohol into acetic acid was not pure and comprised various bacterial species. The genus *Acetobacter* was first proposed by Beijerinck (1899).

Bacterial taxonomy has traditionally been based on morphological, biochemical, and physiological criteria. The first to propose a classification of acetic acid bacteria based on these criteria was Visser't Hooft (1925). Asai (1935) formulated the proposal to classify acetic acid bacteria into two genera: *Acetobacter* and *Gluconobacter.* Later, Frateur (1950) proposed a classification based essentially on five physiological criteria: catalase activity, production of gluconic acid from glucose, oxidation of acetic acid to carbon dioxide and water, oxidation of lactic acid to carbon dioxide and water, and oxidation of glycerol into dihydroxyacetone. Based on these criteria, he proposed the subdivision of *Acetobacter* into four groups: *peroxydans, oxydans, mexosydans,* and *suboxydans* (reviewed by Barja et al., 2003).

The history of the taxonomic criteria applied to bacterial species is provided in the different editions of Bergey's *Manual of Determinative Bacteriology,* which has become a reference for bacterial taxonomy. The eighth edition of this manual (Buchanan & Gibbons, 1974) recognized two genera—*Acetobacter* (able to convert acetate and lactate into carbon dioxide and water and motile through peritrichous flagella or nonmotile) and *Gluconobacter* (unable to oxidize lactate and acetate completely and motile through polar flagella or nonmotile)—and placed the genus *Gluconobacter* within the *Pseudomonadaceae* family. The genus *Acetobacter* was not assigned to any family and was placed with the genera of unknown affiliation. In 1984, a new edition of the manual was published under a different title: *Bergey's Manual of Systematic Bacteriology*. This new edition included among the taxonomic criteria some molecular tests, such as fatty acid composition, electrophoresis of soluble proteins, guanine-cytosine (GC) content, and DNA—DNA hybridization. These techniques suggested that the genera *Gluconobacter* and *Acetobacter* had extensive phylogenetic similarity and, as a result, in this new edition of Bergey's Manual (de Ley et al., 1984) the two genera were included in the family *Acetobacteraceae*, which had previously been included in the division α-*Proteobacteria* (Stackebrandt et al., 1988). The genus *Acetobacter* included four species: *A. aceti*, *Acetobacter pasteurianus*, *Acetobacter liquefaciens*, and *Acetobacter hansenii*. The genus *Gluconobacter* included only one species: *G. oxydans*. The main difference between the two genera continued to be that *Acetobacter* species are able to perform complete oxidation of ethanol—in other words, that ethanol can be oxidized to acetic acid and then to carbon dioxide and water—whereas *Gluconobacter* is unable to oxidize acetic acid completely to carbon dioxide plus water.

The taxonomy of microorganisms has been continually revised and reorganized, mainly based on data obtained using molecular techniques such as DNA—DNA or DNA—RNA hybrization and analysis of 16S rDNA. The *Acetobacteraceae* family has not been exempt

from this process of reordering genera and species. Nine new genera of acetic acid bacteria must be added to the two mentioned earlier: *Acidomonas* (Urakami et al., 1989), *Gluconacetobacter* (Yamada et al., 1997), *Asaia* (Yamada et al., 2000), *Kozakia* (Lisdiyanti et al., 2002), *Saccharibacter* (Jojima et al., 2004), *Swaminathania* (Loganathan & Nair, 2004), *Neoasaia* (Yukphan et al., 2005), *Granulibacter* (Greenberg et al., 2006), and *Tanticharoenia* (Yukphan et al., 2008). In addition, a number of new species have been identified. Thus, the family *Acetobacteraceae* currently comprises 11 genera and 56 species (58 if we include newly proposed species that have yet to be accepted) of acetic acid bacteria (see Table 9.1). *Acetobacter* and *Gluconacetobacter*, which contain 19 and 16 species,

**TABLE 9.1** Genera and Species of Acetic Acid Bacteria (June 2009)

| Genus | Species |
|---|---|
| *Acetobacter* | *Acetobacter aceti* |
| | *Acetobacter pasteurianus*[5] |
| | *Acetobacter pomorum* |
| | *Acetobacter peroxydans* |
| | *Acetobacter indonesiensis* |
| | *Acetobacter tropicalis* |
| | *Acetobacter syzygii* |
| | *Acetobacter cibinongenesis* |
| | *Acetobacter orientalis* |
| | *Acetobacter orleaniensis* |
| | *Acetobacter lovaniensis* |
| | *Acetobacter estuniensis* |
| | *Acetobacter malorum* |
| | *Acetobacter cerevisiae* |
| | *Acetobacter oeni* |
| | *Acetobacter nitrogenifigens* |
| | *Acetobacter senegalensis* |
| | *Acetobacter ghanensis* |
| | *Acetobacter fabarum* |
| *Acidomonas* | *Acidomonas methanolica* |
| *Asaia* | *Asaia bogorensis* |
| | *Asaia siamensis* |
| | *Asaia krugthepensis* |
| | *Asaia lannensis* |

*(Continued)*

**TABLE 9.1**   Genera and Species of Acetic Acid
Bacteria (June 2009) (*cont'd*)

| Genus | Species |
|---|---|
| *Gluconacetobacter* | **Gluconacetobacter liquefaciens** |
|  | *Gluconacetobacter diazotrophicus*[4] |
|  | **Gluconacetobacter xylinus**[5] |
|  | **Gluconacetobacter hansenii** |
|  | **Gluconacetobacter europaeus** |
|  | **Gluconacetobacter oboediens** |
|  | **Gluconacetobacter intermedius** |
|  | *Gluconacetobacter sacchari* |
|  | **Gluconacetobacter entanii** |
|  | *Gluconacetobacter johannae* |
|  | *Gluconacetobacter azotocaptans* |
|  | *Gluconacetobacter swingsii* |
|  | *Gluconacetobacter kombuchae* |
|  | *Gluconacetobacter nataicola* |
|  | *Gluconacetobacter rhaeticus* |
|  | *Gluconacetobacter saccharivorans* |
|  | *Gluconacetobacter persimmonis*[1] |
| *Gluconobacter* | **Gluconobacter oxydans**[2] |
|  | *Gluconobacter frateurii* |
|  | *Gluconobacter assaii* |
|  | *Gluconobacter cerinus* |
|  | *Gluconobacter albidus* |
|  | *Gluconobacter thailandicus* |
|  | *Gluconobacter kondonii* |
|  | *Gluconobacter roseus* |
|  | *Gluconobacter sphaericus* |
|  | *Gluconobacter japonicus* |
|  | *Gluconobacter kanchanaburiensis* |
|  | *Gluconobacter wancherniae* |
| *Granulibacter* | *Granulibacter bethesdensis*[3] |
| *Kozakia* | *Kozakia baliensis* |
| *Neoasaia* | *Neoasaia chiangmaiensis* |
| *Saccharibacter* | *Saccharibacter floricola* |
| *Swaminathania* | *Swaminathania salitolerans* |
| *Tanticharoenia* | *Tanticharoenia sakaeratensis* |

[1]*Newly proposed species that have yet to be accepted.*
*Acetic acid bacteria with sequenced genomes:*
[2]*Prust, C. et al. (2005);*
[3]*Greenberg, D. E. et al. (2007);*
[4]*Bertalan et al. (2009);*
[5]*Azuma (2008).*
Species described in grapes and wine or wine vinegar are shown in
bold.

respectively, show the greatest diversity among the species that have been described. The genus *Acetobacter* differs biochemically from *Gluconoacetobacter* in that it produces ubiquinone-9 (Q-9) rather than ubiqiuinone-10 (Q-10), which is found in *Gluconoacetobacter* species (Yamada et al., 1997). This ubiquinone Q-10 is common to all other genera of acetic acid bacteria (Yamada & Yukphan, 2008). The genus *Acidomonas*, with its single species *Acidomonas methanolica*, was characterized as the only genus able to grow in methanol as a unique carbon source (Urakami et al., 1989), although the recently described *Granulibacter* also has this capacity (Greenberg et al., 2006). The characteristic features of strains assigned to the genera *Asaia* and *Swaminathania* are their production of little or no acetic acid from ethanol and their failure to grow in the presence of more than 0.35% acetic acid. The strains of these genera have mainly been isolated from flowers (Yamada et al., 2000) and rice (Loganathan & Nair, 2004).

The other genera (*Kozakia*, *Saccharibacter*, *Neoasaia*, *Granulibacter*, and *Tanticharoenia*) each contain only a single species. These bacteria, which have generally been described in soils, flowers, and fruits from Asian countries, display very limited production of acetic acid from ethanol and have a very weak capacity to overoxidize acetate or lactate (Cleenwerck & de Vos, 2008). An exception worth mentioning is *Granulibacter*, for which the only species has been isolated in hospital environments in the United States in patients with chronic granulomatous disease, a rare hereditary condition that is characterized by the accumulation of superoxides and hydrogen peroxide, which facilitate the growth of parasites with catalase activity (Greenberg et al., 2006). Some species are also highly osmophilic and able to develop in media containing high concentrations of sugar equivalent to more than 30% glucose (*Acidomonas*, *Asaia*, *Neoasaia*, *Saccharibacter*, and *Tanticharoenia*) (Cleenwerck & de Vos, 2008; Yukphan et al., 2008).

Finally, the genus *Frateuria*, which contains a single species, *Frateuria aurantia*, belongs to the γ-*Proteobacteria*, but it has an oxidative metabolism similar to that of the acetic acid bacteria and is therefore usually considered to be a pseudo-acetic-acid bacteria (Yamada & Yukphan, 2008). It was initially named *Acetobacter aurantia* and characterized on the basis of being positive for ubiquinone Q-8, which is absent from all acetic acid bacteria.

Our understanding of both the taxonomy and the biochemical and physiological characteristics of acetic acid bacteria will be increased substantially when a complete genome sequence becomes available for the different species in this bacterial group. To date, sequences have only been published for *G. oxydans* (Prust et al., 2005), *Granulibacter bethesdensis* (Greenberg et al., 2007), and *Ga. diazotrophicus* (Bertalan et al., 2009). The sequencing of another two species (*A. pasteurianus* and *Gluconacetobacter xylinus*) was reported by the Osaka Institute for Fermentation in Japan at the Second Congress on Acetic Acid Bacteria in Nagoia, Japan, in November 2008, but the sequence is not publicly available.

## 4.1. Isolation

Acetic acid bacteria have traditionally been considered to be fastidious due to their poor recovery from culture media. As a result, various media have been developed for their isolation. In our laboratory, to isolate acetic acid bacteria from grapes, must, and wine, we mainly use GYC agar (5% D-glucose, 1% yeast extract, 0.5% calcium carbonate, and 2% agar), which was first described by Carr and Passmore (1979). The presence of calcium carbonate gives the medium an opaque appearance. The production of acetic acid by the bacteria causes the calcium carbonate to dissolve and a transparent halo to form around the colony. This medium is also adjusted to a pH of 4.5 and is supplemented with natamycin (100 mg/L) to inhibit the growth of yeasts and

fungi and with penicillin (3 U/mL) to inhibit the growth of lactic acid bacteria (Ruiz et al., 2000). Mannitol medium (2.5% mannitol, 0.5% yeast extract, 0.3% peptone, and 2% agar) yields very similar results in the isolation of acetic acid bacteria from enological samples. This liquid medium is particularly useful for the production of biomass, since it is the best medium to support the growth of acetic acid bacteria. The plates are incubated for 2 to 4 d at 28°C under aerobic conditions. Other authors have reported using similar media for the growth of acetic acid bacteria derived from enological samples (Bartowsky et al., 2003; du Toit & Lambrechts, 2002). Bartowsky et al. (2003) also recommend the use of Wallerstein nutrient agar supplemented with 2% ethanol or 10% filter-sterilized wine.

No problems have been reported in the isolation and culture of these bacteria when derived from samples obtained during the winemaking process. However, differences have been observed in bacterial recovery on plates and numbers observed under the microscope, and these have been attributed to the presence of viable but noncultivable bacteria (Millet & Lonvaud-Funel, 2000). In contrast, a number of studies have reported difficulty culturing these bacteria from samples obtained during vinegar production (González et al., 2006a; Ilabaca et al., 2008; Sokollek et al., 1998; Trcek, 2005). This problem has been partially resolved by the introduction of a double layer of agar in the medium (0.5% in the bottom layer and 1% in the top layer) and the use of culture media that simulate the acetifying environment, such as AE medium containing ethanol and acetic acid (Entani et al., 1985).

## 4.2. Identification

Until recently, the classification of bacterial species was based on morphological, biochemical, and physiological criteria. These phenotypic characteristics have also been used to assign isolates to specific genera and species. The more general characteristics used for the phenotypic classification of acetic acid bacteria include cell and colony morphology, Gram staining, and catalase and oxidase activity. As summarized by Cleenwerck and de Vos (2008), other tests that have been useful for the identification of acetic acid bacteria and even the differentiation of genera and species include the following: (1) the production of acetic acid from ethanol; (2) overoxidation of lactate and acetate to carbon dioxide and water; (3) growth in the presence of 0.35% acetic acid; (4) growth in 1% nitric acid; (5) formation of 2-ketogluconic, 5-ketogluconic, and 2,5-ketogluconic acid from glucose; (6) ketogenesis of glycerol; (7) growth on different carbon sources (e.g., methanol); (8) formation of brown water-soluble pigments; (9) formation of γ-pyrones from glucose or fructose; (10) production of acids from sugars; (11) production of cellulose; (12) growth in 30% glucose; (13) presence and position of flagella; and (14) motility. These characteristics can be used to discriminate between different genera and species of acetic acid bacteria, as shown in Table 9.2. Many of the tests yield variable results between different species and even between different strains of the same species. This complicates the identification process significantly and makes it necessary to use a larger number of additional tests in order to reliably identify an isolate.

The same characteristics can be used to identify acetic acid bacterial isolates. Bartowsky et al. (2003) identified wine isolates using a few phenotypic tests that can be considered specific for the acetic acid bacteria found in wine. These tests, which were designed to discriminate between the five species of acetic acid bacteria described in *Bergey's Manual of Systematic Bacteriology* (de Ley et al., 1984), include Gram staining; catalase test; growth in ethanol, sodium acetate, and dulcitol; overoxidation of ethanol; ketogenesis of glycerol; oxidation of lactate; and production of water-soluble brown pigments.

A good alternative to identification based on phenotypic characteristics is the use of genotypic and molecular criteria. Developments in genetics and molecular biology have allowed polymorphism and variability in certain molecules, mainly DNA, to be used for taxonomic characterization and identification. As a result, there has been a marked increase in the number of genera and species of acetic acid bacteria in recent years through the use of molecular techniques such as DNA–DNA hybridization and sequencing of ribosomal genes. Nevertheless, despite their effectiveness for the classification of bacterial species, these techniques are not appropriate for routine use and are too complex to be applied to the processing of large numbers of samples.

## 4.3. Molecular Techniques for the Rapid Identification of Acetic Acid Bacteria

One of the first techniques used in bacterial taxonomy was the analysis of the percentage of GC base pairs. The GC content was already included for the species of *Acetobacteraceae* in *Bergey's Manual of Systematic Bacteriology* (de Ley et al., 1984). However, GC content alone is insufficient for the identification of an isolate. Although the GC content of acetic acid bacteria ranges from 52 to 67%, most species include strains with percentages around the middle of this range (e.g., 59%). Nevertheless, this criterion is used alongside DNA–DNA hybridization and various morphological and physiological characteristics for the classification of new species, and it forms the basis of so-called polyphasic taxonomy (Cleenwerck & de Vos, 2008).

Sequencing is one of the methods proposed for the identification of acetic acid bacteria and generally involves ribosomal genes or the region between the 16S and 23S ribosomal genes (Yamada & Yukphan, 2008). The latter involves intergenic regions known as internal transcribed

**TABLE 9.2** Phenotypic Characteristics of the Different Genera of Acetic Acid Bacteria

| Characteristic | A | G | Ac | Ga | As | K | S | Sa | N | Gr | T |
|---|---|---|---|---|---|---|---|---|---|---|---|
| Production of acetate from ethanol | + | + | + | + | − | + | + | w/− | + | w | + |
| Oxidation of acetate to carbon dioxide and water | + | − | + | + | w | w | w | − | − | w | − |
| Growth in 0.35% acetic acid | + | + | + | + | − | + | + | − | + | nd | + |
| Growth in 1% nitric acid | − | − | + | − | − | − | + | nd | − | nd | − |
| Formation of gluconic acid from glucose | + | + | − | + | + | + | nd | nd | nd | nd | nd |
| Ketogenesis of glycerol | + | + | − | + | w | w | + | − | w | − | + |
| Growth in methanol | − | − | + | − | − | − | − | − | − | + | − |
| Formation of brown, water-soluble pigments | − | − | − | + | − | − | + | − | − | nd | + |
| Formation of γ-pyrones from glucose | − | + | nd | + | − | − | nd | nd | nd | nd | nd |
| Production of acids from fructose | − | + | − | + | + | − | + | + | + | nd | nd |
| Production of cellulose | − | − | − | + | − | − | − | nd | nd | nd | nd |
| Growth in 30% glucose | − | − | + | − | + | w | + | + | + | nd | + |
| Presence and location of flagella | per | pol | pol | per | per | − | per | − | − | − | − |
| Majority ubiquinone | Q9 | Q10 | Q10 | Q10 | Q10 | Q10 | Q10 | Q10 | Q10 | nd | Q10 |
| GC content (%) | 52–64 | 54–64 | 62–63 | 56–67 | 59–61 | 56–57 | 52–53 | 57–60 | 63 | 59 | 66 |

*A = Acetobacter; G = Gluconobacter; Ac = Acidomonas; Ga = Gluconacetobacter; As = Asaia; K = Kozakia; S = Swaminathania; Sa = Saccharibacter; N = Neoasaia; Gr = Granulibacter; T = Tanticharoenia. + = positive; − = negative; nd = not determined; per = peritrichous; pol = polar; w = weak.*
*Adapted from Cleenwerck and de Vos (2008) and Yukphan et al. (2008).*

spacers (ITSs). Although these are transcribed as part of the group of ribosomal genes, they are later eliminated from the final ribosomal RNA molecules. Sequencing of the *RecA* gene has also been proposed (Cleenwerck & de Vos, 2008). However, despite the reduction in the cost of DNA sequencing in recent years, its systematic use in descriptive ecological studies is not practical given the hundreds of samples that may easily be involved. Consequently, approaches are needed that can be used to detect and group different microorganisms using simple, rapid, and inexpensive techniques. These techniques are generally based on the analysis of fragments of ribosomal genes using electrophoretic or similar methods that allow comparison of patterns with those from known samples such as type strains for each species. These techniques are not useful for precise taxonomic identification but, when combined with sequencing of representative samples for each group, they provide a highly accurate means of confirming identification.

Our group has developed various molecular techniques for the rapid and reliable identification and, in some cases, quantification of most species of acetic acid bacteria, particularly those present in grapes, wine, and vinegar. The first method involves restriction analysis of the 16S ribosomal gene following amplification by polymerase chain reaction (PCR) using a protocol known as 16S-ARDRA (Poblet et al., 2000;

Ruiz et al., 2000). The method was fine-tuned as follows (Ruiz et al., 2000):

1) Design of specific primers for the amplification of 16S ribosomal DNA (rDNA) based on sequences from acetic acid bacteria present in sequence databases.
2) PCR amplification of 16S rDNA from acetic acid bacteria. To confirm the specificity, different reference strains of acetic acid bacteria, lactic acid bacteria, and yeasts were used. Only the acetic acid bacteria displayed the characteristic 1450-base-pair amplification product predicted from sequence data.
3) Digestion of the amplification products with different restriction enzymes. Of all the enzymes tested, *Taq*I and *Rsa*I produced the best results for species identification (see Table 9.3). Two groups of species could not be distinguished with any of the enzymes tested: one formed by *Gluconoacetobacter liquefaciens*, *Ga. xylinus*, and *Gluconoacetobacter europaeus* and the other by *Gluconobacter frateurii* and *Gluconobacter asaii*. In order to differentiate between these groups and between other newly described species, a system was proposed involving the sequential use of different restriction enzymes to allow species to be grouped and distinguished from others according to the patterns obtained (González et al., 2006b).

The system described above was designed for the identification of isolates of acetic acid bacteria derived from musts and wines. Isolates from industrial fermentations have also been identified by comparison of their restriction profiles with those obtained from reference strains. All of the isolates had profiles that were identical to those described previously for the different species. The restriction patterns with *Taq*I for some of these wine strains are shown in Figure 9.2. The technique has also been used for the identification of acetic acid bacteria present in grapes (Prieto et al., 2007), wines (González et al., 2004, 2005), vinegars (Ilabaca et al., 2008),

and during the production of traditional balsamic vinegar (Giudici et al., 2003).

Primers have also been designed for amplification of the 16S-23S ITS (Ruiz et al., 2000). These regions usually display greater sequence variability than the ribosomal genes and, therefore, restriction analysis reveals more extensive polymorphism or variability between the strains and species analyzed. However, amplification of the ITS region from the same reference strains followed by digestion with the same restriction enzymes does not result in greater species differentiation. This technique was used by Sievers et al. (1996) to differentiate between two species of *Acetobacter*. Later, Trcek and Teuber (2002) used digestion with the same restriction enzymes to characterize 57 strains of acetic acid bacteria. Those authors were also unable to differentiate between strains of *Ga. xylinus* and *Ga. europaeus*, but they were able to distinguish strains of *Ga. liquefaciens*. The technique has also been used to identify isolates from grapes in Chile (Prieto et al., 2007). Gullo et al. (2006) used a combination of the two previous techniques to identify isolates from traditional balsamic vinegar. The entire region comprising the 16S-23S-5S genes and the ITS regions was amplified. However, the length of the amplicon (approximately 4500 base pairs) can represent a technological barrier in this method.

All of these techniques have been used on plate isolates and are therefore also affected by the problems mentioned earlier in relation to the culture of acetic acid bacteria. The method has occasionally been used on noncultured samples, but with inconclusive results when multiple species were present (Ilabaca et al., 2008). Our group has circumvented the problem of culturing acetic acid bacteria by directly quantifying acetic acid bacteria on different substrates using real-time quantitative PCR. Briefly, this technique allows the amplification process to be monitored continually and a threshold cycle ($C_T$) to be determined, indicating the point at which exponential

**TABLE 9.3**  Size of Restriction Fragments Obtained with TaqI and RsaI Following Digestion of the 16S Ribosomal Gene

| Strain | *Taq*I | *Rsa*I |
|---|---|---|
| G. oxydans LMG 1408[T] | 350 +190 + 175 + 160 + 120 + 120 + 110 | 400 + 400 + 400 + 150 + 90 |
| G. oxydans CECT 360 | 350 + 190 + 175 + 160 + 120 + 120 + 110 | 400 + 400 + 400 + 150 + 90 |
| G. oxydans LMG 1484 | 350 + 190 + 175 + 160 + 120 + 120 + 110 | 400 + 400 + 400 + 150 + 90 |
| G. oxydans LMG 1414 | 350 + 190 + 175 + 160 + 120 + 120 + 110 | 400 + 400 + 400 + 150 + 90 |
| G. frateurii LMG 1365[T] | 350 + 190 + 175 + 160 + 120 + 120 + 110 | 400 + 400 + 300 + 150 + 130 |
| G. asaii LMG 1390[T] | 350 + 190 + 175 + 160 + 120 + 120 + 110 | 400 + 400 + 300 + 150 + 130 |
| A. aceti LMG 1261[T] | 850 + 350 + 210 | 500 + 400 + 300 + 150 + 125 |
| A. aceti CECT 298[T] | 850 + 350 + 210 | 500 + 400 + 300 + 150 + 125 |
| A. aceti LMG 1505 | 850 + 350 + 210 | 500 + 400 + 300 + 150 + 125 |
| A. aceti LMG 1372 | 850 + 350 + 210 | 500 + 400 + 300 + 150 + 125 |
| A. pasteurianus LMG 1262[T] | 500 + 350 + 330 + 210 | 500 + 400 + 300 + 150 + 125 |
| A. pasteurianus LMG 1553 | 500 + 350 + 330 + 210 | 500 + 400 + 300 + 150 + 125 |
| Ga. hansenii LMG 1527[T] | 650 + 350 + 210 + 175 | 500 + 400 + 400 + 150 |
| Ga. liquefaciens LMG 1381[T] | 500 + 350 + 210 + 175 + 160 | 500 + 400 + 400 + 150 |
| Ga. liquefaciens LMG 1347 | 500 + 350 + 210 + 175 + 160 | 500 + 400 + 400 + 150 |
| Ga. xylinus LMG 1515[T] | 500 + 350 + 210 + 175 + 160 | 500 + 400 + 400 + 150 |
| Ga. xylinus LMG 1518 | 500 + 350 + 210 + 175 + 160 | 500 + 400 + 400 + 150 |
| Ga. europaeus DSM 6160[T] | 500 + 350 + 210 + 175 + 160 | 500 + 400 + 400 + 150 |
| Ga. diazotrophicus DSM 5601[T] | 500 + 350 + 210 + 175 + 160 | 500 + 400 + 250 + 150 + 150 |
| Ac. Methanolica LMG 1668[T] | 450 + 190 + 175 + 160 + 120 + 110 | 500 + 400 + 400 + 150 |

A. = Acetobacter; Ac. = Acidomonas; G. = Gluconobacter; Ga. = Gluconacetobacter.
Adapted from Ruiz et al. (2000).

amplification begins. This will be proportional to the amount of DNA present in the original sample, and therefore standard curves can be established to plot the number of cycles against the concentration of DNA (or, in fact, cells) present in the sample. The specificity or identity of the cells or DNA amplified will be determined by the specificity of the primers. We have also designed primers for the simultaneous identification and quantification of acetic acid bacteria (González et al., 2006a) that have been successfully used in wine (Andorrà et al., 2008) and vinegars (Jara et al., 2008). Likewise, specific probes have been designed for use in real-time PCR (TaqMan probes) for the most common species found on grapes and in wine and vinegar (Torija et al., 2009). This method has allowed simultaneous identification and quantification in these substrates and has been successfully used for the identification and

**FIGURE 9.2** *Taq*I restriction profiles following amplification of 16S ribosomal DNA in different strains of acetic acid bacteria isolated during alcoholic fermentation. All strains from the same species displayed the same profile. m = molecular weight marker (100-base-pair ladder; Gibco-BRL).

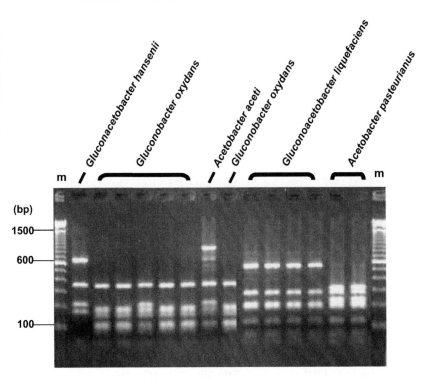

## 4.4. Molecular Techniques for Typing of Acetic Acid Bacteria

quantification of the species present in vinegar made using traditional methods (Jara, 2009). We have also used other techniques that do not require culture, such as cloning of amplified DNA from the 16S gene in *Escherichia coli* and subsequent analysis of transformed colonies by restriction fragment length polymorphism (RFLP) (Ilabaca et al., 2008). In this technique, although the problem of culture is resolved, that of transformation efficiency is introduced, since not all fragments are amplified similarly. Finally, denaturing gradient gel electrophoresis (DGGE) and temperature gradient gel electrophoresis (TGGE) can be used in combination with PCR to separate amplified fragments of the 16S gene according to small sequence differences. This technique has been used successfully for the identification of acetic acid bacteria in wine (Andorrà et al., 2008; López et al., 2003) and vinegars (de Vero et al., 2006; Gullo et al., 2009; Haruta et al., 2006; Ilabaca et al., 2008).

The main objective of any form of microbial classification is to identify isolates in terms of the species to which they belong, this being the fundamental taxonomic unit. However, the discrimination or typing of different strains or genotypes from a given species plays an increasingly important role in industrial applications, since not all strains of the same species generate similar changes in the product. Molecular techniques have the advantage that they allow within-species discrimination. In industrial winemaking, typing of *Saccharomyces* strains has been important in the selection of autochthonous strains for use in industrial starter cultures during alcoholic fermentation (Querol et al., 1992). Although acetic acid bacteria do not have desirable biotechnological properties for use in winemaking, the selection of strains for use in the vinegar industry has been of particular interest (Sokollek et al., 1998). Strain typing has therefore focused on the microbial characterization of the acetification process (Teuber et al., 1987). In the study by Teuber et al. (1987), and also in subsequent studies (Mariette et al., 1991; Sokollek et al., 1998), the plasmid profile of each of the strains was used as a molecular marker. Ohmori et al. (1982) demonstrated the presence of plasmids in strains isolated during vinegar production, and Teuber et al. (1987) considered the plasmid profile to be characteristic of the strain.

Nanda et al. (2001) have also characterized strains of acetic acid bacteria isolated in vinegar, although in this case the samples were from rice vinegar. Two PCR-based methods have been used in this type of characterization: enterobacterial repetitive intergenic consensus (ERIC) PCR and random amplification of polymorphic DNA (RAPD). RAPD involves the use of short oligonucleotides (9—10 nucleotides) that hybridize with random sequences in the bacterial genome; the number and size of the fragments amplified is different for each strain. This technique has also been used recently by Bartowsky et al. (2003) to type strains of acetic acid bacteria isolated from bottled wine with symptoms of being pricked. Our group has used the ERIC-PCR technique alongside another technique known as repetitive extragenic palindrome (REP) PCR to study the strain diversity of acetic acid bacteria during alcoholic fermentation (González et al., 2004). ERIC and REP elements were described as consensus sequences derived from repetitive sequences spread throughout the bacterial chromosome of enterobacter species (Versalovic et al., 1991). However, these sequences have been found in other bacterial groups. The technique involves the use of primers to amplify sequences between these repetitive regions. By employing specific oligonucleotides against the DNA of strains from different species of acetic acid bacteria, characteristic patterns can be obtained

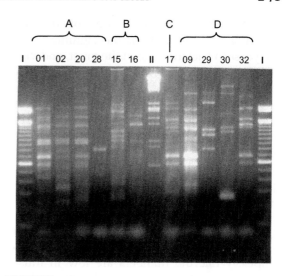

**FIGURE 9.3** Patterns obtained for strains of different species of acetic acid bacteria by enterobacterial repetitive intergenic consensus polymerase chain reaction. A = *Gluconobacter oxydans*; B = *Gluconoacetobacter liquefaciens*; C = *Gluconoacetobacter hansenii*; D = *Acetobacter aceti*.

in each case (see Figure 9.3). Therefore, these techniques can be considered appropriate for analysis of variability beyond the species level in acetic acid bacteria. This technique has also been used for monitoring the growth dynamics of acetic acid bacteria in vinegars (Gullo et al., 2009). A final technique for the typing of acetic acid bacteria involves the use of repetitive $(GTG)_5$ sequences (abundant in all genomes) for the monitoring and characterization of acetic acid bacteria present during fermentation of cocoa beans (Camu et al., 2007).

# 5. GROWTH OF ACETIC ACID BACTERIA IN WINEMAKING PROCESSES

According to the excellent review by Drysdale and Fleet (1988), "further studies are needed to more accurately determine the growth behavior of acetic acid bacteria during the different stages of vinification. Such studies should examine the influences of different conditions of vinification, the significance of the yeast-to-bacteria ratio in the must, and variations in the behavior of different species and strains of acetic acid bacteria." Despite the time that has passed, few studies have been published on this subject. However, a series of conclusions can be drawn from those that have.

## 5.1. Association of Acetic Acid Bacteria With Grapes

The number of acetic acid bacteria present in must is directly proportional to the health of the grapes. While the numbers are low in musts derived from healthy grapes (no more than $10^2-10^3$ colony-forming units [CFU]/mL), they increase by various orders of magnitude in damaged grapes and grapes infected with the fungus *Botrytis cinerea* (Joyeux et al., 1984). Joyeux et al. (1984) also identified the majority species found on the grapes. *G. oxydans* was the main species isolated from healthy grapes and those not infected with *Botrytis*, whereas *A. aceti* and *A. pasteurianus* predominated in unhealthy grapes. As discussed in Section 5.2, *G. oxydans* has a poor tolerance of ethanol. Therefore, synthesis of this compound by damaged grapes or grapes at various stages of rotting may explain the replacement of this species with *Acetobacter* species, which have a much greater alcohol tolerance. In our group, we have carried out counts of acetic acid bacteria in successive harvests and our results support the conclusions of earlier studies. In a harvest in which the grapes were extraordinarily healthy, the bacterial counts in the must did not exceed $10^3$ CFU/mL and more than 80% of the identified colonies belonged to *G. oxydans*. In contrast, after a very wet summer that led to substantial rotting and *Botrytis* infection, the counts in the corresponding musts reached around $10^6$ CFU/mL and the predominant species was *A. aceti* (González et al., 2004,

2005). All of the isolates were genotyped and *G. oxydans* was found to display considerable strain diversity, whereas the diversity was substantially lower in isolates belonging to *A. aceti* (González et al., 2005). We also recovered *Ga. hansenii* from grapes and must. Other authors have also observed the presence of *A. pasteurianus* and *Ga. liquefaciens* in fresh musts, although these species were in the minority (du Toit & Lambrechts, 2002). Finally, the only species of acetic acid bacteria present in a study of Chilean grapes were *G. oxydans* and *A. cerevisiae* (Prieto et al., 2007). Those species displayed a specific distribution according to latitude and only overlapped in a transition zone between the two regions. Interestingly, *A. cerevisiae* had not been described previously in grapes. However, prior to 2002 it was classified as *A. pasteurianus* (Cleenwerck et al., 2002), and it is therefore difficult to determine whether the species is peculiar to Chile or whether it has previously been identified as *A. pasteurianus*.

## 5.2. Growth Dynamics of Acetic Acid Bacteria During Alcoholic Fermentation

Although the growth dynamics of acetic acid bacteria will depend on the initial numbers of each population in the must, most studies agree that these bacteria are unlikely to grow during alcoholic fermentation and that their numbers reduce drastically to $10^1-10^3$ CFU/mL (du Toit & Lambrechts, 2002; González et al., 2004; Joyeux et al., 1984). The massive production of carbon dioxide by the yeast and therefore the establishment of anaerobic conditions in the fermentation medium make it almost impossible for this group of obligate aerobes to proliferate (Drysdale & Fleet, 1988). In terms of the growth of different species, all of these studies agree that the main species present in the must, *G. oxydans*, gradually disappears during the first few days of fermentation and is rarely found in wine. This species is much more competitive in substrates containing high concentrations of sugar but its intolerance of ethanol means that its numbers diminish as alcoholic fermentation advances. The majority species isolated during alcoholic fermentation are *A. aceti* and *A. pasteurianus* (Drysdale & Fleet, 1985; Joyeux et al., 1984), although du Toit and Lambrechts (2002) recently reported a significant presence of the species *Ga. liquefaciens* and *Ga. hansenii*. We have obtained similar results in isolates from alcoholic fermentation (see Figure 9.4) (González et al., 2004, 2005).

Our group has analyzed the effects of some winemaking practices on populations of acetic acid bacteria. We have studied, for instance, the addition of sulfite, inoculation of selected yeasts, and their combined effect in relation to the growth dynamics of different strains and species (González et al., 2005) as well as on the overall population of acetic acid bacteria using independent culture techniques (Andorrà et al., 2008). The predominance of *A. aceti* during alcoholic fermentation was confirmed along with the observation of a considerable reduction in the population of cultivable acetic acid bacteria caused by both addition of sulfite and inoculation of yeast starter cultures (in this case occurring in parallel with the rapid onset of alcoholic fermentation). Notably, analysis of strains revealed that some grape-derived strains survived throughout alcoholic fermentation and some were enriched in the must after it entered the winery. In other words, there is a significant resident population in the winery but this does not solely account for the bacteria recovered at the end of fermentation (González et al., 2005). The use of independent culture techniques showed that both *A. aceti* and *Ga. hansenii* are present throughout alcoholic fermentation and are clearly detectable by DGGE. It should be noted that, since this technique only recovers populations larger than $10^3$ cells/mL, both species must have exceeded this population size during fermentation. This is confirmed by the finding that the total population of acetic acid bacteria measured by quantitative PCR remains close to $10^4$ cells/mL throughout alcoholic fermentation. Notably, neither of these observations was affected by inoculation of yeast starter cultures or addition of sulfite (Andorrà et al., 2008), suggesting that the low recovery of colonies on plates may be due to the induction of viable but noncultivable cell states (Millet & Lonvaud-Funel, 2000).

The effects of standard winemaking procedures such as control of temperature, maceration, micro-oxygenation, etc., during alcoholic fermentation have not been studied in detail, although in general terms the filtration or clarification of must is thought to reduce the population of acetic acid bacteria and therefore the risk of their growth. Nevertheless, once alcoholic fermentation is complete, racking causes aeration of the wine and can lead to proliferation of acetic acid bacteria. Consequently, these populations can reach new titers of around $10^6$ CFU/mL in stored wine (Drysdale & Fleet, 1985), especially if the wine is not stored under anaerobic conditions. Joyeux et al. (1984)

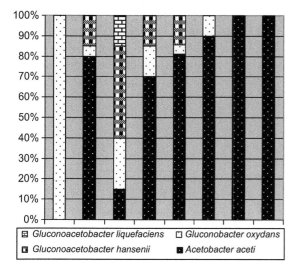

**FIGURE 9.4** Species profiles in must and at the beginning, middle, and end of alcoholic fermentation of a Garnacha grape variety in the 2001 and 2002 vintages (González et al., 2004).

demonstrated that these populations of acetic acid bacteria that survive the alcoholic fermentation process can proliferate rapidly in stored wine when enriched with 7.5 mg/L of oxygen. This growth is accentuated when the storage temperature and pH of the wine are high. Barrel aging of wines can be considered an anaerobic or semi-anaerobic process, and, therefore, acetic acid bacteria would be expected to have little likelihood of proliferating. However, Joyeux et al. (1984) also reported that oxygen penetrates the barrel in quantities of around 30 mg/L per year, which is sufficient for small populations of acetic acid bacteria to survive. Drysdale and Fleet (1985) confirmed this capacity to survive and even grow under semi-anaerobic conditions by isolating acetic acid bacteria from samples taken within the barrel. In addition, the racking performed during aging and even bottling leads to new increases in oxygen concentration and the proliferation of acetic acid bacteria (Millet & Lonvaud-Funel, 2000).

Finally, it is worth mentioning that there has been little analysis of the growth of acetic acid bacteria during malolactic fermentation. Consequently, it is not known whether there is any degree of synergy between the growth of lactic acid bacteria and that of acetic acid bacteria. Alternatively, as occurs with yeasts, the growth of lactic acid bacteria may inhibit that of acetic acid bacteria. The latter possibility may be supported by the observation that malolactic fermentation also causes the release of carbon dioxide, although in proportions that are much lower than those seen during alcoholic fermentation, and therefore creates anaerobic conditions. Nevertheless, the racking processes performed prior to malolactic and after alcoholic fermentation cause sufficient aeration to stimulate the growth of acetic acid bacteria. In an effort to answer some of these questions, populations of lactic acid and acetic acid bacteria were analyzed alongside the concentrations of malic and acetic acid during malolactic fermentation (Guillamón et al., 2003) (see Figure 9.5). Lactic acid bacteria grew to reach $10^8$ CFU/mL within a few days. Interestingly, this growth coincided with substantial growth of acetic acid bacteria. Subsequently, both populations diminished to levels similar to those present at the beginning of the process. Consumption of malic acid began when the population of lactic acid bacteria reached its maximum, and the growth of acetic acid bacteria was also reflected in an increase in the concentration of acetic acid. Therefore, there could be some degree of synergy between lactic acid and acetic acid bacteria, which could grow in parallel during malolactic fermentation. In a similar study, Joyeux et al. (1984) observed

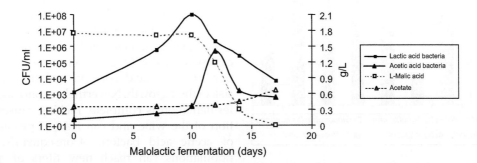

**FIGURE 9.5** Dynamics of population growth in acetic acid and lactic acid bacteria during malolactic fermentation (continuous lines and filled symbols). Consumption of malic acid and production of acetic acid are also shown (dotted lines and empty symbols). CFU = colony-forming units.

that prior growth of *G. oxydans* or *A. aceti* in the must could stimulate malolactic fermentation by *Oenococcus oeni*. However, these data should be confirmed in studies of malolactic fermentation and, in particular, in wines with varying degrees of aeration.

There are some differences of opinion regarding the majority species growing in stored wines or during aging. In Bordeaux wines, *A. aceti* is reportedly present in larger numbers than *A. pasteurianus* (Joyeux et al., 1984). In contrast, Drysdale and Fleet (1985) found that the majority of isolates in Australian wines corresponded to *A. pasteurianus*. In more recent studies, *A. pasteurianus* was found to be the predominant species in South African wines, although on occasion *Ga. liquefaciens* was also isolated in large numbers (du Toit & Lambrechts, 2002). Finally, in all of the studies performed by our group, *A. aceti* was practically the only species isolated at the end of fermentation, during malolactic fermentation, and during storage of wines (González et al., 2004, 2005; Guillamón et al., 2003), although *Ga. hansenii* may also be present in notable quantities (Andorrà et al., 2008). The presence of *A. tropicalis* has also been observed in Austrian wines that were spontaneously fermented and later acetified (Silhavy & Mandl, 2006).

Although any of the acetic acid bacteria that have been described can produce acetic acid, wine spoilage is most commonly linked to *A. pasteurianus*, especially when spoilage occurs in the bottle, where characteristically a *flor* or biofilm can form on the surface of the wine and is easily detected by the presence of a ring of residue (Bartowsky et al., 2003). The development of these rings or biofilms is linked to the formation of air pockets beneath the cork and can be avoided by storing the bottles in a horizontal position (Bartowsky & Henschke, 2008). Recently, a new species of acetic acid bacteria, *Acetobacter oeni*, was described in wines from the Dao region in Portugal that had been spoiled by *Dekkera* species (Silva et al., 2006).

# 6. FACTORS DETERMINING THE GROWTH OF ACETIC ACID BACTERIA: BACTERIAL CONTROL METHODS

The main physicochemical properties that influence the growth of acetic acid bacteria during winemaking are the pH of the must/wine, the temperature, and the concentrations of ethanol, sulfite, and, most importantly, oxygen dissolved in the medium (Drysdale & Fleet, 1988). Although few systematic studies have addressed the effect of these parameters on the growth of acetic acid bacteria, the most relevant conclusions of the studies that have been performed are summarized below.

## 6.1. pH

Optimal growth of acetic acid bacteria occurs at a pH of between 5 and 6 (de Ley et al., 1984). Of course, the pH of wine is much lower and it has been clearly demonstrated that these bacteria can grow in this medium. Thus, although the low pHs found in wine generally inhibit bacterial growth, in the case of acetic acid bacteria, they may be limiting for proliferation but not for survival, since these bacteria have been found in wines with a pH of 3 (Drysdale & Fleet, 1985). Nevertheless, Joyeux et al. (1984) observed that the growth of *A. aceti* was lower in a wine with a pH of 3.4 than in one with a pH of 3.8. Studies have not addressed the pH resistance of different species of acetic acid bacteria isolated from wine. However, certain strains are known to be particularly resistant, since, under aerated conditions such as those found during the production of vinegars, acetic acid bacteria survive and proliferate in highly acidic media with a pH of 2.0 to 2.3, even with limited aeration such as in the traditional Orleans method. Furthermore, the effect of pH is synergistic with that of other growth inhibitors such as ethanol and sulfite. The antimicrobial effect of sulfite increases in highly acidic media. When this compound is added to wine, an equilibrium is formed between molecular $SO_2$ and the bisulfite form ($HSO_3^-$). Although the bisulfite form predominates in wine, only the molecular species has antimicrobial properties. The lower the pH, the greater the proportion of $SO_2$ versus $HSO_3^-$ and, therefore, the greater the antimicrobial effect (Ribéreau-Gayon et al., 2000).

## 6.2. Temperature

The optimal temperature for growth of acetic acid bacteria is between 25 and 30°C (de Ley et al., 1984), although *A. pasteurianus* may have an optimal temperature of around 20°C (Vaughn, 1955). The maximum temperature that can be tolerated is estimated at between 35 and 40°C, depending on the strain and species in question. In fact, a temperature of between 28 and 32°C is maintained during industrial vinegar production, since the reaction can be highly exothermic and there may be considerable increases in temperature that stop the acetification process. In hot countries, resistance to high temperatures is a positive factor for the selection of acetic acid bacteria to be used in vinegar making (Ndoye et al., 2007). In terms of lower temperatures, a significant increase in the population of acetic acid bacteria has been observed in wines stored at 18°C (Joyeux et al., 1984). Growth has also been detected, though at very low levels, in wines stored at 10°C. In fact, populations have been observed to increase from $10^3$ to $10^5$ CFU/mL during standard winemaking practices such as cold maceration prior to inoculation (du Toit & Lambrechts, 2002). Thus, while standard temperatures used during wine storage or barrel aging may slow growth, they do not appear to prevent it. Of course, the higher the temperature, the greater the growth rate.

## 6.3. Ethanol

Ethanol is the main substrate for acetic acid bacteria during growth in wine. However, the degree of ethanol tolerance depends on the species and even the strain. *G. oxydans* is the least tolerant of the species commonly isolated during winemaking. According to *Bergey's Manual* (de Ley et al., 1984), only 42% of *G. oxydans* strains proliferate in a medium containing 5% ethanol. Similarly, strains of *A. aceti*, *Ga. hansenii*, and *Ga. liquefaciens* were unable to grow in medium containing 10% ethanol, and only 20% of *A. pasteurianus* strains grew in this medium. However, it is well known that acetic acid bacteria may be present in wineries in media containing higher concentrations of ethanol, although it is also known that wines containing higher concentrations of ethanol (15% or more) are less likely to become pricked. Nevertheless, strains of acetic acid bacteria have been isolated from wines with ethanol concentrations above 13%; the limit is considered to be 15 or 15.5%, which corresponds to the minimum level recommended for fortified wines. Clearly, then, wild strains must be much more resistant than the collection strains in which these tests are usually carried out.

## 6.4. Sulfite

The most widely used antimicrobial agent in winemaking is also active against acetic acid bacteria. However, the effects of sulfite on the growth and survival of acetic acid bacteria have not been studied in detail (Ribéreau-Gayon et al., 2000). According to Lafon-Lafourcade and Joyeux (1981), the concentrations of sulfite that are generally used in winemaking are insufficient to prevent the growth of acetic acid bacteria. Those authors observed the growth of *A. aceti* in red wines containing 25 mg/L of free $SO_2$. Du Toit et al. (2005) established that 1.2 mg/L free $SO_2$ had an observable effect on the viability of *A. pasteurianus*. On the other hand, complete inhibition of the growth of acetic acid bacteria in grape must was observed following addition of 100 mg/L total $SO_2$ (Watanabe & Ino, 1984). Acetic acid bacteria have a particular capacity to remain in substrates such as wood; therefore, it is recommended that special measures are taken to clean barrels before they are reused. Following an analysis of the effect of treatment with sulfite, potassium carbonate, bleach, and hot water, Wilker and Dharmadhikari (1997) concluded that hot water (85–88°C for 20 min) was the most effective. Use of other preservatives (sorbic, fumaric, or benzoic acid, etc.) proposed as alternatives to sulfite has not been assessed in detail for effect on the survival of acetic acid bacteria. The growing restrictions on the use of sulfite due to new legislation and a certain reluctance on the part of consumers may have undesirable effects, particularly in terms of sensory quality, as a result of a failure to control these microorganisms.

## 6.5. Oxygen

Since acetic acid bacteria are obligate aerobes, oxygen becomes an authentic limiting factor for their growth. However, a number of acetic acid bacteria can still grow despite the anaerobic conditions present during alcoholic fermentation not being favorable for their growth. This explains why any wine exposed to air will rapidly develop a biofilm on its surface that mainly comprises acetic acid bacteria, although yeasts may also grow. Clearly, then, while some oxygen is necessary for wine maturation (Mas et al., 2002), inadequate management of oxygen levels will offer a clear advantage to acetic acid bacteria and cause spoilage. These minimal requirements for oxygen during maturation are usually managed through the use of micro-oxygenation, an increasingly common practice in wineries to accelerate the process of color stabilization in the wines. Nevertheless, although the only study performed to date

revealed that micro-oxygenation stimulated the growth of acetic acid bacteria (du Toit et al., 2006), further studies are required to assess how the process affects these bacterial populations.

## 6.6. Storage and Aging

Standard aging and storage processes also present considerable risks for the growth of acetic acid bacteria and wine spoilage. Aging is usually carried out in oak barrels. Although these may not contain microorganisms when they are new, they are rapidly colonized following exposure to wine (Renouf et al., 2006). The porous nature of the wood that makes it so appropriate for use in wine aging also makes it an appropriate habitat for a range of microorganisms, including acetic acid bacteria. Nevertheless, under aging conditions, the availability of oxygen is considerably reduced and most acetic acid bacteria are found in viable but non-cultivable states (Millet & Lonvaud-Funel, 2000). Although wineries usually employ various washing methods to allow barrels to be reused, only treatment with water at high temperatures (greater than 85°C) appears to be effective for the elimination of acetic acid bacteria (Wilker & Dharmadhikari, 1997). Wine aging may be completed in the bottle, and this offers a new opportunity for the proliferation of acetic acid bacteria. Bottles must be stored horizontally to prevent the formation of air pockets in which acetic acid bacteria can grow (Bartowsky et al., 2003). This occurs particularly in aged red wines, since white wines are increasingly treated by filter sterilization or similar processes that limit the survival of microorganisms significantly (Bartowsky & Henschke, 2008). As mentioned earlier, the survival of acetic acid bacteria in wines may be due to the presence of quinones that can function as terminal electron acceptors and the aeration that occurs during the transfer of liquids between tanks, barrels, and bottles, which represent ideal opportunities for growth of these microorganisms (Joyeux et al., 1984).

# 7. CHANGES OCCURRING IN WINE AS A RESULT OF THE GROWTH OF ACETIC ACID BACTERIA

The growth of acetic acid bacteria in wine is rapidly followed by the production of acetic acid and an increase in volatile acidity, producing what is often referred to as pricked wine. Volatile acidity is considered a defect at levels above 0.4 to 0.5 g/L, depending on the type of wine (Ribéreau-Gayon et al., 2000); it can reach levels up to 1.0 to 1.5 g/L in some sweet wines (botyrized wines or icewines) (Nurgel et al., 2004). However, volatile acidity is directly associated with the formation of ethyl acetate, which appears rapidly during the growth of acetic acid bacteria in wine. Ethyl acetate has a notably low perception threshold and is easily recognized by the smell of glue or nail polish remover (Ribéreau-Gayon et al., 2000). Vinegar and glue are probably the first noticeable aromas in wine spoiled by acetic acid bacteria. However, this is not the only change that occurs as a result of the development of these bacteria. These changes in the medium also depend on the point during vinification at which growth occurs.

## 7.1. Changes in Grapes and Must as a Result of the Growth of Acetic Acid Bacteria

Populations of acetic acid bacteria increase in size over the course of ripening and large populations are present in unhealthy grapes, especially those infected with *B. cinerea*. Nevertheless, any damage to the skin of the grape can provide a route of entry for all types of microorganism, including acetic acid bacteria. Consequently, acid rotting is also considered to be

a combined effect of some yeasts and acetic acid bacteria. These populations of acetic acid bacteria remain high in the must. In both substrates (grapes and must), glucose is the main carbon source for these bacteria. As mentioned earlier, this glucose is directly oxidized to gluconic acid, which is the main compound to accumulate in the medium. Although the presence of high concentrations of this acid in grapes infected with *Botrytis* was until recently considered a consequence of fungal metabolism, the main source is now known to be acetic acid bacteria, which are present in large numbers in botrytized grapes (Barbe et al., 2001). Acetic acid bacteria also metabolize fructose, although in smaller quantities, to form 5-oxofructose.

The capacity of acetic acid bacteria to produce extracellular polysaccharides leads to another significant change due to sugar metabolism (Kouda et al., 1997). Some strains of *A. pasteurianus* and *G. oxydans* produce cellulose fibrils or other polysaccharides that impede filtration of the wines (Drysdale & Fleet, 1988). The main consequence of the production of these polysaccharides during winemaking is the wine filtration difficulties that it generates.

Acetic acid bacteria can also use other carbohydrates such as arabinose, galactose, mannitol, mannose, ribose, sorbitol, and xylose (de Ley et al., 1984). Although these sugars are found at low concentrations in the must, some form part of the residual sugars found in wine as a result of having not been used by yeasts. Consequently, these sugars can also be used by acetic acid bacteria during growth in wine, though ethanol remains the main substrate.

Logically, acetic acid bacteria also produce acetic acid in grapes and must through the metabolism of sugars, although the quantities produced are lower than those of gluconic acid. It is believed that the acetic acid produced is mainly derived from the ethanol produced by the yeast in grapes and must than through the metabolism of hexose sugars.

Another effect of sugar metabolism by acetic acid bacteria is the high capacity of gluconic acid and oxofructose to bind $SO_2$; this reduces the proportion of free $SO_2$ and therefore its antimicrobial and antioxidant capacity. Barbe et al. (2001) reported that maintenance of 50 mg/L of free $SO_2$ in a synthetic must in which *G. oxydans* had grown required 3000 mg/L of total $SO_2$ as a consequence of the high concentrations of gluconic acid (51 g/L), 5-oxofructose (6 g/L), and dihydroxyacetone (2 g/L) formed from glucose, fructose, and glycerol, respectively. The last two components can also bind $SO_2$ efficiently (du Toit & Pretorius, 2002).

## 7.2. Changes in Wine as a Result of the Growth of Acetic Acid Bacteria

The conversion of ethanol into acetic acid in wine is the most widely recognized form of spoilage due to acetic acid bacteria. This ethanol represents the principal carbon source and its conversion into acetic acid is responsible for the generation of pricked wine. The biochemistry of this reaction has been widely studied due to its importance in the production of vinegar. Some strains of acetic acid bacteria can produce up to 150 g/L of acetic acid during vinegar production (Sievers et al., 1997). Such high concentrations are only obtained in highly oxygenated cultures. Although this clearly does not occur during the winemaking process, volatile acidity is easily increased to above 0.8 g/L, a concentration considered detrimental to the quality of wine since it is well above the detection threshold for acetic acid.

As mentioned, ethanol is first oxidized to acetaldehyde, which is then oxidized to form acetic acid. Since acetaldehyde is a metabolic intermediate in this reaction, it is one of the most important products after acetic acid. Its levels increase in wine with decreasing concentrations of dissolved oxygen (Drysdale & Fleet, 1989a). Yeasts also produce acetaldehyde during alcoholic fermentation at concentrations of around 20 to 200 mg/L. Its perception threshold is around 125 mg/L, and at concentrations above 500 mg/L it is considered detrimental to the quality of the wine as a result of the oxidized character it endows (Margalith, 1981). In addition to its impact on the aroma and flavor of the wine, acetaldehyde has a greater capacity to bind $SO_2$ and therefore reduce the levels of free $SO_2$ (Ribéreau-Gayon et al., 2000). As a consequence, wine with a high concentration of acetaldehyde requires higher concentrations of $SO_2$ to achieve good protection of wine during aging and bottling.

After ethanol, glycerol is the main product of alcoholic fermentation and reaches concentrations of between 2 and 25 g/L. This metabolite is important in determining the quality of the wine as it contributes unctuous, syrupy, and viscous characteristics. Acetic acids also oxidize glycerol to dihydroxyacetone, which does not impart these positive characteristics in the organoleptic qualities of the wine. In addition, it binds easily to $SO_2$.

Acetic acid bacteria can also oxidize the different acids present in wine. Drysdale and Fleet (1989b) observed a reduction in malic, tartaric, and citric acids present in wines contaminated with acetic acid bacteria. These acids, and others such as lactic and fumaric acid, would be completely oxidized to carbon dioxide and water via the Krebs cycle (only in those species other than *Gluconobacter* that have a functional Krebs cycle). Some strains of *Acetobacter* and *Gluconobacter*, particularly *A. pasteurianus* strains, can oxidize lactic acid to acetoin, which produces a buttery aroma and flavor (Drysdale & Fleet, 1988).

Ethyl acetate is another compound produced by acetic acid bacteria that has a negative effect on the sensory quality of wine. This ester has a very low perception threshold and is detectable at concentrations as low as 10 mg/L (Berg et al., 1955). The growth of acetic acid bacteria can increase the concentration of ethyl acetate to 140 mg/L in wine and 30 mg/L in must (Drysdale & Fleet, 1989a). Finally, acetic acid bacteria can also oxidize higher alcohols such as isoamyl alcohol, propanol, and 2-phenyl ethanol in wine to their corresponding aldehydes and carboxylic acids (Molinari et al., 1999).

Importantly, most of the acetic acid is introduced into the wine during the stationary and death phases of acetic acid bacteria and not during the growth phase (Kösebalan & Özingen, 1992). Although most acetic acid is produced in an extracellular reaction, a proportion can accumulate inside the cell and is then released when the cells die. Thus, although the numbers of these bacteria increase as a result of aeration (during specific winemaking processes such as pump-over, racking, etc.), the increase in volatile acidity will probably not occur until storage and aging.

Little is known about the effect of acetic acid bacteria on the growth of lactic acid bacteria. Gilliland and Lacey (1964) reported that a strain of *Acetobacter* inhibited the growth of *Lactobacillus* species, and Joyeaux et al. (1984) found that acetic acid bacteria stimulated malolactic fermentation.

Bradley (1965) demonstrated the presence of bacteriophages that were active against *Acetobacter* species. Subsequently, Sellmer et al. (1992) demonstrated that phages could place the viability of acetic acid bacteria cultures at risk during the production of vinegar. It is therefore likely that phages can affect the growth and survival of acetic acid bacteria over the course of wine production.

## 8. INTERACTIONS WITH OTHER MICROORGANISMS IN WINE

Grape must contains a wide range of species of yeast, lactic acid bacteria, and acetic acid bacteria, and their interactions during alcoholic fermentation may be complex (Fleet, 2003). Joyeux et al. (1984) showed that prior growth of *G. oxydans* or *A. aceti* in grape must could result in stuck fermentation. Drysdale and Fleet (1989b) studied the effect of simultaneous inoculation of yeast and bacteria in the must, as usually occurs in the winery. In this case, there was no substantial inhibition of yeast growth; however, the yeasts struggled to consume all of the sugars in the must. *A. pasteurianus* was found to produce greater inhibition of the fermentative capacity of yeasts. The mechanism by which acetic acid bacteria produce this antagonism of yeasts has yet to be elucidated. Acetic acid is an inhibitor of yeasts, but other substances may also be involved in this effect (du Toit & Lambrechts, 2002). Clearly, this inhibition can only occur if there is a considerable delay in the onset of alcoholic fermentation, giving rise to the development of an excess of acetic acid bacteria that can produce these inhibitory substances.

## 9. FINAL RECOMMENDATIONS TO AVOID WINE SPOILAGE DUE TO ACETIC ACID BACTERIA

In summary, although acetic acid bacteria need oxygen, they can survive and even grow to some extent in an almost completely anaerobic medium such as wine. Clearly, then, it is impossible to remove all risk of contamination by these bacteria during winemaking. However, it is possible to balance the need for certain processes against the risk of bacterial growth. The following considerations may help to prevent the undesirable growth of acetic acid bacteria during winemaking:

1) Management of vines and correction of musts should focus on acidification of the must, as this will help to prevent growth of acetic acid bacteria. Reducing the pH of the wine will not only reduce the risk of spoilage during production but also enhance the options available for aging.
2) The health of the grapes will determine the microbial load corresponding to acetic acid bacteria. Reducing the population of acetic acid bacteria at the beginning of fermentation

will reduce the size of the population at the end of the process. These populations are reduced by clearing, filtration, addition of sulfite, and inoculation of yeast starter cultures. Musts obtained from unhealthy grapes should be immediately filtered and inoculated with a yeast starter culture. However, none of these practices alone will guarantee the elimination of acetic acid bacteria.

3) Rapid initiation and progression of fermentation creates anaerobic conditions that will impede the proliferation of acetic acid bacteria.

4) Any process that involves aeration (pump-over, racking, etc.) will provide an opportunity for the growth of acetic acid bacteria and, therefore, favor an increase in volatile acidity. In addition, the recent introduction of micro-oxygenation techniques must be taken into consideration as their effect on the growth of acetic acid bacteria has not yet been analyzed in detail.

5) Maintaining low temperatures during fermentation, aging, and storage limits the growth of acetic acid bacteria. However, only temperatures below 10°C can effectively prevent bacterial growth. Cold maceration alone is also insufficient to eliminate acetic acid bacteria and can even lead to an increase in their numbers (Couasnon, 1999).

6) Although further studies are required, the presence of residual sugars in wine and also malolactic fermentation may increase the likelihood of the growth of acetic acid bacteria.

7) Good winery hygiene is absolutely essential to reduce the risk of contamination with acetic acid bacteria, especially in wines with a lower alcohol content and during barrel aging. It should be remembered that solid porous supports such as wood provide good conditions for the survival of acetic acid bacteria.

# References

Adachi, O., Miyagawa, E., Shinagawa, E., Matsushita, K., & Ameyama, M. (1978). Purification and properties of particulate alcohol dehydrogenase from *Acetobacter aceti*. *Agric. Biol. Chem., 42,* 2331–2340.

Andorrà, I., Landi, S., Mas, A., Guillamón, J. M., & Esteve-Zarzoso, B. (2008). Effect of enological practices on microbial populations using culture-independent techniques. *Food Microbiol., 25,* 849–856.

Asai, T. (1935). Taxonomic studies of acetic acid bacteria and allied oxidative bacteria in fruits and a new classification of oxidative bacteria. *J. Agr. Chem. Soc. Jap., 11,* 50–60.

Azuma. (2008). Genomic comprehension for evolutional flexibility of *Acetobacter pasteurianus*. The 2nd International Conference on Acetic Acid Bacteria. Nagoya, Japan, November 11–14.

Barbe, J. C., de Revel, G., Joyeux, A., Bertrand, A., & Lonvaud-Funel, A. (2001). Role of botrytized grape microorganisms in SO$_2$ binding phenomena. *J. Appl. Microbiol., 90,* 34–42.

Barja, F., Mesa, M. M., Macías, M., Bermudez, I., Cantero, D., & Lopez, J. (2003). Aspectos bioquímicos, moleculares y morfológicos de las bacterias acéticas. In A. Mas, & J. M. Guillamón (Eds.), *Primeras jornadas en I+D+i en la elaboración de vinagre de vino* (pp. 17–20). Tarragona, Spain: Servei de publicacions.

Bartowsky, E. J., & Henschke, P. A. (2008). Acetic acid bacteria spoilage of bottled red wine – A review. *Int. J. Food Microbiol., 125,* 60–70.

Bartowsky, E. J., Xia, D., Gibson, R. L., Fleet, G. H., & Henschke, P. A. (2003). Spoilage of bottled red wine by acetic acid bacteria. *Lett. Appl. Microbiol., 36,* 307–314.

Beijerinck, M. W. (1899). Sur les diverses espèces de bactéries acétifiantes. *Arch. Neerl. des Sciences exactes et Nat.* Haarlem II 180–189.

Belly, R. T., & Claus, G. W. (1972). Effect of amino acids on the growth of *Acetobacter suboxydans*. *Arch. Mikrobiol., 83,* 237–245.

Berg, H. W., Filipello, F., Hinreiner, E., & Webb, A. D. (1955). Evaluation of thresholds and minimum difference concentrations for various constituents of wines. I. Water solution of pure substances. *Food Technol., 9,* 23–26.

Bertalan, M., Albano, R., de Pádua, V., Rouws, L., Rojas, C., Hemerly, A., et al. (2009). Complete genome sequence of the sugarcane nitrogen-fixing endophyte *Gluconacetobacter diazotrophicus* Pal5. *BMC Genom., 10,* 450.

Bradley, D. E. (1965). The isolation and morphology of some new bacteriophages specific for *Bacillus* an *Acetobacter* species. *J. Gen. Microbiol., 41,* 233–241.

Buchanan, R. E., & Gibbons, N. E. (1974). In *Bergey's manual of determinative bacteriology* (8th ed.). Baltimore, MD: Williams & Wilkins.

Camu, N., de Winter, T., Verbrugghe, K., Cleenwerck, I., Vandamme, P., Takrama, J. S., et al. (2007). Dynamics and biodiversity of populations of lactic acid bacteria and acetic acid bacteria involved in spontaneous heap fermentation of cocoa beans in Ghana. *Appl. Environ. Microbiol., 73*, 1809–1824.

Carr, J. G., & Passmore, S. M. (1979). Methods for identifying acetic acid bacteria. In F. A. Skinner, & D. W. Lovelock (Eds.), *Identification methods for microbiologists* (pp. 33–47). London, UK: Academic Press.

Cleenwerck, I., & de Vos, P. (2008). Polyfasic taxonomy of acetic acid bacteria: An overview of the currently applied methodology. *Int. J. Food Microbiol., 125*, 2–14.

Cleenwerck, I., Vandemeulebroecke, K., Janssens, D., & Swings, J. (2002). Reexamination of the genus *Acetobacter*, with descriptions of *Acetobacter cerevisiae* sp. nov., and *Acetobacter malorum*, sp. nov. *Int. J. Syst. Evol. Microbiol., 52*, 1551–1558.

Couasnon, M. B. (1999). Une nouvelle technique: La maceration préfermentaire à froid – Extraction à la neige carbonique. 1re partie: Resultats oenologiques. *Revue des Oenologues et des Techniques Vitivinicoles et Oenologiques, 92*, 26–30.

de Ley, J. (1959). On the formation of acetoin by *Acetobacter*. *J. Gen. Microbiol., 21*, 352–365.

de Ley, J., Gillis, M., & Swings, J. (1984). Family VI. Acetobacteriaceaea. In N. R. Krieg, & J. C. Holt (Eds.), *Bergey's manual of systematic bacteriology, Vol. 1* (pp. 267–278). Baltimore, MD: Williams & Wilkins.

de Vero, L., Gala, E., Gullo, M., Solieri, L., Landi, S., & Giudici, P. (2006). Application of denaturing gradient gel electrophoresis (DGGE) analysis to evaluate acetic acid bacteria in traditional balsamic vinegar. *Food Microbiol., 23*, 809–813.

Deppenmeier, U., Hoffmeister, M., & Prust, C. (2002). Biochemistry and biotechnological applications of *Gluconobacter* strains. *Appl. Microbiol. BioTechnol., 60*, 233–242.

Drysdale, G. S., & Fleet, G. H. (1985). Acetic acid bacteria in some Australian wines. *Food Technol. Aust., 37*, 17–20.

Drysdale, G. S., & Fleet, G. H. (1988). Acetic acid bacteria in winemaking: A review. *Am. J. Enol. Vitic., 2*, 143–154.

Drysdale, G. S., & Fleet, G. H. (1989a). The growth and survival of acetic acid bacteria in wine at different concentrations of oxygen. *Am. J. Enol. Vitic., 40*, 99–105.

Drysdale, G. S., & Fleet, G. H. (1989b). The effect of acetic acid bacteria upon the growth and metabolism of yeast during the fermentation of grape juice. *J. Appl. Bacteriol., 67*, 471–481.

du Toit, W. J., & Lambrechts, M. G. (2002). The enumeration and identification of acetic acid bacteria from South African red wine fermentations. *Int. J. Food Microbiol., 74*, 57–64.

du Toit, W. J., Lisjak, K., Marais, J., & du Toit, M. (2006). The effect of microoxygenation on the phenolic composition, quality and aerobic wine-spoilage microorganisms of different South African red wines. *South Afr. J. Enol. Vitic., 27*, 57–67.

du Toit, W. J., & Pretorius, I. S. (2002). The ocurrence, control and esoteric effect of acetic acid bacteria in winemaking. *Ann. Microbiol., 52*, 155–179.

du Toit, W. J., Pretorius, I. S., & Lonvaud-Funel, A. (2005). The effect of sulphur dioxide and oxygen on the viability and culturability of a strain of *Acetobacter pasteurianus* and a strain of *Brettanomyces bruxellensis* isolated from wine. *J. Appl. Microbiol., 98*, 862–871.

Dutta, D., & Gachhui, R. (2007). Nitrogen-fixing and cellulose-producing *Gluconacetobacter kombuchae* sp. nov., isolated from Kombucha tea. *Int. J. Syst. Evol. Microbiol., 57*, 353–357.

Entani, E., Ohmori, S., Masai, H., & Suzuki, K. I. (1985). *Acetobacter polyoxogenes* sp. nov., a new species of an acetic acid bacterium useful for producing vinegar with high acidity. *J. Gen. Appl. Microbiol., 31*, 475–490.

Fleet, G. H. (2003). Yeast interactions and wine flavour. *Int. J. Food Microbiol., 86*, 11–22.

Frateur, J. (1950). Trial on classification of. *Acetobacter. Cellule, 53*, 287–392.

Gilliland, R. B., & Lacey, J. P. (1964). Lethal action of *Acetobacter* on yeasts. *Nature, 202*, 727–728.

Gillis, M., Kersters, K., Hoste, B., Janssens, D., Kroppenstedt, M., Stephan, M. P., et al. (1989). *Acetobacter diazotrophicus* sp. nov., a nitrogen-fixing acetic acid bacterium associated with sugarcane. *Int. J. Syst. Bacteriol., 39*, 361–364.

Giudici, P., Gullo, M., Pulvirenti, A., Solieri, L., & De Vero, L. (2003). Microflora y la tecnología de producción del vinagre balsámico tradicional. In A. Mas, & J. M. Guillamón (Eds.), *Primeras jornadas en I+D+i en la elaboración de vinagre de vino* (pp. 59–70). Tarragona, Spain: Servei de publicacions.

González, A., Guillamón, J. M., Mas, A., & Poblet, M. (2006b). Application of molecular methods for routine identification of acetic acid bacteria. *Int. J. Food Microbiol., 108*, 141–146.

González, A., Hierro, N., Poblet, M., Mas, A., & Guillamón, J. M. (2005). Application of molecular methods to demonstrate species and strain evolution of acetic acid bacteria population during wine production. *Int. J. Food Microbiol., 102*, 295–304.

González, A., Hierro, N., Poblet, M., Mas, A., & Guillamón, J. M. (2006a). Enumeration and detection of acetic acid bacteria by real-time PCR and nested-PCR. *FEMS Microbiol. Lett., 254*, 123–128.

González, A., Hierro, N., Poblet, M., Rozès, N., Mas, A., & Guillamón, J. M. (2004). Population dynamics of acetic acid bacteria in a red wine fermentation. *J. Appl. Microbiol., 96*, 853–860.

Greenberg, D. E., Porcella, S. F., Stock, F., Wong, A., Conville, P. S., Murray, P. R., et al. (2006). *Granulibacter bethesdensis* gen. nov., sp. nov., a distinctive pathogenic acetic acid bacterium in the family *Acetobacteraceae*. *Int. J. Syst. Evol. Microbiol., 56*, 2609–2616.

Greenberg, D. E., Porcella, S. F., Zelazny, A. M., Virtaneva, K., Sturdevant, D. E., Kupko, J. J., et al. (2007). Genome sequence analysis of the emerging human pathogenic acetic acid bacterium *Granulibacter bethesdensis*. *J. Bacteriol., 189*, 8727–8736.

Guillamón, J. M., González, A., Poblet, M., & Mas, A. (2003). Development of molecular techniques for the analysis of acetic acid bacteria during winemaking. In Lallemand. (Ed.), *Yeast–bacteria interactions* (pp. 45–49). Biarritz, France: Lallemand Technical Meetings.

Gullo, M., Caggia, C., de Vero, L., & Giudici, P. (2006). Characterisation of acetic acid bacteria in traditional balsamic vinegar. *Int. J. Food Microbiol., 106*, 209–212.

Gullo, M., de Vero, L., & Giudici, P. (2009). Succession of selected strains of *Acetobacter pasteurianus* and other acetic acid bacteria in traditional balsamic vinegar. *Appl. Environ. Microbiol., 75*, 2585–2589.

Haruta, S., Ueno, S., Egawa, I., Hashiguchi, K., Fujii, A., Nagano, M., et al. (2006). Succession of bacterial and fungal communities during a traditional pot fermentation of rice vinegar assessed by PCR-mediated denaturing gradient gel electrophoresis. *Int. J. Food Microbiol., 109*, 79–89.

Ilabaca, C., Navarrete, P., Mardones, P., Romero, J., & Mas, A. (2008). Application of culture culture-independent molecular biology based methods to evaluate acetic acid bacteria diversity during vinegar processing. *Int. J. Food Microbiol., 126*, 245–249.

Jara, C. (2009). Desarrollo de métodos de biología molecular para el análisis directo de bacterias acéticas de vinagre. Tarragona, Spain: Doctoral thesis, Universidad Rovira i Virgili.

Jara, C., Mateo, E., Guillamón, J. M., Torija, M. J., & Mas, A. (2008). Evaluation of several methods for isolating high quality DNA in acetic acid bacteria in wine and vinegar. *Int. J. Food Microbiol., 128*, 336–341.

Jojima, Y., Mihara, Y., Suzuki, S., Yokozeki, K., Yamanaka, S., & Fudou, R. (2004). *Saccharibacter floricola* gen. nov., sp. nov., a novel osmophilic acetic acid bacterium isolated from pollen. *Int. J. Syst. Evol. Microbiol., 54*, 2263–2267.

Joyeux, A., Lafon-Lafourcade, S., & Ribéreau-Gayon, P. (1984). Evolution of acetic acid bacteria during fermentation and storage of wine. *Appl. Environ. Microbiol., 48*, 153–156.

Kösebalan, F., & Özingen, M. (1992). Kinetics of wine spoilage by acetic acid bacteria. *J. Chem. Tech. Biotechnol., 55*, 59–63.

Kouda, T., Naritomi, T., Yano, H., & Yoshinaga, F. (1997). Effects of oxygen and carbon dioxide pressures on bacterial cellulose production by *Acetobacter* in aerated and agitated culture. *J. Ferment. Bioeng., 82*, 124–127.

Lafon-Lafourcade, S. (1985). Rôle des microorganismes dans la formation de substances combinant le $SO_2$. *Bull. OIV, 652*, 590–604.

Lafon-Lafourcade, S., & Joyeux, A. (1981). Les bactéries acétiques du vin. *Bull. OIV, 608*, 803–829.

Lisdiyanti, P., Kawasaki, H., Widyastuti, Y., Saono, S., Seki, T., Yamada, Y., et al. (2002). *Kozakia baliensis* gen. nov. sp. nov., a novel acetic acid bacterium in the α-Proteobacteria. *Int. J. Syst. Evol. Microbiol., 52*, 813–818.

Loganathan, P., & Nair, S. (2004). *Swaminathania salitorans* gen. nov., sp. nov., a salt-tolerant, nitrogen-fixing and phosphate-solubilizing bacterium from wild rice (*Porteresia coarctata* Tateoka). *Int. J. Syst. Evol. Microbiol., 54*, 1185–1190.

López, I., Ruiz-Larrea, F., Cocolin, L., Orr, E., Phister, T., Marshall, M., et al. (2003). Design and evaluation of PCR primers for analysis of bacterial populations in wine by denaturing gradient gel electrophoresis. *Appl. Environ. Microbiol., 69*, 6801–6807.

Maestre, O., Santos-Dueñas, I. M., Peinado, R., Jiménez-Ot, C., García-García, I., & Mauricio, J. C. (2008). Changes in amino acid composition during wine vinegar production in a fully automatic pilot acetator. *Proc. Biochem., 43*, 803–807.

Margalith, P. Z. (Ed.). (1981). *Flavour microbiology*. Springfield, IL: Charles Thomas Publishers.

Mariette, I., Schwarz, E., Vogel, R. F., & Hammes, W. P. (1991). Characterization by plasmid profile analysis of acetic acid bacteria from wine, spirit and cider acetators for industrial vinegar production. *J. Appl. Microbiol., 71*, 134–138.

Mas, A., Puig, J., Lladó, N., & Zamora, F. (2002). Sealing and storage position effects on wine evolution. *J. Food Sci., 67*, 1374–1378.

Millet, V., & Lonvaud-Funel, A. (2000). The viable but non-culturable state of wine microorgamisms during storage. *Lett. Appl. Microbiol., 30*, 126–141.

Molinari, F., Gandolfi, R., Aragozzini, F., Leon, R., & Prazeres, D. M. F. (1999). Biotransformations in two-liquid phase systems. Production of phenylacetaldehyde by oxidation of 2-phenylethanol with acetic acid bacteria. *Enzyme Microb. Technol., 25*, 729–735.

Muraoka, H., Watab, Y., Ogasawara, N., & Takahashi, H. (1983). Trigger damage by oxygen deficiency to the acid production system during submerged acetic acid fermentation with *Acetobacter aceti*. *J. Ferment. Technol., 61*, 89–93.

Nanda, K., Taniguchi, M., Ujike, S., Ishihara, N., Mori, H., Ono, H., et al. (2001). Characterization of acetic acid bacteria in traditional acetic acid fermentations of rice vinegar (Komesu) and unpolished rice vinegar (Kurosu) produced in Japan. *Appl. Environ. Microbiol., 67*, 986–990.

Ndoye, B., Cleenwerck, I., Engelbeen, K., Dubois-Dauphin, R., Guiro, A. T., van Trappen, S., Willems, A., & Thonart, P. (2007). *Acetobacter senegalensis* sp nov., a thermotolerant acetic acid bacterium isolated in Senegal (sub-Saharan Africa) from mango fruit (*Mangifera indica L.*). *Int. J. Syst. Evol. Microbiol., 57,* 1576–1581.

Nurgel, C., Pickering, G. J., & Inglis, D. L. (2004). Sensory and chemical characteristics of Canadian ice wines. *J. Sci. Food Agric., 84,* 1675–1684.

Ohmori, S., Uozumi, T., & Beppu, T. (1982). Loss of acetic acid resistance and ethanol oxidizing ability in an *Acetobacter* strain. *Agric. Biol. Chem., 46,* 381–389.

Pasteur, L. (Ed.). (1868). *Etudes sur le vinaigre*. Paris, France: V. Masson.

Poblet, M., Rozès, N., Guillamón, J. M., & Mas, A. (2000). Identification of acetic acid bacteria by restriction fragment length polymorphism analysis of a PCR-amplified fragment of the gene coding for 16S rRNA. *Lett. Appl. Microbiol., 31,* 63–67.

Polo, M. C., & Sánchez-Luengo, A. A. (1991). Las bacterias acéticas. In C. Llaguno, & M. C. Polo (Eds.), *El vinagre de vino* (pp. 25–68). Madrid, Spain: EBCOMP.

Prieto, C., Jara, C., Mas, A., & Romero, J. (2007). Application of molecular methods for analysing the distribution and diversity of acetic acid bacteria in Chilean vineyards. *Int. J. Food Microbiol., 115,* 348–355.

Prust, C., Hoffmeister, M., Liesegang, H., Wiezer, A., Fricke, W. F., Ehrenreich, A., et al. (2005). Complete genome sequence of the acetic acid bacterium *Gluconobacter oxydans*. *Nat. Biotechnol., 23,* 195–200.

Querol, A., Barrio, E., Huerta, T., & Ramón, D. (1992). Molecular monitoring of wine fermentations conducted by dry yeast strains. *Appl. Environ. Microbiol., 58,* 2948–2952.

Raj, K. C., Ingram, L. O., & Maupin-Furlow, J. A. (2001). Pyruvate decarboxylase: A key enzyme for the oxidative metabolism of lactic acid by *Acetobacter pasteurianus*. *Arch. Microbiol., 176,* 443–451.

Renouf, V., Perello, M. C., Strehaiano, P., & Lonvaud-Funel, A. (2006). Global survey of the microbial ecosystem during alcoholic fermentation in winemaking. *J. Int. Sci. Vigne Vin, 40,* 101–116.

Ribéreau-Gayon, P., Dubordieu, D., Donèche, B., & Lonvaud, A. (Eds.). (2000). *Handbook of enology, the microbiology of wine and vinifications*. Chichester, UK: John Wiley & Sons.

Ruiz, A., Poblet, M., Mas, A., & Guillamon, J. M. (2000). Identification of acetic acid bacteria by RFLP of PCR-amplified 16S rDNA and 16S-23S rDNA intergenic spacer. *Int. J. Syst. Bacteriol., 50,* 1981–1987.

Saeki, A., Taniguchi, M., Matsushita, K., Toyam, H., Theeragool, G., Lotong, N., et al. (1997a). Microbiological aspects of acetate oxidation by acetic acid bacteria, unfavorable phenomena in vinegar fermentation. *Biosci. Biotech. Biochem., 61,* 317–323.

Saeki, A., Theeragool, G., Matsushita, K., Taniguchi, M., Toyama, H., Lotong, N., et al. (1997). Development of thermotolerant acetic acid bacteria useful for vinegar fermentation at higher temperatures. *Biosci. Biotech. Biochem., 61,* 138–145.

Sellmer, S., Sievers, M., & Teuber, M. (1992). Morphology, virulence and epidemiology of bacteriophage particles isolated from industrial vinegar fermentations. *System. Appl. Microbiol., 15,* 610–616.

Sievers, M., Lorenzo, A., Gianotti, S., Boesch, C., & Teuber, M. (1996). 16–23S ribosomal RNA spacer regions of *Acetobacter europaeus* and *A. xylinum*, tRNA genes and antitermination sequences. *FEMS Microbiol. Lett., 142,* 43–48.

Sievers, M., Stockli, M., & Teuber, M. (1997). Purification and properteies of citrate synthetase from *Acetobacter europaeus*. *FEMS Microbiol. Lett., 146,* 53–58.

Silhavy, K., & Mandl, K. (2006). *Acetobacter tropicalis* in spontaneously fermented wines with vinegar fermentation in Austria. *Mitteilungen Klosterneuburg, 56,* 102–107.

Silva, L. R., Cleenwerck, I., Rivas, R., Swings, J., Trujillo, M. E., Willems, A., et al. (2006). *Acetobacter oeni* sp. nov., isolated from spoiled red wine. *Int. J. Syst. Evol. Microbiol., 56,* 21–24.

Sokollek, S. J., Hertel, C., & Hammes, W. P. (1998). Cultivation and preservation of vinegar bacteria. *J. Biotechnol., 60,* 195–206.

Stackebrandt, E., Murray, R. G. E., & Trüper, H. G. (1988). *Proteobacteria classis* nov., a name for the phylogenetic taxon that includes the purple bacteria and their relatives. *Int. J. Syst. Evol. Microbiol., 38,* 321–325.

Teuber, M., Sievers, M., & Andresen, A. (1987). Characterization of the microflora of high acid submerged vinegar fermenters by distinct plasmid profiles. *Biotech. Lett., 9,* 265–268.

Torija, M. J., Mateo, E., Guillamón, J. M., & Mas, A. (2009). Design and optimisation of TaqMan-MGB probes for the identification and quantification of acetic acid bacteria species. *Food Microbiol. doi.* 10.1016/j.fm.2009.10.001.

Trcek, J. (2005). Quick identification of acetic acid bacteria based on nucleotide sequences of the 16S-23S rDNA internal transcribed spacer region and of the PQQ-dependent alcohol dehydrogenase gene. *Syst. Appl. Microbiol., 28,* 735–745.

Trcek, J., & Teuber, M. (2002). Genetic and restriction analysis of the 16S-23S rDNA internal transcribed spacer regions of the acetic acid bacteria. *FEMS Microbiol. Lett., 208,* 69–75.

Urakami, T., Tamaoka, J., Suzuki, K., & Komogata, K. (1989). *Acidomonas* gen. nov., incorporating *Acetobacter methanolicus* as *Acidomonas methanolica* comb. nov. *Int. J. Syst. Bacteriol., 39,* 50–55.

Valero, E., Berlanga, T. M., Roldan, P. M., Jiménez, C., García, I., & Mauricio, J. C. (2005). Free amino acids and volatile compounds in vinegars obtained from different types of substrate. *J. Sci. Food Agric., 85*, 603–608.

Vaughn, R. H. (1955). Bacterial spoilage of wines with special reference to California conditions. *Adv. Food Res., 6*, 67–108.

Versalovic, J., Koeuth, T., & Lupski, R. J. (1991). Distribution of repetitive DNA sequences in eubacteria and application to fingerprinting of bacterial genomes. *Nucleic Acids Res., 19*, 6823–6831.

Visser't Hooft, F. (1925). Biochemical investigations on *Acetobacter*. Delft, the Netherlands: Thesis, Delft Technical University.

Watanabe, M., & Ino, S. (1984). Studies on bacteria isolated from Japanese wines. Part 2. Growth of the *Acetobacter* sp. A-1 during the fermentation and the storage of grape must and red wine. *Yamanashiken. Shokuhin. Kogyo. Shydojo. Kenkyu. Hokoku, 16*, 13–22.

Wilker, K. L., & Dharmadhikari, M. R. (1997). Treatment of barrel wood infected with acetic acid bacteria. *Am. J. Enol. Vitic., 48*, 516–520.

Yamada, Y., Hoshino, K. I., & Ishikawa, T. (1997). The phylogeny of acetic acid bacteria based on the partial sequences of 16S ribosomal RNA: The elevation of the subgenus *Gluconacetobacter* to the generic level. *Biosci. Biotech. Biochem., 61*, 1244–1251.

Yamada, Y., Katsura, K., Kawasaki, H., Widyastuti, Y., Saono, S., Seki, T., et al. (2000). *Asaia bogorensis* gen. nov., sp. nov., an unusual acetic acid bacterium in the α-Proteobacteria. *Int. J. Syst. Evol. Microbiol., 50*, 823–829.

Yamada, Y., & Yukphan, P. (2008). Genera and species in acetic acid bacteria. *Int. J. Food Microbiol., 125*, 15–24.

Yukphan, P., Malimas, T., Muramatsu, Y., Takahashi, M., Kaneyasu, M., Tanasupawat, S., et al. (2008). *Tanticharoenia sakaeratensis* gen. nov., sp. nov, a new osmotolerant acetic acid bacterium in the α- Proteobacteria. *Biosci. Biotechnol. Biochem., 72*, 672–676.

Yukphan, P., Malimas, T., Potacharoen, W., Tanasupawat, S., Tanticharoen, M., & Yamada, Y. (2005). *Neoasaia chiangmaiensis* gen. nov., sp. nov., a novel osmotolerant acetic acid bacterium in the α- Proteobacteria. *J. Gen. Appl. Microbiol., 51*, 301–311.

CHAPTER

# 10

# Filamentous Fungi (*Botrytis cinerea*)

*Jesús M. Cantoral* [1], *Isidro G. Collado* [2]

[1] Laboratorio de Microbiología Enológica , Facultad de Ciencias, Universidad de Cádiz, Spain and
[2] Departamento de Química Orgánica, Facultad de Ciencias, Universidad de Cádiz, Spain

## OUTLINE

## 1. INTRODUCTION

The grapevine is the most extensively planted fruit crop in the world. The total surface area dedicated to vines is estimated at 8 million hectares, of which 62% are in Europe. These vines produce 674 million quintals of grapes annually. Almost half of the total production (45%) comes from Europe. Of this total volume, 184 million quintals are sold as fresh grapes, 12.4 million are used to make raisins, and the rest is used to make grape juice (and wine). Of the 282 million hectoliters of wine produced in the world every year, 170 million are produced in Europe. The total annual consumption of wine is 237 million hectoliters, with the rest of production being allocated to vinegars and distilled beverages (data for 2005 from the International Organisation of Vine and Wine [OIV], http://www.oiv.int).

Grapes are grown in both tropical and temperature climates, although the majority of grapevines are located in temperate regions, mostly in Europe. The main producers are Spain, France, Italy, Russia, Turkey, and Portugal, followed by the United States and countries from the southern hemisphere, namely Australia, South Africa, Chile, and Argentina. Grapevine diseases can have serious consequences such as reduced crop yields, diminished quality, and increased production and harvesting costs. The diseases are the result of the interaction between the vine (the sensitive host) and a live pathogenic organism (called a biotic or infectious pathogen, or simply a pathogen). Most of these organisms are bacteria, fungi, nematodes, and viruses.

In this chapter, we will discuss some of the diseases produced by filamentous fungi, namely powdery mildew, downy mildew, gray mold, phomopsis cane and leaf spot, and eutypa dieback, as well as the most serious and most common symptoms and control measures (Ministry of Agriculture, Fisheries and Food, Government of Spain, 2004; Pearson & Goheen, 2007). In Section 3, we will take a closer look at the case of *Botrytis cinerea* based on the considerable experience our research group has acquired in this area in recent years.

## 2. MAIN DISEASES AFFECTING GRAPEVINES

### 2.1. Powdery Mildew

Powdery mildew, which is known by different names in different regions, affects grapevines all over the world, even in tropical regions. The disease is believed to have originated in North America, from where it was probably spread to Europe by trade. It is caused by the filamentous fungus *Uncinula necator* (Schw.) Burr., discovered by Schweinitz in North America in 1834. The anamorph state is called *Oidium tuckeri* (Berk.). It is an obligate parasite of various genera belonging to the family *Vitaceae* (Bulit & Lafon, 1978). The fungus produces black spherical bodies called cleistothecia in its sexual structures; these bodies are formed by the fusion of male and female hyphae (Pearson & Gadoury, 1978) and contain between four and six asci with four to seven ascospores each. The hyphae of the fungus develop multilobed appressoria on which pegs used to penetrate the host develop. Powdery mildew can infect all the green tissues of the grapevine. The fungus only penetrates epidermal cells, into which it introduces globular structures known as haustoria that are used to absorb nutrients. Diseased plants develop a whitish grey powder caused by the presence of abundant conidiophores (Sall, 1980).

Both sides of leaves of all ages are susceptible to infection, which manifests as a fine white powder. When young leaves become infected, they shrink and become distorted, and the infection of clusters close to inflorescences can cause poor fruit set and considerable reductions in crop yields. Infected shoots develop a dark green patch that may turn brown and then black. Berries are prone to infection until they reach a sugar content close to 8%. Once infected, however, they continue to be affected by the production of fungal spores until a sugar content of 15% is reached. If berries become infected before they have reached their full size, the epidermal cells die but the flesh continues to grow, causing the berry to split open and either dehydrate or rot. Infection by *B. cinerea* is common in such cases. When almost-ripe grapes (particularly red ones) become infected, they cannot be sold as fresh grapes or used to make wine as they have a negative impact on flavor.

Powdery mildew stunts growth and reduces crop yields. In particular, it affects the quality of grapes and the plant's resistance to cold. *U. necator* fungi can overwinter as hyphae in dormant buds or as cleistothecia on the surface of canes (the main shoots). Mycelia and conidia survive from one season to the next in the plant tissues. The fungus develops within the bud, where it remains latent until the following season. The conidia, in turn, reproduce in large numbers on infected shoots and spread rapidly to nearby parts.

The infection cycle begins in spring, when temperatures exceed 15°C. When moistened by rain, the cleistothecia open and release ascospores, which germinate and infect green tissues, forming colonies that produce conidia for secondary spread. Temperature is the most important environmental factor in terms of fungal growth, with optimal development occurring between 20 and 27°C. Rain can disrupt the spread of infection as it washes away the conidia and breaks up the mycelium. A relative humidity of between 40% and 100% is sufficient for the conidia to germinate. Low indirect sunlight favors the development of powdery mildew, and bright sunlight can inhibit the germination of conidia.

Disease control is normally achieved with copper and organic fungicides such as benomyl, dinocap, and sterol biosynthesis inhibitors. Sulfur was the first fungicide used to successfully treat powdery mildew and it is still the most widely used treatment because of its efficacy and cost. Copper and organic fungicides are also available commercially but their use is not as widespread as that of sulfur, which has the disadvantage of being less active in wet environments. Cultivation strategies can reduce disease severity and increase the effectiveness of chemical treatments. It is beneficial, for example, to plant vines in such a way that they will be exposed to good air currents and plenty of sunlight. Vine training systems designed to allow good air circulation through the canopy and prevent excessive shade are also a highly effective infection control system. Good air circulation can also be achieved with suitable pruning.

## 2.2. Grapevine Downy Mildew

Grapevine downy mildew is caused by the fungus *Plasmopara viticola* (Berk & Curt.) Berl. & de Toni, which is an obligate parasite of several genera of the family *Vitaceae* (Lafon & Bulit, 1981). It grows intracellularly in infected grapevine tissue, where it forms tubular hyphae with globular haustoria. The fungus generally overwinters as oospores in fallen, dead leaves, although it can also survive as a mycelium in buds and persistent leaves. Oospores, which survive better in the upper layers of damp soil, germinate in the spring and produce a sporangium from which primary zoospores are dispersed by rain. This process requires a relative humidity of 95 to 100% and at least 4 h of darkness; the optimal temperature is between 18 and 22°C. The sporangia are wind-dispersed to leaves, where they germinate. The resulting zoospores swim through water and encyst near the stomata. Because host penetration occurs exclusively through the stomata, only plant structures with these pores are susceptible to infection. The optimal temperature for the growth of *P. viticola* is 25°C but rain is the main factor responsible for epidemics. The worst cases of downy mildew are seen in years with a wet winter followed by a wet spring and a hot summer with intermittent rain storms (Langcake & Lovell, 1980).

Grapevine downy mildew is more common in regions in which the plant growth phase coincides with warm, wet weather. The growth of this fungus is limited in areas with little rainfall in the spring and summer and in vineyards located in more northerly regions, where sufficiently high temperatures are not reached in spring. The fungus attacks all the green parts of the vine, particularly the leaves. It appears as a characteristic oily patch on the upperside of the leaves (which become a dull green or yellowish color) and a downy white growth on the lower sides. The infection of leaves is a very important source of inoculum for the infection of berries. Infected leaves generally fall off, leading to a reduction in the accumulation of sugar in the fruit and an increase in the vulnerability of buds in winter. Young berries are very susceptible to infection. When infected, they turn a grayish color and become covered in gray powder caused by sporulation. Although berries become less prone to infection as they ripen, the infection can spread inwards and attack the older berries. In autumn, mosaic-like symptoms can appear on old leaves.

Preventive cultivation strategies consist of ensuring proper ground drainage, reducing sources of inoculum in winter, and pruning the tips of infected shoots. The timing of treatment is extremely important. The fungus needs wet weather and a temperature of between 15 and 25°C to grow. Optimal growth conditions are thus rain, fog, or showers followed by hot, sunny days. Fungicides are the best means of control (Lafon, 1985). Nonsystemic chemical fungicides such as dithiocarbamates and copper salts, which are used in spring, are preventive only and only protect the treated surfaces. Cymoxanil and chlorothalonil are specific nonsystemic mildew fungicides and are curative if used in the first few days following infection. Aluminium and phenylamides are two of the most widely used systemic fungicides to treat grapevine downy mildew.

## 2.3. Black Rot

Black rot is caused by *Guignardia bidwellii* (Ellis) Viala & Ravaz (anamorphic state, *Phyllosticta ampelicida* [Engleman] Van der Aa) (Sivanesan & Holliday, 1981). The disease originated in North America and was probably spread to

other countries through contaminated material. This fungus can cause crop losses of between 5 and 80%, depending on the severity of the epidemic, which, in turn, depends on the infective capacity of the inoculum, the weather conditions, and the susceptibility of the host plants.

The main symptom on leaves is the appearance of small, dark, circular patches in spring and early summer, about 1 to 2 weeks after infection. The lesions, which are initially cream-colored, become progressively darker and eventually acquire a reddish-brown color on the upper side of the leaf. Pycnidia appear as small black spots (59–196 μm in diameter) in the center of the patches, and long, black canker lesions measuring between several millimeters and 2 cm in length develop on the edges of young leaves. Infected berries develop small white spots and after a few days begin to dehydrate, shrivel, and wither. The infection can affect whole clusters.

The fungus overwinters in fallen mummies (shriveled, diseased grapes) or in old shoots still on the vine. Ascospore release begins in spring, shortly after bud break (opening of buds); this process is favored by frequent rainfall as the ascospores need water in order to germinate. The ascospores produce lesions on the leaf and infect inflorescences and young fruit. Fruit can become infected at any time between midway through the flowering period until the grapes start to change color. Adult leaves and ripe fruit are not susceptible to infection. Pycnidia grow in dry and newly rotted grapes and, once mature, release conidia when dampened by rain, leading to the risk of infection of leaves, inflorescences, and young fruit. Infection starts to decline at the end of July and disappears at the end of August (Spotts, 1980).

Chemical control measures involve the use of preventive fungicides such as manel and ferbam. Treatment should be started when the shoots are 10 to 16 cm long and maintained until the grapes reach a sugar content of 5%. Curative fungicides such as triadimefon are also used and should be applied as soon as the disease is detected.

## 2.4. Phomopsis Cane and Leaf Spot

Phomopsis cane and leaf spot is caused by the fungus *Phomopsis viticola* (Sacc.) Sacc. (synonyms, *Fusicoccum viticola* Reddick and *Cryptosporella viticola* Shear [telemorphic form]). It can affect all of the green parts of the grapevine. While symptoms are similar in the different parts of the plant, the extent of damage varies, with the main shoots being the most affected (Punithalingam, 1979). This disease has particularly worrying consequences in areas with frequent rainfall in spring, as the spread of disease is facilitated when the vines remain wet for several days after bud break.

As the fungus grows, it causes the appearance of black spots or patches (which eventually crack) on green wood at the base of the buds. A bulge forms at the base of newly sprouted shoots; this cracks longitudinally and signs of wood strangulation become visible underneath. This strangulation makes the canes more fragile. In autumn, the bark develops whitish patches and black spots and, in winter, the vine becomes severely damaged as numerous canes start to fall off. Leaves are also prone to attack and develop dark patches that mostly affect the petioles. The veins are rarely affected. The disease also attacks cluster stems, causing partial or total dehydration of the grapes.

One recommended disease control measure is to burn all pruning debris as this can provide a home for overwintering fungi. The use of dichlofluanid, folpet, mancozeb, maneb, or metriam in winter destroys the pycnidia on the canes prior to bud break. These fungicides can also protect young shoots when applied after bud break. The careful application of the product on spurs and canes that need treatment is more efficient than the use of spray guns (Bugaret, 1986).

## 2.5. Eutypa Dieback

Eutypa dieback is caused by the ascomycete fungus *Eutypa lata* (Persoon: Fries) Tulasne and C. Tulasne. T. It is one of the most worrying diseases for grape growers because of the devastating economic losses it can cause. First detected in Australia in 1973, the disease affects many vineyards around the world and is considered to be the most serious threat to vineyard longevity. In Spain, it was discovered for the first time in 1979 and is becoming an increasingly serious problem affecting all of Spain's grape-growing regions.

The ascospores of the fungus infect and colonize the xylem through pruning wounds and then spread to the cambium. After an incubation period of at least 3 years, infected pruning wounds become surrounded by canker and the first symptoms appear in the green parts of the plant. The main symptoms are stunted growth, withering of new branches, necrosis at the margins of leaves, dryness of inflorescences, and the death of one or more branches. There are different strains of the fungus, with varying levels of virulence.

As mentioned above, the disease can have serious economic consequences. The most sensitive grapevine varieties are Sauvignon Blanc, Cabernet Sauvignon, Ugni-blanc, Cinsault, and Chenin, and the most tolerant are Merlot and Semillon (Deswarte et al., 1994). While eutypa dieback has a direct impact on crop yield, it can also affect the quality of wine made from grapes from infected vineyards. It is one of the most serious grapevine diseases known and is becoming an increasing concern for growers. No means have yet been found for eliminating the fungus once it has infected the plant. Precautionary control measures involve destroying infected trunks or areas, pruning, and treating lesions with fungicides.

A toxin isolated in culture medium from *E. lata* was found to be toxic for grapevines (Fallot et al., 1997). This compound,

4-hydroxy-3-(3-methyl-3-butene-1-ynyl) benzaldehyde (eutypine), has been found in all infected vines (raw sap, leaves, inflorescences, herbaceous stems, etc.) but has yet to be isolated in a healthy plant. In vitro assays have shown that eutypine rapidly leads to the development of symptoms on cut grapevine leaves and causes structural alterations similar to those described for infected grapevine leaves in vivo, demonstrating that it is involved in the manifestation of eutypa dieback symptoms. In grapevine cells, eutypine is metabolized into eutypinol (4-hydroxy-3-[3-methyl-3-butene-1-ynyl] benzyl alcohol), which is not toxic for the vine. This biotransformation is catalyzed by a NADPH reductase (eutypine reductase) (Colrat et al., 1999). Tolerant grapevine varieties have a greater capacity to metabolize eutypine than their more sensitive counterparts. The discovery of this mechanism of action opens new perspectives for the development of efficient tools to manage this worrying fungal disease. Preventive measures consist of burning dead vines and of pruning and burning diseased shoots and other plant parts. Pruning wounds can be treated with carbendazim, thiophanate-methyl, or triadimefon paste.

# 3. B. CINEREA AS A MODEL FOR STUDYING GRAPEVINE FUNGAL DISEASES

Species from the phytopathogenic fungal genus *Botrytis* constitute a serious threat to a wide variety of crops. *B. cinerea* (perfect state, *Botryotinia fuckeliana*) is a particularly virulent variant and attacks many types of crops including grapevines, causing characteristic necrotic patches on leaves, stems, and fruits, and forming a grayish powdery mold known as gray mold (Snowdon, 1990) (see Figure 10.1). In viticulture, where the disease is commonly referred to as botrytis bunch rot, the fungus can have particularly serious consequences as it can reduce crop yield and alters the organoleptic properties of wine. The frequency and intensity of attacks have made *B. cinerea* one of the most feared diseases in the agricultural community. The development of rational control programs is of major environmental importance (Rebordinos et al., 2003) and is a priority for the agrochemical industry with far-reaching consequences for viticulture.

## 3.1. Grapevine Infection by *B. cinerea*

The quality of wines made with grapes infected by *B. cinerea* is diminished by the reduction in monosaccharide content (glucose and fructose) and the accumulation of metabolites (glycerol and gluconic acid) and enzymes that catalyze the oxidation of phenolic compounds. These wines do not age well either as they are

FIGURE 10.1 Gray mold caused by *Botrytis cinerea* on a cluster of grapes.

susceptible to oxidation and bacterial contamination (Bulit & Dubos, 1988; Coley-Smith et al., 1980). Although *B. cinerea* can cause serious damage in the winemaking industry, it is also responsible for excellent wines such as Sauternes (France), Tokaji (Hungary), and Trockenbeerenauslesen (Germany, Austria) (Coley-Smith et al., 1980). This benevolent form of *B. cinerea*, known as noble rot, proportionally consumes more acid than sugar, giving rise to smooth, sweet wines with a good body and a pleasant bouquet.

*Botrytis* is a parasite that first establishes itself in the weaker, damaged parts of the host before spreading to the rest of the plant. It can also, however, attack plants that are already infested by other pests or pathogens. *B. cinerea* infects its hosts by physical or chemical penetration (Isaac, 1992). In the case of physical penetration, the fungus exploits natural openings on the plant such as the stomata or small wounds that can appear on the surface of leaves or fruits. These openings are relatively unprotected and are therefore vulnerable to penetration. In both cases, for direct penetration to occurs, the fungus needs to attach itself to the host tissue via an anchor system (appressorium) attached to the germination tubes (branched hyphae). As the fungus is not capable of growing by hyphal branching alone, it secretes substances that destroy or prepare the plant tissue for hyphal penetration.

## 3.2. Chemical Penetration

Once the fungus is near the cell wall, it launches a biochemical attack on the plant tissue and cells to aid the spread of infection. A large number of complex interactions take place during the cell invasion process and the development of symptoms (see Figure 10.2). The fungus uses two chemical weapons at this point: high-molecular-weight enzymes that break down the cell wall and membrane and lead to tissue maceration, and low-molecular-weight toxins that kill the plant cells as the hyphae advance through the host tissue. Because the production of these chemical weapons appears to be essential for the pathogenicity of *B. cinerea*, the interruption of this activity may render the pathogen harmless.

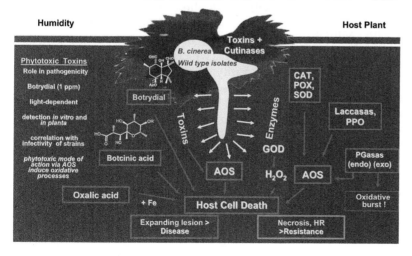

FIGURE 10.2 *Botrytis cinerea* infection process. CAT = catalase; POX = peroxidase; SOD = superoxide dismutase; GOD = glucose oxidase; PPO = polyphenoloxidase; AOS = active oxygen species.

*Botrytis* infections are characterized by the appearance of necrotic lesions and rot accompanied by the secretion of different types of important substances, mainly toxins (Coley-Smith et al., 1980; Collado et al., 2000, 2007; Elad et al., 2004) and enzymes that break down the cell wall (polygalacturonases, pectin lyases, cellulases) (Prins et al., 2000) and the cell membrane (phospholipases, lipases) (Prins et al., 2000; Shepard & Pitt, 1976). The chemical action of the toxins and enzymes secreted during the infection process leads to the production of reactive oxygen species (ROS) (Deighton et al., 1999; Govrin & Levine, 2000; von Tiedemann, 1997). ROS have been detected in all the infectious processes studied to date and have become the focus of much research in recent years (Deighton et al., 1999; Govrin & Levine, 2000; Prins et al., 2000; von Tiedemann, 1997). It has also been demonstrated that ethylene is frequently produced in plant tissues in response to attack by *B. cinerea* (Elad, 1995).

Enzymes play an important role in the infection process. While pectic enzymes are instrumental in the degradation of tissues as they facilitate access to other enzymes, cell-wall-degrading enzymes alter osmotic pressure, causing cell death (Elad, 1995).

Recent studies have uncovered completely new aspects of the mechanisms used by necrotrophic fungi such as *B. cinerea* to attack and invade their hosts. There is now evidence, for example, that the fungus produces ROS during the infection process through the chemical action of toxins and enzymes. ROS are produced by both fungi and hosts when the cell walls are broken down, triggering a series of free-radical reactions (Deighton et al., 1999; Govrin & Levine, 2000; Prins et al., 2000; von Tiedemann, 1997). The infection mechanism by which gray mold is produced is thus complex and involves both external factors, which are necessary for the process to begin, and a series of chemical reactions that damage the host cell wall and membrane and aid the spread of hyphae through the tissue.

Figure 10.2 shows a simplified, summarized version of the different molecular processes and chemical reactions that take place during *B. cinerea* infection. Some of these processes have not yet been studied in detail and much remains to be learned before we can fully understand the infection mechanism employed by this necrotrophic fungus. Several aspects, however, have recently been clarified, such as the role played by toxins in the infection mechanism, which we will discuss in further detail in Section 3.3.5.

## 3.3. Strategies for the Analysis of *B. cinerea*

The serious damage caused by this plant pathogen calls for physiological, biochemical, and molecular studies that will shed light on the mechanisms underlying *B. cinerea* infection. Most of the studies performed to date have focused on identifying the enzymes involved in the infectious process or the metabolites produced during infection. In recent years, however, considerable efforts have been made in the area of molecular biology (Rebordinos et al., 1997), though many questions remain about the molecular mechanisms underlying the pathogenicity of the fungus and the molecular basis of resistance to fungicides. Greater knowledge of these aspects will help in the design of effective control strategies. Much also remains to be learned about the genetics of the fungus, which is difficult to study because of the rarity with which the sexual stage of *Botrytis* is seen in nature (see Figure 10.3). In order to fill these important gaps in our knowledge, our research groups have characterized several strains of *B. cinerea* isolated in different grape-growing areas of Spain, with particular focus on the following aspects:

1. *Morphology.* By modifying in vitro culture conditions and observing the fungus in each of the stages of its life cycle, we were able to perform crosses between sexually

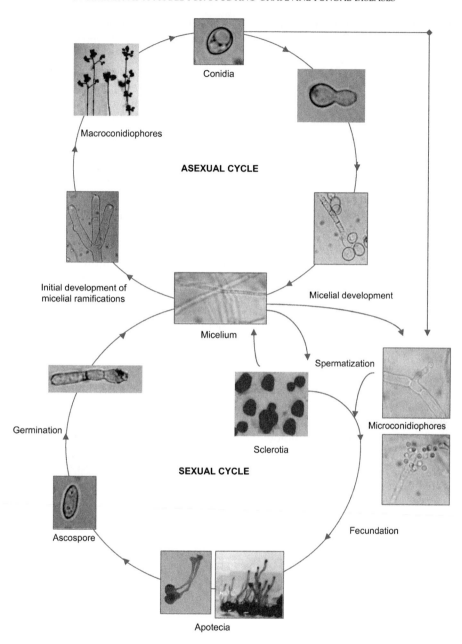

**FIGURE 10.3** Life cycle of *Botrytis cinerea* (perfect state, *Botryotinia fuckeliana*).

compatible strains and study offspring and level of pathogenicity.

2. *Determination of karyotype by pulsed field gel electrophoresis (PFGE).* This revealed polymorphism among the strains analyzed, although several strains had an identical electrophoretic profile. We also developed a novel autonomous transformation system in *B. cinerea*, and cloned, sequenced, and characterized the *gdhA* gene, which is involved in the metabolism and regulation of nitrogen.

3. *Proteomic analysis.* In this first-ever proteomic study of *B. cinerea*, we optimized the entire process, from the extraction of proteins to their identification. We also performed the first differential proteomic study of *B. cinerea*, comparing proteomes from two strains with different virulence. On the basis of our findings, we selected proteins that were expressed in both of the strains analyzed and proteins that were overexpressed in the most virulent strain for further analysis.

4. *Isolation and characterization of compounds from culture broths.* The compounds were found to be toxins or toxin derivatives, some of which displayed considerable biological, phytotoxic, and antibiotic activity.

5. *Role of toxins in infection mechanism of B. cinerea.* We found that both families of toxins we had isolated were made redundant during infection by the most virulent strains.

6. *Pathways involved in the biosynthesis of toxins.* We characterized several of the genes involved in the secondary metabolism of the fungus and several of the main enzymes involved in the biosynthetic pathways of the above toxins.

### 3.3.1. Morphological Characteristics of B. cinerea *Strains and Relationship with* Pathogenicity

Our group isolated and studied several strains of *B. cinerea* from different agricultural regions as well as strains from the Spanish Type Culture Collection (CECT) and strains donated by Dr. F. Faretra (University of Bari, Italy) for crosses. The first step was to morphologically characterize the strains (according to production of resistance structures and growth rate) and to analyze levels of pathogenicity. Morphological differences were observed between the strains, with variations seen according to the time in culture mostly attributable to heterokaryosis. Strains with the least infective capacity did not produce sclerotia. Infective capacity was low in strains that mostly retained a conidial morphology and higher in those with a very fast growth rate. The size of the conidia, determined by scanning electron microscopy (see Figure 10.4), was another highly variable character, as was the number of nuclei in the conidia (determined by fluorescence microscopy). No relationship was found between these variables and pathogenicity (Vallejo, 1997).

The optimization of the method used to obtain *B. cinerea* apothecia under controlled laboratory conditions has resulted in greater

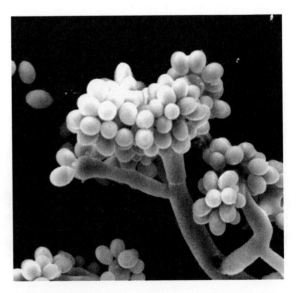

**FIGURE 10.4**  Images of *Botrytis cinerea* conidia obtained by scanning electron microscopy.

knowledge of the fungus thanks to genetic studies of its progeny. The first studies designed to determine sexual compatibility led to the conclusion that this is controlled by a single gene, *MAT 1*, with two alleles, *MAT-1* and *MAT-2*. Not many studies of this type have been performed, however, because of the difficulty of and time required to obtain apothecia, which largely explains why knowledge of the genetic aspects of this fungus is still very limited.

We performed numerous crosses between sexually compatible reference and grapevine strains and studied genotype segregation of sequences encoding ribosomal RNA (rRNA), as rRNA contains valuable information on basic species structure and gene function. rRNA can also be used to perform taxonomic and systematic analyses to establish intra- and interspecific relationships. The segregation of the above sequences in single-spore progeny of two crosses between *B. cinerea* strains was analyzed using an rDNA hybridization probe labeled with digoxigenin on nylon filters containing chromosome bands corresponding to electrophoretic profiles for single-spore segregations obtained by PFGE (Vallejo et al., 2002).

In a subsequent study, in vitro bioassays on *Vitis vinifera* (grapevine) and *Phaseolus vulgaris* (common bean) leaves were used to determine the infective capacity of the progeny (Vallejo et al., 2003). The virulence of both parent strains and progeny were analyzed by inoculating the leaves with a suspension of conidia (of a known concentration) and observing the development of symptoms over the course of days. We found that the pathogenicity trait had been passed from the parent strains to their progeny.

### 3.3.2. Molecular Analysis of B. cinerea: Karyotype Polymorphism and Cloning and Characterization of the gdhA Gene

Molecular biological tools are important for gaining a greater understanding of *B. cinerea* and its mechanisms of action. Restricted

fragment length polymorphism (RFLP) analysis of numerous strains of *B. cinerea* isolated in infected grapes has revealed considerable genetic variability and the presence of the transposable elements *Boty* and *Flipper* in different populations of this fungus. PFGE, in turn, has revealed considerable chromosome polymorphism between different strains of the fungus (Vallejo et al., 1996) as well as the presence of minichromosomes, which may correspond to what are referred to as supernumerary or B chromosomes. These chromosomes are primarily characterized by their variable length, absence in certain strains, and abnormal segregation in sexual crosses. Their exact function, however, remains to be clarified.

Karyotypic analysis by PFGE has also revealed extensive polymorphism between different strains of *B. cinerea* (see Figure 10.5), with between four and eight resolvable bands measuring between 1.88 and 3.86 Mb detected. The minimum genome size of these strains was calculated to range between 11.22 and 22.92 Mb (Rebordinos et al., 2000). In that study, however, we did not find an association between any of the bands and level of pathogenicity, although the high phenotypic variability could be explained by chromosome polymorphism and the heterokaryotic nature of the fungal cells. The study of nonpathogenic mutants created using agents such as ultraviolet radiation will possibly contribute to a greater understanding of the infection mechanisms of *B. cinerea*.

One of the most useful and powerful tools to emerges in molecular biology in recent years is cloning. Thanks to this technique, numerous genes involved in *B. cinerea* pathogenicity and plant-pathogen interactions have been identified. Although numerous enzymes associated with the infection process in this fungus have also been identified, no conclusive evidence has yet been obtained that any are directly responsible for the damage caused by *B. cinerea*.

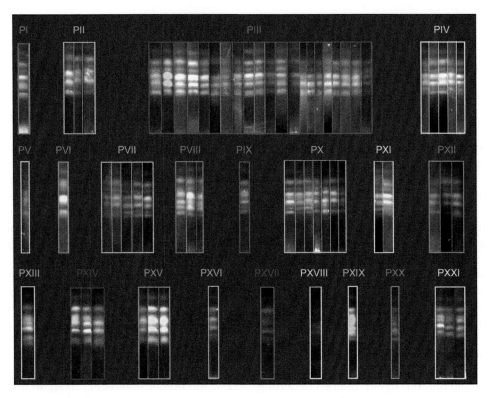

**FIGURE 10.5** Polymorphism in the karyotype of 68 strains of *Botrytis cinerea* (pulsed field gel electrophoresis).

The shortage of nutrients such as nitrogen appears to be linked to pathogenicity and several morphological aspects of *B. cinerea*. The loss of regulatory factors involved in the assimilation of ammonia, for example, significantly reduces the virulence of certain plant and animal fungi. Given the absence of studies on nitrogen and carbon metabolism in *B. cinerea* and the enormous importance of understanding the interactions between these two pathways and their possible association with pathogenesis, our group decided to clone the *gdhA* gene, which encodes the NADPH-dependent glutamate dehydrogenase (GDH) enzyme involved in the synthesis of the essential amino acid glutamate and, as such, responsible for the direct assimilation of ammonia. Analysis of the regulatory mechanisms for this gene, the creation of mutants, and above all the effect of different substrates on the *gdhA* gene promoter may help to establish a direct relationship between nitrogen metabolism and pathogenicity in *B. cinerea*.

The *gdhA* gene was cloned in *B. cinerea* by first screening a genomic library by hybridization with a heterologous probe from *Aspergillus awamori* containing part of the coding sequence of the gene (Santos et al., 2001). This yielded a 3.48 kb DNA fragment from which 2351 nucleotides were sequenced and found to contain an open reading frame (ORF) of 1350 base pairs coding for a 450-amino-acid protein (this sequence is available from the EMBL Nucleotide Sequence Database under accession no. 093934). The size of the monocistronic transcript was estimated at 1.7 kb. The gene was

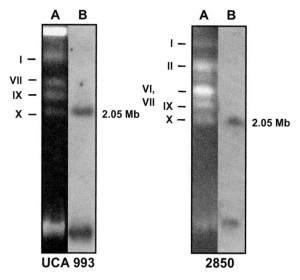

**FIGURE 10.6** Karyotype (A) of two strains (UCA 993, 2850) and localization of *gdhA* gene (B) in chromosome X (numbering of bands according to Santos et al., 2001).

located on chromosome X (see Figure 10.6), whose karyotype was analyzed by PFGE; the gene complemented two *Aspergillus nidulans* mutants and restored NADP-dependent GDH activity (Santos et al., 2001). Expression analyses in the same study indicated that the gene was subject to strong regulation by carbon and nitrogen.

Our group was the first to implement an innovative (nonintegrative) transformation system in *B. cinerea* using the plasmid pUT737; the system had a transformation efficiency of 25–40 transformants/µg of DNA and these transformants maintained their capacity for four generations without selective pressure (Santos et al., 1996).

### 3.3.3. Proteomics Study of B. cinerea

Important advances have been made in recent years in the area of proteomics and its application to the study of biological, metabolic, pathological, and other processes. Indeed, the current age of molecular biology is referred to as the post-genomics era. The proteomic analysis of filamentous fungi is, however, still in its infancy, and most of the studies conducted in this area have been of species of industrial interest. The characterization of the first proteome of *B. cinerea* by Fernández-Acero et al. (2006) represented a breakthrough in terms of optimizing the process from start to finish as their study reported all the stages involved in the analysis, from protein extraction to identification. Other techniques can be used to identify factors involved in the infection processes of different organisms, but most involve the selection of a candidate gene in a previous screening step and the subsequent analysis of its influence on infection processes via expression analysis or directed mutagenesis. Proteomics studies, in contrast, do not require an a priori selection of genes but rather focus on analyzing the expression of whole sets of proteins under given conditions.

Fernández-Acero et al. (2006), using two-dimensional gel electrophoresis and Coomassie blue staining, detected between 380 and 400 protein spots with a molecular weight of between 15 and 85 kDa and an isoelectric point of between 5.4 and 7.7. They selected 22 of these spots for identification by matrix-assisted laser desorption ionization time of flight or electrospray ionization ion trap mass spectrometry. Among the proteins found were different forms of malate dehydrogenase (MDH) and glyceraldehydes-3-phosphate dehydrogenase (GADPH), cyclophilin, and other proteins of unknown function.

In a later study, the proteomes from two strains of *B. cinerea* with different virulence were compared and 28 proteins that were either expressed in both strains or overexpressed in the more virulent strain were selected for analysis (Fernández-Acero et al., 2007). The most relevant proteins in terms of their possible involvement in the infection process were (1) NADPH-dependent MDH, (2) GAPDH, (3) metE/metH, and (4) cyclophilin. The first,

MDH, was found in three clusters and overexpressed in the most virulent strain. This substance catalyzes the transformation of malate to oxaloacetate, the main precursor of oxalic acid, which has been described as a pathogenicity factor. High levels of oxalic acid reduce the pH of cultures, which is a prerequisite for the production of the toxins botrydial and dihydrobotrydial by *B. cinerea*. These data support the hypothesis that MDH is involved in the pathogenicity of this fungus. It was recently demonstrated that these NADPH-dependent oxidases play a role in the differentiation and pathogenicity of *B. cinerea* (Segmüller et al., 2008), a finding that seems to be in agreement with the proteomic data available. The second protein, GAPDH, was found only in the most virulent strain. The role of this enzyme in the pathogenicity of various organisms has been widely reported (Alderete et al., 2001). GAPDH might, thus, in addition to its metabolic function, have a role in the infection cycle of *B. cinerea*, as was recently reported for the hexokinase Hxk1 (Rui & Hahn, 2007). The third protein, the transcriptional regulator metE/metH, was specific to and overexpressed by the most virulent strain. This protein is involved in the synthesis of methionine, a pathway that has been widely used in the design of fungicides. The variability observed between the different strains of *B. cinerea* might be the molecular basis of the different fungicide resistance phenotypes described for this fungus. The fourth protein, cyclophilin, which is associated with protein assembly and regulation, was found only in the most virulent strain. Cyclophilin has been described as a virulence factor in different fungi, including *B. cinerea*, in relation to tissue invasion and colonization processes, confirming the usefulness of proteomics to identify components involved in infection mechanisms. The protein might also be involved in cell signaling as it forms a complex with calmodulin, which is involved in various cell signaling cascades.

The first proteomic map of *B. cinerea*, published recently (Fernández-Acero et al., 2009), contains over 300 identified proteins that have been functionally classified into molecular and biological groups using the PANTHER (protein analysis through evolutionary relationships) database (http://www.pantherdb.org). This information is freely available at the website of the Swiss Institute of Bioinformatics through the ExPASy Proteomics Server (http://world-2dpage.expasy.org/repository/, accession number 0005). Because cellulose is one of the main components of the cell wall, many of the proteins identified play a crucial role in the pathogenesis of the fungus.

Given the enormous difficulty and complexity associated with proteomic analysis, more recent studies have taken a more simplified approach by dividing samples into more manageable packets called subproteomes (Fernández-Acero et al., 2007). The first of these subproteomes to be studied in *B. cinerea*, the secretome (Fernández-Acero et al., 2009; Shah et al., 2009), provided promising results that reflect the potential of this tool for further investigating pathogenicity factors and therapeutic targets and conducting basic research into *B. cinerea*.

### 3.3.4. Isolation and Characterization of Toxins Secreted by *B. cinerea*

In addition to producing organic acids and high- and low-molecular-weight polysaccharides (from glucose monomers and from mannose, galactose, glucose, and ramnose monomers, respectively) (Coley-Smith et al., 1980; Elad et al., 2004), *Botrytis* fungi also secrete a series of secondary sesquiterpene metabolites during the development of necrotic lesions on the host plant (Colmenares, 2001; Colmenares et al., 2002; Deighton et al., 2001; Rebordinos et al., 1996). Recent studies have shown that these metabolites are toxins that constitute a virulence factor in plant pathogens of this type (Colmenares, 2001; Colmenares et al., 2002; Deighton et al., 2001; Rebordinos et al., 1996).

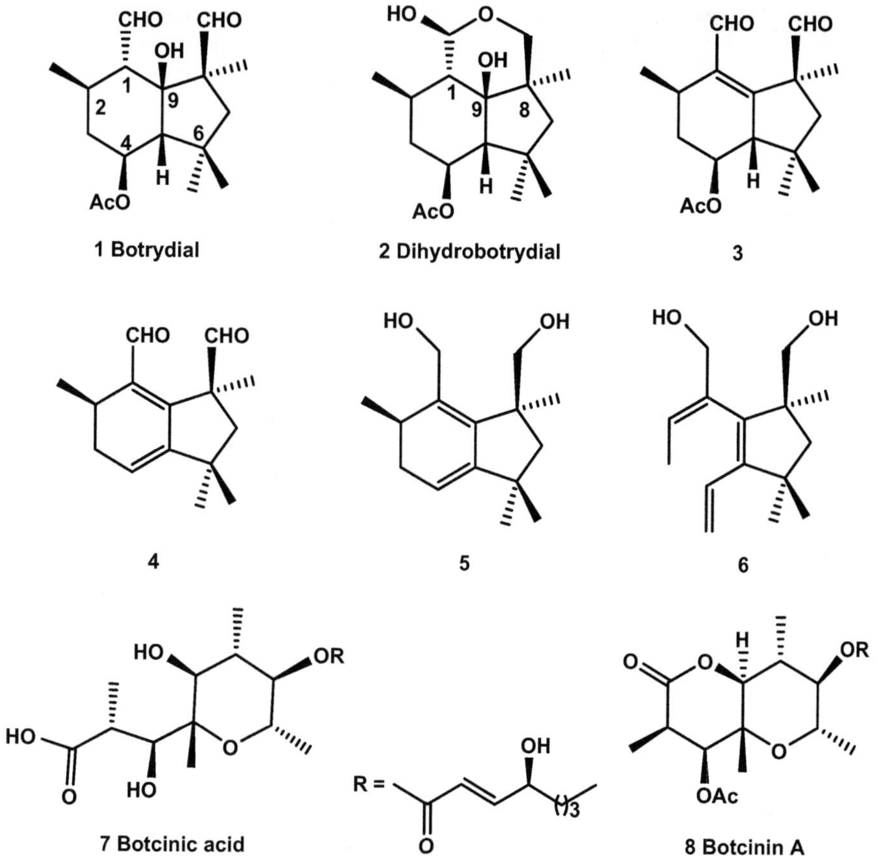

FIGURE 10.7 Structure of secondary metabolites produced by *Botrytis cinerea*.

*B. cinerea* synthesizes a series of metabolites with a botryane skeleton—mainly botrydial and dihydrobotrydial (see Figure 10.7) and derivatives of these—that are responsible for the symptoms associated with this fungus (see Figure 10.8) (Collado et al., 2007; Colmenares, 2001; Colmenares et al., 2002; Durán-Patrón et al., 2000). The in vivo detection of botrydial in *Capsicum annuum* (sweet pepper) plants infected with *B. cinerea* showed that these metabolites, or toxins, are associated with pathogenesis and infection (Deighton et al., 2001). The toxins, which are not host-specific, help the fungus to penetrate the host and colonize the plant tissue, thus increasing the severity of disease.

Two structurally distinct families of toxins synthesized by *Botrytis* fungi have been isolated. The most abundant of these has a botryane skeleton and is a chemical derivative of the plant toxin botrydial. Some of the most relevant toxins isolated are shown in Figure 10.7. A new family of toxins recently isolated from virulent strains of *B. cinerea* and initially called botcinolides have a polyketide backbone. The structures of these toxins were recently revised (Tani et al., 2005, 2006), however, and they have since been renamed botcinins, the most abundant of which is botcinic acid (see Figure 10.7). These toxins act synergically with botrydial and derivatives in

the infection mechanism of *B. cinerea* (Siewers et al., 2005), and it was recently demonstrated that both families have a redundant role in the infection mechanism of aggressive strains (Pinedo et al., 2008).

Diverse biosynthetic studies have revealed detailed information about the biosynthetic pathway to generate botrydial and its derivatives from farnesyl pyrophosphate (see Figure 10.9) (Bradshaw et al., 1977, 1982; Durán-Patrón et al., 2001, 2003). Furthermore, kinetic growth studies of *B. cinerea* that have analyzed botrydial and three of its derivatives have elucidated the biosynthetic pathways of botrydial derivatives secreted by *B. cinerea* (Daoubi et al., 2006). There are two main pathways involved in degrading the most active toxin secreted by the fungus, botrydial, which is used as a chemical weapon during the infection mechanism. Figure 10.10 shows a general diagram of the degradation

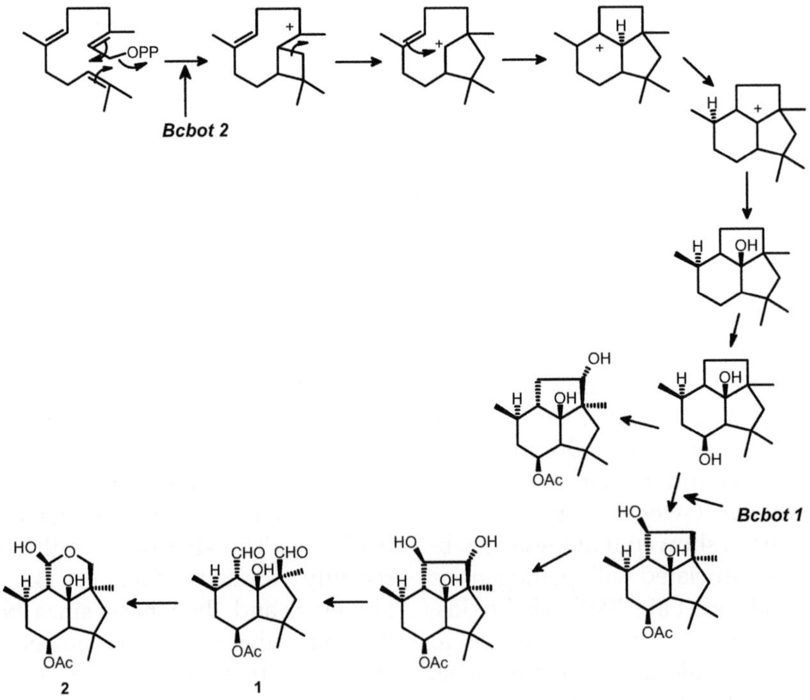

**FIGURE 10.8**  Effect of the toxin botrydial on *Phaseolus vulgaris* leaf.

**FIGURE 10.9**  Biosynthetic pathway of botrydial and derivatives from farnesyl pyrophosphate.

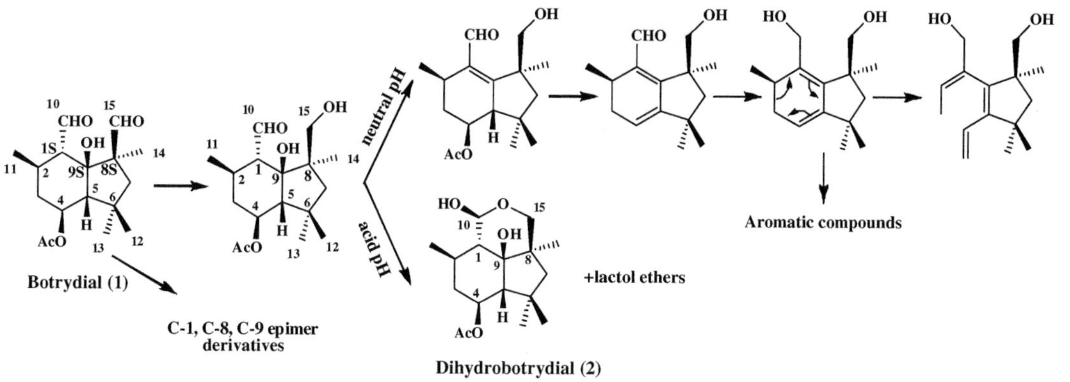

FIGURE 10.10  Detoxification of botrydial and relationship between toxins with a botryane skeleton produced by *Botrytis cinerea*.

process and the relationship between the main metabolites secreted by *B. cinerea*.

Our knowledge of the *B. cinerea* genome, combined with the increasing proteomic data available, will help to elucidate the role played by genes in the different biological processes in the fungus and to identify new molecular targets to be exploited in the design of future fungicides. Recent studies have characterized a cluster of genes involved in botrydial biosynthesis, namely those coding for a sesquiterpene cyclase (*BcBOT2*), an acyltransferase (*BcBOT5*), and three monooxygenases (*BcBOT1, 3,* and *4*) (Pinedo et al., 2008; Siewers et al., 2005) (see Figure 10.11). Recent studies of the *B. cinerea* strains B05-10 ku70 and SAS56 (which produce both families of toxins) and the T4 strain (which produces just one family, namely botrydial and derivatives), resulted in the identification of the first two genes involved in the biosynthesis of botrydial: *BcBOT1* and *BcBOT2*. These genes encoded a P-450 monooxygenase (Siewers et al., 2005) and a sesquiterpene cyclase (STC) (Pinedo et al., 2008), respectively.

The above studies confirmed that *BcBOT1* is involved in one of the final steps in the biosynthesis of botrydial (see Figure 10.9), while *BcBOT2* is involved in the first step of the cyclization of farnesyl pyrophosphate to the botryane skeleton (see Figure 10.9). Presilphiperfolan-8-ol synthase, an enzyme with a key role in the biosynthesis of botrydial and a large number of polycyclic sesquiterpenes of fungal and plant origin with diverse biological activities (see Figure 10.12), has also been characterized (Pinedo et al., 2008).

Finally, the discovery of the polyketide synthases involved in the biosynthesis of botcinins (Collado, Viaud, Tudzinsky groups, unpublished results) will also help in the design of fungicides based on hybrid molecules that behave as double inhibitors of polycyclic sesquiterpenes and polyketide synthases.

### 3.3.5. Evidence on the Role of B. cinerea Toxins in the Infection Mechanism

As indicated previously, *B. cinerea*, which is responsible for the devastating gray mold disease, produces several biologically active

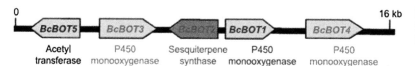

FIGURE 10.11  Cluster of genes involved in the biosynthesis of botrydial (1) and derivatives.

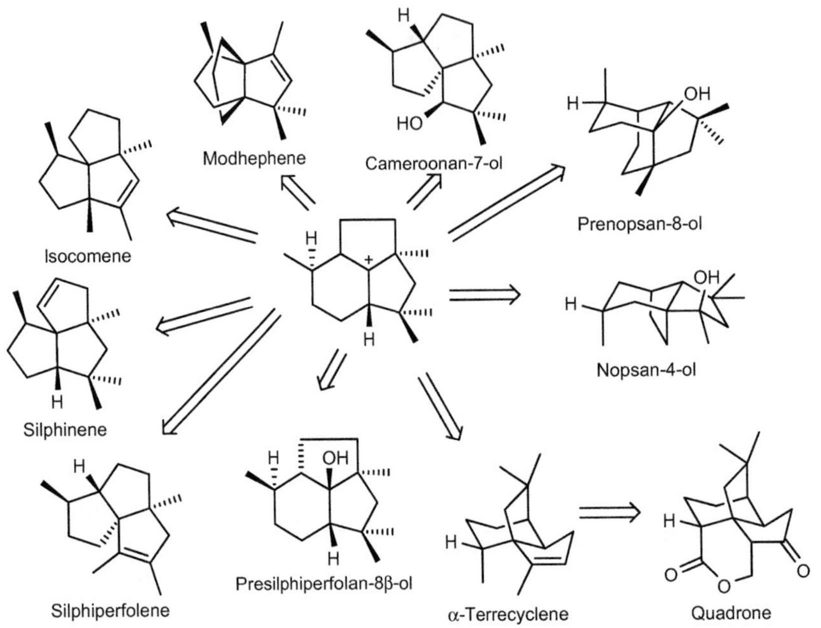

**FIGURE 10.12** The cation presilphiperfolan-8-ol is the precursor of a wide variety of triquinanes and sesquiterpenes derived from fungi and higher plants.

metabolites, several of which (botrydial and compounds 3 and 4 in Figure 10.7) have been found to exhibit high levels of *in vitro* toxicity. Similar results have been obtained *in vivo* during the testing of these metabolites in *B. cinerea*-susceptible and -resistant *P. vulgaris* plants (genotypes N 90598 and N 90563, respectively). The *in vivo* effect, however, occurred much more rapidly, with symptoms becoming evident after just a few hours (see Figure 10.8). Botrydial was capable of reproducing the phytotoxic effect of *B. cinerea* at a concentration of one part per million. At this concentration, the toxin affected 70% of the treated leaves, with symptoms appearing on 4% of the surface area; leaves started yellowing after 60 h on bright sunny days and after 120 h during periods of cloudy weather. The same phytotoxic effect was observed in *Nicotina tabacum*, *Lactuca sativa*, *Fragaria vesca*, *V. vinifera*, and *Citrus limon* leaves, although it is interesting that the symptoms took longer to appear in the last three of these plants.

Similar results have also been found on applying botrydial to tomatoes, peppers, grapes, and strawberries, with chlorosis affecting the pericarp of all the fruit and evident depletion and collapse in the treated areas (Colmenares, 2001; Colmenares et al., 2002). Tests designed to detect botrydial during in vivo infection have been conducted to gain further insights into the role played by this toxin in the infection mechanism. Deighton et al. (2001), in tests conducted in *C. annuum* and on *P. vulgaris* and *Arabidopsis thaliana* leaves, for example, detected and isolated botrydial from the inoculated areas during the early stages of infection, proving that botrydial is produced in the plant during the infection process. Although the authors concluded that the toxin was associated with pathogenesis, they were unable to confirm whether or not it was a primary factor in the infection process.

Various experiments involving the inoculation of *B. cinerea* onto *P. vulgaris* leaves previously treated with botrydial have clearly shown that the toxin facilitates fungal penetration and tissue colonization. Taken together, these results indicate that botrydial is a nonspecific toxin that affects a wide range of plant species; it can also be considered a virulence factor as it appears to be involved in the development and spread of the disease. Whether or not botrydial is a pathogenicity factor is the focus of ongoing studies involving mutants (Collado, unpublished results). Furthermore, recent experiments have shown an interesting relationship between toxin production and virulence. The fungus is resistant to its own toxin thanks to a detoxification mechanism that reduces the compound to other inactive, nontoxic products. Daoubi et al. (2006) recently discovered two detoxification mechanisms regulated by pH (see Figure 10.10). *B. cinerea* toxins have a light-dependent mechanism of action. Phytotoxicity studies consisting of exposing treated areas to light, darkness, and alternating light and darkness clearly showed that the mechanism of action of the most active and abundant toxin, botrydial, was light-dependent and generated ROS via a type 1 photodynamic reaction involving oxygen activation (Colmenares, 2001). When exposed to light, botrydial may induce lipid peroxidation in plant cells followed by membrane changes that give rise to chlorosis, death, and collapse in the affected zone.

It should be noted that, in response to silencing of the *BcBOT2* gene in a *B. cinerea* mutant whose parent strain, *Bc* 05-10, produced botcinin acid and botcinin A (see Figure 10.7), the fungus shifted its secondary metabolism to an overproduction of these two compounds. No such shift was seen in the T4 mutant strain, which does not produce botcinin A.

Virulence tests on *Vicia fabae* and *Lycopersicum esculentum* leaves with *bcbot2* mutants have provided insights into the role played by botrydial in *B. cinerea* virulence. The results showed that the toxin is necessary for infection

in the T4 strain but not in the more virulent B05-10 strain. One possible explanation is that the B05-10 mutant compensates for the loss of botrydial by producing high quantities of botcinic acid and botcinin A, thus retaining similar levels of virulence to the parent strain. The T4 mutants, in contrast, whose parent strains did not produce botcinic acid or botcinin A (see Figure 10.7) had lost their infective capacity as they did not produce either of the toxins.

The results described in this section are particularly relevant in terms of their importance for the design of new fungicides targeting biological pathways involved in the biosynthesis of toxins that play an important role in the mechanism of infection by *B. cinerea*. Targeting these toxins could be the key to controlling *B. cinerea* and its pathogenicity.

## Acknowledgments

The research by our group described in this chapter was funded by grants from the Spanish Ministry of Science of Technology (1FD97-0668-C06, AGL2000-0635-C01 and -C02, AGL2002-04388-C01 and -C02, AGL2003-06480-C02 and -C01, AGL2005-07001-C02, AGL2006-13401-C01 and -C02, and AGL2009-13359-C01 and -C02) and the Autonomous Government of Andalusia (Proyecto de Excelencia grant PO7-FQM-02689).

We thank the following people for their help: Laureana Rebordinos, Inmaculada Vallejo, Francisco Javier Fernández-Acero, María Carbú, and Carlos Garrido (applied microbiology group) and Rosario Hernández-Galán, Antonio José Macías-Sánchez, Josefina Aleu Casatejada, and Mª Rosa Durán-Patrón (organic chemistry group).

## References

Alderete, J. F., Millsap, K. W., Lehker, M. W., & Benchimol, M. (2001). Enzymes on microbial pathogens and *Trichomonas vaginalis*: Molecular mimicry and functional diversity. *Cellular Microbiology, 3,* 359–370.

Bradshaw, A. P. W., Hanson, J. R., & Nyfeler, R. (1982). Studies in terpenoid biosynthesis. The fate of the mevalonate hydrogen atoms in the biosynthesis of the sesquiterpenoid, dihidrobotrydial. *J. Chem. Soc. Perkin Trans., I,* 2187–2192.

Bradshaw, A. P. W., Hanson, J. R., & Siverns, M. (1977). Use of induced 13C-C13 coupling in terpenoid biosynthesis. *J.C.S. Chem. Comm., 819.*

Bugaret, Y. (1986). Données nouvelles sur l'epidémiologie de l'excoriose et leurs conséquences pour la lutte. *Phytoma, 375,* 36—41.

Bulit, J., & Dubos, B. (1988). *Compendium of grape diseases.* St. Paul, MN: APS Press.

Bulit, J., & Lafon, R. (1978). Powdery mildew of the vine. In D. M. Spencer (Ed.), *The powdery mildew* (pp. 525—548). New York, NY: Academic Press.

Coley-Smith, J. R., Verhoeff, K., & Jarvis, W. R. (1980). *The biology of botrytis.* London, UK: Academic Press.

Collado, I. G., Aleu, J., Hernández-Galán, R., & Durán-Patrón, R. (2000). *Botrytis* species: An intringuing source of metabolites with a wide range of biological activities. Structure, chemistry and bioactivity of metabolites isolated from *Botrytis* species. *Curr. Org. Chem., 4,* 1261—1286.

Collado, I. G., Macias-Sánchez, A. J., & Hanson, J. R. (2007). Fungal terpene metabolites: Biosynthetic relationships and the control of the phytopathogenic fungus *Botrytis cinerea. Nat. Prod. Rep., 24,* 674—686.

Colmenares, A. J. (2001). *Obtencion de nuevos metabolitos del Fitopatógeno* Botrytis cinerea. *Evidencias sobre el papel de las toxinas en la fitopatogenicidad del hongo.* Cadiz, Spain: Doctoral thesis. Universidad de Cádiz.

Colmenares, A. J., Aleu, J., Durán-Patrón, R., Collado, I. G., & Hernández-Galán, R. (2002). The putative role of botrydial and related metabolites in the infection mechanism of *Botrytis cinerea. J. Chem. Ecol., 28*(5), 997—1005.

Colrat, S., Latché, A., Pech, J. C., Bouzayen, M., Fallot, J., & Roustan, J. P. (1999). Purification and characterization of an NADPH-aldehyde reductase from *Vigna radiata* which detoxifies eutypine, a toxin produced by *Eutypa lata. Plant Physiol., 119,* 621—626.

Daoubi, M., Duran-Patron, R., Hernandez-Galán, R., Benharref, A., Hanson, J. R., & Collado, I. G. (2006). The role of botrydienediol in the biodegradation of the sesquiterpenoid phytotoxin botrydial by *Botrytis cinerea. Tetrahedron, 62*(35), 8256—8826.

Deighton, N., Muckenschnabel, I., Colmenares, A. J., Collado, I. G., & Williamson, B. (2001). Botrydial is produced in plant tissues infected by Botrytis cinerea. *Phytochemistry, 57,* 689—692.

Deighton, N., Muckenschnabel, I., Goodman, B. A., & Williamson, B. (1999). Lipid peroxidation and the oxidative burst associated with infection of Capsicum annum by *Botrytis cinerea. Plant J., 20,* 485—492.

Deswarte, C., Rouquier, P., Roustan, J. P., Dargent, R., & Fallot, J. (1994). Ultrastructural changes produced in plantlet leaves and protoplasts of *Vitis vinifera* cv. Cabernet Sauvignon by eutypine, a toxin from *Eutypa lata. Vitis, 33,* 185—188.

Durán-Patrón, R., Colmenares, A. J., Hernández-Galán, R., & Collado, I. G. (2001). Some key metabolic intermediates in the biosynthesis of botrydial and related compounds. *Tetrahedron, 57,* 1929—1933.

Durán-Patrón, R., Colmenares, A. J., Montes, A., Hanson, J. R., Hernández-Galán, R., & Collado, I. G. (2003). Studies on the biosynthesis of secobotryane skeleton. *Tetrahedron, 59*(33), 6267—6271.

Durán-Patrón, R., Hernández-Galán, R., & Collado, I. G. (2000). Secobotrytriendiol and related sesquiterpenoids: New phytotoxic metabolites from *Botrytis cinerea. J. Nat. Prod., 63,* 182—183.

Elad, Y. (1995). *Modern fungicides and antifungal compounds* (pp. 217—233). Andover, UK: Intercept Ltd.

Elad, Y., Williamson, B., Tudzynski, P., & Delen, N. (2004). *Botrytis: Biology, pathology and control.* Dordrecht, the Netherlands: Kluwer Academic Publishers.

Fallot, J., Deswarte, C., Dalmayrac, S., Colrat, S., & Roustan, J. P. (1997). L'eutypiose de la vigne. Isolement d'une molécule synthétisée par *Eutypa lata* et toxique pour la vigne. *C.R.A. Sci. Paris, 320,* 149—158.

Fernández-Acero, F. J., Carbú, M., Garrido, C., Vallejo, I., & Cantoral, J. M. (2007). Proteomic advances in phytopathogenic fungi. *Current Proteomics, 4,* 79—88.

Fernández-Acero, F. J., Colby, T., Harzen, A., Cantoral, J. M., & Schmidt, J. (2009). Proteomic analysis of the phytopathogenic fungus *Botrytis cinerea* during cellulose degradation. *Proteomics, 9,* 2892—2902.

Fernández-Acero, F. J., Jorge, I., Calvo, E., Vallejo, I., Carbú, M., Camafeita, E., et al. (2006). Two-dimensional electrophoresis protein profile of the phytopathogenic fungus *Botrytis cinerea. Proteomics, 6,* 88—96.

Govrin, E. M., & Levine, A. (2000). The hypersensitive response facilitates plant infection by the necrotrophic pathogen *Botrytis cinerea. Curr. Biol., 10,* 751—757.

Isaac, S. (1992). *Fungal—plant interactions* (pp. 147—175). New York, NY: Chapman & Hall.

Klober, A. (1992). *Obstau-Weinbau, 17,* 1—92.

Lafon, R. (1985). Les fongicides viticoles. In I. M. Smith (Ed.), *Fungicides for crop protection, Vol. 1* (pp. 191—198). Croydon, UK: British Crop Protection Council, Monogr. 31.

Lafon, R., & Bulit, J. (1981). Downy mildew of the vine. In D. M. Spencer (Ed.), *The downy mildews* (pp. 601—614). New York, NY: Academic Press.

Langcake, P., & Lovell, A. (1980). Light and electron microscopical studies of the infection of *Vitis* spp. by *Plasmopara viticola,* the downy mildew pathogen. *Vitis, 19,* 321—337.

Ministry of Agriculture. (2004). Fisheries and Food, Government of Spain. In *Los parásitos de la vid. Estrategias de protección razonada* (4th ed.). Madrid, Spain: Ediciones Mundi-Prensa.

Pearson, R. C., & Gadoury, D. M. (1978). Cleistothecia, the source of primary inoculum for grape powdery mildew. *Phytopathology, 77*, 1509–1514.

Pearson, R. C., & Goheen, A. C. (1996). Plagas y enfermedades de la vid. In The American Phytopathological Society. (Ed.), Madrid, Spain: Mundi-Prensa.

Pinedo, C., Wang, C.-M., Pradier, J.-M., Dalmais, B., Choquer, M., Le Pecheur, P., et al. (2008). Sesquiterpene synthase from the botrydial biosynthetic gene cluster of the phytopathogen *Botrytis cinerea. ACS Chemical Biology, 3*(12), 791–801.

Prins, T. W., Tudzynski, P., von Tiedemann, A., Tudzynski, B., Have, A., Hansen, M. E., et al. (2000). Infection strategies of Botrytis cinerea and related necrotrophic pathogens in fungal pathology. In J. Kronstad (Ed.), (pp. 123–145). Dordrecht, the Netherlands: Kluwer Academic Publishers.

Punithalingam, E. (1979). Phomopsis viticola. Descriptions of pathogenic fungi and bacteria, no 635. *UK*. Kew: Commonwealth Mycological Institute.

Rebordinos, L., Cantoral, J. M., Prieto, M. V., Hanson, J. R., & Collado, I. G. (1996). The phytotoxic activity of some metabolites of *B. cinerea. Phytochemistry, 42*(2), 383–387.

Rebordinos, L., Santos, M., Vallejo, I., Collado, I. G., & Cantoral, J. M. (1997). Molecular characterization of the phytopathogenic fungus *B. cinerea*. In *Recent research developments in phytochemistry, Vol. 1* (pp. 293–307). Trivandrum, India: Research Signpost.

Rebordinos, L., Vallejo, I., Collado, I. G., & Cantoral, J. M. (2003). Control of agricultural diseases and pests: The case of *B. cinerea*. In J. L. Barredo (Ed.), *Useful microorganisms for health care, foods and enzyme production* (pp. 269–287). Trivandrum, India: Research Signpost.

Rebordinos, L., Vallejo, I., Santos, M., Collado, I. G., Carbu, M., & Cantoral, J. M. (2000). Análisis genético y relación con patogenicidad en *Botrytis cinerea. Rev. Iberoam. Micol, 17*(1), S37–S42.

Rui, O., & Hahn, M. (2007). The *Botrytis cinerea* hexokinase, *Hxk*1, but not the glucokinase, *Glk*1, is required for normal growth and sugar metabolism, and for pathogenicity on fruits. *Microbiology, 153*, 2791–2802.

Sall, M. A. (1980). Epidemiology of grape powdery mildew: A model. *Phytophatology, 70*, 338–342.

Santos, M., Rebordinos, L., Gutierrez, S., Cardoza, R. E., Martín, J. F., & Cantoral, J. M. (2001). Characterization of the *gdh*A gene from the phytopathogen *B. cinerea. Fungal Genet. Biol., 34*, 193–206.

Santos, M., Vallejo, I., Rebordinos, L., Gutierrez, S., Collado, I. G., & Cantoral, J. M. (1996). An autonomously replicating plasmid transforms *Botrytis cinerea* to phleomycin resistance. *FEMS Microbiol. Lett., 137*, 153–158.

Segmüller, N., Kokkelink, L., Giesbert, S., Odinius, D., van Kan, J., & Tudzynski, P. (2008). NADPH oxidases are involved in differentiation and pathogenicity in *Botrytis cinerea. Mol. Plant Microbe In., 21*, 808–819.

Shah, P., Atwood, J. A., Orlando, R., El Mubarek, H., Podila, G. K., & Davis, M. R. (2009). Comparative proteomic analysis of *Botrytis cinerea* secretome. *J. Proteome Res., 8*(3), 1123–1130.

Shepard, D. V., & Pitt, D. (1976). Purification of a phospholipase from *Botrytis* and its effects on plant tissues. *Phytochemistry, 15*, 1465–1470.

Siewers, V., Viaud, M., Jiménez-Teja, D. G., Collado, I., Gronover, C. S., Pradier, J.-M., et al. (2005). Functional análisis of the cytochrome P450 monooxygenase gene bcbot1 of *Botrytis cinerea* indicates that botrydial is a strain-specific virulence factor. *Microbial Plant Molecular Interaction, 18*(6), 602–612.

Sivanesan, A., & Holliday, P. (1981). Guidnardia bidwellii. Descriptions of pathogenic fungi and bacteria, no 710. *UK*. Kew: Commonwealth Mycological Institute.

Snowdon, A. L. (1990). *A colour atlas of post harvest diseases and disorders of fruits and vegetables*. London, UK: Wolfe Scientific.

Spotts, R. A. (1980). Infection of grape by *Guignardia bidwellii* – Factors affecting lesion development, conidial dispersal, and conidial populations on leaves. *Phytopathology, 70*, 252–255.

Tani, H., Koshino, H., Sakuno, E., Cutler, H. G., & Nakajima, H. (2006). Botcinins E and F and botcinolide from *Botrytis cinerea* and structural revision of botcinolides. *J. Nat. Prod., 69*, 722–725.

Tani, H., Koshino, H., Sakuno, E., & Nakajima, H. (2005). Botcinins A, B, C and D, metabolites produced by *Botrytis cinerea*, and their antifungal activity against *Magnaporthe grisea*, a pathogen of rice blast disease. *J. Nat. Prod., 68*, 1768–1772.

Vallejo, I. (1997). *Caracterización de diferentes cepas de* Botrytis cinerea *por técnicas microbiológicas, genéticas y de biología molecular*. Cadiz, Spain: Doctoral thesis. Universidad de Cádiz.

Vallejo, I., Carbu, M., Muñoz, F., Rebordinos, L., & Cantoral, J. M. (2002). Inheritance of chromosome-length polymorphism in the phytopathogenic ascomycete *Botryotinia fuckeliana* (*Botrytis cinerea*). *Mycol. Res., 106*, 1075–1085.

Vallejo, I., Carbu, M., Rebordinos, L., & Cantoral, J. M. (2003). Virulence of *Botrytis cinerea* strains on two grapevine varieties in south-western Spain. *Biologia, 58*(6), 1074–1076.

Vallejo, I., Santos, M., Cantoral, J. M., Collado, I. G., & Rebordinos, L. (1996). Chromosomal polymorphism in *B. cinerea. Hereditas, 124*, 31–38.

von Tiedemann, A. (1997). *Physiol. Mol. Plant P., 50*, 151–166.

CHAPTER

# 11

# Production of Wine Starter Cultures

*Ramón González[1], Rosario Muñoz[2], Alfonso V. Carrascosa[3]*

[1] Instituto de Ciencias de la Vid y del Vino (CSIC-UR-CAR), Logroño, Spain,
[2] Instituto de Ciencia y Tecnología de los Alimentos y Nutricíon (ICTAN, CSIC), Madrid, Spain and
[3] Instituto de Investigación en Ciencias de la Alimentación (CIAL, CSIC-UAM), Madrid, Spain

OUTLINE

## 1. INTRODUCTION

The quality of fermented foodstuffs and beverages is determined in part by the microorganisms used in their preparation. The secondary character of wine, for instance, is determined by sensory characteristics that arise from the direct action of microorganisms on the substrate. Consequently, the exploitation of organisms such as the yeasts and lactic acid bacteria responsible for alcoholic and malolactic fermentation, respectively, is a constantly expanding branch of biotechnology.

The effectiveness of starter cultures in the wine industry is based largely on the microbiological control that can be achieved during winemaking. Good cultivation practices that limit contamination of the fruit with molds or acetic or lactic acid bacteria prior to harvesting are key to obtaining a must that can be correctly fermented. Likewise, good manufacturing practices that include appropriate winery hygiene programs favor the development of the inoculated microorganisms and reduce the microbial competition that they encounter, given that both musts and wine are nonsterile substrates. The implementation of quality-control systems for hygiene such as the Hazard Analysis Critical Control Point approach is by far the best guarantee of success when using starter cultures in the wine industry.

Recent years have seen an increasing emphasis on the importance of wine quality. Thus, obtaining a wine with characteristics that can be clearly distinguished from others or with organoleptic properties that remain consistent year after year tends to be associated with increased competitiveness. To a large extent, consumers determine which wine is sold, and this must be defined in terms of measurable parameters, be that through the use of instruments or by tasting, in order to produce wines correctly.

The use of starter cultures may not meet expectations if the goals are not first understood in terms of measurable characteristics. Consequently, the first step in any wine production process (see Figure 11.1) is to clearly define the type of wine to be produced. This step is more important than usually considered since it guides both the preparation and choice of the starter culture.

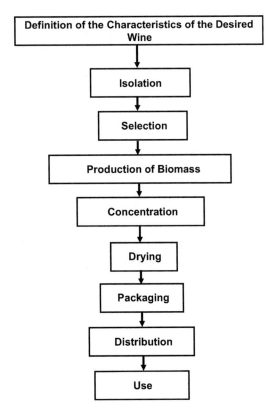

FIGURE 11.1  Stages in the production of starter cultures for use in winemaking.

## 2. YEASTS

### 2.1. Historical Notes

The process of grape-must fermentation for the production of wine has been described since antiquity and has been carried out for thousands of years by trusting in the action of the microorganisms naturally present on the grapes and in the winery. However, the birth of microbiology as a scientific discipline led to improved control of the process as it enabled the identification of the microorganisms responsible for fermentation and offered an opportunity for winemakers to employ pure yeast cultures. Koch's postulates must have been consciously or unconsciously present in the minds of the first microbiologists to propose that the results of wine fermentation might depend on the strain of yeast used. Nevertheless, it should be remembered that under industrial production conditions grape must is not a sterile substrate and the inoculated yeasts therefore compete with other yeasts and bacteria, particularly during the initial phases of fermentation (see Chapter 1).

At the beginning of the last century, the La Claire Institute in France began the isolation, maintenance, and small-scale production of pure strains of yeast (up to 100). These were used to establish *pied de cuve* cultures for inoculation of musts in a process that was gradually scaled up until it was possible to inoculate industrial fermentation tanks (Kraus et al., 1983). Commercially produced cultures that would be ready for inoculation in large tanks without prior preparation and scaling up of *pied de cuve* cultures were first proposed in the 1950s. At this time, the work of German, Californian, and Canadian researchers demonstrated that yeast cells produced under aerobic conditions were perfectly suitable for the fermentation of must (reviewed in Kraus et al., 1983). As we will see later, prior growth under aerobic conditions also offers some advantages in terms of the fermentative capacity of industrially produced cells.

The industrial production of wine yeasts as we know it today began in the 1960s, when various producers of baker's yeast began to produce wine yeasts using a similar procedure. This product had the usual characteristics of pressed baker's yeast, including a 70% moisture content. However, although it was relatively easy to use, it had a short shelf life. It should be remembered that the use of wine yeasts is essentially seasonal and can be difficult to coordinate with production. Around 1964, the first efforts were made to develop a much more stable product, active dried yeast, as a solution to this problem.

The first report of a selected yeast being used in Spain dates back to 1958 and makes reference to various microbiological studies of grape harvests performed since 1956 in Spanish winegrowing regions (Iñigo, 1958). The study advocated the use of selected yeasts to avoid excessive volatile acidity in Spanish wines. The author recommended the sequential inoculation of strains with weak, intermediate, and strong fermentative capacity using *pied de cuve* culture to imitate the natural process occurring during spontaneous fermentation and thus allow the desired wines to be obtained in a controlled manner. It was concluded that the final volatile acidity was lower with a stepwise inoculation program than when fermentation was performed

with a single pure strain of *Saccharomyces cerevisiae*. To explain the observation that inoculation of *Saccharomyces ellipsoideus* some days after inoculation of *Candida pulcherrima* increased the fermentative capacity of *S. ellipsoideus*, it was suggested that the products of autolysis of the first inoculated strain (proteins and vitamins) acted as growth factors for the true fermentative strain. Nowadays, in fact, this effect is achieved by the addition of products of yeast autolysis (yeast cell walls, etc.).

The idea of using selected yeasts was brought to Spain by the Italian researcher Tommaso Castelli, who in 1955 considered Gino de Rossi the founder of the work he had undertaken since 1933. The aim was to obtain a collection of microorganisms from different winegrowing regions according to the logic that those that were most abundant during the different phases of fermentation (initial, tumultuous, and final) would be potentially the most interesting for selection. Castelli claimed that fermentation using pure cultures did not work miracles but allowed greater consistency, increased production of alcohol, improved yield (alcohol to sugar ratio), and reduced volatile acidity, although with the disadvantage of being less effective in the warmest and coolest regions, since temperature-control systems were not widely available. The collaboration between Castelli and Iñigo led to the first microbiological studies of Spanish harvests in the regions of La Mancha and La Rioja (Castelli & Iñigo, 1957, 1958).

## 2.2. Isolation and Selection

Yeast inoculation offers a series of advantages over spontaneous fermentation for the production of wine. These advantages depend on the type of vinification and the characteristics of the yeast strain used. However, most notable among them are a more rapid onset and progression of fermentation and a reduction in volatile acidity associated with greater consistency in the quality of the product between different tanks and vintages. Specific contributions to the organoleptic quality of the wine are also offered according to the characteristics of the yeast strain. In addition to these characteristics, the selection criteria applied to potential new yeast strains should take into account three main elements: the conditions under which fermentation will take place (characteristics of the must and other environmental factors such as fermentation temperature), the quality and character of the wine obtained, and the survival of the strain during industrial production of active dried yeast.

The process of selection of new yeast strains for use in winemaking begins with the isolation of natural yeasts from different substrates (generally from wine during the final phase of fermentation but also, possibly, from grapes, fresh must, or lees). A factor that should be taken into account when isolating new strains from must and in particular from wine is whether the winery has previously used commercial starter cultures, since in this case there is a high risk of repeatedly isolating a known yeast strain for which use may be restricted by intellectual property rights. As a result, the isolation of new strains should be undertaken in newly constructed wineries. Since during wine fermentation there is a selective enrichment in strains of the genus *Saccharomyces*, a spontaneous fermentation that has yielded good results in terms of the kinetics of fermentation and the quality of the product should be an ideal source for the isolation and selection of new industrial strains. However, one should not ignore other substrates that are less rich in strains of *Saccharomyces*, such as must or grapes, as a source of new strains with interesting properties. The selection of new strains is increasingly based on very specific criteria, such as fermentation at low temperatures or specific contributions to secondary aroma, and these are more difficult to identify without using a wide range of isolates.

The selection procedure generally begins with a large number of strains (often several hundred). From an organizational and economic perspective, it is desirable for the initial phases of the process to employ easily evaluated selection criteria that allow less interesting strains to be identified and ruled out more quickly as this will help to reduce the number of strains that need to be analyzed during the subsequent phases of the process. More expensive tests and analytical methods can be left until the final stages, when only a small number of strains need to be analyzed.

A number of criteria have been proposed for the selection of new yeasts for use in winemaking, not all of which are relevant for the specific conditions and characteristics of each wine. Below we describe some of the most widely applied.

**Fermentation power**—understood as a mixture of rapid fermentation, short lag phases, and almost complete consumption of the sugars present in the must—was the first criterion used to select wine yeasts (Kraus et al., 1983). The original purpose of inoculating selected yeasts was to ensure that the fermentation process was not excessively long and also to prevent stuck fermentation. This was sufficient to guarantee that wines of appropriate quality could be obtained year after year without spoilage due to the growth of undesirable microorganisms as a result of poor fermentation kinetics. Good fermentation power is obviously related to the capacity of the strain to overcome the stresses associated with wine fermentation, such as the hyperosmotic environment present during the initial phases of the process or the elevated concentrations of ethanol and scarcity of assimilable nitrogen and growth or survival factors during subsequent phases.

Fermentation power estimated under laboratory conditions, using either natural musts or culture media designed to mimic their composition, tends to be a poor predictor of the behavior of strains under industrial production conditions. As a result, various authors have proposed additional selection criteria based on the identification of factors that limit the survival of yeasts during the vinification process. For instance, Ivorra et al. (1999) observed a negative correlation between the resistance of strains to various stress factors and the likelihood of stuck fermentation. In that study, analysis of expression levels for a number of genes led the authors to propose *HSP12* expression as a marker of resistance to stress in the corresponding strain. Subsequently, Zuzuarregui and del Olmo (2004a) used multivariate analysis as a basis for suggesting **resistance to ethanol** and **oxidative stress** as selection criteria for strains with good fermentation behavior. Interestingly, the response to oxidative stress would naturally be expected to have more relevance for the behavior of the strain during production of active dried yeast than as a predictor of its fermentation power. The same authors showed a correlation between the expression level of certain stress-response genes and the fermentative capacity of the wine strains studied (Zuzuarregui & del Olmo, 2004b). A conclusion that can be drawn from their study is that, although good induction of stress-response genes is necessary for the strain to perform well under adverse conditions, excessive or prolonged expression may also be prejudicial. Increased expression levels of *GPD1* have also been observed in various commercial strains during the initial phase of fermentation (Zuzuarregui et al., 2005). This may improve the response to **osmotic stress** generated by the high concentration of sugars in the must, since the enzyme encoded by *GPD1* is necessary for intracellular accumulation of glycerol.

After ethanol, one of the main causes of sluggish or stuck fermentation tends to be the lack of assimilable nitrogen caused by modern winemaking practices. The capacity to continue fermentation under conditions of limited nitrogen depends on the characteristics of the strain and is highly related to the activation of stress response mechanisms associated with entry into the stationary phase (see Chapter 1). Consequently, **nitrogen demand** has also been proposed as a selection criterion for industrial *Saccharomyces* strains (Manginot et al., 1998).

Recently, Marks et al. (2008) performed a transcriptomic analysis of *S. cerevisiae* during alcoholic fermentation and defined a "fermentation stress response" (FSR). According to those authors, this response will be different from, although obviously overlap with, the response to the individual stress factors described earlier. Of the genes found to participate in the FSR, 62% had not previously been linked to the global stress response and 28% had not yet been functionally annotated in sequence databases.

As described in Chapters 1 and 6, **sulfite resistance** is a common characteristic of *S. cerevisiae* wine strains that can affect their capacity to complete fermentation. Pérez-Ortín et al. (2002) and Yuasa et al. (2004) reported that most of the *S. cerevisiae* wine strains analyzed in their studies had at least one copy of the *SSU1-R* allele, which confers greater resistance to sulfite than the wild-type *SSU1* allele. This is probably due to continued selection pressure (see Chapter 6), given that the use of sulfite is almost as old as wine production itself. Since it is unlikely that the use of sulfites in winemaking will be abandoned in the short or medium term (Romano & Suzzi, 1993), resistance to sulfites will continue to be a character to take into consideration during the selection process, especially when selection is not limited to *S. cerevisiae*, the only species identified to date that contains the *SSU1-R* allele.

Compared with other yeasts found in must or used as starter cultures, strains of *S. cerevisiae* are considered to be relatively heat-tolerant. Thermotolerance, or **resistance to thermal stress**, has been an important characteristic in wine strains, particularly before the introduction of temperature control systems, since the temperatures reached during exothermic fermentation could be suboptimal for growth. As with other characters linked to fermentation power, thermotolerance is associated with the expression of stress-response genes. It is no coincidence that some of the first stress-response genes were characterized on the basis of their expression in response to thermal shock (heat shock proteins) (Craig, 1985). Resistance to thermal stress can thus be a criterion for selection, partly because it is linked to fermentative capacity (Ivorra et al., 1999) and partly because heat tolerance may be an important factor when yeasts are dried during preparation for industrial use.

The ability to control fermentation temperature has also made it possible to develop new styles of vinification. For instance, cold fermentation has become increasingly popular since it confers some very interesting aromatic properties on the wines produced. This occurs as a result of greater retention of the volatile substances formed during fermentation, the effect of temperature on yeast metabolism, and the characteristics of the cryotolerant strains used in winemaking, which tend to belong to (or be hybrids of) the species *S. bayanus*. Consequently, **cryotolerance** has become an additional selection criterion for yeasts destined for use in fermentations performed below 15°C.

Another hereditary characteristic that can influence the capacity of a strain to induce complete fermentation of grape must is the **killer factor**. Three killer toxins have been described in *S. cerevisiae*: K1, K2, and K28 (Magliani et al., 1997). These toxins are encoded by double-stranded satellite RNAs (M1, M2, and M28, respectively) that are found in the cytoplasm of the producing strains and encapsulated in icosahedral particles similar to viruses. In turn, these RNAs depend on other double-stranded RNA viruses known as L-A for their replication and encapsidation. The M RNAs are responsible for the synthesis of the corresponding toxin and for the immunity of

the producing strain to the toxin produced by that strain or by other cells that produce a killer factor of the same type. The toxin is able to kill strains that do not produce a killer factor or that produce a killer factor of a different type. In principle, the use of killer strains of *S. cerevisiae* as starter cultures should contribute to favoring the establishment of the inoculated strain during fermentation. However, there is no consensus regarding the true relevance of the killer factor under natural conditions. The killer factor K2 would seem to be of most interest due to its activity and stability at the pH found in grape must and wine (Heard & Fleet, 1987). However, strains expressing a wide range of killer factors have been constructed by genetic engineering in an effort to improve their survival capacity or their ability to eliminate undesirable yeasts and thereby prevent wine spoilage (see Chapter 7).

Despite its overall importance, good fermentation power is not in itself an adequate criterion for the selection of winemaking strains. Many of the additional selection criteria are based on the positive or negative influence that a strain may have on the sensory qualities of the wine produced. Yeast metabolism has a notable influence on **secondary aroma**. This refers to the contribution to wine aroma made by microorganisms present during fermentation. Chapter 4 describes the most relevant compounds involved in the formation of secondary aroma by yeasts of the genus *Saccharomyces* and other genera. The contribution to secondary aroma may differ among *Saccharomyces* strains and can therefore be used as a target for selection according to the quality of the aroma and its appropriateness in a given type of wine. Some of the most interesting metabolites in this respect are esters of acetic acid and ethyl esters. The levels and relative abundance of the different esters that contribute to this secondary aroma are the result of the activity of groups of enzymes with antagonistic effects: alcohol acetyltransferases, which catalyze the synthesis of these compounds, and esterases, which catalyze their hydrolysis. Genetic engineering has been used to construct strains of *Saccharomyces* in which the production of esters is increased by deleting genes that encode esterases and overexpressing genes that encode alcohol acetyltransferases (see Chapter 7).

The primary or **varietal aroma**, which is essentially derived from the grape, can also be affected by the strain of yeast used (see Chapter 4) through the action of hydrolytic enzymes that release terpenes from their glycosylated precursors or facilitate the extraction of aromas and colors from the cell wall of the grapes. Consequently, the production of hydrolytic enzymes with these activities has also been proposed as a selection criterion.

Another product of yeast metabolism with a positive contribution to many of the sensory properties of wine is **glycerol**. Furthermore, increased glycerol levels go hand in hand with reduced concentrations of alcohol. Consequently, the ever-growing demand for wines with lower alcohol content has led to a growing interest in strains that produce higher levels of glycerol and lower levels of ethanol. Recombinant yeast strains have been produced in which more glycerol is produced at the expense of ethanol. However, the high concentrations of acetic acid produced by these strains made it necessary to incorporate additional genetic modifications to prevent excessive volatile acidity (see Chapter 7).

**Volatile acidity** (due to excess acetic acid) is one of the most easily detected wine flaws. Although excessive volatile acidity can often be attributed to the uncontrolled growth of acetic acid bacteria, there can also be an appreciable, strain-dependent contribution of *S. cerevisiae* metabolism (Delfini & Cervetti, 1991), making this an important selection criterion.

Excess **hydrogen sulfide** can also be an important defect in finished wine. Given that the sensory threshold for hydrogen sulfide is extremely low, differences in the levels produced by different industrial strains of *S. cerevisiae* may act as major determinants of the quality of the finished wine. The main source of sulfur for the formation of hydrogen sulfide can be either sulfate or sulfite, and depletion of nitrogen sources has been identified as one of the main determining factors for its formation. Various genes have been linked to the production of hydrogen sulfide, including *MET17* and *NHS5*, which can suppress the formation of hydrogen sulfide when overexpressed in yeast (Spiropoulos & Bisson, 2000; Tezuka et al., 1992).

Another undesirable product of yeast metabolism is urea. Although it does not influence the sensory quality of wine, its presence in an aqueous, alcohol-containing medium such as wine can over time give rise to the formation of **ethyl carbamate**, a toxic compound that is also a suspected carcinogen. Many countries have produced legislation on the maximum permitted levels of ethyl carbamate in imported wines. In wine, urea is mainly derived from the action of arginase produced by *Saccharomyces* yeasts. This is the first enzyme in the catabolic pathway of arginine and catalyzes its hydrolysis to give rise to urea and ornithine. One of the strategies recommended by the United States Food and Drug Administration (FDA) to minimize the formation of ethyl carbamate is the selection of strains with low urea production. Recombinant strains of sake yeast have even been produced in which the gene *CAR1*, encoding arginase, has been deleted to allow the generation of sake completely lacking ethyl carbamate (Kitamoto et al., 1991). Coulon et al. (2006) recently used a different strategy to the same end, constructing a wine yeast that was able to eliminate urea from the wine through the constitutive expression of the gene *DUR1,2*, which encodes urea amidolyase. The same group used a similar strategy to produce the yeast ECMo01, which has been approved by the FDA and Health Canada for commercial use.

In all protocols for the selection of industrial yeasts, **tasting** is one of the most difficult and at the same time most important steps in identifying the most interesting strains. This is the phase that allows confirmation of various indicators obtained in earlier tests. For instance, the potential contribution of the strain to primary and secondary aroma can now be confirmed by the detection of undesirable flavors or aromas. Likewise, tasting can be used to determine whether the finished wine has the typical characteristics of the type of wine that was intended to be produced.

Specific selection criteria may also be used according to the specific style of vinification. As mentioned earlier, cryotolerance would be a criterion for cold fermentation, but autolytic and flocculation capacity can also be considered for the production of traditional-method sparkling wines (Cava and Champagne); *flor* formation, ethanol tolerance, and other necessary characteristics for the production of biologically aged wines; and autolytic capacity for the production of wines aged on lees. All of these factors are discussed in more detail in other chapters.

Finally, successful marketing of wine yeast strains will depend largely on their behavior under industrial production conditions, particularly in terms of genetic stability, growth on molasses, and survival and metabolic activity following drying and rehydration.

## 2.3. Production of Biomass

The first companies to attempt the commercial production of wine yeast strains were producers of baker's yeast and, in fact, as mentioned earlier, in the 1960s wine strains were still sold as pressed yeast. Nowadays, the initial phases in the production of active dried yeast are very similar to those involved in the production of pressed yeast, and the same factors must be taken into consideration.

The proliferation phase (or production of biomass) naturally begins with a pure culture that must have been maintained under appropriate conditions to ensure both purity and genetic stability (see Chapter 12). The scaling-up process allows multiplication of the yeast from the few hundred million cells typically present in the starting culture by gradually increasing the culture volume through fermenters of increasing capacity (from 5 to 250 L), until fermenters with a capacity of hundreds of thousands of liters are reached. Maintaining the purity of the culture is a factor that must be taken into account throughout scale-up and in all phases of the production process, although the presence of other yeasts and bacteria is normal in the final product. An indicator of quality in active dried yeast is that the presence of bacteria and yeasts other than those belonging to the genus *Saccharomyces* does not exceed 0.01% of the concentration of cells surviving rehydration.

As in all microbial growth processes, the function of the culture medium is to guarantee adequate provision of the nutrients required for growth. For the industrial production of yeasts, the best carbon source is cane or beet molasses. This is an inexpensive substrate that is very rich in sucrose, a sugar that is easily assimilated by *S. cerevisiae* thanks to genes that encode various forms of invertase. Under production conditions, it is calculated that invertase activity allows sucrose to be hydrolyzed some 300 times faster than the resulting glucose and fructose are assimilated by the cells (Sánchez, 1988). Consequently, the initial paucity of monosaccharides in the substrate does not represent a limitation for the use of molasses. In addition to sucrose, molasses usually represents an adequate source of other essential nutrients for yeast growth, including some minerals, oligoelements, and vitamins. However, molasses does not generally represent a good source of nitrogen or phosphorus. Therefore, in addition to dilution, it must also be supplemented with these nutrients, usually in the form of ammonium salts. Since molasses is a byproduct of another industrial activity and no standardization processes are applied at source, it is also necessary to confirm the composition of each batch of molasses so that in each case the appropriate quantities of nutrients can be added. The presence of potential growth inhibitors such as sulfites, organic acids, and nitrites must also be taken into consideration and any defects rectified prior to use. The pH of the molasses is also checked prior to use as a substrate for fermentation and normally adjusted to a pH of 5, which tends to be optimal for the growth of *S. cerevisiae* on this substrate.

Production of biomass is carried out under aerobic conditions in order to achieve two main goals. The first is to obtain the greatest possible yield from the process, expressed as the quantity of biomass produced for a given quantity of molasses. The incomplete oxidation of glucose that occurs during fermentation leads to a net energy yield per mole of sugar consumed that is lower in the case of fermentation (56 kcal/mol) than in respiration (688 kcal/mol). In order to ensure minimal fermentative metabolism and to maximize sugar consumption by respiration, it is essential to maintain good aeration; however, the Crabtree effect must also be taken into account. This metabolic phenomenon seen in many yeasts, including *S. cerevisiae*, leads part of the sugar consumed to be converted into ethanol via the fermentative pathway. This occurs when glucose concentration exceeds a relatively low threshold (0.1−0.5 g/L), even in the presence of sufficient quantities of oxygen. The Crabtree effect is caused by the generation of high intracellular concentrations of pyruvate in the presence of glucose. This favors pyruvate degradation via a pathway involving pyruvate decarboxylase—which has a high loading capacity and a high $K_m$—rather than the pyruvate dehydrogenase complex, which leads directly to acetyl-coenzyme

A (CoA). Since the subsequent reactions that would allow transformation of the acetaldehyde formed by pyruvate decarboxylase into acetyl-CoA are limited, this ultimately favors the formation of ethanol even under aerobic conditions (Potma et al., 1989; Pronk et al., 1996). Figure 11.2 shows a schematic diagram of the relative metabolic flows occurring in the Crabtree effect.

Despite adequate aeration, the high initial concentration of sugars means that the yeast initially adopts a fermentative metabolism following inoculation of the molasses. The sugar concentration also places the yeast under osmotic stress. As the culture grows, the sugars begin to be depleted and controlled feeding is required. To minimize the Crabtree effect, however, it is necessary to maintain the sugar concentrations at low levels. A "fed-batch" process is therefore used in which molasses is added little by little as it is consumed by the yeast, thus minimizing the production of ethanol without halting growth. To optimize fed-batch feeding, feedback systems are usually employed. In those that work best, the addition of the substrates is regulated by the respiratory

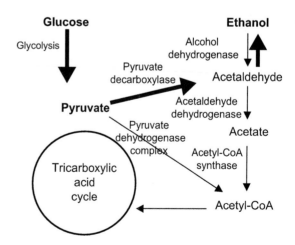

**FIGURE 11.2** Relative metabolic flows associated with the Crabtree effect in *Saccharomyces cerevisiae*.

quotient or RQ (moles of carbon dioxide formed per mole of oxygen consumed) (Aiba et al., 1976; Cooney et al., 1977; Wang et al., 1977). These control systems analyze the flow and composition of gases entering and leaving the fermenter and regulate the provision of the substrate according to the value obtained. As we will see below, the molecular responses to osmotic and oxidative stress are the main adaptations required of the yeast during production (Pérez-Torrado et al., 2005).

Another goal during the production of biomass under aerobic conditions is to allow the yeast cells to synthesize sterols. These compounds are survival factors that cannot be synthesized under anaerobic conditions. Consequently, if the inoculum contains limiting sterol reserves, its capacity to complete fermentation will depend on the characteristics of the must (see Chapter 1).

Finally, the production of biomass should take into account factors that affect the capacity of the yeasts to survive the drying process and to recover their viability and fermentative capacity following rehydration and inoculation into the must. The ability of the yeasts to survive the drying and rehydration process will depend on the expression of stress-response genes, either directly or via the synthesis of storage compounds such as trehalose and glycogen (François & Parrou, 2001). Mobilization of these storage compounds through the action of hydrolytic enzymes is also important for the recovery process (Lillie & Pringle, 1980; Novo et al., 2003; Thevelein, 1994). Trehalose is thought to play an additional role as a protective agent, helping to maintain the integrity of cell membranes and stabilizing the native conformation of proteins (Felix et al., 1999; Leslie et al., 1994). The specific genotype of each strain will thus be important in determining its ability to respond to these treatments and, therefore, to be sold as active dried yeast.

It is also important to take into consideration the culture conditions responsible for induction of stress-response genes and genes involved in the synthesis of storage compounds. These processes are normally repressed under conditions of rapid growth and are induced upon entry into the stationary phase. As a result, the final step in the production of biomass is carried out in the presence of limiting nitrogen while maintaining sufficient levels of the carbon source to allow the synthesis of glycogen and trehalose. The concentration of phosphorus is also controlled since it plays a role in determining the rate of division and the stability of the cells during dehydration (Sánchez, 1988).

Nitrogen limitation plays a role in the entry of cells into the stationary phase. This is a perfectly regulated cellular process brought about by the lack of an essential nutrient (Werner-Washburne et al., 1993). Despite the lack of nutrients, the cells are able to complete the cycle of cell division that has been initiated, meaning that all the cells in the stationary phase lack buds and have the same DNA content. The absence of cell division in no way implies metabolic inactivity, since, as discussed in Chapter 1, most of the fermentation process occurs without any increase in the number of viable cells.

Actively dividing cells pass through the four different phases—G1, S, G2, and M—that make up the cell cycle. In G1, which is normally initiated following cell division, each of the resulting cells (mother and daughter) increases in size, with a corresponding synthesis of macromolecules (proteins, cell wall polysaccharides, RNA, etc.). During the S phase, there is a duplication of the genetic material such that at the end of this phase the cells contain twice the content of DNA and nuclear proteins. G2 involves de novo synthesis of macromolecules and morphological changes required for cell division or mitosis. The M or mitosis phase involves separation of the chromosomes between the dividing cells and corresponds to cell division proper, which gives rise to two cells, mother and daughter. Currently, the most widely accepted view is that, when cells stop

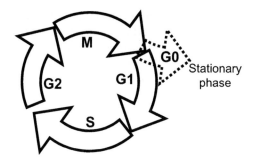

**FIGURE 11.3** Phases of the *Saccharomyces cerevisiae* cell cycle. M = mitosis; G0 = adaptation phase during entry into the stationary phase; G1 = growth phase; S = synthesis phase; G2 = preparation phase prior to division.

dividing and enter the stationary phase, they exist in a phase outside of the normal cell cycle known as G0 (Figure 11.3) (Werner-Washburne et al., 1993). In this phase, the cells acquire a series of adaptations that help them to survive in adverse conditions. Examples of these adaptations include thickening of the cell wall, accumulation of storage polysaccharides, expression of a variety of stress-response genes, and activation of mechanisms to allow recycling of cell components, particularly proteins, via pathways such as ubiquitin-dependent proteolysis or autophagy (see Chapter 2).

The entry and maintenance of cells in the stationary phase involve the expression of various genes with different specific functions, including resistance to the different types of stress mentioned above. These changes in expression are regulated by both transcriptional and post-transcriptional mechanisms. The response to the lack of nutrients that leads ultimately to entry into the stationary phase is regulated by a complex network of signal transduction pathways (Gray et al., 2004) based on protein kinases (some of which can be considered nutrient sensors while others function at later points in the signal transduction pathway). These include the TOR pathway, the cyclic AMP-dependent, protein kinase A pathway regulated by Ras (RAS/cAMP/PKA), the

protein kinase C pathway, and the SNF1 pathway. Cells that have not entered the stationary phase and are therefore actively proliferating have very little chance of surviving the subsequent treatments involved in the production of active dried yeast.

Although the entire production process has been optimized to obtain maximum yield in the production of biomass, there has been little characterization of the molecular mechanisms used by yeast to adapt to the changing conditions that this entails. One of the main difficulties faced in such studies is the simulation of industrial processes under laboratory conditions. Recently, this has begun to be resolved through the use of micro fermenters with similar biomass yield and microbial growth conditions to those found in industrial situations, making it possible to use molecular techniques to understand and improve the behavior of yeasts during industrial production (Pérez-Torrado et al., 2005).

The shift from fermentative to respiratory metabolism leads to the inhibition of mitochondrial metabolism and causes changes in the intracellular redox potential that can place the yeast under oxidative stress. The response of *S. cerevisiae* to oxidative stress has been investigated under different conditions, including the addition of agents responsible for the generation of reactive oxygen species (ROS) (hydrogen peroxide, menadione, and various metal ions) and laboratory conditions that produce changes in the cytoplasmic redox potential (Gibson et al., 2008). Various genes have been identified as markers of the response to oxidative stress. Firstly, *TRX2* (which encodes thioredoxin, an enzyme that protects against the toxic effects of hydrogen peroxide) is expressed strongly during the transition in the first fed-batch from fermentative to oxidative growth and acts as one of the best defenses against oxidative damage. *GSH1*, which initiates the synthesis of glutathione (GSH), and *GRE2*, which encodes a reductase associated with oxidative stress, also act as indicators for the molecular response to oxidative stress (Pérez-Torrado et al., 2009). Activation of the gene *GPD1*, which is linked to the synthesis of glycerol, has also been described as a consequence of the osmotic stress that can occur during the early phases of the production of wine yeast, and to a lesser extent in response to oxidative stress (Pérez-Torrado et al., 2005). The key role played by the Yap1p factor in mediating the induction of a set of genes involved in redox protection (Rodrigues Pousada et al., 2004) is well known, as is the activation of factors involved in the general response to stress such as Hsf1p and its homolog Skn7p. Other enzyme systems involved in the elimination of proteins and lipids damaged by oxygen have also been recently described (Toledano et al., 2007).

Growth under conditions involving limited feeding can lead to the temporary absence of nutrients, leading to stress as a result of nutrient depletion. Consequently, it may be appropriate to consider using resistance to these stresses as a selection criterion. Resistance of yeast to these and other types of stress appears to be correlated with their suitability for use in industrial production (Zuzuarregui & del Olmo, 2004a, 2004b; Zuzuarregui et al., 2005), and it may therefore be appropriate to consider using laboratory tests such as those described by a number of authors (Carrasco et al., 2001; Zuzuarregui & del Olmo, 2004a) as selection criteria for use with these types of yeast.

## 2.4. Drying

Once the required biomass has been obtained, the yeasts must be separated from the culture medium, dried, and prepared for sale (Papin, 1988). The first step in this process involves the production of a paste known as a "cream yeast," which contains 150 to 200 g/L of dry material. To obtain this, the yeasts are separated from the culture medium by centrifugation or filtration. The paste is then filtered again, this time using a rotary vacuum filter or a filter press, to produce a paste containing 35% dry material. It is then extruded to form fine filaments (2–4 mm in diameter) that are further dried (to a moisture content of 4–8%) using a counter-current hot-air fluidized-bed dryer. The granules that are produced by breaking these dried filaments are vacuum packaged in the presence of an inert gas or carbon dioxide. It is important to adequately isolate the granules from the oxygen in the air in order to maintain the stability of the product, which should also be maintained at low temperature (4–8°C).

Despite all of the precautions taken and the process having been optimized over a number of years, the numbers of revivable cells present in different batches of active dried yeast can vary substantially according to the yeast strain and the production conditions. Not surprisingly, one of the fundamental characteristics considered during the selection of new strains of wine yeast for commercial use is their capacity to survive the process of drying while maintaining their viability during extended periods of storage.

Tolerance of desiccation facilitates a process known as anhydrobiosis, referring to a state in which metabolism ceases due to the lack of water. Although *S. cerevisiae* tolerates these conditions well, the molecular details of the phenomenon are poorly understood. The loss of water leads to collapse of the cytoskeleton, which affects cell physiology, membrane integrity, etc. Although it has generally been accepted that molecules such as trehalose can act as replacements for water molecules and stabilize the cell during the desiccation process, this model has recently been questioned due to the demonstration that deletion of *TPS1* (encoding trehalose-6-phosphate synthase) in an *S. cerevisiae* mutant did not reduce the tolerance of desiccation compared with the wild-type strain (Ratnakumar & Tunnacliffe, 2006).

Other recent studies have shown that the cellular response to drying involves the regulation of the transcription or translation of genes involved in metabolism and in the synthesis of proteins and lipids (Novo et al., 2007; Rossignol et al., 2006; Singh et al., 2005). Specifically, hydrophilic proteins or hydrophilins are now believed to confer resistance to the process of dehydration and rehydration in yeasts as they do in other living organisms (Tunnacliffe & Wise, 2007).

## 2.5. Use

Dried yeast can be used directly to inoculate grape must. However, in order to ensure the best results, a prior rehydration step is normally employed. Various procedures can be used for the rehydration of dried yeast, and companies usually advise on the best one for their product. One of the most widely used options is warm water, although extended periods under these conditions should be avoided in order to prevent loss of viability caused by the hypo-osmotic medium. Direct rehydration of yeasts in must tends to be slower and leads to death as a result of the high osmotic pressure. Other media used include mixtures of water and must or water and sugar. Under these conditions, the yeast are provided with a carbon source while maintaining an osmotic pressure that is more compatible with the recovery of vitality. When must or diluted must is used, it should be as clean as possible. In other words, it should be reasonably free of microorganisms and inhibitory substances such as fungicides, pesticides, and sulfites. Some companies that produce products for use in winemaking sell media with a specific formulation, generally a commercial secret, that guarantees the best results during the rehydration phase. Rehydration is generally performed at temperatures between 30 and 38°C for 20 to 30 min. The protective capacity of trehalose in maintaining the integrity of the cell membrane can play a fundamental role in the survival of these yeasts during the critical phase of rehydration. The in vitro presence of this disaccharide results in a reduction of around 20°C in the transition temperature from dry gel to liquid crystal (the transition temperature determined for isolated plasma membranes is 60°C) (Leslie et al., 1994). Thus, cells containing sufficient trehalose that are rehydrated at temperatures of around 38°C can avoid the negative effects of passing through a phase transition.

Recent studies addressing the molecular processes that occur during rehydration have revealed a specific role for genes linked to the synthesis of lipids and proteins (Novo et al., 2007; Singh et al., 2005). These studies have begun to employ the concepts of viability and vitality. Viability refers to the capacity of the cell to divide, whereas vitality relates to the capacity to remain metabolically active. Rodríguez-Porrata et al. (2008) recently performed a systematic analysis of the optimal conditions for rehydration using fluorescence microscopy and flow cytometry. They found that one of the principal causes of reduced vitality during rehydration is the loss of intracellular substances and that magnesium is an important element in limiting this loss and maintaining cell vitality. Vaudano et al. (2009) studied the expression of eight genes during the rehydration process using real-time quantitative polymerase chain reaction (PCR). Their results suggested that the yeast reacts immediately to rehydration only when there is a fermentable carbon source present in the medium. Furthermore, the expression of MEP2 was modulated by the concentration of ammonia, suggesting that catabolic repression by nitrogen is active during the rehydration phase.

During the process of rehydration and inoculation into fresh must, the yeast cells must respond to the presence of nutrients by exiting the stationary phase. As in the entry into the stationary phase, exit from this phase involves reprogramming of gene expression driven by signal transduction pathways involving protein phosphorylation. Although the exit from the stationary phase has been much less extensively studied than entry into that phase, it appears that the essential signal involves the availability of a carbon source (whereas the availability of a nitrogen source is insufficient to signal exit from G0) (Granot & Snyder, 1991).

Once fermentation has been initiated with the dried yeast, it is possible to continue using it in the same winery as a *pied de cuve*. Here, the fermenting must is diluted in fresh must. In practice, this is performed by progressive filling of the fermentation tank or by use of the *pied de cuve* to inoculate must at a proportion of 10 to 30% during fermentation. When the yeast is used in this way, some of the advantages relating to the production process or the purity of the yeast may be lost.

## 3. LACTIC ACID BACTERIA

Malolactic fermentation was discovered at the end of the nineteenth century by Müller-Thurgau. He described the transformation of L-malic acid into L-lactic acid and carbon dioxide. Malolactic fermentation occurs naturally around the end of alcoholic fermentation. Sometimes this fermentation can be delayed or may not even occur as a result of an inappropriate temperature, pH, or sulfite or alcohol content, or even due to the presence of phages.

Starter cultures are preparations containing cultures of one or more strains belonging to one or more species of microorganism that are used to inoculate a substrate in order to initiate fermentation. Since the beginning of the 1980s, starter cultures have been available to induce malolactic fermentation. These contain strains of lactic acid bacteria belonging to the species *Oenococcus oeni*, *Lactobacillus plantarum*, and *Lactobacillus hilgardii* as preparations of one or more strains (Hammes, 1990). It has been demonstrated that strains of these species are the most tolerant of the adverse conditions found in wine (Alegría et al., 2004; Guerzoni et al., 1995).

Why inoculate wine with selected bacteria? The induction of malolactic fermentation by inoculation offers a series of advantages, including greater control over the timing and length of malolactic fermentation and also over the strain of lactic acid bacteria that carries out the process. This inoculation allows:

1) *Rapid onset of malolactic fermentation at the most appropriate moment*. If the bacterial population has been adequately controlled, at the end of alcoholic fermentation the wine will contain very few bacteria and, therefore, may require weeks or even months before an adequate spontaneous bacterial population is present. The loss of viable bacteria in the wine during this period and the requirement for increased temperatures in order for malolactic fermentation to occur correctly creates a costly problem for the winery. The use of an inoculum containing $10^6$ cells/mL can help to avoid significant delays.

2) *Maintenance of wine quality*. The bacterial population never comprises a single microorganism. Spontaneous malolactic fermentation is carried out by different strains of *O. oeni* and, often, other bacterial species. Greater variability in this population increases the risk of negative effects on fermentation or of undesirable metabolites being produced.

3) *Control over the type of wine produced*. The use of selected bacterial cultures ensures that the quality of wine sought by the producers can be obtained. This last point is very important since malolactic fermentation is not only a process of deacidification of the wine but also, depending on the strain used, an opportunity to obtain additional advantages by preventing the production of secondary metabolites that can have a negative effect on the wine.

The use of spontaneous malolactic fermentation to some extent makes the results unpredictable, delays the process considerably, and can ultimately prevent the complete degradation of L-malic acid. Consequently, the use of commercial starter cultures is recommended. As occurs with yeasts, the use of malolactic starter cultures means that numerical superiority of the bacteria over potentially competing native strains is achieved immediately through displacement as a result of mechanisms such as competitive exclusion. Compared with spontaneous malolactic fermentation by the microflora present in the must, appropriately prepared commercial starter cultures produce more predictable results in terms of the rate of degradation and the final concentration of L-malic acid, and in the production of metabolites that ensure and improve sensory quality.

Nowadays, many commercial starter cultures are available to induce malolactic fermentation. Most consist of strains of lactic acid bacteria, mainly *O. oeni*, which have a high malolactic activity and a high tolerance of low pH and high ethanol content. Although these starter cultures have been commercialized in various forms, including fresh, frozen, and lyophilized cultures, from a commercial point of view, lyophilized preparations are preferable. This is because fresh starter cultures must be produced and sold directly in the producing regions, or used immediately in the producing wineries. In the case of frozen starter cultures, transport over long distances is complicated by the difficulty of guaranteeing that the required temperature is maintained. Problems are also associated with lyophilized cultures, since there is a marked loss of viability when they are inoculated directly into the wine (Krieger et al., 1993). Consequently, most efforts are now focused on the development of lyophilized malolactic starter cultures that can be directly inoculated into wine without prior treatment.

After isolation, the starter cultures are subjected to various selection stages prior to laboratory-, pilot-, and industrial-scale development before freezing or lyophilization in preparation for commercial use. Experiments are also performed under semi-production and production conditions to assess the quality of the starter cultures. Generally, to assess whether a strain is suitable for use as a starter culture, the survival, time required for malolactic fermentation, and sensory and chemical characteristics of the wine obtained are assessed.

## 3.1. Selection and Identification of Strains

Three groups of selection criteria have been identified that should be met by starter cultures used to induce malolactic fermentation in wine (Buckenhüskes, 1993); these are summarized in Table 11.1. The selection of strains that are well adapted to carrying out malolactic fermentation under specific winemaking conditions is of particular importance since wine production varies from one region to another, as does the pH of the wine and the temperature. Malolactic bacterial strains vary in their tolerance of the different stresses associated with the wine environment (Gindreau et al., 2003; Gockowick & Henschke, 2003; Maicas et al., 1999a). As a result, strains are usually isolated from samples of wine in which active malolactic fermentation is taking place.

It has been known for some time that exposure to stressful conditions such as heat, cold, ethanol, acid pH, etc. can protect against hostile environment conditions (van de Guchte, 2002). This adaptive response requires the activation of certain defense mechanisms, and in this way the bacteria become more tolerant of adverse conditions through exposure to conditions of moderate stress. For instance, acclimatization to cold temperatures can be used to obtain cryotolerant lactic acid bacteria (Panoff et al., 2000). Another exploitable defense mechanism is the accumulation of osmoprotective organic compounds as a response to osmotic

**TABLE 11.1**    Criteria for the Selection of Lactic Acid Bacteria to Induce Malolactic Fermentation in Wine

| 1. | *FIRST-ORDER CRITERIA* |
|----|------------------------|
| 1.1. | Resistance to low pH |
| 1.2. | Resistance to ethanol |
| 1.3. | Tolerance of low temperatures |
| 1.4. | Reduced metabolism of hexose and pentose sugars |
| **2.** | ***SECOND-ORDER CRITERIA*** |
| 2.1. | High viability following propagation in a standardized medium |
| 2.2. | Short propagation time in a standardized medium |
| 2.3. | High production of biomass in a standardized medium |
| 2.4. | Rapid survival kinetics in a standardized medium |
| 2.5. | Rapid degradation of malic acid in a tartaric acid buffer (pH 4.5) and in standardized wine |
| **3.** | ***THIRD-ORDER CRITERIA*** |
| 3.1. | Production of appropriate organoleptic characteristics in the wine |
| 3.2. | Resistance to phages |
| 3.3. | Sulfite resistance |
| 3.4. | No formation of biogenic amines |
| 3.5. | Potential to form diacetyl and acetoin |
| 3.6. | Limited formation of volatile acids |
| 3.7. | No degradation of glycerin |
| 3.8. | No production of extracellular polysaccharides |
| 3.9. | Little formation of D-lactic acid |

*Adapted from Buckenhüskes (1993).*

stress in certain lactic acid bacteria (Baliarda et al., 2003; Romeo et al., 2001) and a survival mechanism in lactic acid bacteria subjected to desiccation (Carvalho et al., 2003). Studies have been performed to analyze the mechanisms underlying the resistance of *O. oeni* to the stressful conditions found in wine (Bourdineaud et al., 2003; Delmas et al., 2000; Guzzo et al., 2000; Jobin et al., 1999a, 1999b; Tourdot-Marechal et al., 2000). In addition, the response of *L. plantarum* to osmotic stress, cold shock, and redox potential has been analyzed along with its behavior following freezing and drying (Carvalho et al., 2002; Molina-Gutierrez et al., 2002; Ouvry et al., 2002; Smelt et al., 2002).

*O. oeni* is the principal microorganism involved in malolactic fermentation under the stressful conditions habitually found in wine. Nevertheless, the inoculation of *O. oeni* starter cultures leads to significant cell death and, consequently, failure of malolactic fermentation. As a result, in order to achieve better control over malolactic fermentation in the winemaking industry, it is essential to understand the mechanisms involved in stress and ethanol tolerance. One of the most widely studied aspects of the response to ethanol stress is the change occurring in the composition of cell proteins, including heat shock proteins (Jobin et al., 1997). Bourdineaud et al. (2003) found that the *O. oeni* gene *ftsH*, which encodes a protease belonging to the ABC family of proteins, is responsible for this stress, since its expression increases at high temperatures and in response to osmotic shock. *O. oeni* cells have also been found to express the 18 kDa protein Lo18 upon exposure to various stresses and during the stationary phase (Guzzo et al., 1997), and this represents a general marker of stress in this bacteria. More recently, Silveira et al. (2004) used proteomic analysis to show evidence of an active adaptive response to ethanol both in cytoplasmic and membrane proteins. Those authors reported that ethanol induces changes in the patterns of cellular proteins expressed by *O. oeni*. They found a variation in the levels of proteins involved in the maintenance of redox balance, suggesting that this process plays an important role in the adaptation to ethanol. Coucheney et al. (2005) measured malolactic and ATPase activity along with expression of the Lo18 protein in three strains of *O. oeni* selected as malolactic starter cultures. The strain of *O. oeni* that showed the highest malolactic activity in complete cells at pH 3 and the highest expression of Lo18 protein also had the highest rates of growth and malic acid consumption. As a result, Coucheney et al. (2005) suggested that these techniques could be more rapid and reliable than the standard techniques used for the selection of strains as malolactic starter cultures.

A procedure involving a turbidostat has also been described for the selection of strains to be used as starter cultures (Nielsen et al., 1996). The procedure involves the use of a 1 L fermenter containing filter-sterilized wine with an initial ethanol concentration of 11.5% (vol/vol) and a pH of 3.4. The fermenter is inoculated with a mixture of 30 or 40 isolates of *O. oeni* that have been precultured in de Man Rogosa Sharpe (MRS) culture medium. The medium is incubated at 18°C with shaking. The biomass in the fermentation is maintained constant at an optical density (600 nm) of 0.10, measured in a spectrophotometer that controls the addition of sterile wine enriched with yeast extract and containing increasing concentrations of ethanol and decreasing pH. After 4 to 6 weeks of culture, the pH and ethanol concentration in the fermenter reach values that prevent bacterial growth. At this point, a sample is taken and cultured in MRS medium, and representative isolates are then obtained. This method allows large numbers of isolates to be subjected to the desired selection conditions at the same time. The gradual increase in selection pressure in the turbidostat allows selection of strains that are better adapted to growth in wine with a low pH and a high ethanol concentration. Following selection of strains in this way, studies are performed to assess their capacity to maintain their adaptation to the adverse conditions found in the wine during the subsequent processes, including concentration, freezing, and lyophilization of the bacteria.

## 3.2. Production of Biomass

The production of bacteria for malolactic fermentation began many years after that of wine yeast. The process is simpler than the method used for yeasts since batch fermentation is used with a smaller number of successive stages. During this process, special care must be taken regarding the cleanliness of the equipment since *O. oeni* grows very slowly and can be easily and rapidly contaminated.

Although the technology used for the commercial production of *O. oeni* is similar to that used for the production of starter cultures used in dairy products, the complex nutritional requirements of these organisms precludes the use of conventional media. This represents the main difference compared with wine yeasts, which are propagated in a medium that is identical to that used for baker's yeast and to which the same principles are applied.

Strains of *O. oeni* are generally stored frozen or lyophilized. Unlike in yeasts, the propagation of *O. oeni* involves a batch process throughout the sequence of fermentations. The first fermentations are carried out in small vessels over 3 to 5 d at temperatures that vary between 20 and 25°C, depending on the strain. Since *O. oeni* grows very slowly, only a small number of scale-ups are used due to the risk of contamination. The final propagation is carried out in conditions of reduced oxygen and in vessels ranging in volume from 500 to 2000 L. The entire process requires 20 to 30 d. Samples are taken at varying intervals for the purposes of quality control and monitoring.

During the preparation of the malolactic starter culture, conditions should be chosen to produce the highest quantity of biomass with the highest viability and malolactic activity. The inoculation of wines or musts with an optimal number of cells to perform malolactic fermentation requires prior production of large quantities of biomass. The growth rate, the biomass produced, and the ability to perform malolactic fermentation can be optimized through the use of an appropriate combination of medium and culture conditions. Studies have been performed to assess the influence of different culture variables such as the composition of the medium, the concentration of malic acid, the pH, and the temperature on the biomass produced and the malolactic activity of cultures of *O. oeni* and some strains of *Lactobacillus* (Champagne et al., 1989; Naomi et al., 1989).

Differences in the sugar composition of the culture medium influence the growth kinetics and production of biomass in *O. oeni*. Maicas et al. (1999b) suggested the use of a combination of two sugars (glucose and fructose) in the culture medium for *O. oeni*. This co-fermentation allows the production of up to eight times more biomass than fermentation with a single sugar (glucose). Controlling pH, which prevents acidification of the culture medium, also leads to a 38% increase in the production of biomass.

MLO medium (medium for *Leuconostoc oenos*) is the most appropriate for the easy and rapid growth of *O. oeni* under controlled laboratory conditions (Maicas et al., 2000). However, in this medium, the cells lose their natural resistance to the adverse environmental conditions found in wine and fail as starter cultures for the induction of malolactic fermentation (Krieger et al., 1993). Growth of the bacteria in an appropriate preculture medium reduces the preparation time required for the starter culture, prepares the cells to survive following storage, and allows their subsequent growth in wine. The cells grown in these media are adapted and are able to perform malolactic fermentation immediately following inoculation in the wine. This also reduces the problems of contamination linked to the use of starter cultures.

It has been reported that the best moment to collect the biomass produced by *O. oeni* is 18 to 24 h after the culture enters the stationary phase. Centrifuged cells obtained at this point in time have a higher rate of survival and greater malolactic activity following inoculation into wine. If the cells are collected sooner, they die quickly and cannot induce malolactic fermentation (Krieger et al., 1993). Nevertheless, the studies of Kole et al. (1982) addressing pilot-scale production of *O. oeni* showed that centrifuged cultures obtained during the middle of the log phase displayed greater viability following lyophilization. In addition, these lyophilized cultures exhibited greater viability when packaged under a nitrogen atmosphere and stored in cold, dry conditions.

## 3.3. Lyophilization, Packaging, and Storage

A major problem in the development of malolactic starter cultures has been the sensitivity of the bacterial cells to damage occurring during lyophilization. The aim is to dehydrate the bacteria in order to maintain a high level of cell viability. The culture should remain stable during storage periods of various months so that it can be commercially prepared in large quantities prior to the harvest and then appropriately stored in the winery until required.

The biomass produced is frozen almost instantaneously and vacuum dried at low temperature. Careful lyophilization leads to the survival of most cells and helps to preserve fermentative capacity. Addition of cryoprotective agents and careful modification of the freezing and drying process can allow a viability of more than 95% to be obtained. The residual water content of the final product is around 4 to 5%.

Zhao and Zhang (2009a, 2009b) studied the influence of lyophilization conditions on the survival of malolactic cultures of *O. oeni*. They analyzed the effects of a cell-washing step, the pH of the resuspension medium, preincubation with sodium glutamate, initial cell concentration, and lyophilization temperature (Zhao & Zhang, 2009a). The cell viability in samples that were not washed in potassium phosphate buffer was significantly lower than that observed in washed samples. Survival was maximal when the pH of the resuspension medium was 7. Cell viability was also increased when cells were preincubated at 25°C prior to freezing. When 2.5% sodium glutamate was used as a protective agent in the suspension medium, the optimal initial cell concentration was $10^9$ colony-forming units (CFU) per mL. Cell viability increased by 21.6% when the thawing temperature was reduced from −20°C to −65°C. However, survival was markedly reduced in cells frozen in liquid nitrogen (−196°C). Zhao and Zhang (2009b) also demonstrated that the survival of *O. oeni* following lyophilization depends on the protective medium, the rehydration medium, and the storage medium used, making it important to choose these elements carefully in order to obtain maximum cell viability. The addition of polysaccharides and disaccharides to the suspension medium significantly increases cell viability. Rehydration in the disaccharide solutions tested, however, led to a significant reduction in cell viability. Viability was reduced after 6 months of storage at 4°C; the loss of viability was dependent upon the protective agents used, sodium glutamate being the most effective.

At the end of the lyophilization process, the biomass is reduced to a powder that is packaged under sterile conditions in gas- and vapor-impermeable polylaminated aluminum to prevent contact between the lyophilized bacteria and oxygen or additional moisture. Exposure to high temperatures for prolonged periods can also kill the cells.

Each batch of lyophilized bacteria must undergo rigorous quality control; this process applies not only to the final product but also to each phase of production. Quality control begins with maintaining a stock of the strain in liquid nitrogen or in lyophilized form to ensure that each industrial production process begins with a pure strain that maintains all of the original characteristics. It has also been reported that strains recently isolated from wine display greater viability and malolactic activity than strains that have been maintained for some time in culture collections and subcultured successively in synthetic media.

The variables measured to assess the quality of the end product are the concentration of viable cells, the capacity to degrade malic acid, and the microbiological quality. Molecular biological techniques are used for the identification of the strains. Commercial preparations generally contain fewer than $10^3$ contaminating bacteria per gram. Quality control is also performed in the marketed product to confirm that it has been appropriately stored.

## 3.4. Use

Direct inoculation of rehydrated cultures into wine leads to significant bacterial cell death that can reduce the cell population from $10^7$ to $10^4$ CFU/mL. To compensate for this loss, the cell density of the inoculum would need to be increased 100-fold. However, such quantities are not economically viable and the lyophilized bacteria must therefore be reactivated. The reactivation medium generally used is grape must without sulfite diluted 50% with water, with a pH adjusted to 4.0 to 4.5 (with calcium carbonate) and containing 3–5 g/L of yeast extract. King and Beelman (1986) demonstrated the importance of using diluted grape must rather than undiluted must for the culture of lactic acid bacteria. Cavazza et al. (1999) demonstrated that, during reactivation of stored cultures in diluted grape must, the time required for malolactic fermentation is considerably reduced. They also showed that the presence of ethanol at a concentration below 2% (vol/vol) acted as a stimulant for the propagation of starter cultures. Hayman and Monk (1982) reported that the best results were obtained with addition of one volume of sterile wine (without sulfite) to five volumes of must. Consequently, addition of wine during propagation increases the acclimatization both to alcohol and to the pH of wine, and therefore increases the number of viable lactic acid bacteria during the final phase of addition to the wine. In addition, various supplement mixtures are commercially available for addition to the media, although these are more frequently used during subsequent malolactic fermentation in wines with low concentrations of nutrients (Pilatte & Nygaard, 1999).

The first malolactic starter culture for direct inoculation into wines was introduced in 1993. Since then, cultures have been described that are specifically adapted to the analytical and organoleptic characteristics of red, white, and rosé wines. The capacity to survive following direct inoculation in wine and the maintenance of this capacity when strains are prepared as lyophilized cultures are of major practical importance in winemaking. Consequently, lyophilized preparations of selected strains of *O. oeni* that display 100% survival following direct inoculation and that induce malolactic fermentation reliably and rapidly under standard vinification conditions are particularly useful. Both the immediate survival and the lag period of the inoculated bacteria are critical factors since they determine the total duration of malolactic fermentation. These commercial lyophilized preparations thus demonstrate that it is possible to produce malolactic starter cultures that do not require the usual reactivation or preadaptation steps prior to use. This eliminates the risk of contamination and reduces the time required during vinification. These cultures are easy to use. The lyophilized product is simply added directly to the wine following alcoholic fermentation. This can be done during racking or, alternatively, with a pump or by recirculation. Certain optimal conditions in the wine to be inoculated have been established for the use of these cultures: it should contain no free $SO_2$, it should not contain added sulfites (with a maximum $SO_2$ of 40 mg/L in red wine and 30 mg/L in white

wine), and it should be maintained at an optimal temperature of 23°C.

The use of commercial malolactic starter cultures is dependent on the effectiveness of the technology used for storage in order to guarantee a high rate of cell survival and a high degree of functionality during the processing stages and following storage and rehydration. Commercial lyophilized malolactic starter cultures generally contain dead cells, undamaged cells, and live damaged cells. It is thus important to monitor the proportions of these cells during treatments prior to storage (freezing and lyophilization), after rehydration, and during initial establishment of the cultures. To this end, it has recently been reported that quantitative data can be obtained in real time (no more than 1 h) using flow cytometry to determine the numbers of metabolically active and dead cells in a sample (Quirós et al., 2009).

## Acknowledgments

We thank the Spanish Ministry of Science and Innovation (MICINN) (grants AGL2006-02558, AGL2008-01052, AGL2009-07894 and Consolider INGENIO 2010 CSD2007-00063 FUN-C-FOOD) and the Comunidad de Madrid (CAM) (S2009/AGR-1469) for financial support.

## References

Aiba, S., Nagai, S., & Nishizawa, Y. (1976). Fed batch culture of *Saccharomyces cerevisiae*: A perspective of computer control to enhance the productivity in baker's yeast cultivation. *Biotechnol. Bioeng., 18*, 1001–1016.

Alegría, E.-G., López, I., Ruiz, J. I., Sáenz, J., Fernández, E., Zarazaga, M., et al. (2004). High tolerance of wild *Lactobacillus plantarum* and *Oenococcus oeni* strains to lyophilisation and stress environmental conditions of acid pH and ethanol. *FEMS Microbiol. Lett., 230*, 53–61.

Baliarda, A., Robert, H., Jebbar, M., Blanco, C., Deschamps, A., & le Marrec, C. (2003). Potential osmoprotectants for the lactic acid bacteria Pediococcus pentosaceus and Tetragenococcus halophila. *Int. J. Food Microbiol., 84*, 13–20.

Bourdineaud, J.-P., Nehmé, B., Tesse, S., & Lonvaud-Funel, A. (2003). The *ftsH* gene of the wine bacterium *Oenococcus oeni* is involved in protection against environmental stress. *Appl. Environ. Microbiol., 69*, 2511–2520.

Buckenhüskes, H. J. (1993). Selection criteria for lactic acid bacteria to be used as starter cultures for various food commodities. *FEMS Microbiol. Rev., 12*, 253–272.

Carrasco, P., Querol, A., & del Olmo, M. (2001). Analysis of the stress resistance of comercial wine yeast strains. *Arch. Microbiol., 175*, 450–457.

Carvalho, A. S., Silva, J., Ho, P., Teixeira, P., Malcata, F. X., & Gibbs, P. (2002). Survival of freeze-dried *Lactobacillus plantarum* and *Lactobacillus rhamnosus* during storage in the presence of protectans. *Biotechnol. Lett., 24*, 1587–1591.

Carvalho, A. S., Silva, J., Ho, P., Teixeira, P., Malcata, F. X., & Gibbs, P. (2003). Effect of various growth media upon survival during storage of freeze-dried *Enterococcus faecalis* and *Enterococcus durans. J. Appl. Microbiol., 94*, 947–952.

Castelli, T. (1955). Biologia e vino. *Humus, 10*, 1–5.

Castelli, T., & Iñigo, B. (1957). Los agentes de la fermentación vínica en la región manchega y limítrofes. *Ann. Fac. Agr. Univ. Perugia, 13*, 17.

Castelli, T., & Iñigo, B. (1958). Los agentes de la fermentación vínica de la Rioja. *Ann. Fac. Agr. Univ. Perugia, 14*, 15.

Cavazza, A., Vavassori, G. L., & Volonterio, G. (1999). Lactic acid bacteria starter cultures in oenology: Direct-vat inoculation or reactivation? *Rivista di Viticoltura e di Enologia, 52*, 41–52.

Champagne, C., Gardner, N., & Doyon, G. (1989). Production of *Leuconostoc oenos* biomass under pH control. *Appl. Environ. Microbiol., 55*, 2488–2492.

Cooney, C. L., Wang, H. Y., & Wang, D. I. (1977). Computer-aided material balancing for prediction of fermentation parameters. *Biotechnol. Bioeng., 19*, 55–67.

Coucheney, F., Desroche, N., Bou, M., Tourdot-Maréchal, R., Dulau, L., & Guzzo, J. (2005). A new approach for selection of Oenococcus oeni strains in order to produce malolactic starters. *Int. J. Food Microbiol., 105*, 463–470.

Coulon, J., Husnik, J. I., Inglis, D. L., van der Merwe, G. K., Lonvaud, A., Erasmus, D. J., et al. (2006). Metabolic engineering of Saccharomyces cerevisiae to minimize the production of ethyl carbamate in wine. *Am. J. Enol. Vitic., 57*, 113–124.

Craig, E. A. (1985). The heat shock response. *CRC Crit. Rev. Biochem., 18*, 239–280.

Delfini, C., & Cervetti, F. (1991). Metabolic and technological factors affecting acetic acid production by yeasts during alcoholic fermentation. *Vitic. Enol. Sci., 46*, 142–150.

Delmas, F., Divies, C., & Guzzo, J. (2000). Biochemical and physiological studies of a small heat shock protein from a lactic acid bacterium. *Sci. Aliments., 20*, 111–117.

Felix, C. F., Moreira, C. C., Oliveira, M. S., Sola-Penna, M., Meyer-Fernandes, J. F., Scofano, H. M., et al. (1999). Protection against thermal denaturation by trehalose on the plasma membrane H$^+$-ATPase from yeast. *Eur. J. Biochem., 266*, 660–664.

François, J., & Parrou, J. L. (2001). Reserve carbohydrates metabolism in the yeast *Saccharomyces cerevisiae. FEMS Microbiol. Rev., 25*, 125–145.

Gibson, B. R., Lawrence, S. J., Boulton, C. A., Box, W. G., Graham, N. S., Linforth, R. S. T., et al. (2008). The oxidative stress response along a lager brewing yeast strain during industrial propagation and fermentation. *FEMS Yeast Res., 8*, 574–584.

Gindreau, E., Keim, H., de Revel, G., Bertrand, A., & Lonvaud-Funel, A. (2003). Use of direct inoculation malolactic starters: Setting efficiency and sensorial impact. *J. Int. Sci. Vig. Vin, 37*, 51–57.

Gockowick, H., & Henschke, P. A. (2003). Interaction of pH, ethanol concentration and wine matrix on induction of malolactic fermentation with commercial "direct inoculation" starter culture. *Austral. J. Grape Wine Res., 9*, 200–209.

Granot, D., & Snyder, M. (1991). Glucose induces cAMP-independent growth-related changes in stationary-phase cells of *Saccharomyces cerevisiae. Proc. Natl. Acad. Sci. USA, 88*, 5724–5728.

Gray, J. V., Petsko, G. A., Johnston, G. C., Ringe, D., Singer, R. A., & Werner-Washburne, M. (2004). "Sleeping Beauty": Quiescence in Saccharomyces cerevisiae. *Microbiol. Mol Biol. Rev., 68*, 187–206.

Guerzoni, M. E., Sinigaglia, M., Gardini, F., Ferruzzi, M., & Torriani, S. (1995). Effects of pH, temperature, ethanol, and malate concentration on *Lactobacillus plantarum* and *Leuconostoc oenos*: Modelling of the malolactic activity. *Am. J. Enol. Vitic., 46*, 368–374.

Guzzo, J., Delmas, J. F., Pierre, F., Jobin, M.-P., Samyn, B., van Beeumen, J., et al. (1997). A small heat shock protein from *Leuconostoc oenos* induced by multiple stresses and during stationary growth phase. *Lett. Appl. Microbiol., 24*, 393–396.

Guzzo, J., Jobin, M. P., Delmas, F., Fortier, L. C., Garmyn, D., Tourdot-Marechal, R., et al. (2000). Regulation of stress response in Oenococcus oeni as a function of environmental changes and growth phase. *Int. J. Food Microbiol., 55*, 27–31.

Hammes, W. P. (1990). Bacterial starter cultures in food production. *Food Biotechnol., 4*, 383–397.

Hayman, D. C., & Monk, P. R. (1982). Starter culture preparation for the induction of malolactic fermentations in wine. *Food Tech. Aust., 34*, 14–18.

Heard, G. M., & Fleet, G. H. (1987). Occurrence and growth of killer yeasts during wine fermentation. *Appl. Environ. Microbiol., 53*, 2171–2174.

Iñigo, B. (1958). Vinificación del mosto de uva con asociación escalar de levaduras seleccionadas. *Revista de Ciencia Aplicada, 63*, 318–324.

Ivorra, C., Pérez-Ortín, J. E., & del Olmo, M. (1999). An inverse correlation between stress resistance and stuck fermentations in wine yeasts. A molecular study. *Biotechnol. Bioeng., 64*, 698–708.

Jobin, M. P., Delmas, F., Garmyn, D., Divies, C., & Guzzo, J. (1997). Molecular characterization of the gene encoding an 18-kilodalton small heat shock protein associated with the membrane of *Leuconostoc oenos. Appl. Environ. Microbiol., 63*, 609–612.

Jobin, M. P., Garmyn, D., Divies, C., & Guzzo, J. (1999). Expression of the *Oenococcus oeni trxA* gene is induced by hydrogen peroxide and heat shock. *Microbiology, 145*, 1245–1251.

Jobin, M. P., Garmyn, D., Divies, C., & Guzzo, J. (1999). The *Oenococcus oeni clpX* homologue is a heat shock gene preferentially expressed in exponential growth phase. *J. Bacteriol., 181*, 6634–6641.

King, S. W., & Beelman, R. B. (1986). Metabolic interactions between *Saccharomyces cerevisiae* and *Leuconostoc oenos* in a model grape juice/wine system. *Am. J. Enol. Vitic., 37*, 53–60.

Kitamoto, K., Oda, K., Gomi, K., & Takahashi, K. (1991). Genetic engineering of a sake yeast producing no urea by successive disruption of arginase gene. *Appl. Environ. Microbiol., 57*, 301–306.

Kole, M., Altosaar, I., & Duck, P. (1982). Pilot scale production and preservation of a new malolactic culture *Leuconostoc oenos* 44.40 for use in secondary wine fermentation. *Biotechnol. Lett., 4*, 695–700.

Kraus, J. K., Reed, G., & Villettaz, J. C. (1983). Levures sèches actives de vinification. 1ere partie: Fabrication et caractéristiques. *Conn. Vig. Vin, 17*, 93–103.

Krieger, S. A., Hammes, W. P., & Henick-Kling, T. (1993). How to use malolactic starter cultures in the winery. *Austral. New Zeal. Wine Indust. J., 8*, 153–160.

Leslie, S. B., Teter, S. A., Crowe, L. M., & Crowe, J. H. (1994). Trehalose lowers membrane phase transitions in dry yeast cells. *Biochim. Biophys. Acta, 1192*, 7–13.

Lillie, S. H., & Pringle, J. R. (1980). Reserve carbohydrate metabolism in *Saccharomyces cerevisiae*: Responses to nutrient limitation. *J. Bacteriol., 143*, 1384–1394.

Magliani, W., Conti, S., Gerloni, M., Bertolotti, D., & Polonelli, L. (1997). Yeast killer systems. *Clin. Microb. Rev., 10*, 369–400.

Maicas, S., Gil, J. V., Pardo, I., & Ferrer, S. (1999a). Improvement of volatile composition of wines by controlled addition of malolactic bacteria. *Food Res. Int., 32*, 491–496.

Maicas, S., González-Cabo, P., Ferrer, S., & Pardo, I. (1999b). Production of *Oenococcus oeni* biomass to induce malolactic fermentation in wine by control of pH and substrate addition. *Biotechnol. Lett., 21*, 349–353.

Maicas, S., Pardo, I., & Ferrer, S. (2000). The effects of freezing and freeze-drying of Oenococcus oeni upon induction of malolactic fermentation in red wine. *Int. J. Food Sci. Technol., 35*, 75–79.

Manginot, C., Roustan, J. L., & Sablayrolles, J. M. (1998). Nitrogen demand of different yeast strains during alcoholic fermentation. Importance of the stationary phase. *Enz. Microb. Technol., 23*, 511–517.

Marks, V. D., Sui, S. J. H., Erasmus, D., van der Merwe, G. K., Brumm, J., Wasserman, W. W., et al. (2008). Dynamics of the yeast transcriptome during wine fermentation reveals a novel fermentation stress response. *FEMS Yeast Res., 8*, 35–52.

Molina-Gutierrez, A., Stippl, V., Delgado, A., Ganzle, M. G., & Vogel, R. F. (2002). *In situ* determination of the intracellular pH of *Lactococcus lactis* and *Lactobacillus plantarum* during pressure treatment. *Appl. Environ. Microbiol., 68*, 4399–4406.

Naomi, P., Chagnaud, P., Arnaud, A., Galzy, P., & Mathleu, J. (1989). Optimization of the conditions for preparing bacterial cultures for malolactic bioconversion. *J. Biotechnol., 10*, 135–150.

Nielsen, J. C., Prahl, C., & Lonvaud-Funel, A. (1996). Malolactic fermentation in wine by direct inoculation with freeze-dried *Leuconostoc oenos* cultures. *Am. J. Enol. Vitic., 47*, 42–48.

Novo, M. T., Beltran, G., Torija, M. J., Poblet, M., Rozes, N., Guillamon, J. M., et al. (2003). Changes in wine yeast storage carbohydrate levels during preadaptation, rehydration and low temperature fermentations. *Int. J. Food. Microbiol., 86*, 153–161.

Novo, M. T., Beltran, G., Torija, M. J., Poblet, M., Rozes, N., Guillamon, J. M., & Mas, A. (2007). Early transcriptional response of wine yeast after rehydration: Osmotic shock and metabolic activation. *FEMS Yeast Res., 7*, 304–316.

Ouvry, A., Wache, Y., Tourdot-Marechal, R., Divies, C., & Cachon, R. (2002). Effects of oxidoreduction potential combined with acetic acid, NaCl and temperature on the growth, acidification and membrane properties of *Lactobacillus plantarum. FEMS Microbiol. Lett., 214*, 257–261.

Panoff, J. M., Thammavongs, B., & Gueguen, M. (2000). Cryoprotectants and cold stress in lactic acid bacteria. *Sci. Aliments, 20*, 105–110.

Papin, J. P. (1988). Les levures sèches en oenologie. Fabrication et contrôles de qualité. In *Application à l'oenologie des progres recents en microbiologie e en fermentation* (pp. 331–371). Montpellier, France: OIV-ENSA.

Pérez-Ortín, J. E., Querol, A., Puig, S., & Barrio, E. (2002). Molecular characterization of a chromosomal rearrangement envolved in the adaptive evolution of yeast strains. *Genome Res., 12*, 1533–1539.

Pérez-Torrado, R., Bruno-Bárcena, J. M., & Matallana, E. (2005). Monitoring stress-related genes during the process of biomasa propagation of *Saccharomyces cerevisiae* strains used for wine-making. *Appl. Environ. Microbiol., 71*, 6831–6837.

Pérez-Torrado, R., Gómez-Pastor, R., Larson, C., & Matallana, E. (2009). Fermentative capacity of dry active wine yeast requires a specific oxidative stress response during industrial biomasa growth. *Appl. Microbiol. Biotechnol., 81,* 951–960.

Pilatte, E., & Nygaard, M. (1999). Development of a specific activator for malolactic bacteria. *Rev. Oenol. Techniq. Vitivin. Oenol., 92,* 31–33.

Potma, E., Verduyn, C., Scheffers, W. A., & van Dijken, J. P. (1989). Enzymatic analysis of the crabtree effect in glucose-limited chemostat cultures of *Saccharomyces cerevisiae. Appl. Environ. Microbiol., 55,* 468–477.

Pronk, J. T., Steensma, H. Y., & van Dijken, J. P. (1996). Pyruvate metabolism in *Saccharomyces cerevisiae. Yeast, 12,* 1607–1633.

Quirós, C., Herrero, M., García, L. A., & Díaz, M. (2009). Taking advantage of the flow cytometry technique for improving malolactic starters production. *Eur. Food Res. Technol., 228*(4), 543–552.

Ratnakumar, S., & Tunnacliffe, A. (2006). Intracellular trehalose is neither necessary nor sufficient for dessication tolerante in yeast. *FEMS Yeast Res., 6,* 902–913.

Rodrigues Pousada, C. A., Nevitt, T., Menezes, R., Azevedo, D., Pereira, J., & Amaral, C. (2004). Yeast activator proteins and stress response: An overview. *FEBS Lett., 567,* 80–85.

Rodriguez-Porrata, B., Novo, M., Guillamón, J., Rozés, N., Mas, A., & Cordero, R. (2008). Vitality enhancement of the rehydrated active dry wine yeast. *Int. J. Food Microbiol., 126,* 116–122.

Romano, P., & Suzzi, G. (1993). Sulfur dioxide and wine microorganisms. In G. H. Fleet (Ed.), *Wine microbiology and biotechnology* (pp. 373–393). Chur, Switzerland: Harwood Academic Publishers.

Romeo, Y., Bouvier, J., & Gutierrez, C. (2001). Osmotic stress response of lactic acid bacteria *Lactococccus lactis* and *Lactobacillus plantarum. Lait, 81,* 49–55.

Rossignol, T., Postaire, O., Storaï, J., & Blondin, B. (2006). Analysis of the genome response of a wine yeast to rehydratation and inoculation. *Appl. Microbiol. Biotechnol., 71,* 699–712.

Sánchez, C. (1988). Levures sèches actives. Fabrication et contrôles de qualité. In *Application à l'oenologie des progres recents en microbiologie e en fermentation* (pp. 315–330). Montpellier, France: OIV-ENSA.

Silveira, M. G., Baumgärtner, M., Rombouts, F. M., & Abee, T. (2004). Effect of adaptation to ethanol on cytoplasmic and membrane protein profiles of *Oenococcus oeni. Appl. Environ. Microbiol., 70,* 2748–2755.

Singh, J., Kumar, D., Ramakrishnan, N., Singhal, V., Jervis, J., Garst, J. F., et al. (2005). Transcriptional response of *Saccharomyces cerevisiae* to desiccation and rehydratation. *Appl. Environ. Microbiol., 71,* 8752–8763.

Smelt, J. P. P. M., Otten, G. D., & Bos, A. P. (2002). Modelling the effect of sublethal injury on the distribution of the lag times of individual cells of *Lactobacillus plantarum. Int. J. Food. Microbiol., 73,* 207–212.

Spiropoulos, A., & Bisson, L. F. (2000). *MET17* and hydrogen sulfide formation in *Saccharomyces cerevisiae. Appl. Environ. Microbiol., 66,* 4421–4426.

Tezuka, H., Mori, T., Okumura, Y., Kitabatake, K., & Tsumura, Y. (1992). Cloning of a gene suppressing hydrogen sulfide production by *Saccharomyces cerevisiae* and its expression in a brewing yeast. *J. Am. Soc. Brew. Chem., 50,* 130–133.

Thevelein, J. M. (1994). Regulation of trehalose mobilization in fungi. *Microbiol. Rev., 48,* 42–59.

Toledano, M. B., Kumar, C., Le, M. N., Spector, D., & Tacnet, F. (2007). The system biology of thiol redox system in *Escherichia coli* and yeast: Differential functions in oxidative stress, iron metabolism and DNA synthesis. *FEBS Lett., 581,* 3598–3607.

Tourdot-Marechal, R., Gaboriau, D., Beney, L., & Divies, C. (2000). Membrane fluidity of stressed cells of *Oenococcus oeni* is involved in protection against environmental stress. *Appl. Environm. Microbiol., 69,* 2512–2520.

Tunnacliffe, A., & Wise, M. J. (2007). The continuing conundrum of the LEA proteins. *Naturwissenschaften, 94* (10), 791–812.

van de Guchte, M., Serror, P., Chervaux, C., Smokvina, T., Ehrlich, S. D., & Magín, E. (2002). Stress responses in lactic acid bacteria. *Anton. Leeuw., 82,* 187–216.

Vaudano, E., Constantini, A., Cercosimo, M., del Prete, V., & García-Moruno, E. (2009). Application of real-time RT-PCR to study gene expression in active dry yeast during the rehydration phase. *Int. J. Food Microbiol., 129,* 30–36.

Wang, H. Y., Cooney, C. L., & Wang, D. I. (1977). Computeraided baker's yeast fermentations. *Biotechnol. Bioeng, 19,* 69–86.

Werner-Washburne, M., Braun, E., Johnston, G. C., & Singer, R. A. (1993). Stationary phase in the yeast *Saccharomyces cerevisiae. Microbiol. Rev., 57,* 383–401.

Yuasa, N., Nakagawa, Y., Hayakawa, M., & Limura, Y. (2004). Distribution of the sulfite resistance gene *SSU1-R* and the variation in its promoter region in wine yeasts. *J. Biosci. Bioeng., 98,* 394–397.

Zhao, G., & Zhang, G. (2009). Influence of freeze-drying conditions on survival of *Oenococcus oeni* for malolactic fermentation. *Int. J. Food Microbiol., 135,* 64–67.

Zhao, G., & Zhang, G. (2009). Influences of protectants, rehydration media and storage on the viability of freezedried *Oenococcus oeni* for malolactic fermentation. *World J. Microbiol. Biotechnol.* doi: 10.1007/s11274–009–0080–9.

Zuzuarregui, A., Carrasco, P., Palacios, A., Julien, A., & del Olmo, M. (2005). Analysis of the expression of some stress induced genes in several commercial wine yeast strains at the beginning of vinification. *J. Appl. Microbiol., 98*, 299–307.

Zuzuarregui, A., & del Olmo, M. (2004a). Analyses of stress resistance under laboratory conditions constitute a suitable criterion for wine yeast selection. *Anton. Leeuw., 85*, 271–280.

Zuzuarregui, A., & del Olmo, M. (2004b). Expression of stress response genes in wine strains with different fermentative behavior. *FEMS Yeast Res., 4*, 699–710.

# 12

# Preservation of Microbial Strains in the Wine Industry

*María Dolores García López, José M. López-Coronado,*
*Laura López-Ocaña, Federico Uruburu Fernández*
Colección Española de Cultivos Tipo (CECT), Universitat de València, Valencia, Spain

## OUTLINE

# 1. INTRODUCTION

Three objectives must be attained to ensure good preservation of microbial strains in microbiology laboratories: the culture must be pure (without contamination during the preservation process), at least 70 to 80% of the cells must survive storage, and the cells must remain genetically stable. The first two objectives are not particularly difficult to achieve when good microbiological technique is used. The third, however, can present difficulties. Consequently, various methods have been developed for the preservation of microorganisms and no single method is applicable to all situations.

# 2. METHODS FOR THE PRESERVATION OF MICROBIAL STRAINS

Many different microorganisms influence the winemaking process. At the earliest stages, vines can be damaged by the growth of phytopathogenic fungi (see Chapter 10). Later, in the winery, yeasts and lactic acid bacteria will transform the grape must into wine (see Chapters 2, 3, 4, 5, and 9) or have a negative influence on the winemaking process (see Chapters 5, 9, and 10), sometimes even growing in unexpected places such as the corks used to seal the bottles (Álvarez-Rodríguez et al., 2002). Analysis and monitoring of these three microbial groups inevitably requires appropriate preservation of pure cultures. Such preservation is perhaps even more important when commercial starter cultures must be stored in the winery so as not to have to request them from the supplier every time they are needed.

In this chapter, we will provide a general description of the most commonly used methods for the preservation of microbial strains. This will be divided into three sections, in each case discussing the advantages and disadvantages of the different approaches. We will then go on to discuss the specific considerations applicable to the preservation of the yeasts, lactic acid bacteria, acetic acid bacteria, and filamentous fungi that are relevant to the wine industry. Further information can be found in Day and Stacey (2007), Hatt (1980), Hill (1981), Hunter-Cevera and Belt (1996), Kirsop (1980), and Kirsop and Doyle (1991).

## 2.1. Long-term Preservation: The Preferred Approach

Long-term preservation methods are considered the most appropriate option wherever possible, since they involve stopping the growth of the microbial cells and keeping them in a viable state. This guarantees maximum genetic stability by preventing the appearance of successive generations. Nevertheless, the possibility that the preparation method itself leads to changes cannot be ruled out. There are two preservation methods belonging to this group: freezing and lyophilization.

### 2.1.1. Freezing

In the first long-term preservation method, the cells are frozen suspended in a liquid medium containing a cryoprotective agent and stored at temperatures below 0°C. As a result, intracellular and extracellular water is maintained in a solid state. The reduced cell metabolism caused by the low temperature and the absence of liquid water prevents growth. Cells preserved in this way are recovered by increasing the temperature prior to use. This is the best method of preservation in almost all respects, although it has the drawback of requiring special apparatus and carries with it the risk that system failure could result in an unintentional increase in temperature during storage. It is also the most inconvenient method for the distribution of strains.

Four factors influence the viability and stability of cells preserved in this way:

1) *Age of the cells.* In most cases, it is best to use mature cells from the beginning of the stationary phase of the growth curve. However, in the case of microorganisms with a stage in their life cycle that prepares them to resist adverse conditions, it is preferable to use cells from this stage. This occurs in microorganisms that sporulate, in some pleomorphic microorganisms, and even in some simpler microorganisms.

2) *Rate of freezing and thawing.* Although there are standardized freezing protocols for use in certain contexts, it is generally best for changes in temperature to be rapid, both during freezing and thawing, in order to minimize the formation of ice crystals. It is normally appropriate, therefore, to thaw cells at 37°C.

3) *Use of cryoprotective agents.* Cryoprotectants are substances that protect against the damage that can occur in microbial cells during freezing, mainly by favoring the vitrification of extracellular water rather than its crystallization, which causes cell damage and loss of viability of the preserved culture. Cryoprotective agents can also stabilize large intracellular molecules but only if the cryoprotectant can cross the cell membrane. Although many compounds can be used as cryoprotectants, the most commonly used is glycerol at a concentration of between 15 and 20%. Dimethyl sulfoxide, skimmed milk, and carbohydrates such as glucose, lactose, sucrose, and inositol can also be used. The choice of cryoprotective agent is influenced by the type of microorganism to be preserved.

4) *Storage temperature.* Storage temperature should be as low as possible in order to prevent intracellular recrystallization of water, which occurs at temperatures between 0 and approximately −130°C. It is best to store the microbial cells in sealed tubes submerged in liquid nitrogen (−195°C) or in liquid nitrogen vapor (−140°C).

Although various types of freezer are available on the market, the most appropriate reach temperatures below −70°C. Those that only reach temperatures of between −20 and −40°C, as is applicable to most of those currently available in microbiology laboratories, are less recommendable. Among other reasons, this is because the high concentration of solutes in the cell suspensions reduces their freezing point and cell damage occurs as a result of the frequent freezing and thawing that occurs under these conditions. Use of a nonionic cryoprotectant such as glycerol reduces the quantity of ice produced and helps to prevent increases in the ionic concentration. For preservation in freezers, the cells are stored in cryotubes (sterilizable plastic tubes that are resistant to freezing and can be hermetically sealed). Batches of tubes are prepared for each strain and then a single tube is used completely each time the culture is required. This avoids the repeated freezing and thawing of the strains. The use of cryoballs is now widespread. However, they have a number of disadvantages that make them inappropriate for use in preserving cells in optimal conditions over long periods of time. The main disadvantage of this method is that the cells are maintained in an extremely thin layer on the surface of the cryoballs and in the absence of cryoprotective agents. This thin layer of cells thaws very rapidly when a cryoball is taken for inoculation and freezes again when the tube is returned to the freezer. Consequently, the cells that remain in the tube are subjected to frequent freeze–thaw cycles, and this problem is further aggravated when the freezers used for storage do not reach temperatures below $-70°C$. Furthermore, since the vials contain numerous cryoballs, the risk of contamination during handling is very high. The method is also particularly harmful to anaerobic microorganisms such as bacteria of the genus *Clostridium*, since the cells on the surface of the cryoballs are exposed to oxygen in the environment and, as a result, their viability is reduced.

### 2.1.2. Lyophilization

Lyophilization is a gentle process in which water is removed from the cells to stop their growth. The genetic stability obtained with this method is high, although not always as high as that seen with freezing, since lyophilization is achieved by sublimation of the ice in the cells. First, the free water in the cells must be frozen and then eliminated by vacuum without increasing the temperature, as this would affect the viability of the microorganism. The process uses an apparatus known as a lyophilizer or freeze drier, and many different models are available on the market. The microbial cells that are preserved in this way are subjected to a more complex treatment, since freezing is followed by sublimation of water. However, it is a highly recommendable method since these lyophilized samples can be maintained at room temperature, thus making storage and distribution much easier.

The factors that must be taken into account in order to achieve good lyophilization are of course the same as those that influence freezing plus additional factors relating to the subsequent dehydration of the cells. However, before discussing dehydration-related factors, we should briefly consider those mentioned earlier in relation to freezing. Freezing can be performed rapidly, by submerging the tubes in liquid nitrogen, or slowly, using freezers with or without programming. As we saw earlier, various cryoprotectants can be used according to the type of microorganism. However, glycerol should not be used when the cells are to be lyophilized due to its high evaporation point and hygroscopic properties, which can lead to highly viscous lyophilized samples. The use of dimethyl sulfoxide is also inappropriate because it is slightly toxic and can cause damage to the microbial cells as a result of concentration following the evaporation of water. As a result, inositol is recommended as a cryoprotective agent for use during the lyophilization of most bacteria and skimmed milk for use with fungi and actinomycetes. Other cryoprotectants may be more appropriate for certain microorganisms, such as glutamate for lactic acid bacteria, mixtures of glucose and liver broth or chopped meat medium (without meat) for anaerobic bacteria, etc.

The factors that specifically influence the efficacy of lyophilization as a means of preservation are as follows:

1) *Type of microorganism.* Some microbes cannot tolerate lyophilization; these are logically microorganisms that contain more intracellular water. Some filamentous fungi, particularly nonsporulating strains, cannot be stored as lyophilized preparations and other methods must therefore be used.
2) *Cell concentration.* It is best to lyophilize cell suspensions at a concentration of between $10^8$ and $10^9$ cells/mL in the case of bacteria and at slightly lower concentrations for filamentous fungi and yeasts.
3) *Temperature during sublimation.* Sublimation should be performed at the lowest possible temperature, always below $-50°C$.
4) *Extent of dehydration achieved.* Dehydration should always be as extensive as possible, although the concentration of solutes may lead to small traces of water that are not harmful.
5) *Presence of oxygen in the tube.* The lyophilized cells are stored in tubes sealed under vacuum

to prevent both rehydration and entry of oxygen, which can damage the cells.

6) *Storage conditions.* The storage temperature must be constant, preferably between 4 and 18°C, and must not fall below 0°C. Lyophilized samples must be stored in the dark.

## 2.2. Short-term, Alternative Preservation Methods

When the strain does not tolerate the treatments required for long-term preservation or the necessary equipment is unavailable for this type of preservation, alternative methods must be used. In such cases, a combination of short-term methods should always be employed—no single short-term method should ever be used in isolation.

### 2.2.1. *Preservation by Periodic Transfer*

In preservation methods based on periodic transfer, the microbial strain is stored as an active culture in the culture medium in which it was grown. However, the strain cannot be stored indefinitely in the same vial. Because the cells remain active, they continue to excrete toxic metabolic byproducts that accumulate and lead to cell aging and death. It is therefore necessary to transfer cells to another vial containing fresh culture medium. This is the worst method in terms of genetic stability, since continued cell growth implies ongoing turnover of generations, and over time the distant descendents of the initial cells may not retain some of the original characteristics. If this method is to be used, it is advisable to delay aging and extend the periods between reinoculation. This can be achieved in various ways. For instance, by reducing the size of the inoculum or reducing the concentration of some nutrients in the culture medium, by using stab inoculation for facultative anaerobes (since growth in the presence of oxygen is more rapid and generally produces toxic byproducts), and by storing cultures at temperatures of between 4 and 8°C. Sometimes the culture is also covered with a layer of sterile mineral oil. This helps to prevent the toxic effects associated with the increase in the concentration of the culture medium caused by evaporation. Highly aerobic microorganisms such as filamentous fungi cannot be stored in completely closed vials. Finally, an additional drawback of periodic transfer is the increased risk of contamination due to handling of the vials over time, as well as the possibility of mites entering the vials.

### 2.2.2. *Preservation by Suspension in Sterile Distilled Water or Seawater*

Suspension in sterile distilled water or seawater is a widely used alternative that maintains a high percentage of viability in a number of different microorganisms, including filamentous fungi, yeasts, and some bacteria. It involves suspending an aliquot of cells from the culture to be preserved in sterile water. The samples can be prepared in cryotubes. In this case, the cell concentration must not exceed $10^4$–$10^5$ cells/mL in the case of bacteria and yeasts. In the case of nonsporulating filamentous fungi, suspensions can be made with small pieces of agar containing the growing fungus. In the case of marine microorganisms, suspensions are prepared in diluted seawater.

Studies performed by the Spanish Type Culture Collection (CECT) to assess the preservation of microorganisms using this method have revealed high percentages of viability, sometimes for periods of more than 15 years. The stability of morphological and physiological characteristics is also good, although this has not been tested for specific characteristics such as virulence, fermentation power, etc.

## 2.3. Other Methods of Preservation

Some methods that are not widely used are required in order to preserve very specific groups of microorganisms that do not tolerate lyophilization or freezing, such as the bacterial genera *Spirillum* and *Rhodospirillum*. The three methods described here are based on halting growth by eliminating the availability of water for the cells.

### 2.3.1. *Drying on Filter Paper*

In this method, a highly absorbent filter paper (Whatman No. 3) is impregnated with a concentrated solution of cells and allowed to dry under sterile conditions. The adsorption to the paper favors the dispersion of the cells in this matrix and the cells do not form impermeable films that hinder drying. Similar principles apply to other substrates used for drying. It is also possible to dry the cells using a procedure known as liquid drying (L-Dry) because it involves the use of a lyophilizer without prior freezing of the cells. The vacuum created by the lyophilizer dries the cells but excessive vacuum must be avoided to prevent rapid evaporation with boiling or too great a reduction in temperature, which would lead to uncontrolled freezing of the cells.

### 2.3.2. *Drying in Earth, Sand, Silica Gel, etc*

Cells can be added to substances such as earth, sand, or silica gel to protect them during drying. Spore-producing microorganisms can be preserved for extended periods using this method.

### 2.3.3. *Drying on Alginate Beads*

The use of alginate beads is an effective procedure in which the cells are placed in an alginate matrix and water is eliminated by sequential immersion in increasingly hypertonic solutions before air drying to achieve a 70% reduction in water content. The alginate beads can be stored in hermetically sealed tubes at a temperature of between 4 and 18°C. They can even be stored at −80°C due to the low water content of the cells and the protection provided by the alginate support. This method has also been used for the preservation of algae and plant cells.

## 2.4. Recovery

Whatever the method used to preserve microbial strains, the cells are placed under stress (particularly during lyophilization). Consequently, the stored cells are not suitable to be used directly. They must first be revitalized or rejuvenated by seeding in nonselective medium; that is, a medium that ensures maximum growth. After this step has been performed, it is possible to work with the cells and culture them in selective media if necessary. Likewise, we should remember that some microorganisms will tolerate a given preservation technique better than others and that special precautions may be necessary for the preservation of certain strains. As mentioned at the beginning of this chapter, there is no single method appropriate for the preservation of all microorganisms, but it is not difficult to identify the most appropriate method in each case.

An important element in the efficacy of the preservation method used is the recovery of the preserved culture. Cultures should be thawed rapidly (37°C water bath), since slow thawing causes recrystallization. Lyophilized or dried cultures must be rehydrated for a few minutes in an appropriate liquid medium. In both cases the reconstituted cells must be inoculated as soon as possible into appropriate culture media.

## 3. PRESERVATION OF YEASTS IN THE WINE INDUSTRY

We hardly need repeat here the role played by yeasts in the production and storage of wine. However, it is worth stressing the importance of their preservation, particularly in the case of starter cultures. These cultures have generally been obtained after many years of work invested in the selection and improvement of the strains.

It is therefore important that they maintain the characteristics for which they were originally selected. As described earlier, the most appropriate methods are therefore long-term preservation, mainly by freezing or lyophilization. In addition to these long-term methods, however, we will describe the short-term preservation of yeast strains, as well as approaches that present greater risks for the maintenance of phenotypic and genotypic stability of the isolates but are useful to ensure availability over the period of time that they are being used.

## 3.1. Long-term Preservation

Long-term preservation is normally used for cultures that can later be employed in the production of the biomass necessary for fermentations. Freezing has proved to be the most effective method. As mentioned, various factors can influence the stability and in particular the viability of the preserved material. Storage temperature is perhaps the most important factor and should be as low as possible in order mainly to prevent the water recrystallization that can occur at temperatures above $-130°C$ and that can result in reduced viability of the frozen material. As would be expected, the lower the temperature, the lower the degree of recrystallization and therefore the greater the length of time that strains can maintain their viability. In general, at $-80°C$, yeasts remain stable for many years (usually more than 5), although this, of course, varies from one strain to the next. To conserve cultures at these temperatures, 15 to 20% glycerol is added by mixing one volume of 20% glycerol with 0.5 volumes of the cell suspension obtained using the protocol described below. When a temperature of $-20°C$ is used, the method is considered to be medium-term storage, since the strains only remain viable for shorter periods of time. This is mainly due to the variations in temperature associated with these freezers that lead to frequent freeze–thaw cycles with a consequent

loss of viability. An alternative to prevent recrystallization during storage of strains at $-20°C$ is to preserve them in a final glycerol concentration of 50%, which will not freeze at this temperature, thus preventing recrystallization of extracellular water. Although recrystallization of intracellular water cannot be avoided, the damage will occur at a slower rate.

The other method of long-term preservation is lyophilization. This technique does not yield such good results with yeasts as it does with bacteria, since the damage inside these larger and more structurally complex eukaryotic cells reduces the viability of the yeast strains to around 5 years, depending on the strain in question. One way of increasing the usable life of lyophilized samples is to store them at 5°C instead of 18°C (optimal storage temperature for lyophilized bacteria and small-spored filamentous fungi). However, as mentioned, lyophilization offers certain advantages over freezing, since the infrastructure required for storage of the preserved strains is less extensive and distribution of samples is much easier.

Another factor to take into account regarding the viability of the frozen or lyophilized cultures is their recovery, which is crucially important, since poor recovery can negate any advantages obtained through the use of a good preservation method. As mentioned at the beginning of this chapter, frozen cultures should be thawed as rapidly as possible to prevent cell rupture and death caused by water recrystallization. The tube containing the cells should be immersed in a 37°C water bath or warmed in the hand until completely thawed. It is important not to refreeze the tube once it has been thawed, since this will inevitably lead to further loss of viability in addition to that caused by the first round of freezing and thawing. Once inoculated in the appropriate medium, any leftover cell suspension obtained from the lyophilized preparation cannot be stored since it is not viable for more than a few hours. In all cases, the culture medium and temperature used to recover the strains from the lyophilized or frozen sample must be taken into consideration. These must be the most appropriate for growth of the microorganism in question without being too nutritionally rich, since the cells may not be able to assimilate nutrients to any great extent following the period of metabolic inactivity to which they have been subjected.

The age of the cells used for preservation of the strain is another important consideration. In most cases, cells develop some natural resistance to adverse environmental conditions, and this resistance can be exploited in order to achieve greater viability following preservation. These natural resistance mechanisms usually appear in cultures towards the end of the exponential growth phase or at the beginning of the stationary phase. However, some strains may have produced toxic waste products by the time they reach this point on the growth curve. In such cases, it would be more appropriate to use younger cultures. In the case of wine yeasts, the cells should be collected after 48 h of growth in an appropriate culture medium (almost always Glucose Peptone Yeast extract Agar [GPYA]; CECT 140 medium).

### 3.1.1. *Preservation Protocols*

For the long-term preservation of yeasts we recommend cryopreservation or lyophilization.

The preparation of the culture (step one) is common to both, as well as to other preservation methods:

1) Grow the cultures until the end of the log phase in the most appropriate medium, depending on the nutritional requirements of the microorganism, and under appropriate incubation conditions of temperature, presence or absence of oxygen, and shaking. If the microorganisms are only able to grow in liquid medium, they will need to be centrifuged before the cell suspension is prepared in order to eliminate the culture medium and concentrate the cells.

#### 3.1.1.1. FREEZING PROTOCOL

2) Adjust the number of cells to a final concentration of $2-6 \times 10^6$ cells/mL (or the desired concentration).
3) Add appropriate cryoprotectant (glycerol) at an optimal concentration (15% for storage at $-80°C$ or 50% for storage at $-20°C$).
4) Aliquot the suspension into cryotubes, hermetically seal them, and maintain them at ambient temperature for 15 to 30 min.
5) Transfer the tubes to their final storage location: $-20°C$ freezer, ultra-freezer ($-80$ or $-145°C$), or liquid nitrogen ($-196°C$). Some authors have advocated reducing the temperature at a rate of between 1 and 3°C/min until $-30°C$, followed by a reduction of 15 to 30°C/min until the final storage temperature is reached.

Once again, it is important to remember that cryovials should only be used once and should not be refrozen after thawing to prevent problems such as loss of viability or contamination.

#### 3.1.1.2. LYOPHILIZATION PROTOCOL

2) Resuspend the cells (after washing if grown in liquid medium) in one part 15% glucose and two parts sterile skimmed milk until a homogeneous suspension is obtained that does not appear saturated with cells under the microscope.
3) Aliquot the suspension into tubes (approximately 200–250 µL per tube) that will later be frozen prior to lyophilization. The literature on the preservation of microorganisms contains a range of opinions on the rate at which samples should be frozen prior to lyophilization. The most widely accepted approach involves slow freezing (around 1°C/min) or rapid freezing by immersion in liquid nitrogen. The most important element in the end, however, is to obtain a frozen cell suspension that will undergo lyophilization for a period of 16 to 18 h.

4) Once the process is finalized, the vials must be hermetically sealed and stored at 5°C in the dark in order to achieve good preservation of the strains.

The same cell suspension obtained in step two could be used for preservation of the yeast in sterile water by adding approximately 0.5 mL of the suspension to 1.0 mL of sterile water. However, this method is not widely used as it is less effective than freezing yet involves essentially the same amount of work.

Further information on the preservation of yeasts can be found in Beech and Davenport (1971).

## 3.2. Short-term Preservation

The most common method used for the short-term preservation of yeast strains is periodic transfer. This is carried out using GPYA slant cultures stored at 5°C. These conditions slow growth and allow the strains to be stored for a few weeks but longer periods of storage are not recommended. This method tends to be used for the maintenance of working stocks. An alternative short-term preservation method is the use of stab cultures in semisolid media.

## 4. PRESERVATION OF BACTERIA

As we have seen in other chapters, the action of certain bacterial species influences the aroma and flavor of wine and can increase the quality of the finished product. It is important, however, to use selected bacterial strains that improve the organoleptic character without introducing biogenic amines or other undesirable compounds. When a strain with the required metabolic characteristics has been selected, it must be preserved without losing those characteristics so that it can be used reliably to prepare starter cultures as part of a controlled production process.

Although various bacterial species can influence the winemaking process, most belong to the lactic acid bacteria: *Lactobacillus brevis*, *Lactobacillus hilgardii*, *Lactobacillus mali*, and *Lactobacillus plantarum*; *Pediococcus damnosus*, *Pediococcus parvulus*, and *Pediococcus pentosaceus*; *Leuconostoc mesenteroides*, and in particular *Oenococcus oeni*. The acetic acid bacteria are also an important group that can appear in wine and cause the conversion of ethanol into acetic acid. There are five genera that can be included in this group: *Acetobacter*, *Acidomonas*, *Asaia*, *Gluconacetobacter*, and *Gluconobacter*. However, usually only some species of *Acetobacter* and *Gluconacetobacter xylinus* are found in wine. Gluconobacter is isolated in grapes and must, though not in wine.

## 4.1. Long-term Preservation

As mentioned, the long-term preservation methods associated with the greatest guarantee of stability in important physiological characteristics are freezing at temperatures below −70°C and lyophilization, and both methods are also the most appropriate for preserving wine strains. However, the preparation of the cells will differ between the different microbial groups (see Tables 12.1 and 12.2).

### 4.1.1. Guidelines for the Preservation of Lactic Acid Bacteria

#### 4.1.1.1. CULTURE MEDIA

Lactic acid bacteria require highly complex culture media containing specific growth factors. MRS medium (CECT 8), specifically designed by de Man, Rogosa, and Sharpe in 1960 for use with lactic acid bacteria, supports the growth of most lactic acid bacteria, although some species may have additional requirements. For instance, *O. oeni* requires the calcium pantothenate present in tomato juice to stimulate its growth (CECT 85 medium). Species of the genus *Pediococcus* isolated from beer or wine grow better if 40% beer or wine, respectively, are added to the medium.

**TABLE 12.1**    Growth Conditions for Wine Bacteria

| Microorganism | Culture medium | pH | Temperature | Aeration | Incubation time | Observations |
|---|---|---|---|---|---|---|
| *Lactobacillus brevis* | MRS | 6.2–6.5 | 30°C | Aerotolerant anaerobe | 24–36 h | |
| *Lactobacillus hilgardii* | MRS | 6.2–6.5 | 30°C | Aerotolerant anaerobe | 48 h | Grows in 15–18% ethanol |
| *Lactobacillus mali* | MRS | 6.2 | 30°C | Aerotolerant anaerobe | 24–36 h | |
| *Lactobacillus plantarum* | MRS | 6.2–6.5 | 30°C | Aerotolerant anaerobe | 24 h | |
| *Leuconostoc mesenteroides* | MRS | 6.2–6.5 | 26°C | Microaerophilic | 24–48 h | |
| *Oenococcus oeni* | MLO | 4.8–5.2 | 26°C | Microaerophilic | More than 3 d | Grows with wine and at high concentrations of ethanol |
| *Pediococcus damnosus* | MRS | 5.8 | 26°C | Anaerobe | More than 48 h | Grows with 40% must or wine |
| *Pediococcus parvulus* | MRS | 6.2 | 26°C | Anaerobe | More than 48 h | |
| *Acetobacter aceti* | MYP/GYC | 5.5–6.0 | 26°C | Aerobe | 24–36 h | |
| *Acetobacter pasteurianus* | MYP | 5.5 | 26°C | Aerobe | 24–48 h | |
| *Gluconacetobacter xylinus* | MYP/GY | 5.5 | 26°C | Aerobe | 24–36 h | |
| *Gluconobacter* | MYP/GYC | 5.5 | 26°C | Aerobe | 24–48 h | |

All of the conditions shown are the most generally applicable for each species but there is always the possibility that individual strains may differ in their physiological behavior. GYC = CECT 287 medium; GY = CECT 217 medium; MLO = CECT 85 medium; MRS = CECT 8 medium; MYP = CECT 10 medium.

### 4.1.1.2. INCUBATION CONDITIONS

Almost all lactic acid bacteria are mesophiles; they are aerotolerant anaerobes, but they can display varying degrees of oxygen sensitivity, and it is therefore advisable to culture them in anaerobic chambers or use recently prepared or degassed medium to ensure the absence of dissolved oxygen. If the culture is static, the bacteria will grow at the bottom, where no oxygen is present, and in most cases further precautions will not be necessary to achieve anaerobic conditions. The cultures should be collected in the exponential phase of growth or at the beginning of the stationary phase to prevent aging or death of the cells caused by their own metabolites.

### 4.1.1.3. PREPARATION OF THE CELLS FOR PRESERVATION

Cells grown in liquid medium are collected by centrifugation and resuspended in cryoprotective solutions at a concentration of approximately $10^8$ cells/mL. As mentioned, these cryoprotective solutions are intended to prevent the formation of ice crystals that can damage the cells and they tend to be aqueous solutions of small molecules (monomeric sugars such as

**TABLE 12.2**  Long-term Preservation of Wine Bacteria

| Microorganism | Freezing | | Lyophilization | | |
| | Incubation time | Cryoprotectant | Freezing and storage temperature | Cryoprotectant | Freezing temperature | Storage[1] |
| --- | --- | --- | --- | --- | --- | --- |
| *Lactobacillus brevis* | 24–36 h | Glycerol + glutamate or milk + glucose | −80°C | Glutamate | −196°C (liquid nitrogen) | 5 to 22°C |
| *Lactobacillus hilgardii* | 24–36 h | Glycerol + glutamate or milk + glucose | −80°C | Glutamate | −196°C (liquid nitrogen) | 5 to 22°C |
| *Lactobacillus mali* | 48 h | Glycerol + glutamate or milk + glucose | −80°C | Glutamate | −196°C (liquid nitrogen) | 5 to 22°C |
| *Lactobacillus plantarum* | 24 h | Glycerol + glutamate or milk + glucose | −80°C | Glutamate | −196°C (liquid nitrogen) | 5 to 22°C |
| *Leuconostoc mesenteroides* | 24–36 h | Glycerol + glutamate or inositol | −80°C | Glutamate or inositol | −196°C (liquid nitrogen) | 5 to 20°C |
| *Oenococcus oeni* | 2–3 d | Glycerol + glutamate | −80°C | Glutamate | −196°C (liquid nitrogen) | 5 to 20°C |
| *Pediococcus damnosus* | 24–48 h | Glycerol + glutamate | −80°C | Glutamate | −196°C (liquid nitrogen) | 5 to 20°C |
| *Pediococcus parvulus* | 24–48 h | Glycerol + glutamate | −80°C | Glutamate | −196°C (liquid nitrogen) | 5 to 20°C |
| *Acetobacter aceti* | 24–36 h | Glycerol or glycerol + inositol | −80°C | Inositol | −196°C (liquid nitrogen) | 5 to 20°C |
| *Acetobacter pasteurianus* | 24–48 h | Glycerol or glycerol + inositol | −80°C | Inositol | −196°C (liquid nitrogen) | 5 to 20°C |
| *Gluconacetobacter xylinus* | 24–36 h | Glycerol or glycerol + inositol | −80°C | Inositol | −196°C (liquid nitrogen) | 5 to 20°C |
| *Gluconobacter* | 16–18 h | Glycerol or glycerol + inositol | −80°C | Inositol | −196°C (liquid nitrogen) | 5 to 20°C |

[1]*All stored in the dark in vacuum-sealed vials.*
Incubation time refers to the recommended age of cells to better tolerate the freezing process.

glucose, inositol, etc.) or complex mixtures such as skimmed milk.

During lyophilization, the protective effect of glucose and small sugars is also to prevent total loss of water from the cell. The viability of the cells during recovery is greater if 1 to 2% moisture has been retained during lyophilization. The reducing effect of glucose can also be beneficial during the preservation of anaerobic microorganisms.

The most widely used cryoprotective agents during lyophilization of lactic acid bacteria are glutamic acid and skimmed milk. These are inexpensive and guarantee good viability, particularly in the case of glutamic acid, which has been confirmed to confer greater survival over long periods. Some investigators recommend the use of other sugars (for instance, adonitol for lactobacilli) but, although they offer certain advantages, they have not been widely tested.

To preserve the cells by freezing, suspensions in glutamic acid or skimmed milk are mixed with glycerol to achieve a final glycerol concentration of 12 to 15%. Good results are also obtained with a mixture of skimmed milk and glucose.

### 4.1.2. Guidelines for the Preservation of Acetic Acid Bacteria

#### 4.1.2.1. CULTURE MEDIA

The species of acetic acid bacteria that have been isolated from wine can grow in very simple media but need a non-nitrogenated carbon source. Peptones and amino acids cannot be used as carbon sources and the options in descending order of preference are glycerol, ethanol, glucose, and mannitol, followed by other carbon compounds. All of the acetic acid bacteria produce acid from glucose but only some species can produce it from mannitol. Peptones and amino acids can be used as nitrogen sources, but ammonium sulfate is also widely used. The bacteria do not require essential amino acids but their growth is stimulated by yeast extract.

The most appropriate culture medium is YMA (CECT 209), and GYC medium (CECT 287) can be used as an alternative for strains that cannot use mannitol. The acid produced from glucose is controlled by dissolution of the calcium carbonate contained in the medium. The optimal pH is 5.5 to 6.3.

#### 4.1.2.2. INCUBATION CONDITIONS

Acetic acid bacteria are mesophilic but they cannot grow at temperatures above 35°C. They are strict aerobes and can therefore be grown on the surface of solid media. The cells to be preserved are collected at the end of the exponential phase or at the beginning of the stationary phase after an incubation period of 24 to 36 h, or even 48 h if the strain requires it.

#### 4.1.2.3. PREPARATION OF THE CELLS FOR PRESERVATION

Cells are collected from a solid medium in cryoprotective solution to obtain suspensions of approximately $10^8$ cells/mL. Meso-inositol (5%) is used as a cryoprotectant for the lyophilization of acetic acid bacteria. When acetic acid bacteria are preserved by freezing, the suspension is mixed with glycerol to obtain a final concentration of 12 to 15%, as with lactic acid bacteria.

### 4.1.3. Protocols for Long-term Preservation

#### 4.1.3.1. FREEZING PROTOCOL

1) Culture the microorganism under the conditions recommended for each species.
2) Prepare a cell suspension containing a cryoprotective agent at an appropriate concentration.
3) Prepare sterile cryotubes (sterilizable screwtop plastic tubes with a rubber seal that are appropriate for freezing) containing 1 mL of 20% glycerol.
4) Add approximately 0.5 mL of cell suspension to each tube to obtain a final concentration of approximately $3 \times 10^7$ cells/mL and 12 to

14% glycerol. Prepare batches of at least five tubes for each strain.

5) Freeze at a temperature of at least −80°C and maintain the samples at this temperature without allowing cycles of freezing and thawing. Remove a single tube each time the culture is to be used and discard any remaining cells.

Under these conditions, the cultures can remain viable for more than 5 years. The viable period is reduced in −40°C freezers. Freezers at −20°C should not be used with this method. Microorganisms can be stored at −20°C if a final concentration of 50% glycerol rather than 12 to 14% glycerol is used in order to prevent freezing of the cell suspension and damage due to recrystallization of free water.

#### 4.1.3.2. LYOPHILIZATION PROTOCOL

The protocol is similar to that used with freezing but the cell suspensions obtained in step two are placed directly into sterile tubes or vials for lyophilization. The CECT recommends that the vials contain a small rectangle of absorbent paper (Whatman No. 3). Following lyophilization, the vials are closed under vacuum and stored in the dark. They can be stored at ambient temperature (18–22°C) or in the cold (5°C).

### 4.1.4. Recovery of the Preserved Cells

The frozen cultures are thawed at 37°C, inoculated into appropriate media as soon as possible, and incubated under appropriate conditions. Lyophilized cultures stored under vacuum in vials or other recipients should be rehydrated in liquid nutrient medium as soon as they are opened and then cultured under appropriate conditions. Viability should be at least 80% and the stability and authenticity of the strain in terms of its essential characteristics should be confirmed. In strains containing plasmids, it must be ensured that these are present and that they multiply.

## 4.2. Short-term Preservation

### 4.2.1. Maintenance by Periodic Transfer

Maintenance by periodic transfer keeps the cells in an active and easily available form but, as mentioned, the method is not advisable for use over extended periods.

Bacteria associated with winemaking are not easy to maintain in live culture without changes occurring. The strong reduction in pH that occurs in the medium causes serious damage to the cells and, although this can be avoided in some cases by adding insoluble calcium carbonate, which will gradually dissolve in the medium as the acids are produced, this is neither sufficient nor advantageous for some strains, and as a consequence the cells must be subcultured regularly to prevent them dying. In other cases, cells age as a consequence of excessive growth in the nutrient-rich media that some of these bacteria require. The microorganisms should therefore only be maintained in live culture for a defined period while they are in use. Once this period has passed, we should once again use cells preserved by lyophilization or freezing.

To maintain the cells by periodic transfer, the growth conditions are as described in Table 12.1. The cultures obtained should be stored at 5 to 8°C and the time that they can survive under these conditions will vary according to the microorganism. Most lactobacilli should be reseeded every 10 to 15 d. Acetic acid bacteria grown in MYA (CECT 63) or GYC (CECT 287) media remain viable for 1 to 2 months.

However, all strains can be altered after a certain number of subcultures, and methods should be tested to extend the length of time between reseeding. These methods are based on reducing cell activity:

1) *Nutrient limitation.* Cells grown in appropriate conditions (as described earlier) are resuspended in distilled water at a concentration of approximately $10^6$ cells/mL. They are stored at 5 to 8°C.

2) *Reducing storage temperature.* As mentioned, the cells can be stored at −20°C without freezing if mixed with 50% glycerol (final concentration).

Alternative preservation methods such as dessication are not recommended for bacteria used in winemaking.

# 5. PRESERVATION OF FILAMENTOUS FUNGI FROM WINE

Various factors must be taken into consideration when choosing a method for the preservation of filamentous fungi. These include the number of strains, the degree of genetic and phenotypic stability required, and the storage period. However, one of the most important is the type of reproduction used by the strain to be preserved. Fungi that reproduce both asexually (through conidia) and sexually (using spores) are relatively easy to preserve, since all of them survive lyophilization well. However, in those species that do not produce spores under laboratory conditions but rather reproduce asexually by fragmentation of hyphae, lyophilization is impossible and preservation via freezing is difficult to achieve. This is because both freezing and lyophilization depend on the presence of structures that confer resistance, at least in eukaryotic cells. These fungi can only be preserved by periodic transfer or suspension in sterile distilled water, a surprisingly simple method that has proved to be one of the best for use in those strains that do not tolerate long-term preservation methods. Obligate parasites such as *Plasmopara viticola* or *Uncinula necator* (which cause mildew) cannot be cultured in laboratory media and must therefore be preserved in the host by freezing pieces of the infected plant in liquid nitrogen (Dahmen, 1983). For more general information on the preservation of filamentous fungi, see Onions (1971) and Smith and Onions (1994).

## 5.1. Methods for the Preservation of Filamentous Fungi

### 5.1.1. *Freezing*

A distinction must be made between sporulating and nonsporulating fungi. The following protocol can be used in nonsporulating strains:

1) Begin with a plate culture of the fungus using the most appropriate medium and temperature for the microorganism.
2) Cut blocks of approximately 0.5 × 0.5 cm containing the growing vegetative mycelium of the fungus.
3) Place the blocks in groups of five in screwtop vials containing 1 mL of 10% (vol/vol) sterile glycerol and seal.
4) Maintain the vials at 4°C for 30 min and then freeze at temperatures below −80°C.

In the case of sporulating fungi:

1) Prepare a suspension of spores in an appropriate cryoprotective solution, which in the case of filamentous fungi is usually sterile skimmed milk.
2) Adjust the concentration to $10^6-10^7$ spores/mL.
3) Aliquot 0.5 mL of the suspension into cryotubes containing 1 mL of 15% (vol/vol) glycerol.
4) Freeze at temperatures below −80°C.

As mentioned earlier for other microbial groups, to recover the fungus, the thawing process should be rapid, with the immersion of the cryotubes in a 37°C water bath for 30 s. Once thawed, the cells are seeded in an appropriate culture medium.

The CECT has studied the effects of different freezing and thawing methods (Juarros et al., 1993).

### 5.1.2. *Lyophilization*

The use of lyophilization is restricted to the preservation of sporulating filamentous fungi.

1) Prepare a suspension of spores in an appropriate cryoprotective solution, which can be sterile skimmed milk or a combination of equal parts sterile skimmed milk and 5% inositol. Adjust the concentration to $10^6-10^7$ spores/mL.
2) Aliquot equal volumes into lyophilization vials.
3) Freeze the tubes and place them in the lyophilizer.
4) Once the spores are lyophilized, recovery involves rehydration of the cells in an appropriate liquid culture medium, which is then used to inoculate the recommended solid medium.

### 5.1.3. Preservation by Subculture or Periodic Transfer

In general terms, preservation by subculture or periodic transfer is the oldest, simplest, and most accessible method with which to preserve small culture collections for relatively short periods. It can be used in both sporulating and nonsporulating filamentous fungi. However, the viability and stability of the cultures is relatively poor compared with those obtained with freezing or lyophilization.

The aim of this method is to maintain pure, active, stable, and immediately recoverable cultures. To this end, it is important to prevent aging or phenotypic or genotypic changes in the culture. This requires a specific protocol to be designed for each microorganism in which a careful study is performed of the specific culture medium to be used (including its water activity), the temperature, light, aeration, pH, and incubation time.

#### 5.1.3.1. GENERAL RECOMMENDATIONS FOR THE PROCEDURE

1) Culture media should be alternated, since the microorganisms can degenerate if maintained consistently in the same medium.
2) The inoculum should be small. Although the intention is for the microorganism to develop fully, it is neither necessary nor desirable to obtain abundant growth, as this will accelerate aging due to accumulation of toxic metabolites.
3) Once the microorganism has begun to grow, the culture should be kept at a temperature of 5 to 7°C, since reducing the storage temperature will help to reduce the metabolic activity and increase the time between subcultures.
4) It is difficult to predict the maximum time between transfers, since this can vary according to the species and even the strain. There are strains of nonsporulating fungi that need to be subcultured every 3 months, each time with the risk of contamination, loss of characteristics, and even loss of the strain itself. In the case of sporulating filamentous fungi, the time between subcultures can be extended to up to 12 months.

### 5.1.4. Preservation in Sterile Distilled Water

Despite its simplicity, preservation in sterile distilled water is highly effective for nonsporulating filamentous fungi. The method was originally described by Castellani (1939, 1967) for the preservation of pathogenic fungi from humans. Boeswinkel (1976) achieved reasonable success using this method to preserve a collection of 650 pathogenic fungi from plants, including representatives of the divisions Oomycota, Ascomycota, Basidiomycota, and imperfect fungi.

The method is used extensively by the CECT for genera belonging to the division Basidiomycota such as *Agaricus*, *Agrocybe*, *Armillaria*, and *Coprynus*, and for some belonging to the division Oomycota, such as the phytopathogenic genera *Pythium* and *Phytophthora*. Using this method, we have succeeded in maintaining viable cultures for more than 5 years and have even recovered strains after 15 years.

### 5.1.4.1. PROCEDURE

1) Grow the fungus on a plate using the most appropriate culture medium and temperature for each microorganism.
2) Cut blocks of approximately $0.5 \times 0.5$ cm containing the growing vegetative mycelium of the fungus.
3) Place the blocks in groups of five in screwtop microvials containing 1 mL of sterile water and hermetically seal them.
4) Store the vials in the dark at a controlled temperature of 18 to 20°C.
5) In order to recover the microorganism, it is sufficient to place the growing region of the fungus in contact with an appropriate culture medium.

## APPENDIX 1

## Culture Media

All of the culture media referred to in this chapter appear with the number assigned to them by the Spanish Type Culture Collection (CECT). Their composition can be found on the CECT webpage (http://www.cect.org).

## Cryoprotectants

1. Glutamic acid (0.067 M in aqueous solution).
2. Drops of 1N NaOH are added to help dissolve the glutamic acid with shaking. The solution is sterilized for 20 min at 1 atmosphere pressure.
3. Glucose (7.5%, wt/vol).
4. Inositol (5%, wt/vol).
5. Skimmed milk (10%).
6. Homogenized, skimmed milk from any high-quality producer without additives or preservatives. Sterilize for 25 min at 112°C without allowing it to caramelize. It can also be prepared from dried skimmed milk.
7. Dried skimmed milk (10%, wt/vol) + glucose (3%, final concentration).
8. Skimmed milk (10%) + 2M adonitol.

## References

Álvarez-Rodríguez, M. L., López-Ocaña, L., López-Coronado, J. M., Rodríguez, E., Martínez, M. J., Larriba, G., et al. (2002). Cork taint of wines: Role of the filamentous fungi isolated from cork in the formation of 2, 4, 6-trichloroanisole by o methylation of 2, 4, 6-trichlorophenol. *Appl. Environ. Microbiol., 68,* 5860–5869.

Beech, F. W., & Davenport, R. R. (1971). Isolation, purification and maintenance of yeasts. In C. Booth (Ed.), *Methods in Microbiology,* Vol. 4 (pp. 153–182). London, UK: Academic Press.

Boeswinkel, H. J. (1976). Storage of fungal cultures in water. *T. Brit. Mycol. Soc., 66,* 183–185.

Castellani, A. (1939). Viability of some pathogenic fungi in distilled water. *J. Trop. Med. Hyg., 42,* 225–226.

Castellani, A. (1967). Maintenance and cultivation of common pathogenic fungi of man in sterile distilled water. Further researches. *J. Trop. Med. Hyg., 70,* 181–184.

Dahmen, H., Staub, T., & Schwinn, F. T. (1983). Technique for long-term preservation of phytopathogenic fungi in liquid nitrogen. *Phytopathology, 73,* 241–246.

Day, J. G., & Stacey, G. N. (2007). *Cryopreservation and freeze-drying protocols.* Totowa, NJ: Humana Press.

de Man, J. D., Rogosa, M., & Sharpe, M. E. (1960). A medium for the cultivation of lactobacilli. *J. Appl. Bacteriol., 23,* 130–135.

Hatt, H. (Ed.). (1980). *American type culture collection methods: I. Laboratory manual on preservation, freezing and freeze-drying.* Rockville, MD: American Type Culture Collection.

Hill, L. R. (1981). Preservation of microorganisms. In J. R. Norris & M. H. Richmond (Eds.), *Essays in applied microbiology* (pp. 2/1–2/31). Chichester, UK: John Wiley & Sons.

Hunter-Cevera, J. C., & Belt, A. (1996). *Maintaining cultures for biotechnology and industry.* London, UK: Academic Press.

Juarros, E., Tortajada, C., García, M. D., & Uruburu, F. (1993). Storage of stock cultures of filamentous fungi at −80°C: Effects of different freezing-thawing methods. *Microbiología SEM, 9,* 28–33.

Kirsop, B. E. (Ed.). (1980). *The stability of industrial organisms.* Kew, UK: Commonwealth Mycological Institute.

Kirsop, B. E., & Doyle, A. (1991). *Maintenance of microorganisms and cultured cells.* London, UK: Academic Press.

Onions, A. H. S. (1971). Preservation of fungi. In C. Booth (Ed.), *Methods in microbiology,* Vol. 4 (pp. 113–151). London, UK: Academic Press.

Smith, D., & Onions, A. H. S. (1994). *The preservation and maintenance of living fungi.* Wallingford, UK: CAB International.

CHAPTER

# 13

# Application of the Hazard Analysis and Critical Control Point System to Winemaking: Ochratoxin A

*Adolfo J. Martínez-Rodríguez, Alfonso V. Carrascosa Santiago*

**Instituto de Investigación en Ciencias de la Alimentación (CIAL, CSIC-UAM), Madrid, Spain**

OUTLINE

## 1. INTRODUCTION

The Hazard Analysis and Critical Control Point (HACCP) system, which emerged in the 1970s as a systematic way of ensuring food hygiene and protecting consumer health, is based on controlling microorganisms that constitute a hazard for consumers (International Commission on Microbiological Specifications for Foods [ICMSF], 1991). The system was original in that it adopted a preventive rather than a retrospective approach. Previous food safety systems had been based on end-product testing aimed at preventing pathogen-contaminated products from entering the food chain and—rather more difficult—assessing the origin and source of the contamination. The retrospective approach, however, could not fully guarantee food safety, given the impossibility of analyzing entire production batches. Tolerable levels of pathogens in foods were therefore established by legislation and food processing sites were routinely inspected by public bodies.

In the HACCP system, the emphasis is on monitoring and maintaining production conditions that prevent undesirable or hazardous microorganisms from contaminating and growing in food, from the raw material stage to the consumer. The system is now widely applied to the control of all kinds of biological, physical, and chemical hazards (Mortimore & Wallace, 2001). Its wide acceptance and successful application led to its incorporation into food safety legislation in the European Union (EU) (Directive 93/43/EEC on the hygiene of foodstuffs) and the United States (Food and Drug Administration [FDA] regulation CPR-123). The United States National Advisory Committee on Microbiological Criteria for Foods issued HACCP guidelines that included generic plans (National Advisory Committee on Microbiological Criteria for Foods [NACMCF], 1992). The Codex Alimentarius Commission adopted the HACCP system at its

20th session, in 1993, and eventually issued guidelines regarding what became known as the twelve tasks (Food and Agriculture Organization [FAO], 1998b) aimed at correctly implementing seven principles. Both the tasks and principles are described in some detail below. Application of the HACCP system has since become a mandatory standard in countries such as Spain, where it is regulated by Royal Decree 2207/1995 of 28 December, establishing hygiene standards for food products.

The HACCP is the most comprehensive system for preventing risks to consumers from foods. Its scientific approach makes it superior to any other system in terms of its efficacy and breadth of coverage of all aspects that can contribute to removing or reducing food hazards for humans. Consequently, it is the only such system underpinned by legislation.

The HACCP system is designed to identify hazards (potentially harmful microorganisms that can affect food), assess risk (i.e., the probability that these hazards are present in the production system), and guide the establishment of appropriate control measures. Successful application of the system largely builds on compliance with health and sanitation standards and on the use of well-established quality management systems such as Good Agricultural Practice (GAP), Good Animal Husbandry Practice (GAHP), Good Storage Practice (GSP), Good Manufacturing Practice (GMP), and Good Hygiene Practice (GHP). It is also compatible with quality assurance systems such as ISO 9000.

## 2. GENERAL CONSIDERATIONS REGARDING THE HACCP SYSTEM

### 2.1. HACCP Principles

Seven different activities, referred to as the seven principles in the Codex Alimentarius Guideline (FAO, 1997), are necessary to establish, implement, and maintain an HACCP system. These seven principles are described below.

### 2.1.1. Principle 1: Conduct a Hazard Analysis

Hazards should be identified and the associated risks assessed at each phase of the production system. Measures for controlling hazards and risks should also be described. Since this book is concerned with wine microbiology, we will focus exclusively on microbiological hazards that affect the quality of wine and have potential harmful effects on human health.

The hazard in this case is the presence, survival, and growth of microorganisms or the production of substances (toxins, metabolites, etc.) in wine at levels that are unacceptable in terms of ensuring the health of the consumer.

### 2.1.2. Principle 2: Determine the Critical Control Points

Critical control points (CCPs) are steps at which essential control measures designed to prevent or eliminate a food safety hazard or to reduce it to an acceptable level are applied. In other words, they are specific production stages where the implementation of appropriate control measures will ensure the elimination or minimization of a specific hazard.

The initial classification of CCPs distinguished between a CCP1, which was an operational or production phase in which a hazard could be eliminated, and a CCP2, which was an operational or production phase in which the hazard was only partially eliminated (i.e., it was minimized but not brought under control) (ICMSF, 1991). Despite the usefulness of this classification, however, it is no longer applied.

### 2.1.3. Principle 3: Establish Critical Limits

Each control measure associated with a CCP should have an associated critical limit that distinguishes between what is acceptable and unacceptable. Critical limits delimit the boundary between safe and unsafe products. These limits are sometimes referred to as absolute tolerance or safety limits. Control parameters used for this purpose should be variables that are directly related to the presence of undesirable microorganisms and can be measured rapidly (e.g., pH and temperature rather than the more time-consuming microbiological tests). Indirect measures can also be used if they are known to be reliably associated with the presence of the microorganism.

### 2.1.4. Principle 4: Establish a Monitoring System

Monitoring is the systematic, scheduled measurement or observation of the parameters established for all the CCPs to check that these are under control; that is, that they are within the critical limits described in Principle 3. Application of this principle requires the definition of monitoring activities and frequencies and the designation of a person with a supervisory role.

### 2.1.5. Principle 5: Establish Corrective Actions

When monitoring activities indicate a deviation from an established critical limit at a CCP, specific corrective actions or procedures need to be implemented to restore control. It is necessary to both establish these actions and designate a person responsible for implementing them and deciding what to do with the affected product.

### 2.1.6. Principle 6: Establish Verification Procedures

To verify the effectiveness of the HACCP system, periodic checks should be performed by the persons responsible for the control operations to evaluate deviations and product disposition and to analyze samples to confirm whether or not the CCPs are under control. The analyses should incorporate tests (including microbiological tests) other than those used for monitoring purposes, even though incubation times may mean that results will not be immediate. These analyses more or less correspond to standard quality-control checks.

### 2.1.7. Principle 7: Establish Documentation Concerning All Procedures and Records that are Appropriate to these Principles and their Application

Application of this principle is essential in that it facilitates verification audits and subsequently ensures that the system is kept up to date and continually evaluated. It is recommendable to create an HACCP manual containing a written description of the application of the HACCP principles (in terms of hazards, risks, critical points, critical limits, corrective actions, etc.) to the process in question. The manual should also be used to keep records of the different operations implemented during routine functioning of the system.

A number of works have been published on the application of the HACCP system to the food industry. As introductory reading, we recommend the books published by ICMSF (1991) (available in Spanish) and Mortimore and Wallace (2001), both of which provide extensive additional information on the principles of the system and its application to specific food sectors.

## 2.2. Prerequisite Programs

Prerequisite programs such as GAP, GSP, GMP, and GHP need to be correctly implemented before the HACCP system can be applied to the production of a plant-based product such as wine (FAO, 2003). The introduction of the HACCP system will be complicated if these programs are not functioning effectively and the outcome will be a cumbersome, over-documented system. In the interest of avoiding confusion, it should be emphasized that, although GAP, GSP, GMP, and GHP cover some of the elements of the HACCP system, they do not replace it, as will become evident below.

### 2.2.1. Good Agricultural Practice

The primary production process should ensure that foods are safe for the consumer. In the case of wine, grape growers should manage production in such a way that crop contamination, pest proliferation, and animal and plant diseases do not pose a threat to food safety. The land used for cultivation should be fit for purpose and not have been previously contaminated with heavy metals, industrial chemicals, or environmental waste, as such hazards will enter the food chain and render the corresponding commodities unfit for human consumption. Where appropriate, GHP should be designed to ensure that the harvested commodity will not represent a food hazard to the consumer. If appropriate, GSP should be applied to ensure that hazards are eliminated during harvesting, after harvesting, and throughout the entire production process.

### 2.2.2. Good Manufacturing Practice

#### 2.2.2.1. DESIGN AND CONSTRUCTION OF FACILITIES

The structure and location of the facilities required to produce wine must meet certain requirements and comply with legislation specific to wineries. The following general issues should be considered:

1. Premises should be designed to minimize the risk of commodity contamination.
2. The design and layout of premises should enable maintenance, cleaning, and disinfection operations that minimize airborne contamination.
3. All surfaces that come into contact with food should be nontoxic and easy to maintain and clean to prevent contamination.
4. When required, there should be adequate means for controlling temperature and humidity.
5. Effective pest control measures should be in place.

### 2.2.2.2. CONTROL OF OPERATIONS

Measures aimed at reducing the risk of contamination of commodities and foods and ensuring that they are safe and fit for purpose should be implemented. These include the following:

6. Adequate temperature, time, and humidity controls
7. Food-grade packaging
8. Potable water supplies
9. Equipment maintenance

### 2.2.2.3. MAINTENANCE AND SANITATION

Procedures and work instructions that ensure an adequate level of maintenance of the premises and effective cleaning, waste management, and pest control practices should exist. These operations will help in the constant monitoring of potential hazards that could cause food contamination.

### 2.2.2.4. PERSONNEL HYGIENE

Measures should be put in place to ensure that food handlers do not contaminate food and that they maintain an appropriate level of personal hygiene and comply with relevant guidelines.

### 2.2.2.5. TRANSPORTATION

Measures should be put in place to prevent deterioration of the commodity during transport. Raw materials or products to be transported should be properly monitored. Examples include products that need to be refrigerated, frozen, or stored at specific humidity levels. Transport means should be kept in good condition and be easy to clean. Containers used for bulk transport should be used exclusively for food.

### 2.2.2.6. TRAINING

All food handlers should be trained in personal hygiene and in the specific operations for which they are responsible, to a level commensurate with their duties. Food handlers should also be overseen by suitably trained supervisors. Ongoing training for food handlers is essential to the success of a food safety management system.

### 2.2.2.7. PRODUCT INFORMATION AND CONSUMER AWARENESS

The end product should be accompanied by sufficient information to ensure that the personnel at the next stage in the food chain will handle, store, prepare, and display the product safely and in a way that does not increase hazards. This is particularly important for foods that are consumed fresh.

## 3. APPLICATION OF THE HACCP SYSTEM TO WINEMAKING

### 3.1. Background

The application of the HACCP system to winemaking is the subject of a number of books (Federación de Industrias de Alimentación y Bebidas [FIAB], 1997; Hyginov, 2000) and articles in specialist journals (Briones & Úbeda, 2001; Kourtis & Arvanitoyannis, 2001; Morassut & Cecchini, 1999). Not all of these publications cover microbiological hazards, and none consider that the consumption of wine may constitute a microbiological health risk.

A fairly common error, which is essentially due to a lack of in-depth knowledge of the HACCP system, is to consider GAP and GMP to be equivalent to HACCP. This downplays the true significance of the HACCP system as a comprehensive approach consisting of many elements whose goal is to gain true control of an entire process. At best, GAP and GMP include just some of the preventive measures provided by an HACCP system. This confusion is typical—many winery managers, wine experts, and even scientists are unaware of the HACCP system.

Wine is contaminated naturally during production. While more typical in foods of animal origin, foodborne pathogens such as *Aeromonas hydrophila*, *Bacillus cereus*, *Campylobacter jejuni*, enterotoxigenic and enterohemorrhagic *Escherichia coli*, *Listeria monocytogenes*, *Salmonella enteritidis*, and *Shigella dysenteriae*, as well as toxin-producing bacteria such as *Clostridium botulinum* and *Staphylococcus aureus* can contaminate wine throughout the production phases, although they are unable to thrive because of the pH levels that characterize this medium (ICMSF, 1996).

The combined effects of the ethanol and pH levels that characterize wine can cause loss of viability of pathogens such as *Salmonella typhimurium*, *Salmonella sonnei*, and enterotoxigenic *E. coli* (Bellido et al., 1996; Sheth et al., 1988; Weisse et al., 1995), and even of viruses such as hepatitis A (Desenclos et al., 1992). Certain polyphenols present in the wine can also inhibit the growth of *S. enteritidis* (Marimón et al., 1998) and *Campylobacter jejuni* (Gañán et al., 2009). Furthermore, it has been known for some time that moderate wine consumption increases gastric secretion and intestinal motility (Bujanda, 2000; Pfeiffer et al., 1992), making it more difficult for pathogens to invade the intestine. It has also been postulated that moderate wine consumption reduces the infectious potential of intestinal pathogens such as *Helicobacter pylori*, the main cause of chronic gastritis and duodenal ulcers (Brenner et al., 1999; Ruggiero et al., 2006). These data show that wine is a functional food and has a role in defending the intestine from pathogens.

While the literature contains no reports of outbreaks of illness caused by toxins in wine, it is acknowledged that wines may contain a range of toxic substances of microbial origin. A maximum limit of 2 µg/L for ochratoxin A (OTA) levels in musts and wines produced after the 2005 harvest, for example, was recently established by the EU (European Union [EU], 2005). Previously proposed by the International Organization of Vine and Wine (OIV) (OIV, 2002), this limit is now mandatory for all EU member states. Since OTA is the only toxic substance of microbial origin in wines that is regulated internationally, this chapter will describe the application of the HACCP system to this mycotoxin.

## 3.2. Applying HACCP to the Control of Ochratoxin A (OTA) in Wine

Although the HACCP system was conceived to improve and ensure hygiene and sanitation in both the agricultural and food processing sectors (ICMSF, 1991), it has mainly been applied in the latter. One of the reasons is that, whereas a processing facility like a winery might have just one owner, the vineyards supplying the winery may have many owners. Hence, fully preventing or eliminating a food hazard or reducing it to an acceptable level is generally more difficult in the primary processing of plant-based foods (FAO, 2003). It is also more difficult to control parameters outdoors than in indoor production facilities.

Nonetheless, since the application of the HACCP system to OTA in wine is closely associated with the grape growing phase, to all effects and purposes, we consider this phase to be an integral part of the winemaking process. Indeed, in the case of mycotoxins such as OTA, it is crucial to implement production controls aimed at protecting grape berries from fungal infection in the vineyards.

### 3.2.1. Mycotoxins in Wine: Ochratoxin A (OTA)

While it has long been acknowledged by many that the presence of OTA in wine is at least a potential hazard, it was not described as such in the wine literature until the late 1990s (Gottardi, 1997).

Although there are over 300 known mycotoxins, only a few are recognized as representing a level of risk that requires the implementation of strict controls (FAO, 2003). The fact that recommended limits for OTA levels in wines have been established is, in itself, a good enough reason for developing an HACCP system for its control.

Mycotoxins are toxic substances of fungal origin. When ingested, inhaled, or absorbed through the skin, they cause illness or death in both humans and animals (Pitt, 1996). They are secondary metabolites that appear to have no specific function in the growth of the species that produce them. They pass to humans through food that has been contaminated by mycotoxigenic filamentous fungi. Some of the mycotoxins that have been found in grapes and grape products are listed in Table 13.1.

Exposure to mycotoxins can result in acute or chronic toxicity and ultimately lead to deleterious effects on a range of body organs and systems, and even death. It is widely believed, particularly in developing countries, that the most important effect of certain mycotoxins is their capacity to block the immune response and reduce resistance to infectious diseases. OTA is a mycotoxin with nephrotoxic, carcinogenic, teratogenic, immunotoxic, and possibly neurotoxic effects (Turner et al., 2009). It has been associated with Balkan endemic nephropathy (Turner et al., 2009), although the evidence for this has been contested in certain sectors (Delage et al., 2003; Soleas et al., 2001; Zimmerli & Dick, 1996). The tolerable daily intake of OTA is very low, ranging from just 0.3 to 0.89 µg/d for a person weighing 60 kg, with acute toxicity likely to occur at a dose of between 12 and 3000 mg for a person of that weight (Rousseau, 2004). The Joint FAO/World Health Organization (WHO) Expert Committee on Food Additives has established a provisional tolerable weekly intake of OTA as 100 ng/kg of bodyweight (FAO, 2002), which corresponds to 14 ng/d/kg of bodyweight.

TABLE 13.1  Mycotoxins Isolated in Grapes and Grape Products

| Mycotoxin | Substrate | Fungus | References |
|---|---|---|---|
| Byssochlamic acid | Grape | *Byssochlamys fulva* | Samson et al. (1996) |
| | | *Byssochlamys nivea* | |
| Citrinin | Must | *Penicillium citrinum* | Vinas et al. (1993) |
| | | *Penicillium expansum* | |
| Patulin | Must | *Byssochlamys fulva* | Frisvad and Thrane (1996) |
| | | *B. nivea* | |
| | | *Penicillium expansum* | |
| Ochratoxin A | Grapes, must, wine | *Aspergillus carbonarius* | Bau et al. (2005); Cabañes et al. (2002); Gallo et al. (2009); Selma et al. (2008) |
| | | *Aspergillus fumigatus* | Battilani and Pietri (2002) |
| | | *Penicillium pinophilum* | |
| | | *Aspergillus tubingensis* | Oliveri et al. (2008) |
| | | *Aspergillus japonicus* | |

*Adapted from Carrascosa (2005).*

After cereals, wine is the next most common source of OTA for humans (Cabañes et al., 2002). OTA was first detected in wine in 1995 (Zimmerli & Dick, 1996). The Codex Alimentarius Commission has even conceded that grapes and grape products were the source of over 15% of OTA intake in Europe (FAO, 1998a).

The EU has approved a regulation establishing tolerable intake limits for OTA in cereals (5 µg/kg), cereal products (3 µg/kg), and raisins (10 µg/kg) (EU, 2002a). An upper limit of 2 µg/L has also been proposed for musts and wines (FAO, 2003). This limit is the same as that recently established in the EU (2005). The mean content of OTA in red wine in Europe is 0.19 µg/L, and the total daily intake of OTA in Europe has been estimated as 171 g (FAO, 1998a), which would correspond to an OTA concentration of between 0.01 and 3.4 µg/L. It should be noted, however, that the presence of this mycotoxin is more common and its concentration greater in warmer, wetter years; in temperate climates; in the south; and in sweet wines made from overripe or raisined grapes; it is generally more common in red wines, followed by rosé and then white wines (Battilani & Pietri, 2002; Burdaspal & Legarda, 1999). Although OTA is detected in over 50% of wines, there are very few cases where the maximum allowable limit of 2 ng/mL is exceeded. Levels also fall with increasing latitude (Mateo et al., 2007).

Studies on the occurrence of OTA in wines indicate that it remains stable in this substrate for at least 12 months (Mateo et al., 2007). They also point to the extremely important role played by factors such as the year of harvest as different weather conditions can result in enormous differences. In one study, for example, it was found that the percentage of wines containing OTA ranged from one year to the next from 86% (with OTA concentrations of 0.056–0.316 ng/mL) to 15% (range 0.074–0.193 ng/mL) (López de Cerain et al., 2002). The authors of a similar study found OTA levels of between <0.01 and 0.76 ng/mL in the wines

they analyzed and estimated a daily intake of 0.01 ng/d/kg of bodyweight (Blesa et al., 2004). In Spain, OTA has been detected at concentrations of up to 11.7 ng/mL and up to 4 ng/mL in the plasma of patients with chronic kidney failure and healthy individuals, respectively (Pérez de Obanos et al., 2001). These values are similar to those reported for other European countries. The European Food Safety Authority's (EFSA) Scientific Panel on Contaminants in the Food Chain (CONTAM) has established a tolerable weekly intake of 120 ng/kg of bodyweight for OTA (European Food Safety Authority [EFSA], 2006). It is estimated that the true intake of OTA in Europe is between 15 and 60 ng/kg bodyweight, and that OTA levels in wines from Africa, America, Australia, and Japan are lower than in European wines (Mateo et al., 2007).

Based on the above data and the results of studies examining OTA levels in wines, it can be concluded that the limit of 2 µg/L recommended by the EU and the OIV is not often exceeded. Given that a level of 3.4 µg/L of OTA is reached in certain kinds of wines produced in the EU, however, it has to be acknowledged that the risk exists—as has been suggested by a number of authors (Mateo et al., 2007; Olivares-Marín et al., 2009)—even if it is moderate given the infrequent presence of this myotoxin. Nonetheless, in order to achieve a true reduction in OTA levels in wine via the application of an HACCP system, the theoretical framework has to assume this level of risk to be unacceptable.

### 3.2.2. Ochratoxin A (OTA)-producing Microorganisms in Wine

While *Aspergillus ochraceus* and *Penicillium verrucosum* produce OTA in cereals (ICMSF, 1996), these fungi are not commonly isolated in grapes or on vines. The presence of OTA in grapes and grape products is attributed fundamentally to *Aspergillus carbonarius* (Figure 13.1) (Bau et al., 2005; Cabañes et al., 2002; Gallo

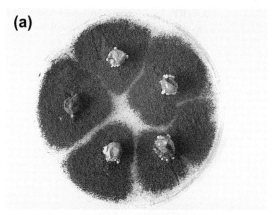

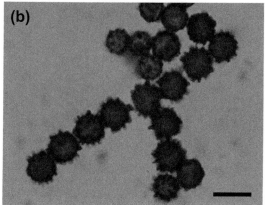

FIGURE 13.1 *Aspergillus carbonarius* from infected grapes grown in solid medium (a) and close-up of conidia (fresh mount) (phase contrast microscopy) (b). Scale bar = 10 μm. Images kindly provided by Dr F.J. Cabañes of the Veterinary Mycology Group at Universitat Autònoma de Barcelona, Spain.

et al., 2009; Gómez et al., 2006; Selma et al., 2008) and, to a lesser degree, to other species from the genus *Aspergillus* section *Nigri* (e.g., *Aspergillus niger*) (Bau et al., 2005; Serra et al., 2003) and to *Aspergillus fumigatus* and *Penicillium pinophilum* (Battilani & Pietri, 2002).

### 3.2.3. The 12 HACCP Tasks Applied to Ochratoxin A (OTA) in Wine

To apply the HACCP principles to the preparation of a food product, it is recommended to perform 12 consecutive tasks (FAO, 1998b). Below we briefly describe each of these tasks both in general terms and in regard to their specific application to OTA in wine.

#### 3.2.3.1. TASK 1: ESTABLISH AN HACCP TEAM

The application of the HACCP system should start with the appointment of a team to perform the tasks necessary to implement the seven principles. The team leader should be familiar with the HACCP system and methods, ensure that the concept is properly applied, be a good listener, and encourage all the team members to become involved. Given that OTA constitutes a microbial hazard, the team should include a microbiologist, preferably with expertise in mycology and mycotoxicology. In order to develop the commodity flow diagram (CFD)—described below—the team should also be able to call on individuals familiar with both viticulture and winemaking processes. To avoid possible conflicts of interest, it is advisable to have wine industry representatives from both the public and private sectors on the team.

As far as OTA is concerned, the scope of the study should cover the entire production chain, from the vineyard to the bottle.

#### 3.2.3.2. TASK 2: DESCRIBE THE PRODUCT

A complete description of the product, including client specifications, is necessary to begin the hazard analysis. A basic generic product description form—like the one shown in Figure 13.2—should be used for this purpose. It should include relevant safety information regarding OTA and data on the limits recommended by law, as well as information on packaging, storage, and recommended temperatures. Where appropriate, labeling information and a sample label should be included. This information will assist the HACCP team in identifying the real hazards associated with the process.

**Product Description Form**

| 1. Product name(s) | Wine (type, appellation, etc.) |
|---|---|
| 2. Description and key characteristics of the end product | pH, alcohol content, $SO_2$ levels, tasting notes, etc. <2 µg/L of OTA |
| 3. Intended use | Human consumption |
| 4. Packaging | |
| 5. Shelf life | |
| 6. Sales point | |
| 7. Labeling instructions | |
| 8. Special storage and distribution conditions | |

Date: _____    Approved by: _____

**FIGURE 13.2** Sample product description form. *Adapted from FAO, (1998b, 2003).*

### 3.2.3.3. TASK 3: IDENTIFY THE INTENDED USE OF THE PRODUCT

The product is intended for human consumption. Given the nature of wine, recommendations should be included on aspects such as safe consumption levels; incompatibilities, if any, with medication; and precautions to be taken by individuals with certain health conditions that would imply limitations on intake.

### 3.2.3.4. TASK 4: DRAW UP THE COMMODITY FLOW DIAGRAM

The CFD describes in detail all the stages involved in production of each kind of wine and the order in which these stages occur.

Examples of CFDs for different wine types are provided in a number of HACCP studies applied to winemaking (see Section 3.2.1). A common feature of these CFDs is to exclude the pre-harvest stages from the winemaking process. This underlying assumption that wine is prepared exclusively in the winery is a genuine conceptual error—as we show in this chapter—and is even contrary to the principles of HACCP, which indicate that each and every stage involved in the preparation of a food product should be included in the system.

In view of the above considerations, we believe that all HACCP plans for the control of

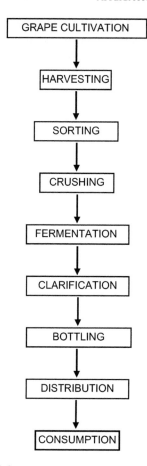

GRAPE CULTIVATION

↓

HARVESTING

↓

SORTING

↓

CRUSHING

↓

FERMENTATION

↓

CLARIFICATION

↓

BOTTLING

↓

DISTRIBUTION

↓

CONSUMPTION

**FIGURE 13.3** Generic commodity flow diagram for winemaking that includes the viticulture stages (growing and harvesting) that should be included in any Hazard Analysis and Critical Control Point system aimed at controlling ochratoxin A.

OTA in wine should take into account the pre-winery phases. To this end, we recommend that each HACCP team should use a generic CFD (like the one shown in Figure 13.3) to guide the creation of a CFD adapted to the wine in question. We also recommend that the model shown in Figure 13.3 be completed by a viticulture expert from the HACCP team, if possible, in collaboration with a representative of the Ministry of Agriculture. The CFD should be verified further by visiting grape production sites and by interviewing enologists, vineyard managers, and winery managers and observing their practices.

It is important to bear in mind that the CFD will form the basis for the hazard analysis and establishment of the CCPs. It is, therefore, generally recommended to provide, along with the CFD, a detailed description of operations with information such as lists of raw materials, additives, the packaging in which these are delivered, storage conditions, activities to be performed throughout the entire process, time and temperature profiles for the different stages, equipment, design characteristics, a blueprint of the facilities, warehousing conditions, customer and distribution problems, etc. (Martínez-Rodríguez & Carrascosa, 2009; Mortimore & Wallace, 2001). For special kinds of wines, such as sweet wines, where the probability of *A. carbonarius* growth is high (Gómez et al., 2006), the CFD should include data on grape over-ripening times, environmental conditions, etc.

The CFD should also cover the post-production stages right up to delivery to the consumer, in order to identify and draw attention to factors that could potentially affect the safety of the product.

### 3.2.3.5. TASK 5: CONFIRM THE COMMODITY FLOW DIAGRAM ON SITE

Once the CFD is completed, the HACCP team should visit the vineyard and the winery to check the data collected against the real operational conditions.

This operation, known as "walking the line," consists of checking, step by step, that all the information regarding materials, practices, controls, etc., has been taken into consideration by the HACCP team. Where appropriate, additional information such as time of harvest, maximum transportation time, transportation conditions to the winery, and temperature at, and duration of, the different stages should be collected and included in the CFD. For special kinds of wines, over-ripening practices at the production site (for botrytized wines) and post-harvest processing (for sweet wines) should be taken into consideration. The site for which the HACCP plan is being designed should be visited as often as necessary to ensure that all information relevant to the hazard in question has been collected.

### 3.2.3.6. TASK 6: IDENTIFY AND ANALYZE HAZARDS (HACCP PRINCIPLE 1)

Success in applying the HACCP system relies on properly identifying and analyzing the hazards that can arise in association with the raw materials and in any of the CFD phases. A correctly applied hazard analysis requires the compilation and evaluation of all data available on the hazard in question and the factors that contribute to its occurrence.

A hazard is any factor that might render a food unsafe for consumption. A microbiological hazard is a hazard caused by a microorganism. We consider microbiological hazards to include hazards established as such on the basis of epidemiological data or widely applied regulations.

Once a hazard has been identified, the associated risk—that is, the probability that it will occur—should be assessed. Like probability, risk is rated between 0 and 1, but it is often described qualitatively as low, medium, or high. Only hazards considered by the HACCP team to constitute unacceptable risks are carried forward to Task 7 (HACCP Principle 2).

Once risk has been assessed, appropriate control measures need to be considered. Control measures are actions or procedures used to bring the identified hazard under control, whether by preventing or eliminating it or by reducing it to an acceptable level. The implementation of control measures requires suitable training of personnel for specific operations already included or to be included in GAP, GMP, and GHP.

*3.2.3.6.1. HAZARD IDENTIFICATION*    As mentioned, this chapter focuses exclusively on OTA as this is the microbiological hazard that is most widely regulated internationally in wine. An unacceptable risk is posed when the permitted level of 2 µg/L of OTA is exceeded. This particular hazard should, thus, be evaluated at each harvest and production phase.

In order to simplify the description of how the HACCP system should be applied to OTA, we will assume that this myotoxin is produced by *A. carbonarius*, even though other species of *A.* section *Nigri* can also produce it, though to a much lesser extent (Carrascosa, 2005; Gómez et al., 2006; Martínez-Rodríguez & Carrascosa, 2009). To treat OTA as a chemical hazard merely because it is a chemical substance would, in our opinion, make it more difficult to identify control measures, most of which will be related to the growth of *A. carbonarius*.

*3.2.3.6.2. IDENTIFYING COMMODITY FLOW DIAGRAM PHASES WHERE OCHRATOXIN A (OTA) CONTAMINATION IS MOST LIKELY TO OCCUR*    Since *A. carbonarius* is an opportunistic pathogen and not highly infectious, the quantity of OTA produced increases with grape skin damage, temperature, and relative humidity (Bellí et al., 2007; Carrascosa, 2005; Kapetanakou et al., 2009; Martínez-Rodríguez & Carrascosa, 2009). *A. carbonarius* generally develops at harvest time, when grapes are most likely to be damaged (Serra et al., 2003). This is therefore the period when OTA levels will be highest in the grape. If the grapes have suffered extensive damage at an earlier stage, the probability of *A. carbonarius* invasion will be higher, as will the risk of higher levels of OTA in the final product. The use of damaged grapes to make wine will thus increase the risk of exceeding maximum recommended levels of OTA (Serra et al., 2005). The winemaking process itself can also favor the growth of *A. carbonarius* and hence the production of OTA (Gómez et al., 2006). For this reason, the CFD should, as far as possible, be adapted to each particular case.

Summing up, the main factors that contribute to biological deterioration by fungi in a vineyard ecosystem are humidity, temperature, and pests. Fungal growth, for example, is greater in more humid, warmer conditions and insects can cause considerable damage to grape skin, leading to the release of nutrients and the spread of fungal mycelia through the pulp of the grape.

OTA generally appears before harvest time (Serra et al., 2003, 2005). First, however, *A. carbonarius* has to develop, and this will only occur in phases with sufficient oxygen supplies, as the fungus is strictly aerobic. These phases occur prior to the crushing of the grapes, as aerobic growth conditions are generally avoided in subsequent winemaking stages to prevent the deterioration of the sensory properties of the final product. *A. carbonarius*, therefore, is likely to develop during the cultivation phase, mainly at the grape ripening stage (Bellí et al., 2007; Cabañes et al., 2002; Kapetanakou et al., 2009).

In other plant-based products, contamination by mycotoxins can also occur in the storage period between harvest and processing. Indeed, storage for periods longer than 48 h at temperatures of 10°C or above is not allowed (FAO, 2003). In the case of winemaking, the normal practice is to crush the grapes immediately after harvesting. If this is not done, the storage control measures described above should be implemented. This is particularly important for sweet wines (Gómez et al., 2006).

Grape ripening generally coincides with the withering of the grape vine, which occurs after veraison. *A. carbonarius* does not appear to be capable of attacking the skin of grapes and invading the pulp (Bellí et al., 2007), which suggests that colonization is strongly favored by pre-existing skin damage (Kapetanakou et al., 2009). The grape berry has two natural barriers to *A. carbonarius*: its thick skin and pH. Only intact skin, however, can prevent invasion,

since *A. carbonarius* can grow at a pH below 4.5, and indeed does when the skin breaks and spores attached to the bloom of the fruit or the vine germinate.

The grape's skin barrier can be weakened by insects (e.g., wasps, mealybugs, fruit and vinegar flies, and pyralid caterpillars), phytopathogenic fungi (which cause diseases such as esca and powdery and downy mildew), birds, and physiological and meteorological conditions (which can cause water stress and other environmental stresses). All these factors contribute to skin damage, thus allowing *A. carbonarius* to access the nutrients in the pulp and begin OTA production.

Attacks on the grapevine by phytopathogenic fungi are more successful when meteorological conditions are propitious. Temperatures of between 20 and 27°C, rainy summers, and damp autumns, for example, all favor the germination of spores and reduce the effectiveness of fungicides (which are most effective in dry conditions).

Damage to grapes during harvesting by rough handling or excessive weight in containers is less likely to lead to the production of OTA as the berries are almost immediately crushed. This kind of damage, however, should receive particular attention in plant-based foods such as cereals that are generally placed in storage until distribution or processing (FAO, 2003).

Further studies are necessary to determine numerous aspects such as the level of grape damage required for *A. carbonarius* to develop, the time it takes for OTA to be produced and the intervening environmental factors, and the relationship between fungal growth and OTA concentrations in wine.

*3.2.3.6.3. POSSIBLE OCHRATOXIN A (OTA) CONTROL MEASURES* OTA control measures must aim to both prevent and reduce OTA contamination. Preventive control measures consist of preventing the development of *A.*

*carbonarius* and, consequently, the synthesis of OTA. The main strategies are those designed to prevent grape skin damage. This requires the implementation of a GAP program involving phytosanitary plans aimed at ensuring optimally healthy plants that have good defenses against possible parasites and measures to prevent water stress and damage from fungi, insects, and birds.

Insecticides (or alternatives such as chemical or biological treatments)—provided they are safe for use with foodstuffs and comply with the legislation underpinning the HACCP system—can be used to protect against moths such as *Lobesia botrana*, *Cryptoblabes gnidiella*, and *Eupoecilia ambiguella*. A good preventive GAP strategy against birds is to eliminate natural shelters or to use optical or acoustic devices to frighten them away. Fungicidal treatments such as sulfur, copper products, and organic fungicides can be used to protect against phytopathogenic fungi (Varga & Kozakiewicz, 2006). These treatments will also protect the grape skin from damage and subsequent invasion by *A. carbonarius*. The effectiveness of biological control measures based on the use of epiphytic yeasts with inhibitory effects on undesirable fungi by competitive exclusion has also been studied (Bleve et al., 2006), although it is not entirely clear whether such measures are truly viable.

Factors that contribute to biological deterioration caused by fungi in vineyards (humidity, temperature, and pests) are uncontrollable since they are dictated by weather conditions. For this reason, preventive measures aimed at significantly reducing OTA levels in grapes and wines need to focus on minimizing damage to the grapes; such measures will include control of insects and phytopathogenic fungi and the elimination of visibly damaged berries before, during, and after harvesting (Bellí et al., 2007; Carrascosa, 2005; Martínez-Rodríguez & Carrascosa, 2009). Indeed, in its Resolution VITI-OENO 1/2005, the OIV issued GAP guidelines that recommend these practices (OIV, 2005).

Other preventive control measures include the use of transgenic grape strains that are resistant to water stress and damage by phytopathogenic fungi (Colova-Tsolova et al., 2001; Kikkert et al., 2001; Vivier & Pretorius, 2000). However, given the time required to adapt the grapevines (3—8 years) and the fact that transgenic foods are more difficult to market, it is likely to be some time before preventive measures of this kind are implemented.

If, despite preventive measures, berries become damaged and fungal growth is detected, the need for OTA reduction measures must be analyzed. To decide whether or not such measures are necessary, it must be determined, firstly, whether *A. carbonarius* invasion has occurred, and, secondly, whether unacceptable levels of OTA are being produced. To test for the presence of *A. carbonarius*, it is necessary to identify and characterize the species present using fast, sensitive, and accurate molecular methods (Oliveri et al., 2008). OTA production should also be analyzed using methods that provide rapid results (Turner et al., 2009; Varga & Kozakiewicz, 2006).

Current recommendations for certain plant-based foods state that fruit damaged by toxigenic fungi should be discarded. To prevent contamination of apple juice by patulin or of corn or copra meal by aflatoxin, for example, the recommendation is to discard 99% of all fruit whose color indicates infection (FAO, 2003). Despite the fact that the presence of *A. carbonarius* is visible (Figure 13.4), no visual selection method has yet been developed to separate healthy and infected fruit prior to the crushing stage. Before removing infected fruit, thus, it is necessary to perform laboratory tests on batches of grapes from contaminated vineyards to test for the presence of *A. carbonarius* and OTA.

Traceability is obviously an important aspect in relation to OTA control measures. It is crucial that the origin of all batches of grapes entering

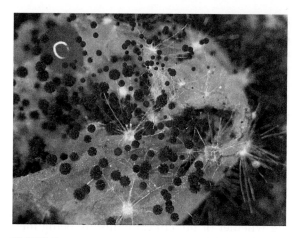

**FIGURE 13.4** Grape infected with *Aspergillus carbonarius*. Photograph kindly provided by Dr Venancio from the Department of Biological Engineering at the University of Minho, Portugal.

the crushing phase is recorded. Each batch should be visually inspected for mold and, where necessary, the presence of *A. carbonarius* or OTA should be confirmed so that contaminated batches can be eliminated.

Once the presence of OTA has been detected, contaminated grapes should not be mixed with uncontaminated grapes (EC, 2002). If an unacceptable level of OTA is subsequently detected in the must, despite the precrushing control measures, detoxification measures such as those recommended for other plant-based foods can be used (Coker, 1997). Understandably, such methods should not compromise either the safety or the sensory properties of the wine. A range of OTA detoxification methods suitable for application to wine have been studied. One method involves the use of activated carbon to reduce OTA levels (Olivares-Marín et al., 2009). It has also been demonstrated that lactic acid bacteria (del Prete et al., 2007) and yeasts (García-Moruno et al., 2005) can adsorb OTA. With reference to yeasts, it has been shown that mannoproteins play an important role in OTA adsorption and that the adsorption capacity of the yeast cell wall

can be enhanced by previous heat treatment (Nuñez et al., 2008). Nonetheless, these and many other proposed methods lead only to a small reduction in OTA levels; they can also interfere with the organoleptic properties of wine and are of questionable viability (Amézqueta et al., 2009).

In brief, pre-harvest control measures should be preventive while those implemented during and after harvesting should aim at reducing OTA levels (principally via the sorting and elimination of damaged berries); finally, detoxification methods should be used if OTA levels are detected after the crushing stage (Carrascosa, 2005; Martínez-Rodríguez & Carrascosa, 2009).

### 3.2.3.7. TASK 7: DETERMINE THE CRITICAL CONTROL POINTS (HACCP PRINCIPLE 2)

The decision tree provided in the Codex Alimentarius (FAO, 1993, 1997) can be used by the HACCP team to help determine the CCPs to be included in the CFD. The CFD should indicate all phases where hazards are likely to arise and all phases with control measures should be considered CCPs. If suitable control measures cannot be established for a particular phase or subsequent phases, the corresponding product should be classified as unfit for human consumption.

In the example we are analyzing, the grape sorting phase in the CFD would be a CCP, because it is a point at which a control measure can be applied; namely, the elimination of berries contaminated by *A. carbonarius* and thus OTA. Although, as a general rule, preventive and reduction measures should be applied during the growing and sorting phases, respectively, adaptation of the HACCP system to a particular site may mean that infected berries are eliminated by pruning during the ripening phase in the vineyard. We have not included the use of detoxification methods as a CCP, given that these methods are still in an experimental stage. However, if such methods or any other new methods are included in the HACCP system, the corresponding phase should logically be considered a CCP.

### 3.2.3.8. TASK 8: ESTABLISH CRITICAL LIMITS FOR EACH CRITICAL CONTROL POINT (HACCP PRINCIPLE 3)

Critical limits are usually established on the basis of readily measurable CCP parameters such as temperature and pH that indicate the presence of a hazard. In the case of OTA, temperature, rainfall, and relative humidity during the cultivation phase are factors that are beyond human control, and grape pH does not necessarily inhibit infection. For this reason, critical limits will need to be based on parameters that are directly related to grape damage, the presence of fungi, and OTA concentration.

No studies to date have adequately documented the relationship between OTA concentration in grape pulp and grape berry damage or fungal growth. There are, however, such studies for other plant-based products that can potentially be affected by mycotoxins. For apple juice, copra cake, copra meal, and pistachios, for example, the guidelines recommend that no more than 1% of infected fruit (defined as a fruit with >10% surface damage) should enter the processing stage. In other words, 99% of all damaged fruit is eliminated, thereby eliminating the mycotoxin or at least reducing concentrations to an acceptable level (FAO, 2003).

If control measures have previously resulted in the detection of *A. carbonarius* or OTA in grapes, we recommend that the same critical limits should be applied initially and subsequently adapted on the basis of information from studies of the winery's processes.

As detoxification tests are deployed on a large scale and conclusive data become available, particularly regarding the effect of detoxification on the sensory quality of wine, and as wineries put these methods to the test, it should become possible to establish critical limits based on data showing the quantity of detoxifying agents required to reduce or eliminate different levels of OTA from contaminated wines. In other fruit sectors, for example, high-pressure water jets are used to remove parts of the fruit damaged by mycotoxigenic fungi, but critical limits need to be set to ensure that the pressure used is sufficient to remove the damaged tissue without causing further damage to the fruit (FAO, 2003).

### 3.2.3.9. TASK 9: ESTABLISH A MONITORING SYSTEM (HACCP PRINCIPLE 4)

Monitoring activities are essential for checking whether or not critical limits are being met at each CCP. The methods used should be both sensitive and rapid to ensure that any loss of control is detected by trained personnel at as early a stage as possible. This is crucial to implementing appropriate corrective measures aimed at preventing or reducing product loss. Monitoring activities include the analysis of samples collected according to a sampling plan based on statistical principles. The most common measurements used to monitor mycotoxin formation in plant-based products are storage time, temperature, and humidity, as these all provide rapid results and allow suitable corrective measures to be taken quickly (FAO, 2003).

Visual inspection methods aimed at determining the level of grape damage and fungal growth in the vineyard should be the first step in the monitoring of OTA in wine. If the critical limits are exceeded, tests should then be performed to determine the presence of *A. carbonarius* and OTA. Tests for the detection of *A. carbonarius* should be performed by experts in the taxonomy of filamentous fungi using rapid molecular microbiological techniques (Oliveri et al., 2008), while OTA tests should be performed by chemical analysis experts, preferably with experience in mycotoxins, using the most validated methods available.

Rapid detection of OTA can be achieved with commercial kits based on immunoaffinity and similar methods (Varga & Kozakiewicz, 2006).

These should be sufficiently sensitive to detect 2 μg/L of OTA (Turner et al., 2009) and should preferably have been validated in must or wine. Other analytical methods are equally valid, although they do not produce results as quickly or require samples to be sent to a laboratory. Such tests will allow batches of grapes to be classified as acceptable or unacceptable on the basis of the level of OTA detected. In other words, OTA tests will help to ensure that the grape sorting phase is performed correctly.

All wineries and vineyard holdings should thus devise an inspection program to periodically check for visible signs of contamination and establish the frequency with which laboratory tests for *A. carbonarius* and OTA should be performed. In all other stages of the production chain, application of GMP—just one element in the HACCP system—should be sufficient to prevent the proliferation of *A. carbonarius* and thus protect against undesirable levels of OTA.

### 3.2.3.10. TASK 10: ESTABLISH CORRECTIVE ACTIONS (HACCP PRINCIPLE 5)

If the monitoring activities in place determine that the critical limits are not being met (indicating that the process is out of control), corrective actions should be implemented. These actions should assume a worst-case scenario yet be based on an evaluation of hazards, risks, severity, and the intended use of the product. Personnel should receive suitable training in the application of corrective measures, which should ensure that control of the CCP is regained and that the affected raw materials or products are isolated and discarded if necessary. Wherever possible, an alarm system should be put in place to warn personnel that a critical limit is being approached. Suitable corrective actions applied at this point should avoid deviation from the critical limits and prevent product loss.

There are two kinds of corrective actions: those aimed at regaining control (e.g., discarding batches of grapes with excessive OTA levels) and those aimed at isolating the product associated with the period when the CCP was out of control and modifying product disposition (by discarding, downgrading, or reprocessing the product; for example, mixing wines to reduce the OTA concentration or, if possible, detoxifying the wine).

If grape sorting prior to crushing is not a routine practice, the most effective corrective action will be to remove damaged grapes or grapes containing OTA detected by the monitoring system to prevent these from entering the winemaking process.

### 3.2.3.11. TASK 11: VERIFY THE HACCP SYSTEM (HACCP PRINCIPLE 6)

One of the most important ways to check the effectiveness of an HACCP system is through a verification audit, which consists of a systematic, independent inspection to check that all actions are being correctly documented (analysis of documentation) and that the system is being implemented as documented (analysis of HACCP records). Accordingly, procedures for validating each CCP should be established and the effectiveness of the overall system checked on the basis of quantitative analyses of OTA content in representative samples taken from batches of grapes prior to crushing and wine after production. Three-monthly audits are recommended for other plant-based products at risk of contamination by mycotoxins (FAO, 2003).

The HACCP system should be verified periodically by an individual designated for this purpose. Microbiological or chemical tests can be used to ensure that the system is under control and that the product meets customer specifications. These tests will enable verification of the suitability of the CCPs and control measures in place and of the scope and efficacy of the monitoring procedures. An internal auditing plan, as well as being an essential tool for verifying the effectiveness of the HACCP system, will also document ongoing efforts to keep the HACCP up to date.

The HACCP system can be verified in the following ways:

10. By taking samples for analysis using a method other than that used for monitoring purposes
11. By talking to personnel, especially the person in charge of monitoring the CCPs
12. By observing operations at the CCPs
13. By commissioning external audits from an independent auditor

It is important to emphasize that the application of a generic HACCP system is not viable as each HACCP system must be adapted to the specific formulation, handling, and preparation methods for the product in question.

Periodic product tests aimed at checking that acceptable limits have not been exceeded should be performed. If limits are exceeded, it should be possible to detect where the system failed and to identify at which point control was lost. In such a case, it may be necessary to change critical limits or to validate and introduce new control measures. Changes should also be made if a study of deviations and product dispositions reveals an unacceptable degree of control at a particular CCP.

If OTA concentrations in the end product exceed the limits established by law, the traceability and record-keeping system will enable the defective batch to be traced and will also indicate the CCP where control was lost. This CCP should, if necessary, be modified.

### 3.2.3.12. TASK 12: KEEP RECORDS (HACCP PRINCIPLE 7)

Record-keeping is essential to the correct application of the HACCP system, as it demonstrates that procedures have been followed appropriately, critical limits have been respected, monitoring has been adequate, and corrective actions have been implemented where necessary. Record-keeping also enables problematic aspects of the system to be documented with a view to implementing continual improvements. The implementation of traceability systems that include batch identification right back to the vineyard enables specific OTA-contaminated batches of product to be located for elimination purposes and so avoids losses associated with the unnecessary elimination of uncontaminated batches. Proper documentation can also provide legal evidence of due diligence regarding food safety management.

Records should at least include all documentation related to processes, GMP and GHP, CCP monitoring, compliance with critical limits, deviations, and corrective actions.

## Acknowledgments

We thank the Spanish Ministry of Science and Innovation (MICINN) (grants AGL2006-02558, AGL2009-07894 and Consolider INGENIO 2010 CSD2007-00063 FUN-C-FOOD) and the Comunidad de Madrid (CAM) (S2009/AGR-1469) for financial support.

## References

Amézqueta, S., González-Peñas, E., Murillo-Arbizu, M., & López de Cerain, A. (2009). Ochratoxin A decontamination: A review. *Food Control, 20*, 326—333.

Battilani, P., & Pietri, A. (2002). Ochratoxin A in grapes and wine. *Eur. J. Plant Pathol., 108*, 639—643.

Bau, M., Bragulat, M. R., Abarca, M. L., Minguez, S., & Cabañes, F. J. (2005). Ochratoxigenic species from Spanish wine grapes. *Int. J. Food Microbiol., 98*, 125—130.

Bellí, N., Marín, S., Coronas, I., Sanchis, V., & Ramos, A. J. (2007). Skin damage, high temperature and relative humidity as detrimental factors for *Aspergillus carbonarius* infection and ochratoxin A production in grapes. *Food Control, 18*, 1343—1349.

Bellido, J. B., Gozález, F., Arnedo, A., Galiano, J. V., Safont, L., Herrero, C., et al. (1996). Brote de infección alimentaria por *Salmonella enteritidis*. Posible efecto protector de las bebidas alcohólicas. *Med. Clin., 107*, 641—644.

Blesa, J., Soriano, J. M., Moltó, J. C., & Mañes, J. (2004). Concentration of ochratoxin A in wines from supermarkets and stores of Valencian Community (Spain). *J. Chrom. A., 1054*, 397—401.

Bleve, G., Grieco, F., Cozzi, G., Logrieco, A., & Visconti, A. (2006). Isolation of epiphytic yeasts with potential for biocontrol of Aspergillus carbonarius and A. niger on grape. *Int. J. Food Microbiol., 108*, 204–209.

Brenner, H., Rothenbacher, D., Bode, G., & Adler, G. (1999). Inverse graded relation between alcohol consumption and active infection with *Helicobacter pylori*. *Am. J. Epidemiol., 149*, 571–576.

Briones, A. I., & Úbeda, J. F. (2001). Elaboración de un plan APPCC en una bodega. *Tecnol., Vin* 89–93, May/June.

Bujanda, L. (2000). The effects of alcohol consumption upon the gastrointestinal tract. *Am. J. Gastroenterol., 95*, 3374–3382.

Burdaspal, P. A., & Legarda, T. M. (1999). Ocratoxina A en vinos, mostos y zumos de uva elaborados en España y en otros países europeos. *Alimentaria, 36*, 107–113.

Cabañes, F. J., Accensi, F., Bragulat, M. R., Abarca, M. L., Castellá, G., Minguez, S., et al. (2002). What is the source of ochratoxin A in wine? *Int. J. Food Microbiol., 79*, 213–215.

Carrascosa, A. V. (2005). APPCC en la elaboracíon del vino: Ocratoxina A. In Microbiologia del vino. A. V. Carrascosa, R. Muñoz, R. González coord. Chap 13, 361–380. Ed. AMV, Madrid.

Coker, R. D. (1997). *Mycotoxins and their control: Constraints and opportunities. NRI Bull. 73.* Chatham, UK: Natural Resources Institute.

Colova-Tsolova, V., Perl, A., Krastanova, S., Tsvetkov, I., & Atanassov, A. (2001). Genetically engineered grape for disease and stress tolerance. In K. A. Roubelakis-Angelakis (Ed.), *Molecular Biology & Biotechnology of the Grapevine* (pp. 411–432). Amsterdam, the Netherlands: Kluwer Publ. Co.

del Prete, V., Rodríguez, H., Carrascosa, A. V., de las Rivas, B., García-Moruno, E., & Muñoz, R. (2007). *In vitro* removal of OTA by wine lactic acid bacteria. *J. Food. Protect., 70*, 2155–2160.

Delage, N., d'Harlingue, A., Colonna Ceccaldi, B., & Bompeix, G. (2003). Occurrence of mycotoxins in fruit juices and wine. *Food Control, 14*, 225–227.

Desenclos, J. C. A., Klontz, K. C., Wilder, M. H., & Jun, R. A. (1992). The protective effect of alcohol on the occurrence of epidemic oyster-borne hepatitis A. *Epidemiol., 3*, 371–374.

European Commission (EC). (2002). Directive 2002/32/EC of the European Parliament and of the Council of 7 May 2002 on undesirable substances in animal feed. *Coun. Stat. Off. J., L140*, 10–22.

European Food Safety Authority (EFSA). (2006). Opinion of the Scientific Panel on Contaminants in the Food Chain on a Request from the Commission related to ochratoxin A in food. Question N° EFSA-Q-2005–154. Adopted April 4, 2006. *EFSA J, 365*. Retrieved August 26, 2010 from <http://www.efsa.europa.eu/en/science/contam/contam_opinions/1521.html>.

European Union (EU). (2002a). Commission regulation EC No 472/2002 of 12 March 2002 amending regulation (EC) No. 466/2001 setting maximum levels for certain contaminants in foodstuffs. *Off. J. Europ. Comm. March, 16*.

European Union (EU). (2002b). Reglamento (CE) no 257/2002 de la Comisión de 12 de Febrero de 2002. *Off. J. Europ. Comm., L 41/12–15*.

European Union (EU). (2003). EU food law news. Contaminants-Mycotoxins-EC Permitted levels. Retrieved August 26, 2010 from <http://www.foodlaw.rdg.ac.uk/news/eu-03068.htm>.

European Union (EU). (2004). EU rules on ochratoxin A extended to coffee, wine and grape juice. *Off. J. Europ. Comm., IP/04/1215*.

European Union (EU). (2005). Reglamento (CE) No 123/2005 de la comisión de 26 de enero de 2005 por el que se modifica el reglamento (CE) no 466/2001 con respecto a la ocratoxina A. *Off. J. Europ. Comm., L 25/3–5*.

Federación de Industrias de Alimentación y Bebidas (FIAB). (1997). *Aplicación del sistema de análisis de riesgos y control de puntos críticos en vinos.* Valencia, Spain: Semana Vitivinícola.

Food and Agriculture Organization (FAO). (1993). Guidelines for the application of the HACCP system. In *Training considerations for the application of the HACCP system to food processing and manufacturing, 11* Geneva, Switzerland: FAO/WHO/FNU/FOS/93.3.

Food and Agriculture Organization (FAO). (1997). HACCP system and guidelines for its application. Annex to CAC/RCP 1–1969, Rev. 3. In *Codex Alimentarius food hygiene basic texts.* Rome, Italy: FAO/OMS.

Food and Agriculture Organization (FAO). (1998a). Position paper on ochratoxin A, Codex Alimentarius Comission, Codex Committee on Food Aditives and Contaminants, thirty-first session. The Hague, the Netherlands, March 22–26, 1999, CX/FAC 99/14.

Food and Agriculture Organization (FAO). (1998a). *Food quality and safety systems – A training manual on food hygiene and the Hazard Analysis and Critical Control Point (HACCP) system.* Rome, Italy: Food Quality and Standards Service Food and Nutrition Division.

Food and Agriculture Organization (FAO). (2002). *Report on meetings of expert committees and study groups. Evaluation of certain micotoxins. EB110/6, 110th session, provisional agenda item 7.1, joint FAO/WHO Expert Committee on Food Additives. Fifty-sixth report.* Geneva, Switzerland, February 6–15.

Food and Agriculture Organization (FAO). (2003). *Manual sobre la aplicación del sistema de análisis de peligros y de puntos críticos de control (APPCC) en la prevención y control de las micotoxinas.* Rome, Italy: Estudio FAO Alimentación y Nutrición no 73.

Frisvad, J.C. & Thrane, U. (1996). Mycotoxin production by food-borne fungi. (5th ed.). In R. A. Samson, E. S. Hoekstra, J. C. Frisvad & O. Filtenborg (Eds.), *Introduction to food-borne fungi* (pp. 251−260). Centraalbureau voor Schimmelcultures, Baarn, the Netherlands.

Gallo, A., Perrone, G., Solfrizzo, M., Epifani, F., Abbas, A., Dobson, A. D. W., et al. (2009). Characterisation of a pks gene which is expressed during ochratoxin A production by *Aspergillus carbonarius*. *Int. J. Food Microbiol., 129*, 8−15.

Gañán, M., Martínez-Rodríguez, A. J., & Carrascosa, A. V. (2009). Antimicrobial activity of phenolic compounds of wine against *Campylobacter jejuni*. *Food Control, 20*, 739−742.

García-Moruno, E., Sanlorenzo, C., Boccaccino, B., & di Stefano, R. (2005). Treatment with yeast to reduce the concentration of ochratoxin A in red wine. *Am. J. Enol. Viticult., 56*, 73−76.

Gómez, C., Bragulat, M. R., Abarca, M. L., Mínguez, S., & Cabañes, F. J. (2006). Ochratoxin A-producing fungi from grapes intended for liqueur wine production. *Food Microbiol., 23*, 541−545.

Gottardi, G. (1997). Come utilizzare in modo semplice il sistema HACCP nel settore enologico. *Enotecnico, 33*, 20−27.

Hyginov, C. (2000). *Elaboración de vinos. Seguridad-calidad-métodos. Introducción al HACCP y al control de los defectos.* Acribia, Zaragoza, Spain.

International Commission on Microbiological Specifications for Foods (ICMSF). (1991). El sistema de análisis de riesgos y puntos críticos. Acribia, Zaragoza, Spain.

International Commission on Microbiological Specifications for Foods (ICMSF). (1996). *Microorganisms in Foods V. Microbiological specifications of food pathogens.* London, UK: Chapman & Hall.

International Organisation of Vine and Wine (OIV). (2002). CST 1/2002 adoptada en 2002 por los Estados miembros de la OIV en ocasión de la Asamblea General de Bratislava (Eslovaquia).

International Organisation of Vine and Wine (OIV). (2005). Resolution VITI-OENO 1/2005. Paris, France, October 25.

Kapetanakou, A. E., Panagou, E. Z., Gialitaki, M., Drosinos, E. H., & Skandamis, P. N. (2009). Evaluating the combined effect of water activity, pH and temperature on ochratoxin A production by *Aspergillus ochraceus* and *Aspergillus carbonarius* on culture medium and Corinth raisins. *Food Control, 20*, 725−732.

Kikkert, J. R., Thomas, M. R., & Reisch, B. I. (2001). Grapevine genetic engineering. In K. A. Roubelakis-Angelakis (Ed.), *Molecular biology and biotechnology of the grapevine* (pp. 387−404). Dordrecht, the Netherlands: Kluwer Academic Publishers.

Kourtis, L. K., & Arvanitoyannis, I. S. (2001). Implementation of Hazard Analysis Critical Control Point (HACCP) system to the alcoholic beverages industry. *Food Rev. Inter., 17*, 1−44.

López de Cerain, A., Gonzáles-Peòas, E., Jimenez, A. M., & Bello, J. (2002). Contribution to the study of ochratoxin A in Spanish wines. *Food Addit. Contam., 19*, 1058−1064.

Marimón, J. M., Bujanda, L., Gutiérrez-Stampa, MªA., Cosme, A., & Arenas, J. (1998). Antibacterial activity of wine against *Salmonella enteritidis* pH or alcohol? *J. Clin. Gastroenterol., 27*, 179−180.

Martínez-Rodríguez, A. J., & Carrascosa, A. V. (2009). HACCP to control microbial safety hazards during winemaking: Ochratoxin A. *Food Control, 20*(5), 469−475.

Mateo, R., Medina, A., Mateo, E. M., Mateo, F., & Jiménez, M. (2007). An overview of ochratoxin A in beer and wine. *Int. J. Food Microbiol., 119*, 79−83.

Morassut, M., & Cecchini, F. (1999). HACCP in oenological industry: A system of self-control not only for hygienic-sanitary hazards. *It. Food Bever. Technol., 16*, 8−12.

Mortimore, S. & Wallace, C. (2001). HACCP: Enfoque práctico. Acribia, Zaragoza, Spain.

National Advisory Committee on Microbiological Criteria for Foods (NACMCF). (1992). Hazard analysis and critical control point system. *Int. J. Food Microbiol., 16*, 1−23.

Núñez, Y. P., Pueyo, E., Carrascosa, A. V., & Martínez-Rodríguez, A. J. (2008). Effects of aging and heat treatment on whole yeast cells and yeast cell walls and on the adsorption of OTA in a wine model system. *J. Food Protect., 71*, 1496−1499.

Olivares-Marín, M., del Prete, V., Garcia-Moruno, E., Fernández-González, C., Macías-García, A., & Gómez-Serrano, V. (2009). The development of an activated carbon from cherry stones and its use in the removal of ochratoxin A from red wine. *Food Control, 20*, 298−303.

Oliveri, C., Torta, L., & Catara, V. (2008). A polyphasic approach to the identification of ochratoxin A-producing black Aspergillus isolates from vineyards in Sicily. *Int. J. Food Microbiol., 127*, 147−154.

Pérez de Obanos, A., López de Cerain, A., Jiménez, A. M., González-Peñas, E., & Bello, J. (2001). Ocratoxina A en plasma humano: Nuevos datos de exposición en España. *Rev. Toxicol., 18*, 19−23.

Pfeiffer, A., Holgl, B., & Kaess, H. (1992). Effect of ethanol and commonly ingested alcoholic beverages on gastric emptying and gastrointestinal transit. *Clin. Investig., 70*, 487−491.

Pitt, J. I. (1996). What are mycotoxins? *Australian Mycotoxin Newsletter, 7*(4), 1.

Rousseau, J. (2004). Ocratoxina en los vinos: Estado de los conocimientos. *Vinidea. Net. Internet Tech. J.,* 1−4.

Ruggiero, P., Tombola, F., Rossi, G., Pancotto, L., Lauretti, L., del Giudice, G., et al. (2006). Polyphenols reduce gastritis induced by *Helicobacter pylori* infection or VacA toxin administration in mice. *Antimicrob. Agents Chemother., 50* (7), 2550–2552.

Samson, R.A., Hoekstra, E.S., Frisvad, J.C., & Filtenborg, O. (1996). *Introduction to food-borne fungi* (5th ed.). Centraalbureau voor Schimmelcultures, Baarn, the Netherlands, p. 322.

Selma, M. V., Martínez-Culebras, P. V., & Aznar, R. (2008). Real-time PCR based procedures for detection and quantification of *Aspergillus carbonarius* in wine grapes. *Int. J. Food Microbiol., 122*, 126–134.

Serra, R., Abrunhosa, L., Kozakiewicz, Z., & Venanci, A. (2003). Black *Aspergillus species* as ochratoxin A producers in Portuguese wine grapes. *Int. J. Food Microbiol., 88*, 63–68.

Sheth, N. K., Wisniewski, T. R., & Franson, T. R. (1988). Survival of enteric pathogens in common beverages: An in vitro study. *Am. J. Gastroenterol., 83*, 658–660.

Soleas, G. J., Yan, J., & Goldberg, D. M. (2001). Assay of ochratoxin A in wine and beer by high-pressure liquid chromatography photodiode array and gas chromatography mass selective detection. *J. Agric. Food Chem., 49*, 2733–2740.

Turner, N. W., Subrahmanyamb, S., & Piletsky, S. A. (2009). Analytical methods for determination of mycotoxins: A review. *Anal. Chim. Act., 632*, 168–180.

Varga, J., & Kozakiewicz, Z. (2006). Ochratoxin in grapes and grape-derived products. *Trends Food Sci. Technol., 17*, 72–81.

Vinas, I., Dadon, J., & Sanchis, V. (1993). Citrinin-producing capacity of Penicillium expansum strains from apple packinghouses of Lerida (Spain). *Int. J. Food Microbiol., 19*, 153–156.

Vivier, M. A., & Pretorius, I. S. (2000). Genetic improvement of grapevine: Tailoring grape varieties for the third millennium. *S. Afr. J. Enol. Vitic., 21*, 5–26.

Weisse, M. E., Eberly, B., & Person, D. A. (1995). Wine as digestive aid: Comparative antimicrobial effects of bismuth salicylate and red and white wine. *BMJ, 311*, 1657–1660.

Zimmerli, B., & Dick, R. (1996). Ochratoxin A in table wine and grape-juice: Occurrence and risk assessment. *Food Add. Contam., 13*, 655–668.

CHAPTER

# 14

# Applied Wine Microbiology

Braulio Esteve-Zarzoso [1], Mireia Martínez [2], Xavier Rubires [3],
María Yuste-Rojas [2], Mireia Torres [2]

[1] Dpt Bioquimica i Biotecnologia, Universitat Rovira i Virgili, Tarragona, Spain,
[2] Miguel Torres, Barcelona, Spain and [3] Pago Jean León, Barcelona, Spain

OUTLINE

# 1. INTRODUCTION

Earlier chapters of this book have explored varying aspects of wine microbiology in exceptional detail. Although the information covered in those chapters provides a good indication of what is actually happening in today's wineries, it is not always readily applicable to the "real world." Small and medium-sized wineries often follow their intuition or continue to do what has worked for them in the past. The aim of this chapter is to provide some general guidance on the practical application of microbiology to the winery but without losing sight of the extreme importance of keeping facilities and equipment as clean as possible and of closely following protocols. Many problems will be avoided if these two principles are followed.

The two main cornerstones of the practices of any winery should be to ensure careful, proper handling throughout the process, from the vineyard to the bottle, and to never lose sight of the fact that wine is intended for human consumption; any departure from these basic premises can cause problems (Garijo, 2008). Another important consideration is that wineries are interested in quick, simple, and affordable solutions that do not significantly impact the quality of the final product. By adhering to the recommendations of the International Organisation of Vine and Wine (OIV) regarding additives and to legislative requirements regarding contaminants (pesticides, heavy metals, toxic substances, etc.) in countries to which the wine is to be exported, wineries will avoid many problems related to the sale of their products.

# 2. MICROBIOLOGICAL CONTROL OF GRAPES

One of the keys to avoiding problems in the winemaking process is to minimize the grape microflora, as this will prevent the development of undesirable microorganisms from the outset.

To do this, it is necessary to carefully inspect and protect the grapes from the moment they start to grow to the moment they are harvested and transported to the winery. The optimal time for microbial growth in grapes is the ripening stage as this is when the grape's protective barrier, the skin, is most likely to be broken, leading to the release of sugars onto the surface of the berry and the proliferation of different types of microorganism.

The most relevant filamentous fungi are those that produce metabolites with a negative effect on wine quality, irrespective of their abundance. The best known grapevine fungus is *Botryotinia fuckeliana* (anamorphic state, *Botrytis cinerea*), which can have both positive and negative effects on the wine. When this fungus grows inside the berries (producing what is known as noble rot), it removes water from the fruit and, thus, increases the concentration of compounds that determine the primary aroma of the wine. When the fungus affects the surface of the grape, however, it is known as gray mold. This form of the fungus produces β-glucans, which ultimately interfere with wine clarification and filtration operations. Other filamentous fungi found on grapes are *Cladosporium*, *Mucor*, and *Rhizopus* species, but they do not have a significant bearing on the fermentation process (Fleet, 1992). Special attention should also be paid to fungi that produce toxins that can be passed into wine. For example, ochratoxin A, which is produced by *Aspergillus* and *Penicillium* species, is becoming an increasing concern in the winemaking community. This metabolite has been detected at different concentrations in a range of wines from different regions (Belli et al., 2005; Solfrizzo, 2008). Chapter 13 describes how to apply a hazard analysis and critical control point plan to control ochratoxin A levels in wine. This procedure is particularly relevant in view of the recent European Union (EU) legislation establishing maximum allowable levels for this toxin in food products.

Numerous studies have analyzed the presence of yeast on the surface of grapes (de Andrés-de Prado et al., 2007; Fleet & Heard, 1992) and many have indicated that *Saccharomyces cerevisiae* is present only in very small numbers on healthy grapes (Martini, 1993; Pretorius, 2000). Most of the yeasts in such cases are aerobic species, the most common of which are those belonging to the genera *Candida*, *Hanseniaspora*, *Kluyveromyces*, *Pichia*, and *Rhodotorula*. The grape microflora, however, varies greatly depending on factors such as geographical location, rainfall, and temperature (Longo et al., 1991; Parrish & Carroll, 1985), fungicide use (Monteil et al., 1986), soil type (Farris et al., 1990; Poulard et al., 1980), vineyard age, grape variety, and harvesting method (Martini et al., 1980; Pretorius et al., 1999; Rosini et al., 1982). Other variations in microflora can be introduced by the sampling procedure used. Vaughan-Martini and Martini (1995), for example, reviewed the differences generated by sampling methods according to whether or not the sample was enriched prior to analysis.

What is certain is that grapes should be handled as gently as possible, as the slightest pressure on the berry could cause the release of juice containing sugars that will favor the growth of the yeast that come into contact with it. Adequate hygiene and sanitation standards in the winery are extremely important as any lapses will lead to the instant proliferation of *S. cerevisiae* yeasts that come into contact with these sugars. Accordingly, efforts should be made to ensure that harvesting equipment and grape reception facilities are kept as clean as possible. To this end, clean containers should be used to transport the grapes to the winery, and grape reception facilities should be hosed down if used continually or cleaned with disinfectants if the arrival of grapes is intermittent.

Bacteria grow in the same conditions as yeast, which explains why acetic and lactic acid bacteria can proliferate in the musts of wineries with poor hygiene conditions. The recommended course of action on the detection of bacteria is to isolate and clean the affected material with suitable disinfectants (alkaline disinfectants followed by water and acid-based disinfectants).

In any case, the best strategy for preventing the proliferation of undesirable microorganisms in the winery is to prevent their growth in the vineyard. Vineyards should thus be designed or adapted to facilitate the application of phytosanitary protection products and prevent overcrowding of grape clusters. Another effective measure is to remove grapes before they become too big as this is when the risk of skin breakage is greatest.

# 3. INOCULATION METHODS

Yeasts, whether naturally present or deliberately added to the must, have been an essential part of the winemaking process since time immemorial. While wine can certainly be made with naturally occurring populations of yeast, the demand for greater control over physical, chemical, and indeed microbiological properties has led to the increasing use of inoculated strains. Spontaneous, or natural, fermentation is performed primarily by *S. cerevisiae* yeast strains (Amerine & Kunkee, 1968), although other species may participate in the process (Torija et al., 2001) and alter the properties of the final product.

To activate fermentation, wineries can use either commercial preparations of yeast (or bacteria in the case of malolactic fermentation) or *pied de cuve* cultures. The use of commercial cultures is now widespread as they are convenient and easy to use, and provide guarantees about the origin of the yeasts. These preparations contain large numbers of viable cells and are used to ensure the rapid establishment of the selected species during fermentation. They may, however, also contain a number of contaminating microbes (Radler & Lotz, 1990).

By seeding the must with known microorganisms, today's winemakers have succeeded in ensuring that fermentation will primarily be conducted by strains with desirable properties. To ensure quality, however, and indeed consistency from one year to the next, it is necessary to control microbial activity during this process (Martini & Vaughan-Martini, 1990).

## 3.1. Direct Inoculation

The success of inoculation, whether performed with active dried yeast or starter cultures, depends largely on correct rehydration, but there are also a number of other factors to bear in mind prior to inoculation. The grape juice must, for example, contain only a small population of resident microorganisms. This can be achieved through static clarification and flotation (which achieve a 50–80% reduction in the native yeast population), centrifugation (60–90% reduction), and vacuum filtration (99% reduction). The inoculation tank must also be clean, as any microorganisms present could proliferate and compete with the inoculated strains. Another important factor is the time that elapses between harvesting and inoculation, as, the longer this time, the greater the proportion of autochthonous yeasts that will thrive and the fewer the nutrients that will be available for the inoculated strains. Furthermore, the temperature of the must should be kept as low as possible as high temperatures can favor the proliferation of naturally occurring populations.

The yeast rehydration process is generally similar across different commercial preparations, with seeding always performed at a density of 20 g/hL. The dry yeast is rehydrated in a volume of very warm water (37°C) at a concentration of 10% weight by volume. This mixture is then shaken gently for 10 min and left to rest for an additional 10 min (maximum 30 min) to allow time for the yeast to become rehydrated. The vessels used for rehydration should be twice the volume of the water used as the mix increases in size during the process. The next step is to vigorously mix the solution until it is uniform and ready to be inoculated. The temperature of the must is critical at this stage as yeast viability can be seriously compromised by shifts in temperature of more than 10°C. To safeguard against problems of this nature, an intermediate thermal conditioning step is recommended.

Using standard commercial preparations, which have a viability of approximately $10^{10}$ colony-forming units (CFU)/g, the above procedure will give rise to a yeast population of $2 \times 10^6$ CFU/mL, which is used to initiate the exponential growth phase (Degre, 1992).

Commercial *Oenococcus oeni* starters are also available for activating malolactic fermentation via direct inoculation. The rehydration phase is also critical in this case, and it is important to follow the manufacturers' instructions.

## 3.2. Preparation of *Pied de Cuve* Cultures and Calculation of Inoculation Rates

The must or wine needs to be inoculated quickly with the selected microorganisms to prevent the growth of unwelcome competitors. This is achieved by omitting the conditioning stage (in which the yeasts acclimatize to the new culture medium) and adding the yeasts directly to the must at the height of their metabolic activity; that is, during the exponential growth phase (Fleet & Heard, 1992).

### 3.2.1. Preparation of a *Pied de Cuve* Culture for Alcoholic Fermentation

A *pied de cuve* culture is simply a continuous fermenter in which the aim is to maintain a large population of yeast in the growth phase. To do this, it is necessary to periodically remove liquid from the fermenter tank and add must (as a source of nutrients) so that the yeast population can continue to proliferate. The liquid that is removed can be used to inoculate the must in the fermentation tank. As mentioned above, the fermentation conditions in the tank to which the active dried yeast has been added must be maintained to ensure that the yeast population is constantly in the exponential growth phase. The temperature of the tank must be maintained at approximately 17°C and it is often advantageous to add nutrients or fermentation activators at the moment of inoculation, particularly in the case of musts with depleted ammoniacal nitrogen supplies.

Strict monitoring of temperature in the *pied de cuve* tank is essential, and population numbers should be monitored using a Neubauer chamber until a level of between $5 \times 10^7$ and $1 \times 10^8$ cells/mL is reached. This is the moment at which the inoculation rate required to achieve an initial culture of $2 \times 10^6$ cells/mL can be calculated. A population level of between $5 \times 10^7$ and $1 \times 10^8$ cells/mL in the *pied de cuve* tank corresponds to an inoculation rate of between 2 and 5% in the fermentation tank. Microscopic observation of the physiological state of yeasts in the *pied de cuve* culture can provide valuable information as a very large population with a very low proportion of actively budding yeasts indicates that the yeasts are nearing the lag phase, whereas a very large population containing a high proportion of actively budding yeasts indicates that the yeasts are in the middle of their exponential growth phase.

### 3.2.2. *Preparation of a* **Pied de Cuve** *Culture for Malolactic Fermentation*

The procedure for preparing a *pied de cuve* starter for malolactic fermentation is much the same as that used for alcoholic fermentation but a longer conditioning stage is required (Champagne et al., 1989). To ensure that the lactic acid bacteria in the *pied de cuve* tank continue to proliferate, it is necessary to add wine from the tank awaiting malolactic fermentation. There are numerous commercially available nutrient preparations that can help to stimulate bacterial growth. The traditional practice in wineries has been to use wine already undergoing malolactic fermentation to inoculate wine that has not yet entered this stage. Calculating the inoculation rate in malolactic fermentation is more complicated than in alcoholic fermentation, however, as there is no quick method available for counting the number of viable cells (Blackburn, 1984). Instead, fermentation kinetics must be used as an indicator for the calculation of the appropriate inoculation rate.

## 4. MOLECULAR METHODS FOR ANALYZING THE MICROORGANISMS USED IN THE WINERY

Advances in technology have greatly improved the control that winemakers have over the yeasts that participate in the fermentation process. Modern molecular methods, in particular, offer rapid results, but other important features are ease of use and affordability (Andorrà, 2008). Of the range of techniques described in the literature, mitochondrial DNA (mtDNA) restriction analysis and random amplification of polymorphic DNA (RAPD) are becoming increasingly common in the winemaking industry.

mtDNA restriction analysis, which is widely used to analyze yeasts, does not require extensive technical skills or a significant outlay of capital, and has the added advantage that it provides rapid results. This means that corrective measures can be taken in the early stages. mtDNA restriction profiles, for example, are available on the same day because the DNA can be isolated directly from the tank, with no need for prior culture.

RAPD, which is used to analyze bacteria, is slightly more complicated as it requires the use of amplification products and thermocyclers, which add somewhat to the cost. A further disadvantage is that the material needs to be correctly identified and kept separate from other material, which is not very practical in a winery laboratory. While RAPD has been described as an ideal method for typing *O. oeni* strains (Reguant & Bordons, 2003), it requires strict adherence to standardization protocols, which again is not very practical in a winery setting. Furthermore, results take several days to be processed.

Samples taken for analysis must be statistically representative, regardless of the type of microorganism being studied. In other words, the information obtained using the sampling technique must provide a true picture of what is happening in the fermentation tank. In the case of alcoholic fermentation, must samples collected 48 h after inoculation will provide sufficient information with which to assess the success of the operation. In the case of malolactic fermentation, however, samples need to be taken at different time points, particularly in the early stages of fermentation, as this is when bacteria proliferate.

# 5. QUALITY CONTROL ANALYSIS OF COMMERCIAL YEASTS AND INOCULATION

Quality control procedures for use with yeast products obtained from commercial suppliers are becoming increasingly important, particularly in view of the wide range of products available on the market. The first test that should be performed when a winery receives a new batch of active dried yeast is to check the viability of the strains following rehydration. This is done by performing serial dilutions in an appropriate culture medium or by staining with vital dyes. These tests are used to check the number of viable yeast that will be inoculated into the must. Viability can vary by up to 30% from one product to the next, depending on the drying methods used and the sensitivity of

the yeasts. The same methods can also be used to check batch uniformity and to test for loss of viability during storage.

The yeast strains should also be characterized, as different strains behave differently during the fermentation process and thus lend different organoleptic properties to the end product. A range of methods exists for identifying and differentiating between yeast strains at the genus and species level (Esteve-Zarzoso et al., 1998). Several of these methods have been described in previous chapters, but the most reliable, rapid, and economic option for wineries is the mtDNA restriction analysis technique described by Querol et al. in 1992 and modified by López et al. in 2001. Furthermore, the test does not require skilled personnel.

Several studies have described how different yeast preparations supplied by different providers with different instructions for use all contained the same strain that had been characterized using different molecular methods (Fernández-Espinar et al., 2001). There have even been cases in which the same strain was found to be sold under different names by different suppliers and at considerably different prices. In certain cases, there was a price difference of 30%, which is by no means insignificant when it comes to determining the price of the final product. In such cases, in addition to mtDNA restriction analysis, it is necessary to karyotype the strains to conclusively demonstrate full uniformity between different commercial products. Uniformity of both mitochondrial and nuclear markers between samples indicates that they correspond to the same strain.

Figure 14.1 shows the chromosomal profile obtained via pulsed field gel electrophoresis (PFGE) for three commercial preparations with identical restriction patterns obtained using restriction enzymes associated with a high level of variability in the profiles obtained (*Alu*I, *Hinf*I, *Rsa*I).

Similar studies have revealed cases in which the same supplier was selling the same strain

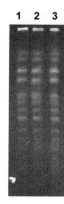

| Ref | Yeast strain | Manufacturer | €/kg |
|-----|--------------|--------------|------|
| 1   | A            | α            | 32   |
| 2   | B            | β            | 25   |
| 3   | C            | χ            | 27   |

**FIGURE 14.1** Pulsed field gel electrophoresis (PFGE) of three commercial yeasts. The figure shows the chromosomal profile obtained by PFGE for three commercial preparations with an identical restriction pattern obtained using restriction enzymes associated with a high level of variability in the profiles obtained (*Alu*I, *Hinf*I, *Rsa*I).

under different names and at different prices, primarily as part of a marketing strategy. The choice of "one strain or another" can, however, influence the price of the final product, which is why several wineries perform quality checks on all the yeast products they receive. These controls are performed yearly as the batches supplied vary from one year to the next.

Another important aspect of winemaking is the analysis of inoculation success. Wines sometimes develop unexpected organoleptic characteristics because fermentation is actually conducted by a native yeast rather than the inoculated strain. To prevent this from happening, it is standard practice to take regular, statistically representative samples throughout the fermentation process to determine, using mtDNA restriction analysis, whether or not the inoculated strain has become established. Studies of this type have shown that not all inoculated strains survive in all cases and that fermentation performed by autochthonous yeasts is much more common than is generally thought (Esteve-Zarzoso et al., 2000).

# 6. MONITORING THE ESTABLISHMENT OF INOCULATED LACTIC ACID BACTERIA

The establishment of inoculated lactic acid bacteria is generally controlled by monitoring the consumption of malic acid and the formation of lactic acid. Nowadays, however, molecular biology techniques can also help to analyze the establishment of these bacteria during malolactic fermentation. Zapparolli et al. (1998) described the use of specific primers for the gene encoding the malolactic enzyme to monitor the establishment of *O. oeni* strains during malolactic fermentation. In a later study, Reguant and Bordons (2003) used multiplex RAPD-polymerase chain reaction (PCR) to characterize *O. oeni* and monitor the population dynamics of the different strains of this species during malolactic fermentation.

# 7. RAPID SOLUTIONS FOR STUCK FERMENTATIONS

The interruption of fermentation that occurs when yeasts stop converting the fermentable sugars in the must is generally known as stuck fermentation. Treatment (which essentially involves the restoration of ideal fermentation conditions) can be difficult, as stuck fermentation has a number of possible causes.

According to Lourens and Reid (2003), the most significant factors that affect yeast viability are osmotolerance; ethanol tolerance; fermentation temperature; availability of nutrients; and presence of medium-chain fatty acids (hexanoic, octanoic, and decanoic acids), which have an inhibitory effect on sugar transport. Other factors include the presence of pesticide traces and the effects of preformentation clarification treatments. Musts that undergo extensive clarification have great difficulty fermenting because they lack sterols and long-chain fatty acids, which are both considered survival factors for yeasts.

## 7.1. Steps to Take in the Event of a Stuck Fermentation (Anchor Yeast, 1994)

1) For every hectoliter of wine, rehydrate 60 to 100 g of active dried *Saccharomyces bayanus* or *S. cerevisiae* strains that are capable of fermentation under difficult conditions and of degrading fructose. If it is not possible to find a yeast with both properties, use two separate types of yeast.

2) When rehydrated, 2 kg of active dried yeast will produce a volume of 20 L. Add 10 L (half the volume of the starter mix) of the problematic wine and wait for fermentation to be activated.

3) The total volume will now be 30 L. Add an additional 15 L of the problematic wine and wait again for fermentation to start. The success of this method will depend on whether or not fermentation is activated after each addition of the problematic wine. If it is, this will mean that the new yeast has adapted successfully to the alcohol level of the new medium.

4) The total volume will now be 45 L. Add equal volumes (e.g., 45, 90, 180, 360 L) of the problematic wine to the new culture until all the stuck wine has been transferred to the new tank. This process takes at least 2 d.

The following aspects should be taken into account:

1) The fermentation temperature should be maintained at between 18 and 22°C for white wine and at between 20 and 25°C for red wine.

2) The lees in the tank with fermentation problems should under no circumstances be

added during the conditioning phase as they may contain toxic substances that could inhibit the growth of the new culture.

3) In red wines in which fermentation has become stuck, the grape skins must be removed if maceration has not been completed to prevent unwanted bacterial contamination.

4) The new tank may be aerated but the old tank may not, as aeration could trigger the growth of acetic acid bacteria.

5) The probability of restarting a stuck fermentation with a residual sugar level of under 10 g/L or an alcohol content of 14% or greater is low. The conditioning phase should be started as soon as there is any indication of stuck fermentation.

## 8. MONITORING CONTAMINATION BY UNDESIRABLE MICROORGANISMS

Yeasts, lactic acid bacteria, and acetic acid bacteria can all alter the quality of wine. Most of these microorganisms are already present in the wine and can grow in this ecological niche. Their proliferation at the wrong time, however, can lead to the production of metabolites that can alter sensory quality or even cause adverse health effects.

### 8.1. Yeast-induced Wine Alterations: Precautionary Measures

Certain yeasts have been associated with refermentation in sweet wines (Enrique et al., 2007), namely strains from the species *S. cerevisiae*, *Zygosaccharomyces bailii*, and *Saccharomyces ludwigii*, all of which have high resistance to sulfur dioxide and ethanol. The best way to prevent refermentation is to ensure appropriate storage conditions; in particular, temperatures of below 15°C, adequate sulphur dioxide levels, protection against the formation of large volumes of air in the tanks, and, of course, careful cleaning of the tanks prior to use.

Biofilm-forming yeasts from the genera *Pichia*, *Candida*, and *Hansenula* that grow on the surface of wines or following contact with air affect young wines and wines with a low alcohol level. To prevent these yeasts from forming in the bottle, in addition to the measures described above, it is necessary to subject the wine to sterile filtration, add 30 mg of free sulfur dioxide, and leave a minimum ullage (air pocket) in the bottle.

Certain wines can develop organoleptic flaws such as unpleasant odors attributable to sulfur compounds during the production process. There are a number of reasons for this. A shortage of nitrogen sources in the must, for example, leads to the production of hydrogen sulfide (Jiranek et al., 1995; Park, 2008), while the genotypes of certain yeasts are associated with an increased capacity to produce sulfites. These yeasts can also produce appreciable amounts of sulfur compounds as byproducts of their metabolism of pesticides.

The acetoin produced by yeasts during fermentation contributes to the bouquet of the wine but it is also the precursor of 2,3-butanediol and diacetyl. While 2,3-butanediol can contribute to aromatic balance, diacetyl is considered a flaw. *Saccharomyces* yeasts produce only small quantities of acetoin, unlike the apiculate yeasts *Kloeckera* and *Hanseniaspora* and *Zygosaccharomyces* species, which produce considerable amounts.

Among the best-studied contaminating wine yeasts are *Brettanomyces* species, which produce four byproducts during growth: esterases, volatile fatty acids (acetic acid), volatile phenols (4-ethylphenol and 4-ethylguaiacol), vinyl phenols (4-vinylphenol and 4-vinylguaiacol), and tetrahydropyridines. These yeasts are more common in red wines, which are rich in cinnamic acid precursors and in which the cinnamate decarboxylase activity of *Brettanomyces* is not inhibited by phenolic compounds. The richest wines—that is, those made from ripe grapes and therefore characterized by high alcohol levels, lower acidity, and longer maceration times—are often those that are richest in assimilable substrates and therefore potentially more amenable to the growth of *Brettanomyces* yeasts. The following points are important for preventing contamination by this yeast:

1. Particular care should be taken to keep the grape reception area clean during harvest time.
2. Sulfur dioxide is the only effective antiseptic against *Brettanomyces* species that is authorized for use in wine. Levels of molecular (active) sulfur dioxide must be controlled as, the higher the pH, the more sulfur dioxide will be needed to maintain adequate levels of molecular sulfur dioxide. Increases in alcohol content and aging temperatures lead to increased molecular sulfur dioxide levels and hence greater protection against *Brettanomyces* yeasts. The growth of these species is inhibited at levels of 0.3 parts per million (ppm) of molecular sulfur dioxide, and levels of over 0.5 ppm lead to the rapid elimination of these yeasts. Levels of molecular sulfur dioxide must be maintained at between 0.5 and 0.8 ppm to protect against *Brettanomyces* species and other organisms that could alter the quality of the wine.
3. Factors that can increase the risk of contamination and subsequent growth of *Brettanomyces* species and other undesirable microorganisms include the inadequate treatment of previously used barrels, the storage of wine in unsuitable conditions (temperature shifts), the presence of air pockets (excessive oxidation), and insufficient racking (frequency of barrel disinfection, etc.). Wood is very porous and can house yeasts in its different layers, which makes them difficult to eliminate.

*Brettanomyces* species and lactic acid bacteria are also likely to grow in new barrels. Several of the phenolic compounds that are extracted from new wood may provide substrates for *Brettanomyces* oxidation or reduction reactions that can give rise to unwanted compounds. In new barrels, there are also large quantities of cellobiose that are degraded by β-glucosidases, giving rise to glucose molecules used by *Brettanomyces* yeasts for growth. Cellobiose forms in barrels subjected to toasting. It is also important to know the origin of any new barrels to be used by the winery.

## 8.2. Spoilage by Lactic Acid Bacteria: Precautionary Measures

Lactic acid bacteria are common in wineries. However, the inoculation of selected strains is actually the best strategy for preventing the proliferation of bacteria capable of producing secondary metabolites that can have harmful health effects (the case of biogenic amines) or diminish the organoleptic quality of the wine. To prevent the growth of unwanted lactic acid bacteria (*Lactobacillus* and *Pediococcus* species) in musts, low sulfite and high pH levels must be avoided (Pfannebecker & Fröhlich, 2008). Inoculation with high concentrations of commercial malolactic starter cultures can be used to displace autochthonous populations already present on winery equipment or in aging barrels. Similar treatment is also necessary for wines that have already completed malolactic fermentation as bacteria that survive through aging can produce biogenic amines during secondary metabolism.

Malolactic fermentation, while common in red wines, is not desirable in certain wines (especially young whites). To prevent the growth of lactic acid bacteria in such cases, the wine should be filtered and treated with sulfites after alcoholic fermentation.

## 8.3. Spoilage by Acetic Acid Bacteria: Precautionary Measures

Acetic acid bacteria contaminate grapes, musts, and wines (Bartowsky & Henschke, 2008). To prevent the proliferation of these microorganisms in stored wine, the corresponding tanks and barrels should be filled to the maximum and the wine treated with additional sulfites as the levels added during the production process are not sufficient to prevent growth. A total level of 100 ppm of sulfur dioxide in the must is necessary for this purpose. The optimal temperature for the growth of acetic acid bacteria is 25 to 30°C, although activity has been detected at temperatures of close to 10°C in certain wineries. Extra caution should be taken to monitor residual populations of these bacteria in the wine as they represent a permanent risk of spoilage. One solution is to refill tanks and barrels regularly.

The excessive growth of acetic acid bacteria on grapes can lead to changes in the must that can interfere with the growth of yeast during alcoholic fermentation and the course of malolactic fermentation. Examples are organoleptic changes (caused by the production of undesirable metabolites) and physical alterations to the wine (e.g., some species are capable of producing polysaccharides that interfere with filtration). Aeration is another important factor that should be controlled as acetic acid bacteria preferentially grow in the presence of low levels of oxygen. The slightest aeration following alcoholic fermentation, for example, can lead to the growth of $10^2-10^3$ CFU/mL of these bacteria.

# 9. MICROBIOLOGICAL ASPECTS OF THE PRODUCTION OF TYPICAL WINES (SHERRY, CAVA)

The types of microorganism used to produce a particular wine will depend on the end product. Even though the base wine might be the same, different strains of yeast, each with unique characteristics, will be used to conduct alcoholic fermentation depending on whether the aim is to produce a standard wine, or Cava or Fino, for example.

Table 14.1 shows the main properties a yeast should have, depending on the wine being made. Below we describe some of the properties sought in yeasts used to make Cava and Fino wines. It should, however, be borne in mind that all wineries have their own practices that lend their wines their distinctive character. A key issue in all wineries, however, is the viability of the starter cultures used. It is therefore important to ensure a high rate of viability following rehydration as this is the key to successful inoculation.

## 9.1. Fino Wines

The process used in the fermentation of white base wines to make Fino wines is similar to that used when activating alcoholic fermentation by inoculation. In the former case, however, the wine is subsequently fortified and left to age (Ibeas et al., 1997; Martínez et al., 1997; Moreno-Arribas & Polo, 2008). Although there is controversy regarding the molecular differences between the four races of *flor* yeasts used to age Fino wine, Esteve-Zarzoso et al. (2004) detected a *flor* yeast strain with a very similar karyotype to that described by Ibeas et al. (1997) in a different winery in the same region. This may be because of the high levels of ethanol and acetaldehyde present or because of the oxidative metabolism of these yeasts. Both studies concluded that, while all wineries have

**TABLE 14.1** Main Properties Sought in Fermentation Yeasts Shown by Type of Wine

|  | Fermentation of white wine | Fermentation of red wine | Secondary fermentation in Cava | Biological aging of Fino wines |
|---|---|---|---|---|
| Fermentation temperature | Tolerance of low temperatures preferable | Moderate resistance | Moderate resistance | Moderate resistance |
| Color | No production of colored metabolites | No degradation of color | No production of colored metabolites | No production of colored metabolites |
| Volatile acidity | Low production | Low production | Low production | Low/moderate production |
| Degradation of malic acid | High | Low | High | High |
| Ethanol tolerance | Moderate/high | High | Moderate/high | High |
| Filtration | Necessary | Desirable | Necessary | Not desirable |
| Formation of biofilm | Undesirable | Undesirable | Undesirable | Necessary |
| Production of aromas | Desirable | Desirable | Desirable | Not necessary |
| Production of/ tolerance of acetaldehyde | Low | Low | Low | High |
| Fermentation kinetics | Moderate | Moderate | Low | Low |

their own distinctive strains, each barrel is a world of its own.

## 9.2. Cava

Because Cava is produced using two separate fermentation steps, efforts should be taken to ensure that the yeasts that participate in each of these steps are compatible. The inoculation of yeasts that participate in secondary fermentation (which occurs in the bottle) must be carefully conducted to ensure that the yeasts are perfectly adapted to the high alcohol levels present. The importance of these measures cannot be overemphasized. The use of a yeast with poor flocculation ability requires longer filtration times. If a filtration step is not used, higher quantities of bentonite will be necessary, resulting in a greater absorption of the components of the wine. If the yeasts are not properly acclimatized, the secondary fermentation will take longer as the yeasts will need more time to consume the sugars added for secondary fermentation.

## 10. MICROBIOLOGICAL QUALITY CONTROL OF THE FINAL PRODUCT

The advances made in classical microbiology methods have mainly involved the development of specific culture media and biochemical tests capable of discriminating, both quantitatively and qualitatively, between different microorganisms. Although widely used, these techniques have the disadvantage that reliable results (e.g., from colony counts) are not available for 2 to 3 d (Pless et al., 1994). This can increase the price of the final product as stocks of wine are accumulated while the results of quality checks on the final product are being processed. Methods designed to overcome this problem include bioluminescence imaging, immunoassays (d'Aoust & Sewell, 1986;

Emsweiler-Rose et al., 1984), molecular techniques based on DNA probes (Fitts, 1985; Flowers et al., 1987), and quantitative measurement systems based on PCR and electrical impedance.

The need for rapid microbial detection systems in the food industry is closely linked to the benefits that such systems could offer companies in terms of improving production processes and time to market. The quantitative and qualitative analysis of microorganisms by electrical impedance is based on the indirect measurement of metabolic activity. These measurements can then be used to calculate theoretical population numbers long before the colony would have become visible on solid culture medium (Deak & Beuchat, 1993).

Electrical impedance is based on the principle that the molecules in the culture medium (proteins, carbohydrates, etc.) are electrically neutral or only weakly ionized. The activity of microorganisms, however, converts these molecules into numerous smaller molecules with a greater electrical charge and electrical mobility (amino acids, lactates, etc.). These changes can be measured by submerging two electrodes in the culture medium (Futschik et al., 1988).

Classical quantification of microbial load requires approximately 3 d but faster results are needed in certain industrial applications. Electrical impedance, in this sense, has been described as a rapid, reliable means of monitoring the presence of viable microorganisms in food. Martínez et al. (2004) showed that the indirect measurement of carbon dioxide production by electrical impedance was an ideal way of detecting and quantifying yeast populations in wine samples; the correlation coefficient between this and the classical microbiology method (plate count on specific solid media) was 0.98, with the added advantage that results (confirmation of the absence of microorganisms) were available in under 21 h. In the case of aerobic bacteria, electrical impedance was also more suitable for detection and quantification purposes, with a correlation coefficient of 0.99 between this method and the plate count method, and results available in under 15 h (Martínez et al., 2004).

## 11. NEW CHALLENGES FACING THE SCIENTIFIC COMMUNITY: GENETICALLY MODIFIED ORGANISMS (GMOS)

From a strictly scientific viewpoint, genetically modified organisms (GMOs) undoubtedly offer many health and agricultural advantages (Fleet, 2008).

Considering that the full sequence of the *S. cerevisiae* genome is available and that this yeast can be modified with relative ease using the increasingly sophisticated genetic modification tools that are now widely available, research centers around the world are in a position to create *à-la-carte* yeasts to meet the requirements of the base must or to achieve particular characteristics in the end product.

Within the scope of a strictly scientific, enological project, researchers in South Africa have developed genetically modified yeasts capable of improving the quality of wine (Pretorius, 2000). It would be advantageous for winemakers to avail of yeasts capable of increased production of exogenous pectolytic or glycosidic enzymes; bacteria with a low production of histamines; yeasts with low nutritional requirements, low methanol production, or extracellular release of mannoproteins; heat-sensitive microorganisms; etc.

In any case, given the enormous importance attached to the presence of transgenic products in food, it should not be forgotten that the Spanish state bulletin (BOE, L268/24) published Regulation (EC) No 1830/2003 on the traceability and labeling of GMOs and the traceability of food and feed products produced from GMOs, establishing strict regulations regarding the approval, labeling, and monitoring of GMOs. This legislation came into force on November 7, 2003, with a 6-month adaptation period for all operators.

From a commercial perspective, the obligatory labeling of the presence of GMOs or elements from GMOs in the final product delivered to the consumer, combined with the existence of strong media and public resistance, places food companies interested in benefiting from the advantages of GMOs in a difficult position.

The world of wine is strongly traditional. Major modifications that require considerable changes in age-old practices are not well viewed by critics or consumers, which makes it very difficult to start using GMOs. Although these organisms have enormous potential, they are not yet fully accepted by the international community.

The future of GMOs in winemaking will fundamentally depend on public opinion and ultimately the end consumer. Despite the advances that could be made, media campaigns have put a stop to the possible industrial breakthroughs in this field. Historically, basic science has always been one step ahead of the rest of society and today's winemakers need to be ready to deliver products with genetically enhanced characteristics to the market should a change of opinion come about.

## References

Amerine, M. A., & Kunkee, R. E. (1968). Microbiology of winemaking. *Ann. Rev. Microbiol., 22*, 323–358.

Anchor Yeast. (1994). *The cellarmaster's wine yeast and fermentation handbook.*

Andorrà, I., Landi, S., Mas, A., Guillamón, J. M., & Esteve-Zarzoso, B. (2008). Effect of oenological practices on microbial populations using culture-independent techniques. *Food Microbiol., 25*(7), 849–856.

Bartowsky, E. J., & Henschke, P. A. (2008). Acetic acid bacteria spoilage of bottled red wine – A review. *Int. J. Food. Microbiol., 125*(1), 60–70.

Belli, N., Marín, S., Duaigües, A., Ramos, A. J., & Sanchis, V. (2005). Ochratoxin A in wines, musts and grape juices from Spain. *J. Sci. Food Agric., 84*(6), 591–594.

Blackburn, D. (1984). Present technology in the use of malolactic bacteria. *Practical Winery, 5,* 64–72.

Champagne, C., Gadner, N., & Doyon, G. (1989). Production of *Leuconostoc oenos* biomass under pH control. *Appl. Environ. Microbiol., 55,* 2488–2492.

d'Aoust, J. Y., & Sewell, A. M. (1986). Detection of salmonellae by the enzyme immunoassay (eia) technique. *J. Food. Sci., 51,* 484–507.

de Andrés-de Prado, R., Yuste-Rojas, M., Sort, X., Andrés-Lacueva, C., Torres, M., & Lamuela-Raventós, R. M. (2007). Effect of soil type on wines produced from V*itis vinifera* L. Cv. Grenache in commercial vineyards. *J. Agric. Food Chem., 55*(3), 779–786.

Deak, T., & Beuchat, L. R. (1993). Comparision of condutimetric and traditional plating techniques for detecting yeast in fruit juice. *J. Appl. Bacteriol., 75,* 546–550.

Degre, R. (1992). Selection and commercial cultivation of wine yeast and bacteria. In G. H. Fleet (Ed.), *Wine microbiology and biotechnology* (pp. 421–447). Chur, Switzerland: Harwood Academic Publishers.

Emsweiler-Rose, B., Gehle, W. D., Johnston, R. W., Okred, A., Moran, A., & Bennet, B. (1984). An enzyme immunoassay technique for the detection of salmonellae in meat and poultry products. *J. Food Sci., 49,* 1010–1020.

Enrique, M., Marcos, J. F., Yuste, M., Martínez, M., Vallés, S., & Manzanares, P. (2007). Antimicrobial action of synthetic peptodes towards wine spoilage yeasts. *Int. J. Food Microbiol., 118*(3), 318–325.

Esteve-Zarzoso, B., Fernández-Espinar, M. T., & Querol, A. (2004). Authentication and identification of *Saccharomyces cerevisiae* flor yeast races involved in sherry ageing. *Anton. Leeuw., 85,* 151–158.

Esteve-Zarzoso, B., Gostíncar, A., Bobet, R., Uruburu, F., & Querol, A. (2000). Selection and molecular characterisation of wine yeasts isolated from the El Penedès area (Spain). *Food Microbiol., 17,* 553–562.

Esteve-Zarzoso, B., Manzanares, P., Ramón, D., & Querol, A. (1998). The role of non-*Saccharomyces* yeasts in industrial wine making. *Int. Microbiol., 1,* 143–148.

Farris, G. A., Budroni, M., Vodret, T., & Deiana, P. (1990). Sull'origine dei lieviti vinari i lieviti dei terreni, delle foglie e degli acini di alcun vigneti sardi. *L'Enotecnico, 6,* 99–108.

Fernández-Espinar, M. T., López, V., Ramón, D., Bartra, E., & Querol, A. (2001). Study of the authenticity of commercial wine yeast strains by molecular techniques. *Int. J. Food Microbiol., 70,* 1–10.

Fitts, R. (1985). Development of DNA–DNA hybridization test for the presence of salmonellae in foods. *Food Technol., 39,* 95.

Fleet, G. H. (1992). The microorganisms of winemaking. Isolation enumeration and identification. In G. H. Fleet (Ed.), *Wine microbiology and biotechnology* (pp. 1–25). Chur, Switzerland: Harwood Academic Publishers.

Fleet, G. H. (2008). Wine yeasts for the future. *FEMS Yeast Res., 8*(7), 979–995.

Fleet, G. H., & Heard, G. M. (1992). Yeast-growth during fermentation. In G. H. Fleet (Ed.), *Wine microbiology and biotechnology* (pp. 27–54). Chur, Switzerland: Harwood Academic Publishers.

Flowers, R. S., Mongola, M. A., Curiale, M. S., Gabis, D. A., & Silliker, J. H. (1987). Comparative study of a DNA hybridization method and the conventional culture procedure for detection of salmonellae in foods. *J. Food. Sci., 52,* 781–785.

Futschik, K., Pfützner, H., Doblander, A., & Asperger, H. (1988). Automatical registration of microorganism growth by a new impedance method. *Abst. Int. Meet. Chem. Eng. & Biotechnol., Achema, 88,* 3.

Garijo, P., Santamaría, P., López, R., Sanz, S., Olarte, C., & Gutiérrez, A. R. (2008). The occurrence of fungi, yeast and bacteria in the air of a Spanish winery during vintage. *Int. J. Food Microbiol., 125*(2), 141–145.

Ibeas, J. L., Lozano, I., Perdigones, F., & Jiménez, J. (1997). Dynamics of flor yeast populations during the biological ageing of sherry wines. *Am. J. Enol. Vitic., 48,* 75–79.

Jiranek, V., Langridge, P., & Henchke, P. A. (1995). Regulation of hydrogen sulphide liberation in wine-producing *Saccharomyces cerevisiae* strains by assimilable nitrogen. *Appl. Environ. Microbiol., 61,* 461–467.

Longo, E., Cansado, J., Agrelo, D., & Villa, T. G. (1991). Effect of climatic conditions on yeast diversity in grape musas from northwest Spain. *Am. J. Enol. Vitic., 42,* 141–144.

López, V., Querol, A., Ramón, D., & Fernández-Espinar, M. T. (2001). A simplified procedure to analyse mitochondrial DNA from industrial yeasts. *Int. J. Food Microbiol., 68,* 75–81.

Lourens, K., & Reid, G. (2003). Stuck fermentation management. *Austral. New Zeal. Grape. Wine. Feb.* 72–74.

Martínez, M., Torres, M., Bobet, R., Cantarero, X., Álvarez, J. F., & Rubires, X. (2004). Application de l'impédance électrique pour le contrôle microbiologique des vins embouteillés. *Rev. Fran. Oenol., 205,* 34–36.

Martínez, P., Pérez-Rodriguez, L., & Benitez, T. (1997). Evolution of flor yeast population during the biological ageing of fino Sherry wine. *Am. J. Enol. Vitic., 48,* 160–168.

Martini, A. (1993). Origin and domestication of the wine yeast Saccharomyces cerevisiae. *J. Wine Res., 4,* 165–176.

Martini, A., Federici, F., & Rosini, G. (1980). A new approach to the study of yeast ecology of natural substrates. *Can. J. Microbiol., 26,* 856–859.

Martini, A., & Vaughan-Martini, A. (1990). Grape must fermentation: Present and past. In J. F. T. Spencer, &

D. M. Spencer (Eds.), *Yeast technology* (pp. 105–123). Berlin, Germany: Springer-Verlag.

Monteil, H., Blazy-Mangen, F., & Michel, G. (1986). Influence des pesticides sur la croissance des levures des raisins et des vins. *Sci. Alim., 6*, 349–360.

Moreno-Arribas, M. V., & Polo, M. C. (2008). Occurrence of lactic acid bacteria and biogenic amines in biologically aged wines. *Food Microbiol., 25*(7), 875–881.

Park, S. K. (2008). Development of a method to measure hydrogen sulfide in wine fermentation. *J. Microbiol. Biotechnol., 18*(9), 1550–1554.

Parrish, M. E., & Carroll, D. E. (1985). Indigenous yeast associated with muscadine (*Vitis rotundifolia*) grapes and musts. *Am. J. Enol. Vitic., 36*, 165–169.

Pfannebecker, J., & Fröhlich, J. (2008). Use of a species-specific multiplex PCR for the identification of pediococci. *Int. J. Food Microbiol., 128*(2), 288–296.

Pless, P., Futschik, K., & Shoff, E. (1994). Rapid detection of salmonellae by means of a new impedance-splitting method. *J. Food. Protec., 57*, 369–376.

Poulard, A., Simon, L., & Cuinier, C. (1980). Variabilite de la microflore levurienne de quelques terroirs viticoles du pays. *Nantais. Conn. Vig. Vin, 14*, 219–238.

Pretorius, I. S. (2000). Tailoring wine yeast for the new millennium: Novel aproaches to the ancient art of winemaking. *Yeast, 16*, 675–729.

Pretorius, I. S., van der Westhuizen, T. J., & Augustyn, O. P. H. (1999). Yeast diversity in vineyards and wineries and its importance to the South African wine industry. *S. Afr. J. Enol. Vitic., 20*, 61–74.

Querol, A., Barrio, E., & Ramón, D. (1992). A compartive study of different methods of yeast strain characterization. *Syst. Appl. Microbiol., 15*, 439–446.

Radler, F., & Lotz, B. (1990). The microflora of active dry yeast and the quantitative changes during fermentation. *Wein Wissenschaft, 45*, 114–122.

Reguant, C., & Bordons, A. (2003). Typification of *Oenococcus oeni* strains by multiplex RAPD-PCR and study of population dynamics during malolactic fermentation. *J. Appl. Microbiol., 95*(2), 344–353.

Rosini, G., Federici, F., & Martini, A. (1982). Yeast flora of grape berries during ripening. *Microb. Ecol., 8*, 83–89.

Solfrizzo, M., Panzarini, G., & Visconti, A. (2008). Determination of Ochratoxin A in grapes, dried vine fruits and winery byproducts by high-performance liquid chromatography with fluorometric detection (HPLC-FLD) and immunoaffinity cleanup. *J. Agric. Food Chem., 56*(23). 11 081–11 086.

Torija, M. J., Rozès, N., Poblet, M., Guillamón, J. M., & Mas, A. (2001). Yeast population dymanics in spontaneous fermentations: Comparision between two different wine producing areas over a period of three years. *Anton. Leeuw. Int. J.G., 79*, 345–352.

Vaughan-Martini, A., & Martini, A. (1995). Facts, myths and legends on the prime industrial microorganism. *J. Ind. Microbiol., 14*, 514–522.

Zapparolli, G., Torriani, S., Pesente, P., & Dellaglio, F. (1998). Design and evaluation of malolactic enzyme gene targeted primers for rapid identification and detection of *Oenococcus oeni* in wine. *Lett. Appl. Microbiol., 27*, 243–246.

# Index

Printed and bound by CPI Group (UK) Ltd, Croydon, CR0 4YY

08/05/2025

01864827-0008